MW00714369

Protein Electrophoresis in Clinical Diagnosis

Protein Electrophoresis in Clinical Diagnosis

David F Keren

Medical Director, Warde Medical Laboratory, Ann Arbor, MI

Department of Pathology, St. Joseph Mercy Hospital, Ann Arbor, MI

Clinical Professor of Pathology, The University of Michigan Medical School,
Ann Arbor, MI

A member of the Hodder Headline Group
LONDON

First published in Great Britain in 2003 by
Arnold, a member of the Hodder Headline Group,
338 Euston Road, London NW1 3BH

http://www.arnoldpublishers.com

Distributed in the United States of America by
Oxford University Press Inc.,
198 Madison Avenue, New York, NY10016
Oxford is a registered trademark of Oxford University Press

© 2003 Arnold

All rights reserved. No part of this publication may be reproduced or
transmitted in any form or by any means, electronically or mechanically,
including photocopying, recording or any information storage or retrieval
system, without either prior permission in writing from the publisher or a
licence permitting restricted copying. In the United Kingdom such licences
are issued by the Copyright Licensing Agency: 90 Tottenham Court Road,
London W1T 4LP.

Whilst the advice and information in this book are believed to be true and
accurate at the date of going to press, neither the author[s] nor the publisher
can accept any legal responsibility or liability for any errors or omissions
that may be made. In particular (but without limiting the generality of the
preceding disclaimer) every effort has been made to check drug dosages;
however it is still possible that errors have been missed. Furthermore,
dosage schedules are constantly being revised and new side-effects
recognized. For these reasons the reader is strongly urged to consult the
drug companies' printed instructions before administering any of the drugs
recommended in this book.

British Library Cataloguing in Publication Data
A catalogue record for this book is available from the British Library

Library of Congress Cataloging-in-Publication Data
A catalog record for this book is available from the Library of Congress

ISBN   0340 812133

1 2 3 4 5 6 7 8 9 10

Commissioning Editor: Serena Bureau
Development Editor: Layla Vandenbergh
Project Editor: Zelah Pengilley
Production Controller: Deborah Smith
Cover Design: Stewart Larking

Typeset in 10 on 13 pt Sabon by Phoenix Photosetting, Chatham, Kent
Printed and bound in Great Britain by Butler & Tanner Ltd, Frome, Somerset

What do you think about this book? Or any other Arnold title?
Please send your comments to feedback.arnold@hodder.co.uk

To my wonderful family

Contents

Preface

This text presents the use of protein electrophoresis of serum, urine, and cerebrospinal fluid in clinical diagnosis. It is a revision of two previous books on this subject with several substantive and many trivial changes. The title has been changed from *High-Resolution Electrophoresis and Immunofixation: Techniques and Interpretations* to the present one in recognition of the fact that the information in this book may be useful to individuals who interpret a wide variety of electrophoretic gels (high-resolution or not).

There have been several significant changes in the field since the second edition of *High-Resolution Electrophoresis and Immunofixation: Techniques and Interpretation* was published in 1994 by Butterworth-Heinemann. That text was largely written in 1993, making it 10 years old at the time of publication of this book. At the time the second edition was written there was no high-resolution technique available that provided automated or semi-automated methods for laboratories with a large clinical volume of testing. There are now several products presented, with examples. Some of them are gel-based, others use capillary zone electrophoresis. Some automated gel-based systems have achieved an excellent degree of resolution that allows efficient performance of high-quality techniques by even moderately sized institutions. At the time of the second edition, capillary zone electrophoresis itself was quite new with no methods available that had been approved for use in the clinical laboratory by the Food and Drug Administration. I discussed it only briefly. Now two such instruments are already in place in many laboratories and there have been several publications about the advantages, artifacts and limitations of these techniques. Improved resolution especially of protein bands has

resulted in better detection of bisalbuminemia by laboratories using capillary zone electrophoresis. That technique also has enhanced our ability to detect genetic variants such as α_1-antitrypsin inhibitor deficiencies (PiZZ and PiSZ) as well as benign variants.

In addition to technical advances in the general detection and quantification of specific proteins, the techniques available to identify the monoclonal proteins have also changed significantly. Immunosubtraction using capillary zone electrophoresis was not even mentioned in the second edition. Now, immunosubtraction by automated technology is used in many laboratories that employ capillary zone electrophoresis. It is highly efficient, but suffers from a lack of sensitivity and flexibility when compared with immunofixation. This is discussed in the present text. For laboratories using some of the newer semi-automated gel-based techniques, semi-automated immunofixation is now available. Beyond the instruments, there are new reagents available such as the 'Penta' (pentavalent) reagent that one company supplies as a potential screen for monoclonal proteins. Immunoselection was also just touched on in the second edition. Now that γ heavy chain disease is more readily recognized, a broader discussion of immunoselection is provided.

Advances have also been made in reagents available for measurement of monoclonal free light chains by nephelometry in both serum and urine. Recent publications indicate that measurement of free light chains in serum may be useful not only to follow patients with known monoclonal free light chains (Bence Jones proteins), but also to assist in the difficult diagnoses of amyloid (AL) and even non-secretory myeloma. This is discussed in Chapters 6 and 7.

In the second edition, high-resolution agarose gels were preferred for detecting oligoclonal bands in the cerebrospinal fluid from patients whose differential diagnosis included multiple sclerosis. Now, the recommendation is the use of isoelectric focusing or an immunofixation technique to identify the bands as immunoglobulins. One company currently offers a semi-automated immunofixation method for detecting oligoclonal bands (O-bands) in cerebrospinal fluid on unconcentrated fluid. There have also been improvements in the detection of leakage of cerebrospinal fluid into nasal and aural fluids; these are discussed in Chapter 8.

Beyond technical issues, there have been improvements in our knowledge of proper utilization and interpretation of protein electrophoresis when searching for the presence of a monoclonal gammopathy. In 1998, the College of American Pathologists Conference XXXII convened a panel of experts to provide recommendations for the clinical and laboratory evaluation of patients suspected of having a monoclonal gammopathy. These guidelines were published in 1999 and provide an important framework that emphasizes the partnership of clinicians with laboratory workers.

Urine evaluation has improved with the newer technologies and has been complicated by some current procedures. For example, individuals receiving a pancreas transplant may have the exocrine pancreas drainage empty into the urinary bladder. As discussed in Chapter 7, this results in γ-migrating bands that may be mistaken for monoclonal gammopathies. They are exocrine pancreas secretions.

The method I suggested in the first edition and expanded in the second for tailored reporting with detailed sign-outs has been improved by feedback from our clinicians. Our current sign-outs are presented in tabular form and readers are encouraged to use them if they wish.

A couple of concepts were removed from this edition. The technique of two-dimensional electrophoresis, while useful in research has not caught on in the clinical laboratory and is not discussed in this volume. Further, the indirect immunofluorescence techniques reviewed in the previous texts has been replaced by immunohistochemical and flow cytometry studies.

Finally, in this text I recommend the use of the term monoclonal free light chain (MFLC) to replace the term Bence Jones proteins. It is both a practical and a technical improvement. Some individuals confuse the presence of an intact immunoglobulin monoclonal gammopathy in the urine with a Bence Jones protein. It is not. Bence Jones protein is a monoclonal free light chain and has much greater significance to the patient's diagnosis and prognosis than an intact immunoglobulin monoclonal gammopathy in the urine. The term MFLC says exactly what we found (i.e. a monoclonal free light chain). So it is technically correct and removes any possible ambiguity. This has the advantage that individuals will no longer use a hyphen that Dr Henry Bence Jones never used on any of the papers or books that he wrote.

I hope you enjoy this book, or at least find it useful. Your comments and questions help to provide a focus for making this text more relevant to the use of electrophoresis in clinical diagnosis. Please feel free to contact me via e-mail: kerend@wardelab.com.

David F. Keren, MD
Ann Arbor, Michigan, USA
January 12, 2003

Acknowledgements

I wish to thank my family and especially my wife for their great patience with me as I worked through this task.

I am deeply grateful to the many people who helped in the development of the materials for this and previous editions of this book. They allowed me to use their clinical material or illustrations to give the reader a first-hand view of the potential uses and limitations of electrophoretic techniques in the clinical laboratory. Figures, Tables or specific cases were kindly provided by Dr R. S. Abraham, Dr Francesco Aguzzi, Dr Arranz-Peña, Dr Gary Assarian, Cynthia R. Blessum, Dr Xavier Bossuyt, Dr Arthur Bradwell, Dr. Stephen O. Brennan, Chrissie Dyson, Dr Beverly Handy, Margaret A. Jenkins, Carl R. Jolliff, Dr Jerry Katzmann, Dr Robert H. Kelly, Dr Joseph M. Lombardo, Dr A. C. Parekh, Dr Jeffrey Pearson, Dr Arthur J. Sloman, Dr Lu Song, Dr Tsieh Sun, Dr E. J. Thompson, Dr Adrian O. Vladutiu and Dorothy Wilkins.

Drs John M. Averyt, John L. Carey, III, and Jeffrey S. Warren provided helpful suggestions in reviewing selected portions of this book. Their input kept many outrageous statements out of the final copy, and corrected several errors of omission.

Special thanks are due to Ron Gulbranson and Debbie Hedstrom who prepared many of the electrophoretic gels, electropherograms, and special studies that are in the present book. They have had endless patience with me while I cluttered their work area with capillary zone electropherograms and gels containing interesting cases.

Finally, I also thank the many individuals who have taken the time to write, telephone or speak to me personally about some aspect of my work. They have framed many questions about concepts or omissions from previous work that I hope to clarify in this text.

David F. Keren, MD
Ann Arbor, Michigan, USA
January 12, 2003

1

Protein structure and electrophoresis

The term 'electrophoresis' refers to the migration of charged particles in an electrical field. The success of electrophoresis in separating serum, urine and cerebrospinal fluid proteins into clinically useful fractions results from the heterogeneity of the charges of these molecules. It is useful, therefore, to review briefly the structural features that result in the observed migration.

PROTEIN STRUCTURE

The major structural and functional molecules produced by cells are proteins. They are important in host defense, cell structure, movement, and as regulatory molecules. Proteins are composed of individual units called amino acids; the general structure is shown in Fig. 1.1. As the name implies, each amino acid contains an acidic carboxyl group (-COOH) and a basic amino group (-NH$_2$). The R groups (sequences specific for each amino acid) attached to the alpha-carbon can be neutral, acidic, or basic. Unless the R group is a hydrogen atom (glycine), the structure around the alpha carbon is asymmetrical. Therefore, amino acids can exist as one of two stereoisomers that are mirror images.

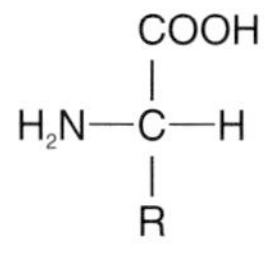

COOH
|
H$_2$N—C—H
|
R

Figure 1.1 General structure of amino acids.

These are the D- and L-forms (Fig. 1.2). In proteins, the L-form is almost always present.[1]

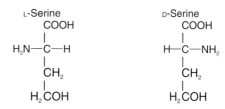

Figure 1.2 The L and D forms of serine.

When in solution, at the typical intracellular pH, many amino acids behave as both acids and bases. This results in the formation of a *zwitterion*, a molecule in which both the amino and carboxyl ends are ionized, yet the molecule is electrically neutral (Fig. 1.3). Since this molecule is electrically neutral, that is, *isoelectric*, it does

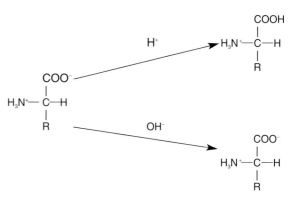

Figure 1.3 Zwitterion and effect of acidic and basic solutions on amino acid charge.

not migrate in an electrical field. Of course, the charge on each amino acid depends on both the R group and the pH of the solution in which it is dissolved.

By altering the pH of an aqueous solution, the charge on an amino acid can be changed. In acidic solution, the amino acid accepts a proton on its carboxylate group, resulting in a net positive charge on the molecule (Fig. 1.3). The positively charged cation thus formed will migrate toward the negative pole (cathode) in an electrical field. Conversely, in basic solution the ammonium group gives up a proton, leaving the amino acid with a net negative charge (Fig. 1.3). This negatively charged anion migrates toward the positive pole (anode) in an electrical field. The pH of the solution and the nature of the R group (Fig. 1.4) have an important effect on the migration of individual amino acids. For example, both aspartic acid and glutamic acid at pH 7.0 or greater are always non-protonated (negatively charged) and consequently are referred to as aspartate and glutamate. At neutral pH, arginine, histidine and lysine are all protonated (positively charged). These differences in charge are important in determining the migration of proteins during electrophoresis. Further, in genetic variants such as α_1-antitrypsin deficiency, the substitution of a charged amino acid for a neutral amino acid, or the converse, will alter the migration of the variant form that allows us to detect the abnormality.

The solubility of proteins also is related to their amino acid composition. For example, tyrosine, threonine and serine have hydroxyl moieties in their R group, and readily form hydrogen bonds with water. Similarly, asparagine and glutamine have amide R groups that are hydrophilic. In contrast, several amino acids are hydrophobic. Large numbers of these hydrophobic amino acids (phenylalanine, tryptophan, valine, leucine, isoleucine, proline, alanine or methionine) in a protein render it, or portions of it, relatively insoluble in aqueous solutions. Functionally, hydrophobic amino acids serve as key constituents of proteins, sometimes as the parts of proteins that interact with the lipid (hydrophobic) membranes of cells. At neutral pH,

most hydrophobic amino acids exist as zwitterions (Fig. 1.3). However, when large numbers of hydrophobic amino acids occur in other molecules, such as occasional products of malignant plasma cell clones, they may exhibit properties that cause solubility problems. Hydrophobic amino acid content is not the only factor that relates to solubility problems. For example, the complex cold-precipitating property of cryoglobulins (see Chapter 6) is not thought to be related to an excess of hydrophobic amino acids.[2]

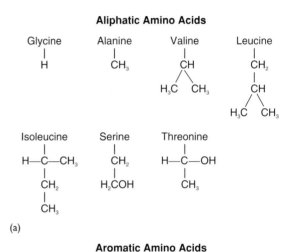

(a)

(b)

Figure 1.4 (a–b) R-Group structure of amino acids.

Basic Amino Acids

Lysine	Histidine	Arginine
CH₂	CH₂	CH₂
CH₂		CH₂
CH₂		CH₂
CH₂		NH
NH₃⁺		C=NH₂⁺
		NH₂

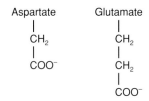

Acidic Amnio Acids

Aspartate	Glutamate
CH₂	CH₂
COO⁻	CH₂
	COO⁻

(c)

Polar Amino Acids

Aspargine   Glutamine

Immuno Acids

COOH — CH   Proline

(d)

Figure 1.4 (c–d) R-Group structure of amino acids.

Peptide bonds and polypeptides

PRIMARY STRUCTURE

Amino acids may be linked by *peptide bonds*: the linkage of the carboxyl group of one amino acid to the amino group of the next. During the process, a water molecule is produced and the covalent peptide bond is formed (Fig. 1.5). The resulting molecule has the charge characteristics inherent in the R groups, together with the carboxyl and amino terminal groups. When a group of amino acids is linked through peptide bonds in a linear array, it is referred to as a *polypeptide*. All polypeptides have the same constituents on each end, a *C*-terminal (with the free carboxyl group) and the *N*-terminal (with the free amino group) (Fig. 1.5). This linear sequence of amino acids is termed the primary conformational structure of protein molecules.

When relatively few amino acids are included in a polypeptide, they are called oligopeptides. There are dramatic differences in the sizes of proteins from tiny hormone oligopeptides like vasopressin (nine amino acids) to massive molecules such as immunoglobulin M (IgM), which contain 10 polypeptide chains folded in a regular array containing over 6000 amino acids.

SECONDARY STRUCTURE

The primary structure folds into a regular secondary structure along one dimension in a simple structure, such as an α-helix, random coils or β-sheet. The secondary structure is held together mainly by hydrogen bonds that form between peptide bonds.[1]

Figure 1.5 During peptide bond formation, a molecule of water is given off. The resulting molecule has an *N*-terminal end with an amino group, and a *C*-terminal end with a carboxyl group.

TERTIARY STRUCTURE

More complex folding into a three-dimensional structure occurs as a result of formation of hydrogen bonds, ionic bonds, disulfide bonds, van der Waals forces and hydrophobic interactions. The hydrophobic interactions are the major forces holding the tertiary structure in place.[3] Because proteins usually are present in an aqueous environment, the hydrophobic portions of the structure tend to reside in the interior of the tertiary structure.[1]

QUATERNARY STRUCTURE

Some proteins have a quaternary structure that consists of complexes of polypeptide monomers. For example, although a single molecule of IgG is considered one protein, it contains two identical light polypeptide chains (220 amino acids each) and two identical heavy polypeptide chains (440 amino acids each) (Fig. 1.6). In these more complex proteins, the same forces involved in tertiary structures stabilize its folded structure (Fig. 1.6).[1,3] An important part of understanding protein structure is to realize that the tertiary and quaternary structures are not rigid. Portions of the amino acid chain can move under certain circumstances.[1] Further, this movement is often key to the function of the molecule, such as in α_1-antitrypsin (see Chapter 4) where movement of a large domain is key to its inhibition of enzymes.[4]

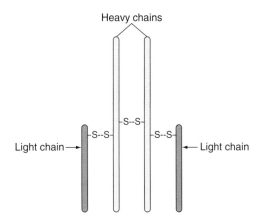

Figure 1.6 General structure of immunoglobulin G (IgG). Two identical heavy chains and two identical light chains are held together by disulfide bonds.

GLYCOSYLATION

Proteins that have carbohydrate groups attached covalently during synthesis or secretion are said to be *glycosylated*. Some proteins exist in both glycosylated and non-glycosylated forms. The existence of these molecules is of importance in understanding electrophoretic patterns and in diagnosing conditions such as leakage of cerebrospinal fluid from a skull injury (where identification of desialated transferrin versus sialated transferrin is key) (see Chapter 8).[5] Functionally, the attached carbohydrate groups protect the protein from digestion or help stabilize the conformation of the protein and may affect its clearance.[6] The attached carbohydrate groups can also affect the charge of the protein and, consequently, its migration during electrophoresis.

POST-TRANSLATION MODIFICATION

Although the amino acid sequence is determined by the DNA sequence, post-translational phosphorylation, N-terminal acetylation, acylation of sidechains, C-terminal α-amidation of glycine, sulfation of tyrosine groups, and γ-carboxylation of glutamic acid may all occur, resulting in changes in function and migration of the proteins involved.[1,7–10] Such post-translational modifications of protein explain why monoclonal protein samples produce several points (or blobs) when analysed by two-dimensional electrophoresis instead of the single band seen on immunofixation or serum protein electrophoresis.[11]

Because the composition of amino acids and carbohydrates in a protein is unique, each protein has a specific charge and migration pattern under defined conditions during electrophoresis. Protein molecules, like their constituent amino acids, have their overall charge determined by the pH of the solvent; consequently, there is a specific pH at which the negative and positive charges balance, and at which the protein will not migrate. The pH at which the positive and negative charges of a given protein balance is referred to as its *isoelectric point* (pI). The pI is constant and highly specific for a given protein molecule.[1]

When a protein is dissolved in a solution that is acidic relative to that protein's pI, it will gain protons and migrate toward the cathode. In solutions that are basic relative to that protein's pI, the protein will donate protons and migrate toward the anode. Although there are charge effects of sulfhydryl groups for proteins in the serum, urine, and cerebrospinal fluid, the amount of charge due to free sulfhydryl groups is negligible with regard to electrophoretic migration.

ELECTROPHORETIC TECHNIQUES IN CLINICAL LABORATORIES

Moving boundary electrophoresis

Analysis of proteins for clinical purposes began in the middle of the nineteenth century. Schmidt used the term globulin in 1862 to describe proteins that were insoluble in water.[12] By the early twentieth century, serum proteins were broadly divided into albumin and globulins, depending on the precipitation of 'globules' after the addition of sodium sulfate to serum proteins. This process would result in formation of a white residue, albumin (*alba* Latin for white), after the salt was removed by dialysis and the water was evaporated.[13] The albumin to globulin ratio was an early, crude index for evaluating liver pathology.

Studies of the electrophoretic mobility of proteins were first carried out by Arne Tiselius in the 1930s employing a liquid medium.[14] For these studies, Tiselius devised a U-shaped electrophoretic cell and employed a Schlieren band optical system to detect the degree of refraction of light by proteins as they moved through the tube. He found that there was a difference in the refractive index of light at boundary interfaces between major protein fractions as they moved by the light detection device. Thus, this technique was called moving boundary electrophoresis. For his technique, Tiselius dissolved a specific volume of a protein mixture in buffer, and carefully layered the solution on the electrophoresis tube below the same buffer. The buffer was placed in contact with electrodes, and the sensitive optical band method was used to monitor the progress of protein fractions during electrophoresis.

When the current was turned off at the end of the run, a mixing of the boundaries occurred and only the fractions at the extreme cathodal and anodal ends of the tube could be collected in a relatively purified form. Therefore, originally, this moving boundary electrophoresis was used to test the success of protein purification that had been performed by other means. Although crude by today's standards, such early techniques were sufficiently sensitive to allow Michaelis to determine isoelectric points of the marginally purified enzymes he had for study.[15] This moving boundary technique was also instrumental in the original definition of the major fractions of human serum proteins as albumin, α_1-, α_2-, β-, and γ-globulin.

Zone electrophoresis

Zone electrophoresis involved the use of a solid support medium with staining by a protein dye to visualize the major protein bands. One of the key practical problems with moving boundary electrophoresis was its inability to achieve a complete separation of electrophoretically adjacent major protein fractions. Also, the refractive index that was used to quantify the proteins in moving boundary electrophoresis was limited in its discrimination of subtle differences. The development of zone electrophoresis made it possible to overcome these difficulties by providing a stable support medium in which proteins could migrate, be stained and quantified. Zone electrophoresis, then, offered the important feature of stabilizing the migration of the proteins that moving boundary electrophoresis could not achieve.

The first major supporting medium for electrophoresis was filter paper. Studies using this support medium were begun as early as 1937. However, it was not until the early 1950s that these techniques were simplified and rigorously defined for practical use in clinical laboratories.[16]

The use of filter paper as a support medium introduced new variables into electrophoresis. As with the moving boundary technique, the migration of individual proteins depended on the pI of the molecule, the pH of the buffer, electrolyte concentration of the buffer, and amount of current applied. However, the texture of the filter paper was found to be another important factor because it offered substantially more resistance to the movement of the proteins than was the case in the free-moving boundary system.

Early in the development of paper electrophoresis, Kunkel and Tiselius observed that the texture of paper products differed, and that this resulted in different migration of proteins depending on the brand and lot of paper used.[17] Although the actual distance that human serum albumin would migrate may be greater on one brand of paper than on another brand, the relationship between albumin and the subsequent major fractions was relatively constant, and could be used to create a correction factor specific for that preparation of filter paper. It was also observed that the type of paper affected the migration of smaller molecules less than larger molecules.[18]

The use of paper as a support medium also introduced the effect known as *electroosmosis* or *endosmosis* as another factor influencing the migration of proteins. The support medium (filter paper, in the present case, but also cellulose acetate and agarose) contains anionic groups that possess a negative charge relative to the buffer solution. Obviously, each support medium is stationary and cannot migrate; however, the positively charged ions in the buffer solution flow toward the negative electrode (cathode).

An understanding of endosmosis is important in order to relate the migration of proteins to their surface charge. For example, most electrophoretic systems that are used in the study of human serum, cerebrospinal fluid, or urine proteins require an alkaline buffer with a pH of 8.6. At this pH, almost all serum proteins, including the γ-globulins, will have a negative charge. Yet, they do not all migrate toward the anode. Most of the γ-globulins and some β-globulins may migrate toward the cathode (Fig. 1.7). This is because of movement of the cationic ions in the buffer toward the cathode. Depending on the amount of negative charge of the support medium, there will be an equal, but opposite (positive) charge of the buffer in the adjacent support medium. This will pull the molecules that have a weaker negative charge, owing to their lower pI, toward the cathode (Fig. 1.8). These molecules are not moving toward the cathode because they have a positive charge under the conditions of the assay; rather, they are just 'weak swimmers' (weak negative charge) caught in a futile attempt to swim against the flow of a strong river.

There were some major problems with paper electrophoresis that limited its application. Paper electrophoresis was slow, requiring several hours (often overnight) in order to achieve adequate separation of major protein fractions. Furthermore, it was opaque (frustrating early densitometric scanners), gave poor resolution, and had significant problems with non-specific protein absorption.[19] To quantify proteins from individual

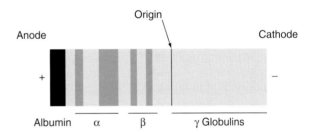

Figure 1.7 Schematic of electrophoretic pattern demonstrates that the γ-globulins migrate toward the cathode under typical conditions which use buffer with a pH of 8.6.

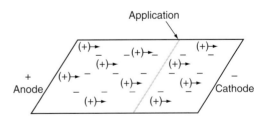

Figure 1.8 Endosmosis. Side view of electrophoresis illustrates negative charge of support medium and flow of positively charged buffer (+) to the cathode. Such flow affects migration of proteins in the support medium.

bands, they would be cut out, eluted and subjected to a protein assay. Therefore, it is not surprising that a search was conducted for better support media for protein electrophoresis.

Cellulose acetate and agarose became popular stabilizing media for the clinical laboratories in the 1960s and 1970s. With these media, electrophoresis could be performed in less than an hour, and the clarity of the media facilitated densitometric scanning to estimate the protein concentration of the major fractions.

For several years, cellulose acetate electrophoresis was the most popular method for performing routine serum protein electrophoresis.[20] Cellulose acetate electrophoresis has advantages over paper electrophoresis: only minimal adsorption of serum proteins occurs upon application, and a sharper separation of the major serum protein bands is obtained much more quickly than by paper electrophoresis. However, the resolution on traditional cellulose acetate systems is inferior to that obtained with most agarose gel electrophoresis systems (see below).[21] Cellulose acetate has been popular in the clinical laboratory because of its simplicity, reproducibility, reliable quantification of protein fractions by densitometry, and relatively low cost.[22]

The successful replacement of paper electrophoresis by cellulose acetate electrophoresis was facilitated by the demonstration that accurate estimates could be made of each major protein fraction by densitometric scanning. In 1964, Briere and Mull demonstrated that densitometric scanning of serum proteins separated by cellulose acetate electrophoresis gave the same measure of the major protein fractions as did elution and spectrophotometric measurement (Table 1.1).[23] Unfortunately, the relatively poor resolution of most commercially available cellulose acetate membranes limited the sensitivity of the technique. Subtle abnormalities such as heterozygotes for α_1-antitrypsin deficiency and small monoclonal gammopathies (especially those in the α_2- or β-regions) were often undetectable. The insensitivity of some earlier low-resolution methods was demonstrated by a College of American Pathologists Survey report.[24] It disclosed that many systems had a lower detection rate of a subtle monoclonal gammopathy that was picked up by most participants using electrophoretic systems with better resolution (Table 1.2).

Agar is a polysaccharide product that is produced commercially by boiling red algae, filtering out the larger impurities, and removing the water-soluble impurities by freeze-thawing.[25-27] After precipitation in ethanol, the mixture consists mainly of 1–4 linked 3,6-anhydro-α-L-galactose and 1–3 linked β-n-galactose.[28] The final agar gel is a chemically complex structure. For practical purposes, it contains varying quantities of agarose and agaropectin. Agaropectin has a relatively high sulfate, pyruvate, and glucuronate content, which imparts a strong negative charge to the gel and results in considerable endosmotic flow; pure agarose has few anionic groups.[29]

Most commercial preparations of agar gels used for electrophoresis today contain relatively pure preparations of agarose. This minimizes nonspecific adsorption of some proteins (such as β-lipoprotein and thyroglobulin) to the agar and the amount of endosmotic flow.[30,31] Although purified agarose preparations substantially reduce the

Table 1.1 Comparison of densitometric scans on cellulose acetate with eluted protein concentration[a]

Method	Albumin (g/dl)	α_1 (g/dl)	α_2 (g/dl)	β (g/dl)	γ (g/dl)
Densitometry	4.5 ± 0.36	0.27 ± 0.08	0.62 ± 0.10	0.64 ± 0.12	0.95 ± 0.27
Elution	4.8 ± 0.34	0.22 ± 0.06	0.52 ± 0.09	0.59 ± 0.13	1.01 ± 0.25

[a]Data from Briere and Mull.[23]

Table 1.2 Detection of small monoclonal gammopathy by different electrophoretic systems[a]

Company	Monoclonal absent	Monoclonal present	Per cent correct
Worthington Panagel agarose	0	8	100
Beckman SPE2 agarose	1	17	94
Beckman SPE1 agarose	155	310	66
Helena agarose	65	127	66
CIBA Corning agarose	89	93	51
Beckman cellulose	3	2	40
Helena cellulose	169	91	35
Helena REP agarose	78	41	34
Gelman cellulose	17	6	28

[a]Data expresses the number of laboratories using the indicated technique. From CAP survey 1991.[24]

endosmotic effect, some manufacturers use substantial quantities of agaropectin to promote endosmosis. Some endosmotic flow is desirable for electrophoresis of serum proteins because it pulls the γ-globulins cathodally. With these systems, most serum monoclonal gammopathies migrate cathodally as do oligoclonal bands seen in cerebrospinal fluid from patients with multiple sclerosis. By moving these important bands away (cathodal) from the origin, these systems minimize the effect that minor distortions, often present at the point of application, have on interpretation of γ-region abnormalities. Distortions at the point of sample application can be especially problematic when one is dealing with a cryoglobulin that often precipitates at the origin (see Chapter 6).

Capillary zone electrophoresis

In the last decade, capillary zone electrophoresis (CZE) has been developed for use in clinical laboratories.[32–42] This technique is a liquid-based system that bears some similarities to the early Tiselius system in that no permanent gel is produced in the process, although available systems may create virtual gel images (see below).[42]

For CZE, a small volume of sample (1–5 μl) is aspirated into a thin fused silica capillary 25–50 μm in diameter (Fig. 1.9). The strong negative charge on the interior of the capillary, together with the narrow lining, provides a large net negative surface area. Under conditions of electrophoresis, this sets up a strong endosmotic flow of cations toward the cathode. In this system, the pull of this endosmotic flow is stronger than the pull of the anode for the

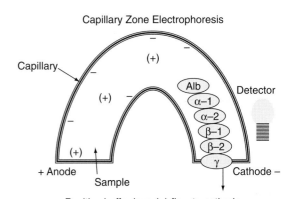

Figure 1.9 Capillary zone electrophoresis, schematic. The sample is aspirated into the anodal end of the 25–50 μm diameter fused silica capillary. The negative charge (−) on the interior of the capillary sets up a strong endosmotic flow of cations (+) toward the cathode. An ultraviolet detector evaluates absorbance at 200–215 nm by the fractions indicated.

anionic proteins being evaluated. The proteins then migrate toward the cathode, but are variously impeded in their migration, based upon the negative charge of the proteins. Thus, the electrical field separates the proteins by charge. An ultraviolet detector that evaluates the absorbance at 200–215 nm (in various systems) determines the protein concentration (Fig. 1.10). Since peptide bonds absorb at 214 nm, these systems provide quantitative measurements of the various proteins that are not influenced by the presence of carbohydrate groups. However, this method of measurement suffers from the disadvantage that other substances that absorb at this wavelength will produce bands that may mimic monoclonal proteins or genetic variants. Radiocontrast dyes have been the most notorious causes of confusion with the CZE system (see Chapter 2).[43]

EARLY CLINICAL APPLICATIONS OF ELECTROPHORESIS

It was soon recognized that when tissues responsible for the synthesis or excretion of proteins were altered by disease, the resulting serum would produce distinctive electrophoretic patterns that could be helpful in diagnosis (Table 1.3). For example, it was known as early as 1940 that in the

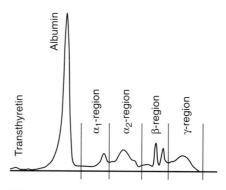

Capillary Zone Electrophoresis

Figure 1.10 Capillary zone electropherogram of serum performed on a Beckman Paragon CZE 2000.

nephrotic syndrome the serum contained markedly decreased levels of albumin and γ-globulin with increased levels of α_2-globulin.[44] This was due to loss of albumin and γ-globulin through the damaged glomeruli, with increased synthesis and retention of the large molecules in the α_2-region (α_2-macroglobulin and haptoglobin). At the same time, it was recognized that the urine from these patients contained the albumin lost from the serum as well as many other serum proteins. In reversible conditions, such as minimal change nephropathy, then termed lipoid nephrosis, a return to the normal serum electrophoretic pattern was noted after resolution of the renal disease.[45]

Table 1.3 Early clinical use of zone electrophoresis

Clinical diagnosis	Electrophoretic pattern	Reference
Nephrotic syndrome	Decreased albumin, decreased γ, increased α	Longsworth and MacInnes[44]
Liver disease	Decreased albumin, decreased β,[a] increased γ	Wajchenberg et al.[46]
Myeloma	Increased γ, decreased albumin	Reiner and Stern [49]
Agammaglobulinemia	Decreased γ	Bruton [51]
Active systemic lupus erythematosus	Increased γ	Coburn and Moore [54]
Multiple sclerosis, neurosyphilis	Increased CSF γ	Kabat et al.[56]

[a]Decreased beta in massive liver necrosis.

A small decrease in serum albumin was quickly recognized as a relatively nonspecific occurrence found in a variety of conditions that cause metabolic stress and as a feature of the acute-phase reaction pattern (discussed later). However, the level of serum albumin in patients with liver disease gave clinically useful information because it was significantly correlated with the amount of tissue damage.[46] Further, it was noted that patients with severe liver disease had a broad elevation of γ-globulin,[47] although the immunological significance of that observation would not be understood for several years.

Detection of monoclonal gammopathies in serum and urine is one of the most important uses of clinical protein electrophoresis. Many different types of electrophoretic patterns can result from products of the neoplastic B-lymphocyte and plasma cell proliferation that occur in chronic lymphocytic leukemia and multiple myeloma, respectively. Most frequently, patients with multiple myeloma have markedly elevated γ-globulin regions with a restriction in the migration (Fig. 1.11), although abnormalities may be found anywhere from the $α_1$- to the γ-globulin region.

The older, paper zone electrophoresis techniques

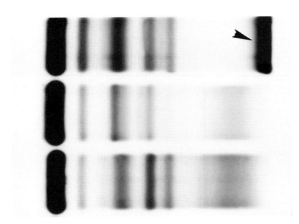

Figure 1.11 Three electrophoretic patterns are shown with the anode at the left and the cathode at the right. The location of the anode can be assumed to be on the same side as the albumin band. Note the prominent area of restriction in the slow γ-region of the top pattern (arrow). This pattern is typical of patients with multiple myeloma. (Paragon SPE2 gel system stained with Paragon Blue.)

were occasionally too insensitive to detect the abnormal serum protein in some of these patients.[48] Using these early techniques, some studies reported as many as 22 per cent of patients with multiple myeloma had no significant abnormality in the serum.[49] Some of their patients had light chain disease with monoclonal free light chains (MFLC, formerly termed Bence Jones proteins) in their serum and, more frequently, in their urine. Since immunoglobulin light chains are relatively small molecules (25 kDa as monomers to 50 kDa as dimers), they would pass into the urine and only a minimal, or no monoclonal restriction was seen with the older zone serum protein electrophoresis techniques. Many of their patients might have been expected to have decreased γ-globulins; this decrease is subtle, probably below the ability of the early methods to discern. The MFLC were detected best by electrophoresis of urine (see Chapter 7).[50]

A major step forward in understanding the basic immunology of the human immune system involved the early application of serum protein electrophoresis by Colonel Ogden Bruton at the Walter Reed Army Medical Center.[51] One of his patients was a boy with a history of recurrent pyogenic infections. When protein electrophoresis was performed on this child's serum, it was discovered that the γ-globulin region was absent (in reality it was very low, but undetectable with the zone electrophoretic techniques available in 1952). By using γ-globulin replacement therapy, Bruton was able to successfully treat this individual. Because we understand the inheritance pattern of this condition, today it is referred to as X-linked agammaglobulinemia.[52] Early workers also recognized the existence of other forms of immunodeficiency diseases associated with low γ-globulin levels on electrophoresis of serum. When seen in infants, they often represented transient hypogammaglobulinemia of infancy; in young adults, the common variable immunodeficiency syndrome was the most frequent cause of isolated hypogammaglobulinemia.[53]

Broad increases in the γ-globulin region were found to be associated with the immune response to infectious agents and occasionally were reported in autoimmune diseases. For instance, Coburn and Moore[54] found that patients with clinically active systemic lupus erythematosus had elevated levels of γ-globulin. These findings preceded by 5 years the demonstration by Hargraves et al.[55] of the autoimmune phenomenon called the LE (lupus erythematosus) cell, and represent one of the earliest laboratory observations suggesting the complicity of γ-globulin in the pathogenesis of this disease.

The availability of zone electrophoresis encouraged many investigators to examine a wide variety of fluids and extracts. Not surprisingly, considering its ready availability and long history of use in the diagnostic laboratory, urine was one of the first fluids studied. Urine from patients with nephrotic syndrome contained considerable albumin and some globulins in amounts that gave an inverse correlation with the serum concentrations of these proteins.[44] Monoclonal free light chains had long since been described by Henry Bence Jones, but with the advent of zone electrophoresis it became apparent that they had greater heterogeneity than previously thought. Although MFLC were always globulins, they migrated anywhere from the α- through the γ-region (Fig. 1.12). This finding raised important questions about the structure of these molecules, which were previously assumed to be homogeneous by virtue of their peculiar thermo-precipitating characteristics (see Chapter 7).[50]

Because of the difficulty involved in diagnosing

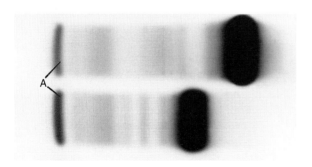

Figure 1.12 Two electrophoretic patterns of concentrated urine from patients with multiple myeloma. Both samples have relatively little albumin (A), but large restrictions in the slow γ- (top sample) and fast γ- (bottom sample) regions. These are both monoclonal free light chains (MFLC) which have significantly different migrations. (Paragon SPE2 gel stained with Paragon Blue.)

many central nervous system disorders, it was logical to use electrophoresis to examine cerebrospinal fluid for any new clues that might aid clinicians. Analysis of concentrated cerebrospinal fluid from patients with multiple sclerosis and neurosyphilis showed markedly elevated γ-globulin content.[56] Similar elevations in the γ-globulin were found in central nervous system infections.

Despite these and many other observations, the clinical applications of zone electrophoresis were limited to obvious extreme elevations or reductions of major protein components. Many diseases were associated with more subtle alterations of proteins that were beyond the limitations of the early zone electrophoretic methods. Better resolution of protein bands, simpler methods to quantify the fractions, and greater sensitivity were required for protein electrophoresis to aid in the diagnosis of these conditions.

With the earlier methods, described above, it was arguable whether agarose gels could provide better resolution of the major protein bands than cellulose acetate. The heterogeneity of agar preparations and the ready availability of pure, commercially prepared cellulose acetate caused much wider usage of the latter. In addition, cellulose acetate allowed more rapid electrophoresis and could be dried, stained, and cleared more easily than agar.[29]

Around 1970, however, reports appeared that demonstrated advantages of the careful, high-resolution agarose electrophoresis system described by Wieme.[31] One could achieve diagnostically useful results on agarose with techniques and equipment well within the capabilities of the clinical laboratory.[57] The availability of more highly purified agarose preparations, which minimized endosmotic flow and offered good optical clarity and unimpeded migration because of their porosity, was credited with some of the improvement in resolution.[58] By the mid-1970s, using the modifications described in the next section, a few agarose and cellulose acetate methodologies had evolved that facilitated the consistent demonstration of up to 12 distinct protein fractions. It is certainly debatable whether one needs to evaluate so many proteins for clinical purposes. Most of the proteins demonstrated by electrophoresis are readily available for specific assay by nephelometry or other immunoassay techniques. The better quality of the more recent electrophoretic techniques has, however, improved the ability to detect subtle monoclonal proteins, as documented by the results of an independent study of electrophoretic results by a College of American Pathologists' survey sample (Table 1.2). Systems with the better overall resolution had higher detection rates of the small M-protein than systems with poorer resolution.

REFERENCES

1. Jones AJS. Analysis of polypeptides and proteins. *Adv Drug Del Rev* 1993;**10**:29–90.
2. Levo Y. Nature of cryoglobulinaemia. *Lancet* 1980;**1**:285–287.
3. Johnson AM, Rohlfs EM, Silverman LM. Chapter 20. Proteins. In: Burtis CA, Ashwood, ER, eds. *Tietz textbook of clinical chemistry*. Philadelphia: WB Saunders, 1999.
4. Carrell RW, Lomas DA. Alpha1-antitrypsin deficiency – a model for conformational diseases. *N Engl J Med* 2002;**346**:45–53.
5. Zaret DL, Morrison N, Gulbranson R, Keren DF. Immunofixation to quantify beta 2-transferrin in cerebrospinal fluid to detect leakage of cerebrospinal fluid from skull injury. *Clin Chem* 1992;**38**:1908–1912.
6. Hotchkiss A, Refino CJ, Leonard CK, et al. The influence of carbohydrate structure on the clearance of recombinant tissue-type plasminogen activator. *Thromb Haemost* 1988;**60**:255–261.
7. Furie B, Furie BC. Molecular basis of vitamin K-dependent gamma-carboxylation. *Blood* 1990;**75**:1753–1762.
8. Huttner WB. Tyrosine sulfation and the secretory pathway. *Annu Rev Physiol* 1988;**50**:363–376.
9. Eipper BA, Mains RE. Peptide alpha-amidation. *Annu Rev Physiol* 1988;**50**:333–344.
10. James G, Olson EN. Fatty acylated proteins as components of intracellular signaling pathways. *Biochemistry* 1990;**29**:2623–2634.
11. Harrison HH, Miller KL, Abu-Alfa A, Podlasek SJ. Immunoglobulin clonality analysis. Resolution of ambiguities in immunofixation electrophoresis results by high-resolution, two-dimensional electrophoretic analysis of paraprotein bands eluted from agarose gels. *Am J Clin Pathol* 1993;**100**:550–560.
12. Joliff CR. Analysis of the plasma proteins. *J Clin Immunoassay* 1992;**15**:151–161.
13. Harrison HH, Levitt MH. Serum protein electrophoresis: basic principles, interpretations, and practical considerations. *Check Sample (ASCP)* 1987;**7**:1–16.
14. Tiselius A. A new apparatus for electrophoretic analysis of colloidal mixtures. *Trans Faraday Soc* 1932;**33**:524–531.
15. Tiselius A. Introduction. In: Bier M, ed. *Electrophoresis, theory, methods and applications*. New York: Academic Press, 1959.
16. Durrum EL. A microelectrophoretic and microionophoretic technique. *J Am Chem Soc* 1950;**72**:2943–2948.
17. Kunkel HG, Tiselius A. Electrophoresis of proteins on filter paper. *J Gen Physiol* 1951;**35**:39–118.
18. McDonald HJ. Polyvinylpyrrolidine: the electromigration characteristics of the blood plasma expander. *Circ Res* 1953;**1**:396–404.
19. Johansson BG. Agarose gel electrophoresis. *Scand J Clin Lab Invest Suppl* 1972;**124**:7–19.
20. Kohn J. A cellulose acetate supporting medium for zone electrophoresis. *Clin Chim Acta* 1957;**2**:297–304.
21. Cawley LP, Minard B, Penn GM. *Electrophoresis and immunochemical reactions in gels. Techniques and interpretations*. Chicago: ASCP Press, 1978.
22. Kohn J. Small-scale membrane filter electrophoresis and immunoelectrophoresis. *Clin Chim Acta* 1958;**3**:450–454.
23. Briere RO, Mull JD. Electrophoresis of serum protein with cellulose acetate. A method for quantitation. *Am J Clin Pathol* 1964;**34**:547–551.
24. College of American Pathologists. Survey Report EC-07. 1991.

25. Serwer P. Agarose gels: properties and use for electrophoresis. *Electrophoresis* 1983;4:375–382.

26. Serwer P. Improvements in procedures for electrophoresis in dilute agarose gels. *Anal Biochem* 1981;112:351–356.

27. Serwer P, Hayes SJ. Exclusion of spheres by agarose gels during agarose gel electrophoresis: dependence on the sphere's radius and the gel's concentration. *Anal Biochem* 1986;158:72–78.

28. Rees DA. Structure, conformation, and mechanism in the formation of polysaccharide gels and networks. *Adv Carbohydr Chem Biochem* 1969;24:267–332.

29. Nerenberg ST. *Electrophoretic screening procedures*. Philadelphia: Lea and Febiger, 1973.

30. Guo Y, Xinhui L, Yong F. The effects of electroendosmosis in agarose electrophoresis. *Electrophoresis* 1998;19:1311–1313.

31. Wieme R. *Agar gel electrophoresis*. Amsterdam: Elsevier, 1965.

32. Jenkins MA, O'Leary TD, Guerin MD. Identification and quantitation of human urine proteins by capillary electrophoresis. *J Chromatogr B Biomed Appl* 1994;662:108–112.

33. Jenkins MA, Kulinskaya E, Martin HD, Guerin MD. Evaluation of serum protein separation by capillary electrophoresis: prospective analysis of 1000 specimens. *J Chromatogr B Biomed Appl* 1995;672:241–251.

34. Jenkins MA, Guerin MD. Quantification of serum proteins using capillary electrophoresis. *Ann Clin Biochem* 1995;32:493–497.

35. Jenkins MA, Guerin MD. Optimization of serum protein separation by capillary electrophoresis. *Clin Chem* 1996;42:1886.

36. Jenkins MA, Guerin MD. Capillary electrophoresis as a clinical tool. *J Chromatogr B Biomed Appl* 1996;682:23–34.

37. Jenkins MA. Clinical applications of capillary electrophoresis. Status at the new millennium. *Mol Biotechnol* 2000;15:201–209.

38. Jenkins MA. Three methods of capillary electrophoresis compared with high-resolution agarose gel electrophoresis for serum protein electrophoresis. *J Chromatogr B Biomed Sci Appl* 1998;720:49–58.

39. Katzmann JA, Clark R, Sanders E, Landers JP, Kyle RA. Prospective study of serum protein capillary zone electrophoresis and immunotyping of monoclonal proteins by immunosubtraction. *Am J Clin Pathol* 1998;110:503–509.

40. Keren DF. Capillary zone electrophoresis in the evaluation of serum protein abnormalities. *Am J Clin Pathol* 1998;110:248–252.

41. Bienvenu J, Graziani MS, Arpin F, et al. Multicenter evaluation of the Paragon CZE 2000 capillary zone electrophoresis system for serum protein electrophoresis and monoclonal component typing. *Clin Chem* 1998;44: 599–605.

42. Jolliff CR, Blessum CR. Comparison of serum protein electrophoresis by agarose gel and capillary zone electrophoresis in a clinical setting. *Electrophoresis* 1997;18:1781–1784.

43. Bossuyt X, Mewis A, Blanckaert N. Interference of radio-opaque agents in clinical capillary zone electrophoresis. *Clin Chem* 1999;45:129–131.

44. Longsworth LG, MacInnes DA. Electrophoretic study of nephrotic sera and urine. *J Exp Med* 1940;71:77–86.

45. Lenke SE, Berger HM. Abrupt improvement of serum electrophoretic pattern in nephrosis after ACTH-induced diuresis. *Proc Soc Exp Biol Med* 1951;78:366–369.

46. Wajchenberg BL, Hoxter G, Segal J, et al. Electrophoretic patterns of the plasma proteins in diffuse liver necrosis. *Gastroenterology* 1956;30:882–893.

47. Franklin M, Bean WB, Paul WD, Routh JI, de la Hueraga J, Popper H. Electrophoretic studies in liver disease. I. Comparison of serum and plasma electrophoretic patterns in liver disease, with special reference to fibrinogen and gamma globulin patterns. *J Clin Invest* 1951;30:718–728.

48. Moore DH. Clinical and physiological applications of electrophoresis. In: Bier M, ed. *Electrophoresis theory, methods and applications*. New York: Academic Press, 1959.

49. Reiner M, Stern KG. Electrophoretic studies on the protein distribution in the serum of multiple myeloma patients. *Acta Haematol* 1953;9:19–29.

50. Moore DH, Kabat EA, Gutman AB. Bence Jones proteinemia in multiple myeloma. *J Clin Invest* 1943;22:67–75.

51. Bruton OC. Agammaglobulinemia. *Pediatrics* 1952;9:722–727.

52. Noordzij JG, de Bruin-Versteeg S, Comans-Bitter WM, et al. Composition of precursor B-cell compartment in bone marrow from patients with X-linked agammaglobulinemia compared with healthy children. *Pediatr Res* 2002;51:159–168.

53. Gitlin D. Low resistance to infection: relationship to abnormalities in gamma globulins. *Bull NY Acad Med* 1955;31:359–365.

54. Coburn AF, Moore DH. The plasma proteins in disseminated lupus erythematosus. *Bull Johns Hopkins Hosp* 1943;73:196–214.

55. Hargraves MM, Richmond H, Morton R. Presentation of two bone marrow elements. The 'tart' cell and the 'LE' cell. *Mayo Clin Proc* 1948;23:25–28.

56. Kabat EA, Moore DH, Landow H. An electrophoretic study of the protein components in cerebrospinal fluid and their relationship to the serum proteins. *J Clin Invest* 1942;21:571–577.

57. Rosenfeld L. Serum protein electrophoresis. A comparison of the use of thin-layer agarose gel and cellulose acetate. *Am J Clin Pathol.* 1974;62:702–706.

58. Elevitch FR, Aronson SB, Feichtmeir TV, Enterline ML. Thin gel electrophoresis in agarose. *Tech Bull Regist Med Technol* 1966;36:282–287.

2

Techniques for protein electrophoresis

PRINCIPLES OF PROTEIN ELECTROPHORESIS

When proteins migrate in an electrical field, the extent of their migration and the degree of the resolution of each band depend on several factors. Two key factors that affect the migration of any protein are its pI (see Chapter 1) and the pH of the buffer. The pI of any given protein is constant and dependent on its amino acid and carbohydrate content. However, the charge that the protein expresses is determined by the pH of the solution in which it is dissolved. For example, a protein such as fibrinogen has a pI of 5.5. In an electrophoresis buffer with a pH of 8.6, it donates protons to the buffer and is left with a net negative charge. However, in a solution with a pH of 4.0, it would accept a proton and has a net positive charge.

The amount of resolution in protein electrophoresis can be improved from the five-band method to better discriminate between separate but closely migrating major protein components. Important factors in achieving improved resolution include optimizing the velocity of migration, minimizing passive diffusion, and avoiding interactions of proteins with the supporting medium. Each can be influenced by adjusting the variables in the electrophoretic system. Excellent detailed discussions of these factors can be found in Wieme,[1] and Briere and Mull.[2]

The speed with which a protein migrates in an electrical field (electrophoretic mobility) under defined conditions of pH, ionic strength, temperature, and voltage is characteristic for that protein. The formula defining the variables involved in calculating the electrophoretic mobility (μ) of a protein is

$$\mu = \frac{d}{Et}$$

where d is the distance traveled from the origin in centimeters, E is the strength of the electrical field in V/cm, and t is the duration of electrophoresis in seconds. Because the strength of the electrical field is inversely proportional to its length, that is, V/cm, a shorter support medium will permit faster separation of proteins.

Increasing the voltage results in a faster separation of proteins; unfortunately, it also results in more heat generation, which is deleterious to resolution of individual bands. The amount of heat generated (in joules) when the electric current passes through the apparatus can be calculated by:

$$\text{Heat generated} = \frac{xE^2}{A}$$

where x is the specific conductance of the apparatus, E is the strength of the electric field in V/cm, and A is the mechanical heat equivalent. From this, it follows that heat production increases exponentially as the voltage is increased.

This excessive heat production plays havoc with good resolution of electrophoretic bands. One of the major effects of heat is to increase the thermal agitation and hence the diffusion of the protein molecules. Diffusion broadens the width of a band, thereby decreasing the resolution. Heat production can also decrease the viscosity of the medium. Although this does permit a more rapid electrophoretic migration (μ) of the proteins through the gel, it is more than counterbalanced by an even greater increase in diffusion, with a resulting decrease in resolution. Before closed systems were common, the heat generated further complicated resolution by causing enough evaporation to change ionic strength.

The ionic strength of the buffer is also an important factor in the resolution of individual protein bands. As the concentration of the salt ions in a buffer increases, the velocity of electrophoretic migration decreases for each protein being assayed. There is no effect, however, on the relative migration of serum proteins as a result of ionic strength. The effect of ionic concentration on the migration of proteins in the electric field is largely the result of interaction of the buffer ions with the surface charges on the protein.

Consider a buffer in which we increase the concentration of NaCl. At the typical pH 8.6 of agar gel electrophoresis, human serum albumin has a negative surface charge. The positive sodium ions are attracted to the negative charges on albumin and diminish its effective net negative charge in the solution (Fig. 2.1). Further, positively charged ions, now in immediate proximity to the albumin, are attracted to the cathode during electrophoresis and tend to retard the progress of albumin toward the anode. This accumulation of positive charges in the buffer around the negatively charged albumin is known as the diffuse double layer. This is why it is important to control evaporation with resultant concentration of ions in the buffer during electrophoresis.[1]

Another factor limiting the effective separation of protein bands is adsorption of the molecules to the agar gel itself. Because of the negative charges possessed by the relatively purified agarose solutions used today, pH < 5.0 is impractical. Below this pH, serum proteins would have a positive charge and would precipitate in the gel.

Depending upon buffer strength, voltage, heat dissipation, purity and thickness of the gel, available electrophoretic systems display from five to as many as 12 protein bands, which encompass more than 95 per cent of the total mass of serum proteins.[3] A comparison of the resolutions generally available is shown in Fig. 2.2. Better resolution is an important part of detecting monoclonal gammopathies.[4,5]

To improve the detection of monoclonal gammopathies, the Protein Commission of the Societa Italiana di Biochimica Clinica published guidelines for criteria for performance of sensitive electro-

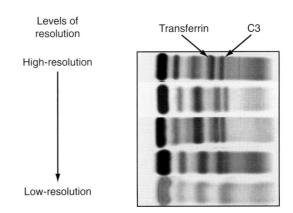

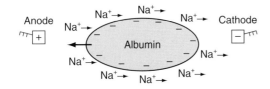

Figure 2.1 At the pH of the typical serum protein electrophoresis gel (8.6), albumin has a strong negative charge. The positive ions in the buffer, in this case sodium, are attracted to the negative charges on albumin and diminish its migration toward the anode.

Figure 2.2 This composite figure illustrates the resolution of different samples run on different electrophoretic analyses. Examining the β-region bands (transferrin and C3) helps to evaluate the resolution available by the different techniques.

phoresis procedures. Aguzzi et al.[5] summarized these recommendations as follows:

1. It should be possible to see the faint transthyretin (prealbumin) band in the serum of all healthy persons.
2. It should be possible to detect, if present, the heterozygosity of α_1-antitrypsin.
3. It should be possible to recognize the two main components of the α_2-zone (haptoglobin and α_2-macroglobulin).
4. The two main components of the β-zone (transferrin and C3) should be clearly resolved as two distinct bands.
5. In the γ-zone, it should be possible to recognize the presence of small monoclonal components (< 1 g/l, i.e. < 100 mg/dl) and possibly the oligoclonal pattern.
6. Zonal terminology and densitometric reporting should not be used, and results should be expressed in terms of qualitative and semiquantitative variations of specific proteins.

Similarly, recently published conclusions from a conference on guidelines for clinical and laboratory evaluation patients with monoclonal gammopathies recommends the use of systems that provide resolution sufficient to separate β_1 (transferrin) and β_2 (C3) proteins to improve detection of monoclonal gammopathies.[4] Measurements of proteins present in concentrations smaller than 10 mg/dl require immunoassay techniques.[6]

ELECTROPHORESIS ON AGAROSE

The basic principles of electrophoresis apply to both the manual and the automated systems (see below). The method of Wieme, as modified by Johansson, is commonly used with agarose.[7] A 1 per cent concentration of agarose is used in 0.075 M, pH 8.6 barbital buffer containing 2 mM calcium lactate. The calcium ions are especially useful for improving the resolution in the β-region. Commercially available agarose electrophoresis

kits usually have a uniform thin (about 1 mm) layer of agarose on an inert plastic support.

The specimen must be applied to the agarose surface in a very narrow band. In manual systems, excess moisture is removed from the surface of the gel by blotting with filter paper. The blotting is needed to help the proteins diffuse into the gel and to prevent excessive lateral movement at the point of application. An inadequately blotted gel will have distortions in all the bands (Fig. 2.3). A plastic template with uniform narrow slits for sample application is firmly layered onto the blotted gel. The template should be applied evenly to the surface of the agarose so that no air pockets are present; these may distort the application of the sample. In most systems, 3–5 µl of sample are placed over each slit and allowed to diffuse into the gel for 5–7 min. Consistency and attention to detail in sample application are extremely important in manual techniques because the final bandwidth and configuration are determined by the initial application. For example, in Fig. 2.4, a small drop of serum fell on the top pattern prior to electrophoresis, affecting the pattern.

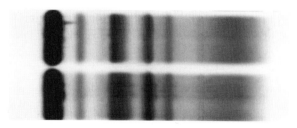

Figure 2.3 Serum protein electrophoretic strip of two sera stained from an inadequately blotted gel. Note the distortion (irregularities) in all bands of these samples. The occurrence of the distortion in all bands indicates that this is an artifact due to an application problem (in this case insufficient blotting). (Paragon SPE2 system stained with Paragon Violet.)

Occasional gel preparations have distortions arising from their initial preparation, or from problems with storage. A distortion in the gel may give a pattern like that in Fig. 2.5. The top sample shows a normal migration. However, the albumin bands of the next three samples show a distinctive

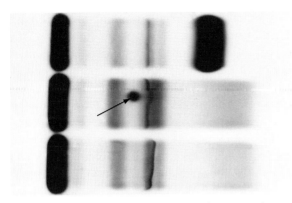

Figure 2.4 The middle sample in this electrophoretic strip has a dark elliptical spot overlying the α_2- to β_1-region (arrow). This represents a drop of serum falling on the gel (about in the α_1-region) prior to electrophoresis. This did not interfere with the interpretation of this sample, which has a subtle β-region band. However, such distortions can be problematic in more subtle cases. They also make very ugly pictures. (Paragon SPE2 system stained with Paragon Violet.)

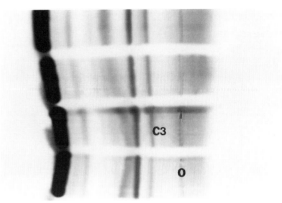

Figure 2.5 The top sample in this gel is a normal serum protein electrophoresis pattern. The next three samples show gel distortion manifest by a bowing of the albumin and α_1-region of the second sample, and of albumin, α_1- and α_2-regions of the bottom two samples. Note that the origin artifact labeled O in the bottom sample, and the C3 band in the β_2-region of the bottom two samples do not show any distortion. This indicates that the application itself was not the problem. (Panagel SPE2 system stained with Amido Black. Note the lighter staining pattern than that in Fig. 2.3 stained with Paragon Violet.)

bowing. The origin artifact (O) and C3 in the β-2 region do not have this distortion. This indicates the problem was not in the application, rather in the gel itself toward the anode.

On the semi-automated, gel-based systems, application devices are used along with automated washing to streamline the technical process. The Helena Rep Unit (Helena Laboratories, Beaumont, MI, USA) has been available for several years and the Sebia Hydrasis (Sebia, Issy-les-Moulineaux, France) has become available more recently. The Sebia Hydragel $\beta_1-\beta_2$ 15/30 method provides serum and urine gel results with crisp separation of the $\beta_1-\beta_2$ region as mentioned above. However, when processing urine samples the technologist needs to be careful of samples that contain solid matter (cells, crystals, etc.). This may interfere with the wicking of the sample onto the gel. If the sample does not wick properly, no sample is applied to the gel. The electrophoretic result on a urine containing a large amount of protein will then appear as though no protein were present. Because of this, I recommend comparing the total protein on the

urine with the electrophoretic pattern to be sure the sample applied. This is not a problem in serum both because the lack of visible bands is an obvious problem for any serum, but even in concentrated normal urine, albumin may not be seen (see Chapter 7).

After sample application is complete, some mechanism for cooling the gel is used. With some systems, the gels are cooled by convection. In other systems, gels are oriented such that their plastic backing is in direct contact with a cooling block (typically kept at 4°C prior to electrophoresis). The cooling block must be properly prepared and stored; if it is not at the proper temperature, the heat generated from the voltage applied to these samples will produce the effects described above and poor resolution of bands will result. A Peltier cooling device is used to control the heat in some of the automated systems.

The amount of buffer in the reservoir is another important factor. If there is too much buffer in the reservoir, the migrating γ-region, upon reaching the buffer, will form an artifactual slow γ-band (Fig. 2.6). Usually, this is obvious because all of the γ-regions have bands. However, when one sample has a relatively large amount of polyclonal, slow migrating γ-globulin (as may occur in patients with chronic active hepatitis), a slow γ restriction may be seen that could be mistaken for a monoclonal gammopathy.

Most agarose systems run with an electrical field of about 20 V/cm (a setting of 200 V for each 10-cm length of agarose) and a current of about 100–120 mA. Under these conditions, the typical run lasts 30–50 min. When electrophoresis has been completed, the proteins are fixed with an acid fixative. Some systems still use picric acid for this step but most new methods do not. (Note that picric acid can become explosive when stored for long periods of time.[8] Good laboratory technique, including checking the bottle for the expiry of the reagent and for crystallization around bottle caps, is especially important with this reagent.) After fixing the proteins, the gel is dried with a gentle stream of hot air for 5–10 min (in our laboratory, we just play recordings of my old lectures!).

For examination of serum and urine proteins, we prefer staining the gel with Amido black. Some commercial suppliers have their own versions of similar dye. Whereas both Coomassie Brilliant Blue and Amido Black give similar patterns, the Amido Black has less background between major bands, which makes interpretation of serum and urine electrophoretic patterns more straightforward.[7,9] Part of the difference is that Coomassie Brilliant Blue is more sensitive than Amido Black and stains small protein molecules at these sites. For cerebrospinal fluid, where sensitivity can be a problem even after concentrating a sample 80-fold, Coomassie Brilliant Blue is preferred. Coomassie Brilliant Blue-stained gels give better results for black and white photography.

ELECTROPHORESIS ON CELLULOSIC MEDIA

Cellulose acetate has been available as a supporting media for protein electrophoresis since the late 1950s.[10] The clarity of the background made this a considerable improvement over filter paper. Cellulose acetate has the advantage of uniform porosity. As with agarose, a wide variety of systems are available with cellulosic media. The membrane is obtained by dissolving, in a volatile organic solvent, the product of mixing carbonic anhydride with cellulose.[11] The resulting membranes provide consistent five-band resolution. However, by using gelification to prevent the membranes from drying, gelled cellulose acetate is created, with improved resolution of protein bands.[11] Preparations of gelled cellulose acetate (Cellogel; Chemetron, Milan, Italy) can separate serum proteins into the same fractions seen with the high-resolution agarose methods.[12–15] Unlike agarose electrophoresis, cooling is not required to provide optimal resolution.[11] Furthermore, immunofixation analysis may also be performed on these gels. Under certain circumstances, the Cellogel strips may be reused.[16]

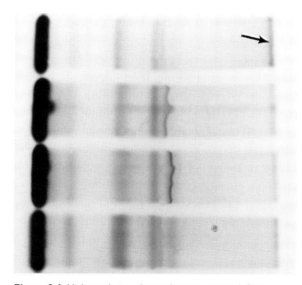

Figure 2.6 High-resolution electrophoretic strip with four samples that all appear to have a small, slow γ-restriction (arrow in restriction of top lane only). If such a band were present in only one sample, a monoclonal gammopathy should be suspected. However, when it occurs in more than one, it represents an artifact. In this case, the artifact results from excess buffer in the reservoir. (Paragon SPE2 system stained with Paragon Blue. Note lighter staining than seen in Fig. 2.3 which used the same system with Paragon Violet.)

For electrophoresis results with high-resolution, cellulose acetate gelatinized membranes are equilibrated for 10 min in a Tris–glycine–salicylic acid buffer solution provided for Cellogel by Chemetron. The Cellogel strip is blotted between two filter paper sheets and placed on an electrophoretic bridge that is then introduced into the electrophoretic chamber. A linear sample applicator (designed to deliver 1 µl) is used to apply undiluted serum 3 cm from the cathodic end of the gel. The samples are electrophoresed for 25 min at 300 V. The samples are then fixed and stained with a variety of stains including Coomassie Brilliant Blue, Ponceau Red, Nigrosin, Schiff, gold, and immunological reagents.[11] Samples are then destained in methanol–water–acetic acid (47.5:47.5:5). Excellent resolution has been reported by Aguzzi and Rezzoni (using the Cellogel system), examples of which (provided by Dr Aguzzi) are used in later chapters.[17]

CAPILLARY ZONE ELECTROPHORESIS (CZE)

The Paragon CZE 2000 system provides automated, high-quality electrophoretic separation of proteins with excellent resolution of transferrin and C3 (Fig. 2.7).[18–24] In a similar manner, the Sebia Capillarys System (Sebia, Inc.) provides an electropherogram (Fig. 2.8). As a supplement to the electrophrogram, the Paragon CZE 2000 (Beckman–Coulter, Fullerton, CA, USA) converts the electropherograms into virtual gel images (Fig. 2.9).[25] On CZE systems, the electropherogram is the basic information while the virtual gel image is generated from that data. This is the opposite of gel-based electrophoresis where the stained gel is the basic information and the densitometric scan is a line drawing generated from that information. Both the Paragon CZE 2000 and the Sebia Capillarys CZE systems provide high quality electrophoresis with minimal investment of the technologist's time.

The Paragon CZE 2000 was the first automated CZE apparatus approved by the Food and Drug

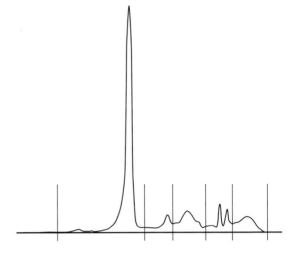

Figure 2.7 Electropherogram of normal serum (Paragon CZE 2000).

Administration for clinical evaluation of serum and urine proteins. While serum is sampled undiluted, evaluation of urine requires manipulations that include concentration and desalting before analysis may be performed. To perform

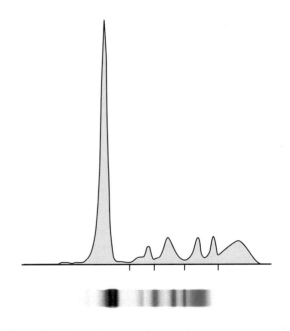

Figure 2.8 Electropherogram of a normal serum.

electrophoresis on serum, a sample volume of 10 μl (plus dead space) is needed. It is placed into a bar-coded tube in one of 10 sectors. Each sector can accommodate seven samples. Data from each sector can be viewed as a virtual gel sector analysis (Fig. 2.9). After the samples are aspirated, they pass into seven 20-cm long uncoated fused silica capillaries (inner diameter 25 μm). Software controls both the physical operation of the instrument

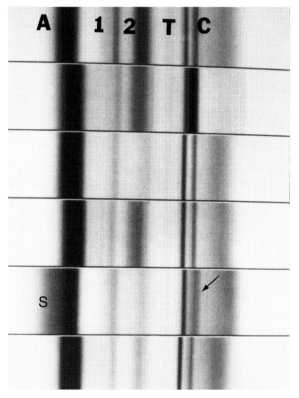

Figure 2.9 Virtual gel image photographed directly from the video monitor of the Paragon CZE 2000. A is just the left of the albumin band, I is in the α_1-region, 2 is at the start of the α_2-region, T is just left of the transferrin band and C is to the right of the C3 band. In the second lane (counting from the top), note that the C3 band appears dark and broad. This reflects the presence of an IgA κ monoclonal gammopathy migrating in this location. Note also the fifth lane from a patient that was receiving heparin. The broad diffuse protein slur (s) anodal to the albumin band reflects α_1-lipoprotein that often migrates here in serum from heparinized patients (see Chapter 4). Also, there is a very subtle diffuse band overlying the C3 region (arrow). This was a subtle IgA κ monoclonal gammopathy.

as well as the display and output of results.[20] The apparatus uses pressure injection to sample undiluted serum from the sector. By use of a lower ionic strength buffer to carry the sample than the ionic strength of the running buffer, a stacking effect (like airplanes queuing up for landing) is produced.[26] The individual samples travel into one of the six capillaries. At the alkaline pH of the Beckman buffer system (pH 9.9 by this laboratory's pH meter), endosmotic flow is created by high voltage and the narrow bore of the negatively charged silica capillary. This propels the proteins toward the cathode where a deuterium lamp emits ultraviolet light to all seven channels.[26] Separation results from the individual isoelectric points, tertiary structure and charge of the proteins under the conditions of the electrophoresis. This light passes through a 214 nm interference filter and ultraviolet silicon detectors sense the absorbance. The software automatically provides delimits for the five major protein fractions, however, the operator can readjust these as needed. Also, the operator can define limits of M-proteins for measurement (Fig. 2.10). A copy of this measurement is stored with the patient's file in order to measure the M-protein in the same manner on subsequent samples.

In the serum protein electrophoresis mode, the instrument has a throughput of 42 samples per hour. Automated immunosubtraction may be performed on the Paragon CZE 2000 (see Chapter 3). For the immunosubtraction mode, five samples are processed in 1 h. Results are displayed as the electropherogram and measurements of the percentage of major protein fractions.

The software allows the operator to concentrate on specific areas of the electropherogram by use of the zoom feature. With the zoom feature selected, an area of the curve may be indicated for closer study. This may be useful in examining small distortions in the β- or γ-region for the presence of M-proteins (Fig. 2.11).

Recently, the Sebia Capillarys system received Food and Drug Administration (FDA) approval for clinical use.[27] As in the Beckman instrument, the inner diameter is 25 μm, however, the Capillarys

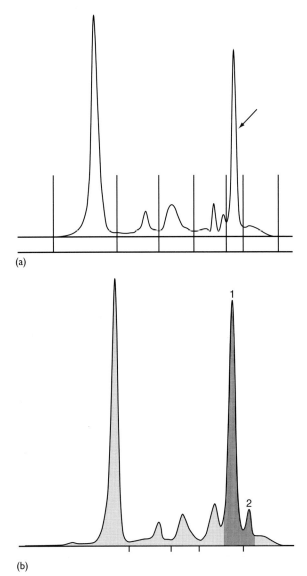

(a)

(b)

Figure 2.10 (a) Measurement of a monoclonal gammopathy on a Paragon CZE 2000 electropherogram is illustrated. The monoclonal gammopathy is indicated by the arrow and the limits were set by the interpreter. (b) Measurement of a biclonal gammopathy on a Sebia Capillarys electropherogram. The two monoclonal peaks are stained more darkly than the rest of the pattern and are labeled 1 and 2.

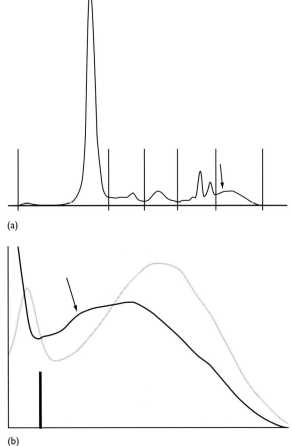

(a)

(b)

Figure 2.11 (a) The arrow indicates a slight irregularity in the fast γ-region of this electropherogram (Paragon CZE 2000). (b) The designated area was expanded to allow a better look at the fast γ-region. The rounded area indicated by the line turned out to be polyclonal. (Paragon CZE 2000.)

uses eight 17-cm long capillaries. For this system, serum is diluted 1:10 and the peptide bonds are interrogated at 200 nm. To set up the strong endosmotic flow, the Capillarys system uses a pH 10 buffer. Further, the assay is performed at 35°C, which may prevent the clumping of cryo-globulins.

Bossuyt et al.[28] compared the reference intervals for the major serum protein fractions by cellulose acetate, agarose, and CZE. They found significant differences in the ranges for all fractions other than γ-globulins in men and women (Table 2.1). For all fractions there were no significant differences between cellulose (Microzone System; Gelman Sciences, Ann Arbor, MI, USA) and agarose (Paragon SPE kit) gels. However, for CZE

Table 2.1 Serum protein reference intervals in men and women[a]

Fraction	Cellulose (G/dl)	Agarose (G/dl)	CZE (G/dl)
Albumin	4.20–5.31	4.19–5.36	4.17–5.23
α_1	0.12–0.25	0.13–0.27	0.26–0.45
α_2	0.38–0.67	0.38–0.70	0.34–0.64
β_1	0.61–1.00	0.65–1.14	0.58–0.95
γ	0.53–1.30	0.49–1.21	0.53–1.32
B. Women			
Albumin	3.94–4.96	4.01–5.11	3.74–4.98
α_1	0.14–0.26	0.14–0.28	0.26–0.51
α_2	0.44–0.69	0.41–0.69	0.39–0.64
β_1	0.55–0.95	0.65–1.00	0.55–0.87
γ	0.53–1.30	0.49–1.21	0.53–1.32

[a]Data modified from Bossuyt et al.[28] Range in for all three instruments expressed as 95% confidence intervals. CZE, capillary zone electrophoresis.

(Paragon CZE 2000), the lower and upper ranges for α_1-globulin were twice as high as the ranges for the gel-based systems. The α_2- and β-globulin fractions for CZE were slightly lower than those from cellulose and agarose gels. Similar results were reported when Katzmann et al.[23] compared CZE (Paragon CZE 2000; Beckman Instruments) with agarose gel electrophoresis (Helena REP system). In their study, there was a 46 per cent increase in the α_1-globulin fraction in the CZE samples compared with the agarose. They also reported a 36 per cent decrease in α_2-globulin fraction and a 10 per cent decrease in β-globulin fraction.[23]

The increase in α_1-globulin likely reflects the increased ability of CZE to detect both α_1-lipoprotein and α_1-acid glycoprotein (orosomucoid) compared with cellulose and agarose gels. The high sialic acid content of orosomucoid interferes with binding of protein dyes, whereas CZE detects proteins via peptide bond absorbance that is not influenced by this factor.[28,29]

In contrast to the Bossuyt et al.[28] study, however, Katzmann et al.[23] noted a 21 per cent decrease in γ-globulin by CZE compared with their agarose

method. This difference likely relates to the differences in the gels and stains used by the agarose commercial products, since the CZE instrument used was the same in both cases. Despite minor differences in the fractions from CZE to gel-based techniques, Petrini et al.[30] found good agreement between interpretation of results when they compared CZE to a high-resolution cellulose acetate electrophoresis on one thousand sera.

Pediatric reference ranges

Reference ranges for the five major protein fractions in a pediatric population was established by Bossuyt et al.[31] (Table 2.2). They divided the population into four groups by ages: 1–2 years, 3–4 years, 5–9 years, 10–14 years. No differences were found between boys and girls for any of these age groups in any of the fractions. The α_2-globulin fraction values were lower in the older children (5–14 years) than in the younger children (1–4 years) because of higher values of α_2-macroglobulin in the latter.[31] Not surprisingly, the γ-globulin fraction was higher in the older groups

Table 2.2 Pediatric reference ranges for capillary zone electrophoresis (Paragon CZE 2000)

Age (years)	n	Albumin[a]	α_1	α_2	β_1	γ
1–2	33	63.5 (54.7–70.4)	6.3 (4.2–8.5)	10.8 (7–15.6)	9.4 (7.5–11.6)	10.1 (4.7–16.0)
3–4	44	61.5 (53.9–70.4)	6.4 (4.8–8.1)	10.4 (7.6–15.2)	9.5 (7.4–11.6)	11.6 (7.1–17.8)
5–9	70	62.2 (52.6–66.3)	6.2 (4.2–7.6)	9.9 (7.4–13.5)	9.4 (7.9–11.3)	12.2 (8.5–18.7)
10–14	48	61.2 (54.1–69.1)	5.9 (4.4–8.0)	8.9 (6.8–11.4)	9.7 (8.5–12.9)	13.7 (8.8–17.6)

[a]Data from Bossuyt et al.[31] with permission. Results expressed as fraction percentages as median, 95% confidence limits in parenthesis.

than in the younger because of the increased amount of immunoglobulin G (IgG) in older children. Even in the 3–4 year group the γ-globulin fraction was higher than in the 1–2 year group.

Specimens that are hemolysed, lipemic and plasma are not recommended for analysis by CZE. Further, radiocontrast dyes create peaks anywhere from the α_2- to the γ-globulin region.[32,33] Because of this, any restriction suggestive of an M-protein that has not previously been characterized must be proven to be an M-protein by immunofixation or immunosubtraction before reporting it as such. As shown in Fig. 2.12, the electropherogram cannot distinguish between this artifact and a true M-protein. With version 1.5 software for the Paragon CZE 2000, fibrinogen was not seen even in samples from heparinized patients.[24] However, version 1.6 of this software now shows the fibrinogen band.

The crisp resolution on CZE has proven to be an excellent vehicle to detect M-proteins.[22,34–38] In studies by Katzmann et al.[23] and Bossuyt et al.,[39] the sensitivity of CZE to detect M-proteins was 93 per cent and 95 per cent, respectively. In contrast, the same two studies detected M-proteins in only 86 per cent and 91 per cent of their samples by the agarose techniques they used. It should be noted, however, the agarose techniques used were of relatively low resolution (five-band patterns that did not show two bands in the β-region). In an earlier publication, Katzmann et al.[40] found good linearity of the comparison of M-protein peaks on CZE versus agarose. Bienvenu et al.[20] found that the

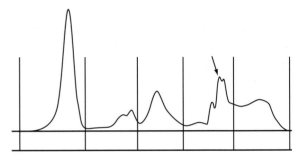

Figure 2.12 Electropherogram (Paragon CZE 2000) from patient that had received a radiocontrast dye. The arrow indicates the position of the dye peak between the anodal small transferrin band and the cathodal C3 band to which it is linked. The size of the dye peak depends on the dose and the time since it was given.

detection limit for an M-protein was < 0.5 g/l (.05 g/dl). A similar level of sensitivity was reported by Smalley et al.[41]

Whereas overall the high-resolution available from CZE has been an advantage, there have been a few reports of problems detecting small or unusual M-proteins.[21,42–44] Capillary zone electrophoresis failed to detect IgA, IgD and IgM M-proteins that were present in concentrations < 3.2 g/l (.32 g/dl) in the study by Bossuyt and Marien.[43] In early studies, Jenkins and Guerin[21] found that six M-proteins with high pI values (between 6.9 and 8.3 for IgM and > 8.5 for IgG monoclonal proteins) and extreme cathodal migration could not be detected. By increasing the ionic strength of the boric acid buffer from 50 mmol, pH 9.7, to 75 mmol, pH 10.3, they were able to detect all of these problem M-proteins. Henskens et al.[42]

reported a similar occurrence with an IgM M-protein. In another case, 2-mercaptoethanol pretreatment was needed before the concentration of a β-migrating IgM M-protein could be accurately determined by CZE (Fig. 2.13).[44]

As with agarose and cellulose acetate electrophoresis, the most challenging M-proteins are those that are of relatively low concentration that also migrate in the β-region. Transferrin and C3 may obscure these small M-proteins.[43] The recent modification of the Paragon system to their version 1.6 system may help to detect some of these M-proteins.

One of the most significant ongoing problems with CZE is the presence of false positive bands that occur when radiocontrast dyes are present.[32,33,45] With the increasing use of these dyes, this problem occurs weekly in our laboratory. Blessum et al.[45] successfully removed an artifactual

α_2-region peak caused by sodium meglumine ioxitalamate (Telebrix, Guerbet Cedex, France) by desalting the sample (Fig. 2.14). For their procedure, they used D-Salt Dextran plastic desalting columns, 5 kDa cutoff from Pierce (Pierce Biotechnology, Rockford, IL, USA). However, Arranz-Peña et al.[32] reported obstruction in some of their capillaries after trying that procedure. They removed the interference by adding 0.2 g of activated charcoal to 1 ml of serum, vortexing for 20 s, and centrifuging at 200 g for 5–10 min at room temperature. Sometimes, two or three centrifugations were needed to clear the serum. In Table 2.3 is the listing of radiocontrast dyes that were tested by Arranz-Peña et al. complete with the region where false positives are known to occur.[32] In addition to radiocontrast dyes, other molecules that absorb at 200–215 nm will create unusual bands on CZE. For example, Bossuyt and co-workers[46,47] reported that the antibiotic piperacillin-tazobactam (Tazocin; Wyeth Lederle)

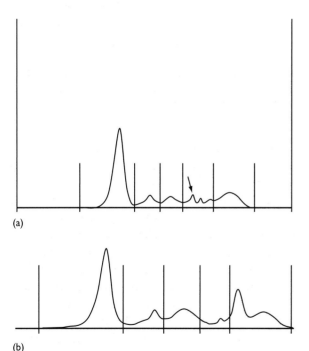

Figure 2.13 (a) Electropherogram (Paragon CZE 2000) with small β-region peak (arrow). The amount of the peak was far too small for the 1500 mg/dl that was measured for the IgM. (b) Electropherogram (Paragon CZE 2000) of same case after treatment with 2-mercaptoethanol. The M-protein is more prominent and now located in the γ-region.

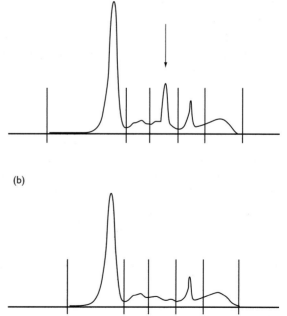

Figure 2.14 Electropherograms (Paragon CZE 2000) of a patient serum sample before (a) and after (b) desalting. Figure from Blessum et al.[45] Used with permission.

Table 2.3 Location in the electropherogram and ultraviolet (UV) maxima for radio-opaque media[a]

Location	Radio-opaque	Interfering substance	UV maximum (nm)
Prealbumin	Bilisegrol	Meglumine iotroxate	237
α_2-Globulin (anodal)	Gastrografin and Urograffin	Sodium meglumine amidotrizoate	237
	Uroangiografin	Meglumine amidotrizoate	238
α_2-Globulin (middle)	Telebrix	Ioxitaalamic acid	240
	Xenetix	Iobitridol	242
α_2-Globulin (cathodal)	Iopamiro	Iopamidol	242
	Omnitrast and Omnipaque	Iohexol	242
	Ultravist	Iopromide	242
β_2-Globulin (anodal)	Hexabrix	Sodium-meglumine ioxaglate	243
β_2-Globulin (middle)	Optiray	Ioversol	244
	Iomeron	Iomeprol	244

[a]Data from Arranz-Peña et al.,[32] with permission.

will produce a small peak in the β-globulin region and the sulfamide sulfamethoxazole produces a small peak at the anodal edge of the albumin fraction.

In our laboratory, I have seen numerous small deflections that likely represent radiocontrast dyes or other interferences. The most convenient way I have found to rule out a monoclonal gammopathy is to perform an immunofixation with the Penta (pentavalent) system (anti-G, A, M, K, and L in the same reagent) on all suspicious cases (see Chapter 3). This simple semi-automated immunofixation is negative with contrast dyes and other non-M-protein bands. Laboratories that use CZE should encourage their clinicians to wait at least a week to order serum protein electrophoresis on patients that have received radiocontrast dyes.[32]

DENSITOMETRIC SCANNING

Densitometry provides objective information, however, it should not be used without direct visual inspection of the gel because it may miss subtle α_1-antitrypsin variants, small monoclonal gammopathies and oligoclonal patterns.[5,9] Although the eye can detect variations in migration more readily than most available clinical laboratory densitometers, differences in density of staining may be objectively noted by the densitometer. Therefore, the objective information obtained from densitometric scans of protein gels helps to draw the attention of the observer to a subtle quantitative abnormality that otherwise might have been missed. Unfortunately, the precision of densitometry is poor in examining the smallest serum fraction (α_1-globulin), although there may be considerable variation in other fractions [48] (Table 2.4). Because of this, some consider clinical densitometry of protein electrophoretic fractions to be a semiquantitative procedure.[48]

The values obtained by densitometric scanning of electrophoretic gels will differ depending on the dilution of serum, the stain, the densitometer and the electrophoretic system used for the analysis. The dilution of the sample used can be particularly important when estimating the concentration of an M-protein by densitometry.[9,49] For example,

Table 2.4 Precision of densitometric quantification[a]

Fraction	Mean CV%	CV% range
Albumin	2.9	1.3–5.1
α_1	9.5	4.4–20.2
α_2	5.1	1.8–9.2
β	5.3	1.6–14.0
γ	6.7	2.5–27.2

[a]Data from Kahn and Strony [48] using 10 measurements of the relative concentration in 30 samples.

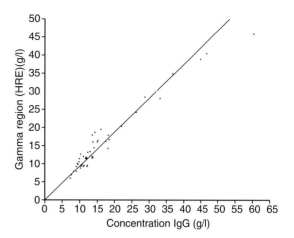

Figure 2.15 Correlation of γ-region concentration, as determined by densitometric scans of serum protein electrophoresis gels, with total IgG as determined by nephelometry. Samples were serum from patients with known γ-migrating monoclonal gammopathies. The excellent correlation demonstrates why we follow these patients with serum protein electrophoresis and densitometric scans only rather than repeating immunoglobulin quantification and/or immunofixation. With β-migrating monoclonal proteins, however, the usual transferrin, β₁-lipoprotein and C3 bands interfere with the accuracy of scanning small monoclonal proteins.

in normal serum samples the concentration of IgG determined by nephelometry was compared with the densitometric scan of the total gamma region and it was found that better linearity was achieved at a 1:4 dilution of serum than when neat solution was used.[9] When protein is too concentrated, the dye binding is not linear in the larger fractions such as albumin or monoclonal proteins.[9,49,50]

When using densitometry, linearity is better on gels stained with Amido Black, and some of the analogous commercial reagents, than with Coomassie Brilliant Blue, although the latter is more sensitive. Sun[51] noted that Amido Black is also superior to Ponceau S for estimating the major protein fractions. Silver stain, although very sensitive, does not provide the type of linearity of these more conventional stains. Electrophoresis strips stained with Amido Black provide a reasonable estimate of human serum albumin, transferrin, and γ-globulin.[52] There is excellent agreement between the densitometric scans and nephelometric measurements of IgG using the serum of patients having γ-migrating IgG monoclonal gammopathies.[9] Therefore, the densitometric scan is used to follow these patients (Fig. 2.15). Since most monoclonal proteins migrate in the γ-region, this provides a convenient method for follow-up of these patients. However, when the monoclonal protein migrates in the α- or β-globulin regions, densitometry may be used

but is problematic. The other major proteins such as haptoglobin, α_2-macroglobulin, transferrin, and C3 that are found in these regions interfere with this method.

Differences in the performance of densitometers have been reported. For example, Schreiber et al.[53] noted that albumin values were higher on the Beckman Appraise densitometer than on the Helena EDC. These instruments differed by an average of 2.5 g/l (250 mg/dl) in measuring the γ-regions on the same gels.[53] This confirmed a similar observation by Sun[54] about relatively low albumin values on the Helena densitometer. In the same study, Sun noted that the Gelman densitometer tested produced low γ-globulin values compared with other instruments. Nonetheless, once a laboratory has developed its own normal range, currently available densitometers provide useful information that allows for comparison from one patient to another and for following patients with monoclonal gammopathies. Clinicians should be

aware that these measurements are not standardized and that there is considerable variability in the
measurement of an M-protein from one laboratory
to another.

The correlation between polyclonal immunoglobulin concentration determined by nephelometry
versus the concentration determined by densitometry is far from perfect.[55] Schreiber et al.[53]
compared the concentration of γ-globulins with
nephelometric values of IgG + IgM + 1/2 IgA (they
assumed that about half of the IgA migrated in the
β-region of their system, and hence would not be
included in the γ-fraction). Although the correlation between the two techniques was very good (an
average correlation coefficient of 0.95), the densitometric technique consistently gave lower results
than did nephelometry. This discrepancy became
more pronounced at higher immunoglobulin concentrations.[53] Similar results were reported by
Chang et al.[49] who recommended diluting sera that
contain total protein from 9.1 to 11.4 g/dl 1:10,
whereas total protein > 11.5 g/dl should be diluted
1:20 to provide better linearity. Further, in estimating albumin concentrations, overstaining of
albumin by some dyes may cause its concentration

to be overestimated compared with nephelometric
techniques.[56]

While densitometry suffers from the limitations
of the dye used to stain the protein and the dilution
of the sample, nephelometric techniques also have
inherent problems due to antigen excess effects as
well as the characteristics of antisera used in the
measurement.[57,58] In their studies, Sinclair et al.[57]
and Tichy[59] found that electrophoresis followed by
densitometry was superior to immunochemical
methods for following IgM monoclonal gammopathies.

Stemerman et al.[60] also recommended following
monoclonal gammopathies (IgG in their study) by
densitometric scans of the serum protein electrophoresis patterns. They noted that there was a loss
of linearity above 6 g/dl requiring dilution for
accurate results. Further, they presented a table to
aid the interpreter in determining if a change in the
quantity of a monoclonal gammopathy is significant (Table 2.5). They used Coomassie Brilliant
Blue to improve the sensitivity and scanned the
monoclonal band with the borders delineated
automatically by their Pharmacia LKB 2220
recording integrator.

Table 2.5 Minimal differences in paraprotein measurement that indicate true changes between sera[a]

Initial paraprotein concentration (g/l)	85% Probability of a true difference (g/l)	95% Probability of a true difference (g/l)
0	1.1	1.8
10	1.6	2.5
20	2.0	3.3
30	2.4	4.0
40	2.9	4.7
50	3.3	5.4
60	3.7	6.1
70	7.2	6.8
80	4.6	7.5

[a]Data modified from Stemerman et al.[60]

Table 2.6 Comparison of normal serum protein values by cellulose acetate versus high-resolution electrophoresis densitometry

Serum fraction	Cellulose acetate[a]	High-resolution agarose[b]
Albumin	3.54–5.0[c]	4.11–5.39
α_1	0.21–0.34	0.10–0.24
α_2	0.40–0.75	0.33–0.73
β[d]	0.73–1.07	0.57–1.05
γ	0.66–1.32	0.62–1.33

[a]Gelman ACD Densitometer used for scanning.
[b]Beckman Appraise Densitometer used for scanning.
[c]Results expressed in Gm/dl as 2 SD range.
[d]Although the high-resolution scans can be separated into two β fractions, only one is used for comparison with the five-band pattern.[9]

The M-protein peak on densitometric scans is measured in the same manner described above for CZE. The indicator is placed at the notch just before and just after the M-protein peak. Clearly, no currently available technique is able to ompletely separate the M-protein concentration from polyclonal immunoglobulins that may migrate in the same region. As with capillary zone electropherograms, the β-region is more problematic because of the presence of other major proteins (see Chapter 4). However, this technique provides a reliable method to follow M-proteins in serum and urine.

We perform densitometric scanning on all gel-based serum patterns. Even though gels with high-resolution allow for recognition of up to twelve proteins, our densitometric patterns of these gels use a standard five serum-fraction pattern. This reflects the variation in small bands such as β_1-lipoprotein and α_1-antichymotrypsin, and the variable proportions of C3 and C3 breakdown product (which migrate in different regions – see Chapter 4), depending on the conditions of storage of the sample and *in vivo* activation. A comparison with our former routine five-band cellulose acetate technique showed a lower level of α_1- and α_2-globulins in the Amido Black-stained agarose strips compared with the Ponceau S-stained cellulose acetate (Table 2.6). However, the superior band discrimination by Amido Black made this difference in densitometric information insignificant for the purpose of clinical interpretation.[9]

REFERENCES

1. Wieme R. *Agar gel electrophoresis*. Amsterdam: Elsevier, 1965.
2. Briere RO, Mull JD. Electrophoresis of serum protein with cellulose acetate. A method for quantitation. *Am J Clin Pathol* 1964;34:547–551.
3. Laurell CB. Electrophoresis, specific protein assays or both in measurement of plasma proteins. *Clin Chem* 1973;19:99–102.
4. Keren DF, Alexanian R, Goeken JA, Gorevic PD, Kyle RA, Tomar RH. Guidelines for clinical and laboratory evaluation patients with monoclonal gammopathies. *Arch Pathol Lab Med* 1999;123:106–107.
5. Aguzzi F, Kohn J, Petrini C, Whicher JT. Densitometry of serum protein electrophoretograms. *Clin Chem* 1986;32:2004–2005.
6. Wild D. *The immunoassay handbook*. London: Nature Publishing Group, 2001.

7. Johansson BG. Agarose gel electrophoresis. *Scand J Clin Lab Invest Suppl* 1972;**124**:7–19.

8. Safety note. *Clin Chem* 1980;**26**:804.

9. Keren DF, Di Sante AC, Bordine SL. Densitometric scanning of high-resolution electrophoresis of serum: methodology and clinical application. *Am J Clin Pathol* 1986;**85**: 348–352.

10. Kohn J. A cellulose acetate supporting medium for zone electrophoresis. *Clin Chim Acta* 1957;**2**:297–304.

11. Destro-Bisol G, Santini SA. Electrophoresis on cellulose acetate and Cellogel: current status and perspectives. *J Chromatogr A* 1995;**698**:33–40.

12. Aguzzi F, Jayakar SD, Merlini G, Petrini C. Electrophoresis: cellulose acetate vs agarose gel, visual inspection vs. densitometry. *Clin Chem* 1981;**27**:1944–1945.

13. Merlini G, Piro P, Pavesi F, Epis R, Aguzzi F. Detection and identification of monoclonal components: immunoelectrophoresis on agarose gel and immunofixation on cellulose acetate compared. *Clin Chem* 1981;**27**:1862–1865.

14. Janik B, Dane RG. High-resolution electrophoresis of serum proteins on cellulosic membranes and identification of individual components by immunofixation and immunosubtraction. *J Clin Chem Clin Biochem* 1981;**19**:712–713.

15. Ojala K, Weber TH. Some alternatives to the proposed selected method for 'agarose gel electrophoresis'. *Clin Chem* 1980;**26**: 1754–1755.

16. Destro-Bisol G. Reusing Cellogel strips after visualization of electrophoretically separated isozymes. *Electrophoresis* 1993;**14**:238–239.

17. Aguzzi F, Rezzani A. An advantageous but neglected technique for immunofixation of monoclonal components on cellulose acetate membranes. Standardisation and evaluation. *Giorn It Chim Clin* 1986;**11**:293–299.

18. Jenkins MA. Clinical applications of capillary electrophoresis. Status at the new millennium. *Mol Biotechnol* 2000;**15**:201–209.

19. Landers JP. Clinical capillary electrophoresis. *Clin Chem* 1995;**41**:495–509.

20. Bienvenu J, Graziani MS, Arpin F, et al. Multicenter evaluation of the Paragon CZE 2000 capillary zone electrophoresis system for serum protein electrophoresis and monoclonal component typing. *Clin Chem* 1998;**44**:599–605.

21. Jenkins MA, Guerin MD. Optimization of serum protein separation by capillary electrophoresis. *Clin Chem* 1996;**42**:1886.

22. Jenkins MA, Guerin MD. Capillary electrophoresis as a clinical tool. *J Chromatogr B Biomed Appl* 1996;**682**:23–34.

23. Katzmann JA, Clark R, Sanders E, Landers JP, Kyle RA. Prospective study of serum protein capillary zone electrophoresis and immunotyping of monoclonal proteins by immunosubtraction. *Am J Clin Pathol* 1998;**110**:503–509.

24. Keren DF. Capillary zone electrophoresis in the evaluation of serum protein abnormalities. *Am J Clin Pathol* 1998;**110**:248–252.

25. Jolliff CR, Blessum CR. Comparison of serum protein electrophoresis by agarose gel and capillary zone electrophoresis in a clinical setting. *Electrophoresis* 1997;**18**:1781–1784.

26. Klein GL, Jolliff CR. Chapter 16. Capillary electrophoresis for the routine clinical laboratory. In: Landers JP, ed. *Handbook of capillary electrophoresis*. Boca Raton: CRC Press, 1993.

27. Bossuyt X, Lissoir B, Marien G, et al. Automated serum protein electrophoresis by Capillarys. *Clin Chem Lab Med* 2003;**41**:704–710.

28. Bossuyt X, Schiettekatte G, Bogaerts A, Blanckaert N. Serum protein electrophoresis by CZE 2000 clinical capillary electrophoresis system. *Clin Chem* 1998;**44**:749–759.

29. Dati F, Schumann G, Thomas L, et al. Consensus of a group of professional societies and diagnostic companies on guidelines for interim reference ranges for 14 proteins in serum based on the standardization against the IFCC/BCR/CAP Reference Material (CRM 470). International Federation of Clinical Chemistry. Community Bureau of Reference of the Commission of the European Communities. College of American Pathologists. *Eur J Clin Chem Clin Biochem* 1996;**34**:517–520.

30. Petrini C, Alessio MG, Scapellato L, Brambilla S, Franzini C. Serum proteins by capillary zone electrophoresis: approaches to the definition of reference values. *Clin Chem Lab Med* 1999;**37**:975–980.

31. Bossuyt X, Claeys R, Bogaert G, Said HI, Wouters C, Groven C, Sneyers L, Marien G, Gorus F. Reference values for the five electrophoretic serum protein fractions in Caucasian children by capillary zone electrophoresis. *Clin Chem Lab Med* 2001;**39**:970–972.

32. Arranz-Pena ML, Gonzalez-Sagrado M, Olmos-Linares AM, Fernandez-Garcia N, Martin-Gil FJ. Interference of iodinated contrast media in serum capillary zone electrophoresis. *Clin Chem* 2000;**46**:736–737.

33. Bossuyt X, Mewis A, Blanckaert N. Interference of radio-opaque agents in clinical capillary zone electrophoresis. *Clin Chem* 1999;**45**:129–131.

34. Litwin CM, Anderson SK, Philipps G, Martins TB, Jaskowski TD, Hill HR. Comparison of capillary zone and immunosubtraction with agarose gel and immunofixation electrophoresis for detecting and identifying monoclonal gammopathies. *Am J Clin Pathol* 1999;**112**:411–417.

35. Clark R, Katzmann JA, Kyle RA, Fleisher M, Landers JP. Differential diagnosis of gammopathies by capillary electrophoresis and immunosubtraction: analysis of serum samples problematic by agarose gel electrophoresis. *Electrophoresis* 1998;**19**:2479–2484.

36. Jenkins MA, Kulinskaya E, Martin HD, Guerin MD. Evaluation of serum protein separation by capillary electrophoresis: prospective analysis of 1000 specimens. *J Chromatogr B Biomed Appl* 1995;**672**:241–251.

37. Jenkins MA, Guerin MD. Capillary electrophoresis procedures for serum protein analysis: comparison with established techniques. *J Chromatogr B Biomed Sci Appl* 1997;**699**:257–268.

38. Jenkins MA, Ratnaike S. Five unusual serum protein presentations found by capillary electrophoresis in the clinical laboratory. *J Biochem Biophys Methods* 1999;**41**:31–47.

39. Bossuyt X, Bogaerts A, Schiettekatte G, Blanckaert N. Detection and classification of paraproteins by capillary immunofixation/subtraction. *Clin Chem* 1998;**44**:760–764.

40. Katzmann JA, Clark R, Wiegert E, et al. Identification of monoclonal proteins in serum: a quantitative comparison of acetate, agarose gel, and capillary electrophoresis. *Electrophoresis* 1997;**18**:1775–1780.

41. Smalley DL, Mayer RP, Bugg MF. Capillary zone electrophoresis compared with agarose gel and immunofixation electrophoresis. *Am J Clin Pathol* 2000;**114**:487–488.

42. Henskens Y, de Winter J, Pekelharing M, Ponjee G. Detection and identification of monoclonal gammopathies by capillary electrophoresis. *Clin Chem* 1998;**44**:1184–1190.

43. Bossuyt X, Marien G. False-negative results in detection of monoclonal proteins by capillary zone electrophoresis: a prospective study. *Clin Chem* 2001;**47**:1477–1479.

44. Keren DF, Gulbranson R, Carey JL, Krauss JC. 2-Mercaptoethanol treatment improves measurement of an IgM kappa M-protein by capillary electrophoresis. *Clin Chem* 2001;**47**:1326–1327.

45. Blessum CR, Khatter N, Alter SC. Technique to remove interference caused by radio-opaque agents in clinical capillary zone electrophoresis. *Clin Chem* 1999;**45**:1313.

46. Bossuyt X, Verhaegen J, Marien G, Blanckaert N. Effect of sulfamethoxazole on clinical capillary zone electrophoresis of serum proteins. *Clin Chem* 2003;**49**:340–341.

47. Bossuyt X, Peetermans WE. Effect of piperacillin-tazobactam on clinical capillary zone electrophoresis of serum proteins. *Clin Chem* 2002;**48**:204–205.

48. Kahn SN, Strony LP. Imprecision of quantification of serum protein fractions by electrophoresis on cellulose acetate. *Clin Chem* 1986;**32**:356–357.

49. Chang CY, Fritsche HA, Glassman AB, McClure KC, Liu FJ. Underestimation of monoclonal proteins by agarose serum protein electrophoresis. *Ann Clin Lab Sci* 1997;**27**:123–129.

50. Carter PM, Slater L, Lee J, Perry D, Hobbs JR. Protein analyses in myelomatosis. *J Clin Pathol Suppl* 1975;6:45–53.

51. Sun T. *Interpretation of protein and isoenzyme patterns in body fluids.* New York/Tokyo: Igaku-Shoin, 1992.

52. Uriel L. Interpretation quantitative des resultats après electrophorese en gelose, 1. Considerations générales, application a l'étude de constituants proteiques isolés. *Clin Chim Acta* 1958;3: 234–238.

53. Schreiber WE, Chiang E, Tse SS. Electrophoresis underestimates the concentration of polyclonal immunoglobulins in serum. *Am J Clin Pathol* 1992;97:610–613.

54. Sun T. High-resolution agarose electrophoresis. In: Ritzmann SE, ed. *Protein abnormalities. Vol 1. Physiology of immunoglobulins: diagnostic and clinical aspects.* New York: Alan R. Liss, 1982.

55. Austin GE, Check IJ, Hunter RL. Analysis of the reliability of serial paraprotein determinations in patients with plasma cell dyscrasias. *Am J Clin Pathol* 1983;79:227–230.

56. Mendler MH, Corbinais S, Sapey T, et al. In patients with cirrhosis, serum albumin determination should be carried out by immunonephelometry rather than by protein electrophoresis. *Eur J Gastroenterol Hepatol* 1999;11:1405–1411.

57. Sinclair D, Ballantyne F, Shanley S, Caine E, O'Reilly D, Shenkin A. Estimation of paraproteins by immunoturbidimetry and electrophoresis followed by scanning densitometry. *Ann Clin Biochem* 1990;27: 335–337.

58. Bush D, Keren DF. Over- and underestimation of monoclonal gammopathies by quantification of kappa- and lambda-containing immunoglobulins in serum. *Clin Chem* 1992;38:315–316.

59. Tichy M. A comparison of methods of monoclonal immunoglobulin quantitation. *Neoplasma* 1985;32:31–36.

60. Stemerman D, Papadea C, Martino-Saltzman D, O'Connell AC, Demaline B, Austin GE. Precision and reliability of paraprotein determinations by high-resolution agarose gel electrophoresis. *Am J Clin Pathol* 1989;91:435–440.

3

Immunofixation, immunosubtraction, and immunoselection techniques

Whereas serum protein electrophoresis can detect restrictions that resemble monoclonal gammopathies, it cannot definitively identify a restriction as an M-protein. For that, immunochemical methods together with electrophoresis must be employed. In this chapter, I review the principles of the techniques employed for the identification of restrictions seen on protein electrophoresis as M-proteins. Most laboratories now perform immunofixation electrophoresis (IFE) to identify the restriction as an M-protein. Laboratories using capillary zone electrophoresis (CZE) may be able to perform immunosubtraction (ISUB), depending on which CZE system they are using. Immunoelectrophoresis (IEP) is still used by some laboratories. It does have a couple of advantages over immunofixation, but it is slow, less sensitive and more difficult to interpret than IFE. Lastly, cases of heavy chain disease benefit from performance of immunoselection (ISEL).

PRINCIPLES OF IMMUNOPRECIPITATION

Whether one is performing IFE, ISUB, IEP, or ISEL the basic principles of the precipitin reaction are the same, and understanding them is critically important to making the correct interpretation of a given sample. Immunoprecipitation involves the interaction of antibody molecules with antigen in either a gel or liquid matrix in which the molecules are free to diffuse. The key to precipitation is the multivalent nature of antibodies and antigens (each has two or more sites with which they can interact).[1,2]

In most chemical reactions, when substance A is mixed with substance B to form a product C (let's call it a precipitate), the reaction can be expressed as

$$A + B \rightarrow C$$

As shown in Fig. 3.1, as one increases the amount of substance A while maintaining the amount of

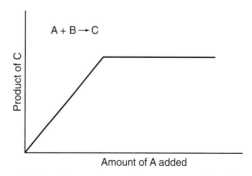

Figure 3.1 As substance B is used up, no more product C is formed by the addition of substance A.

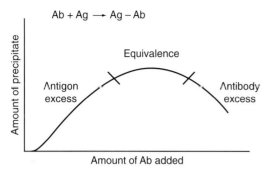

Figure 3.2 This classic immunoprecipitin curve shows that the amount of precipitate (antigen–antibody complex) decreases with the addition of excessive amounts of antibody.

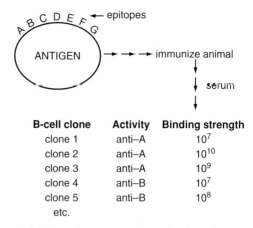

B-cell clone	Activity	Binding strength
clone 1	anti–A	10^7
clone 2	anti–A	10^{10}
clone 3	anti–A	10^9
clone 4	anti–B	10^7
clone 5	anti–B	10^8
etc.		

Figure 3.3 Most antigens are complex molecules with many surface epitopes to which antibodies will form. In turn, several different B-cell clones may respond to each epitope, creating a diverse array of antibody specificities with a variety of binding strengths in the reagent antisera that we use.

substance B constant, the amount of precipitate C will increase up to a point, and then remains constant, despite addition of more substance A. No more precipitate forms as more C is added because all the available substance B has combined with A to form the precipitate. Similarly, in immunological reactions a precipitable product will form when an antibody is added to its antigen. Unlike the chemical reaction shown in Fig. 3.1, however, a decrease in the amount of precipitate occurs when excess antibody is added to a constant amount of antigen (Fig. 3.2). Therefore, there is something about the interaction of antibodies with antigens that differs from the simple chemical reaction shown above.

Antibody–antigen interactions are highly complex because of the many variables that may occur. Any given antigen molecule has many different surface determinants (epitopes) to which antibodies may bind (Fig. 3.3). Each epitope may elicit several different clones of B lymphocytes to differentiate into antibody-secreting plasma cells. The antibodies produced by each clone will differ from one another in structure such that their ability to bind to the epitope will vary from one clone to the next. This strength of antibody binding to a particular epitope is called its affinity.

The binding of antibody molecules to epitopes depends on four types of non-covalent interactions. The forces involved are:

1. *Coulombic forces*, which result from the interaction of oppositely charged groups such as NH_3^+ and COO^-
2. *Hydrogen bonding*, which results from the interaction of a hydrogen atom closely linked to an electronegative atom (such as oxygen), with another electronegative atom.
3. *Hydrophobic bonding*, which is analogous to the effects of oil in water. Even when dispersed, oil will coalesce, excluding the intervening water molecules. Hydrophobic bonding occurs because of the preference of apolar groups for self-association.
4. *van der Waals forces*, which are the interactions that occur in the outer electrons of the reactants. These are relatively weak forces that gain considerably in strength as the distance between antigen and antibody molecules decreases.

None of these interactions has the strength of covalent bonding; therefore, antibody–antigen interactions are readily reversible. The strength of the interaction of a particular antibody with a particular epitope (affinity) depends on the number and strength of the four types of bonds described above.

Understanding the immune precipitin reaction and

the need to consider the concentration of the antibody and antigen is critically important to optimize IFE, ISUB, IEP, and ISEL techniques and to avoid potential errors. Although the following concepts are quite basic, commercial products whose instructions try to 'simplify' the technique occasionally ignore them. Before launching into a discussion of the chemistry involved, I will use the following case to illustrate the relevance of the antibody and antigen concentrations to the amount of protein present in the gammopathy being studied.

Although the protein electrophoresis tract looked like a monoclonal gammopathy – serum protein electrophoresis (SPE) for Fig. 3.4a – the immuno-fixation showed a very broad reactivity with IgG and κ that suggested a polyclonal reactivity. No reaction was seen in the region where the slow γ-band was seen. Serum protein electrophoresis was performed on the sample. As shown in Fig. 3.4b, this gave a slow γ-band similar to the one that detected in the SPE lane of the referred gel. We measured IgG, IgA, IgM, κ and λ by nephelometry. From the data shown in Fig. 3.4c, it was concluded that the serum contained an IgG κ monoclonal gammopathy.

The problem with the referred IFE related to the dilutions of patient serum applied to the gel. We performed IFE, using the same commercial kit as the outside laboratory, but adjusting the dilution of the serum to approximate equivalence for the antibody–antigen interaction (see below). As shown in Fig. 3.4d, our IFE demonstrates that the slow γ-band is an IgG κ monoclonal gammopathy. We ran the κ at two concentrations, 1:10 and 1:20, both of which gave interpretable results. This demonstrates that one must adjust the concentration to be *near* the equivalence range to obtain an interpretable result.

The dilutions of patient's serum on the first IFE gel resulted in antigen excess effect in the region of the monoclonal band. Note also that no precipitate at all is seen in the IgA or λ lanes. Did the technologist forget to place the sample or the antiserum in this tract, or was the dilution used too large, resulting in an undetectable precipitate (in effect, antibody excess)? This is another key problem with using dilutions that do not attempt to account for the approximate concentrations of the immuno-globulins looked for. Why were broad precipitates of IgG and κ on the original IFE anodal to the major band in the slow γ-region? This represented lower concentrations of the monoclonal protein that had migrated behind the major band. The broad migration may have reflected some self-aggregation, heavy glycosylation, or the known microheterogeneity of monoclonal proteins.[3] It stained well, however, because the concentration of the monoclonal band in this region was considerably lower than at the slow γ-region where most of the protein migrated. Because of its high concentration at the slow γ-end, the complexes formed were small and washed away during the wash steps (antigen excess effect), resulting in confusion, or worse, a false negative. Hopefully, this digression on a real case will whet your appetite for the rather dry, but straightforward, discussion of these important basics.

In the immune precipitin reaction, the multi-valency of the antibody and antigen allow for the formation of a lattice (Fig. 3.5). In the zone of antigen excess (Fig. 3.2), the large amount of antigen present makes it likely that only relatively small immune complexes are formed, with the formula $AB(1)AG(2)$. These molecules are too small to precipitate. As more antibody is added to the system, a precipitate (large antibody–antigen latticework) begins to form until a maximal precipitate is noted. This maximal precipitate occurs in the zone of equivalence where the formula is $AB(1)AG(1)$ (Fig. 3.5). With the addition of still more antibody, the epitopes on the surface of the antigen are saturated with antibody molecules and therefore are not available for reaction with other crosslinking antibodies. Here the formula is $AB(x)AG(1)$ where x is the number of epitopes expressed on the surface of a particular antigen. For optimal IFE, a large crosslinked precipitate is needed. Smaller immune complexes usually formed at antigen excess will wash away. In the case shown in Fig. 3.4, there was too much antigen present at the cathodal end of the gel resulting in the band being washed away (antigen excess effect).

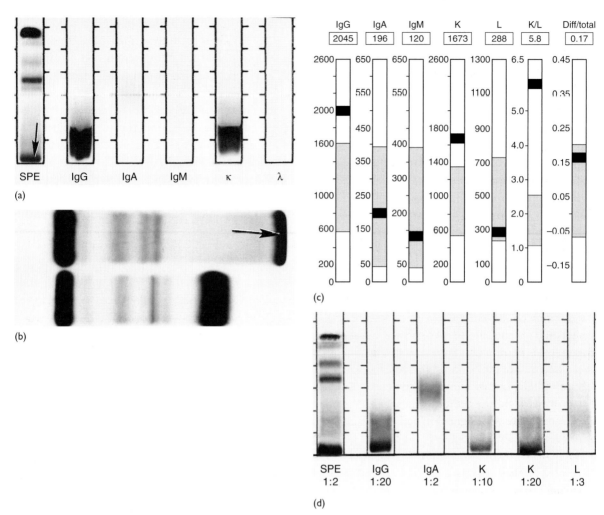

Figure 3.4 (a) Immunofixation of a serum referred to us from another laboratory. Specific antisera were added to the lanes as indicated by the immunoglobulin labels. The dilutions used were not provided. The anode is on the top. Note the sharp band at the extreme cathodal end of the SPE lane (arrow). Why are the bands in the IgG and κ lanes so broad? Why is there no sharp band at the cathodal end? Did the technologist add the sample to the lanes labeled IgA, IgM and λ? (Paragon system stained with Paragon Violet; anode at the top.). (b) The top sample is from the case shown in (a). The distinctive γ-region band is also seen in our gel (arrow). The bottom sample is not from the present case but shows a striking monoclonal gammopathy. (Paragon SPE2 system stained with Paragon Violet.) (c) The immunoglobulin quantifications for the serum from (a) are shown. The normal ranges are shaded for each column. The patient's values are shown in the square at the top of each column and depicted with a black square. K/L is the κ/λratio. Diff/Total is (IgG + IgA + IgM) − (κ + λ). In light chain disease it is often high. In this case, the IgG and κ are elevated along with the κ/λ ratio. This, together with the obvious monoclonal band in (b) is consistent with an IgG κ monoclonal gammopathy. (d) Immunofixation in our laboratory demonstrates the IgG κ monoclonal gammopathy. No IgM was run on this gel. Instead, two concentrations of κ were used to see if the slow γ-band became weaker at greater concentrations. Note that IgA and λ are visible, because the dilutions used were optimized to precipitate the normal polyclonal IgA and λ which were present. Note that the IgA is migrating in the correct position for polyclonal IgA, the β-region. (Paragon Immunofixation system stained with Paragon Violet.)

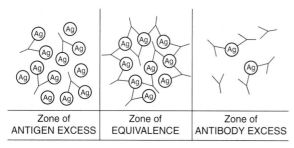

| Zone of ANTIGEN EXCESS | Zone of EQUIVALENCE | Zone of ANTIBODY EXCESS |

Figure 3.5 Multivalency of antibody and antigens is responsible for the classic immunoprecipitin curve (Fig. 3.2).

DOUBLE DIFFUSION IN TWO DIRECTIONS (OUCHTERLONY TECHNIQUE)

Ouchterlony devised a simple way to use the immunoprecipitation reaction to determine antibody reactivity and to identify unknown antigens.[4] He cut wells in an agarose gel and put antibody in one well and antigen in another. These molecules diffuse radially out of the wells (Fig. 3.6). As they diffuse away from the center of the well, their concentration decreases logarithmically. A precipitin band forms somewhere between the two wells at the point at which their concentrations are equivalent. If the precipitate is closer to the antigen well, it indicates that the antibody was more concentrated than the antigen because it had to diffuse further than the antigen (thereby becoming more dilute) before a precipitate could be seen. Other factors such as the size of the molecules and interactions with the gel also affect this reaction but, in general, the beauty

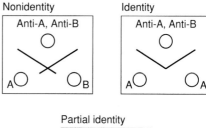

Figure 3.7 Ouchterlony plates using specific antibodies can be used to determine the antigen content of unknown solutions. When different antigens (A and B) are reacted with antibodies to A and B, two lines form, which cross one another (nonidentity). When the antigen in both wells is the same, the two lines meet (identity) because the antibodies to the antigen are absorbed out in the precipitation and do not pass beyond the precipitin line to react with the other antigen. When the antigen is similar (A′), only some of the antibodies to the antigen (A) will be removed. Antibodies to antigen A that are not expressed on A′ will pass through the precipitin band formed by anti-A and A′, and a second line (spur) will form with antigen A. This is the partial identity pattern.

of the antibody–antigen interactions when they diffuse through agarose gel for a distance is that the concentration of the reactants is automatically adjusted to form the precipitate.

Ouchterlony also found that by using a known antigen and antibody he could determine whether an unknown antigen was structurally the same, similar, or dissimilar. For example, immunoprecipitation of an antiserum with activity against both antigens A and B is shown in Fig. 3.7. If antigen A is in one test well and antigen B is in the other, a pattern of non-identity occurs where the two lines cross. The precipitin line does not resemble a solid barrier, rather a grossly visible latticework (Fig. 3.5). Antibodies and antigen that lack specificity for the epitopes on this immune complex readily pass through this latticework.

If antigen A is placed in both wells, a pattern of identity occurs. Here, the lines meet but do not cross one another, because all the antibodies to antigen A react with substance A they do not pass through the

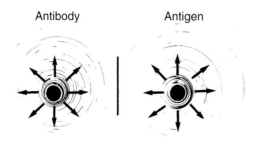

Figure 3.6 As antibody and antigen diffuse from the well, their concentrations decrease logarithmically. The precipitin line forms where the concentrations are equivalent.

lattice. If antigen A is placed in one well and a chemically similar antigen A' (which may lack one epitope that antigen A possesses) is placed in the other test well, a pattern of partial identity results. This is a difficult pattern to identify and to understand. Here, all of the molecules that react with antigen A' are precipitated onto the lattice when they meet antibody to A. However, since antigen A' lacks one epitope that is present on antigen A, antibodies to this unique epitope pass through the lattice formed by anti-A + antigen A' and are available to interact with antigen A. Because there are only relatively few of these molecules and most of the antibodies that react with anti-A have been removed by the interaction with A', antigen A must diffuse slightly further (to become more dilute in concentration) before it is in equivalence to form a precipitate with the fewer remaining anti-A molecules that react with the unique epitope. This explains the 'spur' of antigen A, which is the classic definition of 'partial identity.'

IMMUNOELECTROPHORESIS

A logical development of the immune precipitin reaction was to combine it with electrophoresis to achieve separation by charge and then identification of the molecules by immunological techniques. For many years, IEP was the mainstay for combining these two techniques in both research and clinical laboratories.[5] It is performed by placing the patient's serum into a series of wells in an agarose gel. The sample is electrophoresed to permit separation of the major serum proteins (Fig. 3.8). After electrophoresis, the gel is removed from the electrophoretic apparatus and the troughs are filled with antisera to various specific components of interest, usually including antipentavalent human immunoglobulin (which reacts with the three major heavy and light chain classes), and the remaining troughs are filled with monospecific antisera against IgG, IgA, IgM, κ-, and λ-chains. For comparison of electrophoretic migration patterns, and to be sure that the correct antisera have been placed into the appropriate troughs, a control serum is alternated

with the patient's sample on the IEP strip. These antisera slowly diffuse from the trough into the gel while the protein components slowly diffuse in a radial fashion. Precipitin lines form where the antisera and the specific antigens are at equivalence. Owing to the geometry of the application wells and radial diffusion, the precipitin lines are in the form of arcs. Large quantities of monoclonal proteins are readily identified by IEP by comparing the migration of the control to the patient's serum across a trough containing a specific antiserum (Fig. 3.9).

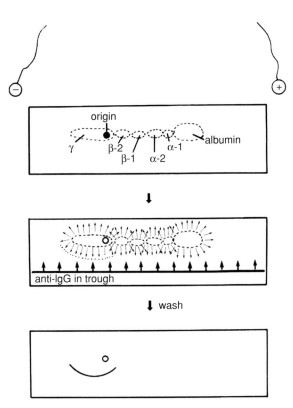

Figure 3.8 Immunoelectrophoresis (IEP) takes advantage of the principles of electrophoresis, gel diffusion (which adjusts concentrations during diffusion), and antibody–antigen precipitin reactions to identify specific proteins. In the example shown, only IgG is precipitated and present after the wash step. The closer the IgG is to the antiserum trough, the greater is its concentration. Large molecules, such as pentameric IgM, can be difficult to examine by this technique because they diffuse slowly through the agarose. Small monoclonal proteins are also difficult to identify because this technique does not offer good resolution of individual proteins.

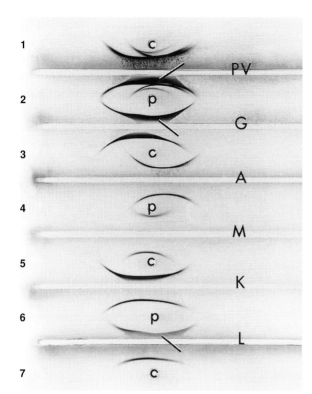

Figure 3.9 Immunoelectrophoresis (IEP) shows a large IgG λ monoclonal gammopathy. Note that control (c) serum alternates with patient (p) serum. Antisera to pentavalent (PV), IgG (G), IgA (A), IgM (M), κ (K), and λ (L) were placed in the troughs. A large arc with excessively anodal migration is seen in the polyvalent, IgG, and λ areas of the patient's sample (indicated in all three locations). The λ arc stains weakly, which is the prozone phenomenon often seen with λ reagents. Diagnosis of such large monoclonal gammopathies is relatively easy by IEP.

Immunoelectrophoresis was a significant advance in the identification of monoclonal proteins associated with multiple myeloma and Waldenström macroglobulinemia. Unlike the antigen excess problem with IFE discussed above, IEP is not as profoundly affected by antigen excess because, as the protein diffuses into the agar, its concentration decreases greatly, allowing the system to adjust automatically to antigen–antibody equivalence and a precipitin arc to be formed. When monoclonal proteins are present in large quantities, IEP is usually able to detect the condition properly because of this diffusion effect.[5] When a large amount of monoclonal protein is present, it needs to diffuse further to decrease its concentration and will have its precipitin arc close to the trough. Occasionally, the center of the arc actually diffuses into the trough (Fig. 3.10). Except in extreme cases, the diagnosis can readily be made without diluting the serum.

Similarly, IEP can detect a second monoclonal protein that results from the presence of a monoclonal free light chain (MFLC, also known as Bence Jones protein) simultaneously with an intact monoclonal immunoglobulin molecule. The MFLC has a different pI, a much smaller molecular weight, and it is usually present at a different concentration than the intact monoclonal immunoglobulin molecule. Because of these factors, MFLC protein has a

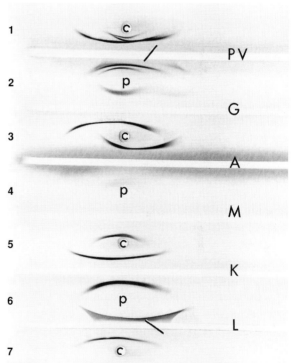

Figure 3.10 Immunoelectrophoresis of urine shows a large λ Bence Jones protein. Note that control (c) is serum so that we can see the position of normal IgG, IgA, and IgM. Normal urine would have too little immunoglobulin to detect with this method. The patient's urine (p) has so much of the monoclonal λ protein that it has substantially diffused into the trough before reaching 'equivalence' and forming a precipitate with the anti-pentavalent (PV) and anti-λ (L) reagents (indicated). Note that the pentavalent reaction is quite weak. It is typical for polyvalent reagents to have relatively weak anti-λ reactivity. Compare this with Fig. 3.9 (symbols are same), in which the dense reaction in the pentavalent reagent is caused by the anti-IgG (G).

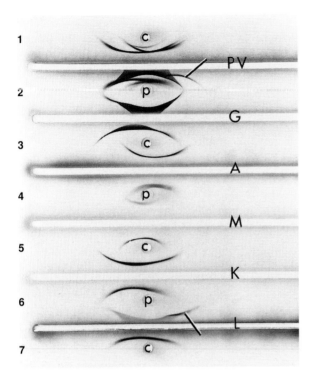

Figure 3.11 Immunoelectrophoresis from the serum of a patient with both a monoclonal IgG λ protein and a free λ monoclonal free light chain (MFLC) in the serum. The free λ light chain reactivity can be seen with the pentavalent reagent, where the extra arc due to the anti-free λ is indicated. The same area is indicated in the anti-λ reaction with the patient's serum in the bottom trough. This area is not seen with the anti-IgG reagent.

different migration than the intact monoclonal gammopathy. Free light chains also express certain epitopes on their surfaces that are not usually expressed on the surfaces of a light chain attached to intact immunoglobulin molecules. Commercial antisera are able to detect these epitopes (Fig. 3.11). Consequently, as with the Ouchterlony patterns of partial identity discussed above, when the intact immunoglobulin molecule reacts with the antiserum it cannot remove antibodies to these hidden determinants. The latter antibodies continue through the precipitin band to react with the free light chain.

Limitations of IEP

Unfortunately, there are several problems with IEP that encouraged the development of newer approaches to the detection of M-proteins. Although IEP is an accurate means to characterize a large M-protein, it is also very slow, largely because of the diffusion step, which requires at least 18 h for optimal results. Further, for greatest sensitivity the gel must be washed and stained before final examination. Immunoelectrophoresis with no complicating factors takes from 24 to 48 h to complete. Immunofixation electrophoresis techniques are more sensitive, semi-automated and can be performed in less than 3 h.

In addition to being slow, IEP is unable to distinguish monoclonality in certain situations.[6–12] The case shown in Fig. 3.12 is a typical example of the 'umbrella effect' problem with IEP. The patient

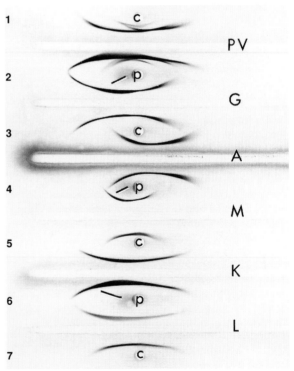

Figure 3.12 Nondiagnostic immunoelectrophoresis in a patient with a fourfold increase in IgM and an obvious γ spike on serum protein electrophoresis (shown on the top of Fig. 3.13). The patient's IgM is pentameric and is barely able to migrate out of the well. The diffuse hazy area indicated near the well for each that contains patient's serum is where the monoclonal protein has deposited. The normal κ and λ arcs represent the patient's normal serum IgG and do not reflect his monoclonal protein (the so-called 'umbrella effect').

was a 60-year-old woman who complained of lethargy and was noted, upon physical examination, to have prominent axillary lymph nodes. Despite an IgM level (1330 mg/dl) greater than three times our upper limit of normal (350 mg/dl) and an obvious spike on serum protein electrophoresis (Fig. 3.13), the IEP (Fig. 3.12) of this patient's serum was non-diagnostic. One could certainly say that the patient had more IgM than the control, but that was already known from the immunoglobulin measurement. Monoclonality

(marked predominance of κ or λ) could not be determined because the patient had normal amounts of IgG. The smaller molecular size of IgG (160 kDa) than IgM (1000 kDa) allows IgG to diffuse more quickly through agarose and to react with antisera to the light chains. This prevents reaction of anti-light chain antisera with the bulkier, more slowly diffusing IgM. This shielding of an IgM monoclonal gammopathy by polyclonal IgG molecules is called the 'umbrella effect,' and is a well-known problem in IEP interpretation.[13,14]

There are several ways that one can still use IEP to make the correct diagnosis in a case with the umbrella effect.[15,16] One can add 2-mercaptoethanol to break disulfide bonds and then repeat the IEP. To perform this procedure, we recommend adding 5 µl of 0.1 M 2-mercaptoethanol to 1 ml of serum and incubating for 1 h at 37°C. This will usually break up aggregates of IgM to allow for visualization of the monoclonal gammopathy by IEP.[15,17-21] Unfortunately, this method is sometimes unsuccessful. Reduction of disulfide bonds works best to identify those monoclonal IgM proteins that are elevated more than fourfold over normal.[15] Other successful methods involve either a sizing column or a charge column to separate the IgM from the IgG, and then repeating the IEP on the separated fraction. These methods are labor-intensive and take considerable time, thereby delaying the diagnosis. Such inefficient procedures are also expensive, which is an even greater problem with modern reimbursement systems that penalize hospitals for a slow diagnosis. While some have argued that the combined cost of such special procedures with IEP is less than that of IFE,[15] this has not been our experience. With IFE, one gets the correct result the first time, in a few hours. With IEP and special procedures, a couple of days are needed, as the first IEP may be unreadable. Finally, we use 50 per cent less of costly specific antisera to perform IFE than to perform IEP and much less technologist time with the advent of semi-automated IFE and ISUB.

Another difficulty with IEP is that its geometry prevents an optimal resolution of individual protein molecules. Therefore, it can be difficult to

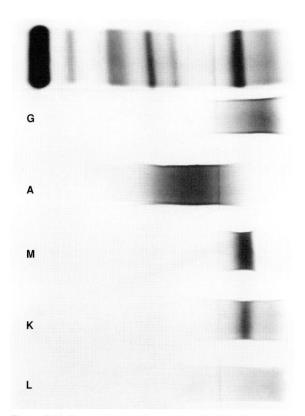

Figure 3.13 Immunofixation of serum from Fig. 3.12. Top sample is the serum protein electrophoresis of the patient's serum run at the same time as immunofixation (anode at left). The interpretation is an IgM κ monoclonal gammopathy. Note that the monoclonal protein has the same electrophoretic migration in the reaction with anti-IgM and anti-κ. Also, by using the serum at the proper dilution (see text), you have a built in control for the antisera. IgA reacts in a broad β-region where IgA normally migrates. Similarly, anti-IgG and anti-λ reagents show proper reactivity. Dilutions used were: IgG, 1:10; IgA, 1:2; IgM, 1:13; κ, 1:14; λ, 1:4. (Panagel system stained with Coomassie Blue; anode is at the left.)

detect biclonal gammopathies, that are more readily detected with IFE, which allows greater discrimination than does IEP between two different monoclonal proteins that have relatively close pIs (see Chapter 1)

Immunoelectrophoresis lacks sensitivity to detect small M-proteins. Some believe that this insensitivity of IEP is an advantage because the larger monoclonal gammopathies detected are more likely to be of clinical significance than smaller monoclonal gammopathies. Small M-proteins, however, may have clinical importance. As discussed in Chapter 6, Kyle has coined the term 'monoclonal gammopathy of undetermined significance' for patients with relatively small amounts of monoclonal gammopathy who lack clinical evidence of multiple myeloma.[22-24] Patients with monoclonal gammopathy of undetermined significance (MGUS) need to be followed indefinitely, because the gammopathy may develop into a malignant process (myeloma or Waldenström macroglobulinemia). Alternatively, small quantities of M-proteins may represent the presence of a malignant B-cell neoplasm. Such small M-proteins may be detected in light chain myeloma, amyloid (AL), chronic lymphocytic leukemia, Burkitt lymphoma, and in well-differentiated lymphocytic lymphoma.[23,25-27] It has become clear that some patients with neurological complaints have monoclonal gammopathies that are related to their clinical symptoms.[28-34]

Another problem with IEP is over-interpretation of polyclonal increases, resulting in misdiagnosis of inflammatory conditions as monoclonal gammopathies. This results when relatively massive polyclonal increases occur in IgG. Often, this produces a distortion of the IgG band (as it requires a larger space to diffuse to reach equivalence than is available). This can produce an appearance similar to the restriction seen in true monoclonal patterns (Fig. 3.14). Since there is twice as much total κ (light chain bound to intact immunoglobulin molecules as well as the small amount of free light chains that normally circulate; see Chapter 7) as λ under normal circumstances, and since the usual κ/λ ratio is 2.0 in most polyclonal increases, the λ-

band may not enter the trough, and thus appear normal (Fig. 3.14).

IMMUNOFIXATION ELECTROPHORESIS

Immunofixation electrophoresis has been a dramatic development for the clinical laboratory.[7,13,35-38] This technique can be performed in about 3–4 h, is not subject to the umbrella effect, can readily detect the small monoclonal gammopathies, is easier to interpret than IEP, uses half as much antisera as IEP, and requires the same equipment as that needed for many of the commercially available protein electrophoresis methods. Further,

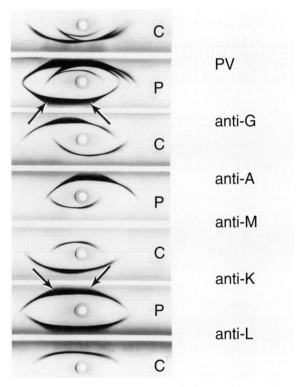

Figure 3.14 Immunoelectrophoresis which was misdiagnosed as an IgG κ monoclonal gammopathy. It is a polyclonal pattern where there is so much IgG and κ that they have spread into the trough and their pattern falsely suggests a restriction (arrows). This should have been caught. The λ actually also spills into the trough, but since it was broader and did not line up with the IgG, it was dismissed. Also, the IgA arc is markedly increased because of a corresponding polyclonal increase in IgA.

semi-automated versions of IFE are currently available that allow for examination of several samples with minimal technologist time. Sensitive new methods even allow for detection of oligoclonal bands in cerebrospinal fluid by IFE technique on a few microliters of unconcentrated sample (see Chapter 8).[39]

Selection of the dilution of patient's serum

One advantage that IEP had over IFE was the slow diffusion step that allowed concentration of the M-protein to adjust to the concentration of the antibody. This avoided the problem of the antigen excess effect that may occur with IFE if the sample is not properly diluted. In IFE, the reaction occurs quickly because there is no diffusion step. This was the step in IEP that allowed the antigen and antibody to adjust to equivalence for maximal precipitation.

To achieve optimum sensitivity and to avoid the antigen excess effect, it is important to use a dilution of the serum that places it close to the equivalence range for the immunoprecipitation reaction. There is also considerable variation in the strength and specificity of commercial antisera. Monos et al.[40] compared several commercial sera with the same M-proteins and found considerable variation in detection of the monoclonal protein. They also noted that an M-protein present in a concentration of 700 mg/dl could be missed if the incorrect dilution of the patient's serum was used with a commercial antisera that had poor reactivity for that particular protein.

There are several ways to determine which dilution of serum to use. The first is to use a standard dilution (suggested by commercial IFE kits). While this is an easy approach, it presents problems in cases with very high or very low concentrations of M-proteins. Monos et al.[40] found that a dilution of 1:10 for IgG, 1:5 for IgA, IgM, and κ-chains and 1:2 for λ-chains allowed them to make the proper diagnosis of a monoclonal gammopathy in most cases. They cautioned, however, that if equivocal results are obtained using these standard dilutions of serum, the analysis should be repeated using a different antiserum.[40] Fortunately, there is usually a relatively large leeway for this dilution, although this varies from one monoclonal protein to another and from one commercial antiserum to another. As shown in Fig. 3.15, the monoclonal gammopathy could be detected by a variety of dilutions around the equivalence region. This would not have happened, however, with all monoclonal proteins, or with all commercial antisera (see Fig. 3.16). When possible, it is a great advantage to know the concentrations of the major immunoglobulin classes before one sets up the IFE. A study by Hadler et al.[41] recommended that the optimum concentration for detection of M-proteins in the system they were using was 28–35 mg/dl. In our laboratory, we have found that with most commercial antisera, the antigen (IgG, IgA, IgM, or κ) needs to be present at a concentration of about 100 mg/dl for best results. Antisera to λ tend to give weaker reactions and one may wish to dilute the patient's serum to about 50 mg/dl for λ.

An example of the weakness of the anti-λ reagents is shown in Fig. 3.16. The serum protein electrophoresis gel shows a prominent slow β-migrating monoclonal gammopathy. By nephelometry, the IgA measured 14 400. Yet, with a 1:10 dilution of IgA, one did not have an antigen excess problem. The monoclonal band stained very strongly, perhaps obscuring the possible second IgA band (likely a multimer of the major band). At 1:100, both the major band and the second band are seen to advantage with the anti-IgA reagent. However, with antisera to λ at 1:10, a problem with antigen excess effect is seen. The center of the major band precipitate has washed away owing to inadequate size of the immune complexes formed by the anti-λ reagent; nevertheless, at 1:10 the same monoclonal protein reacts well with the anti-IgA reagent. Even at 1:50, the center of the λ band has begun to wash away. With most antigen excess effects, one can still make the diagnosis of the type of monoclonal gammopathy, however, in some cases, it can make

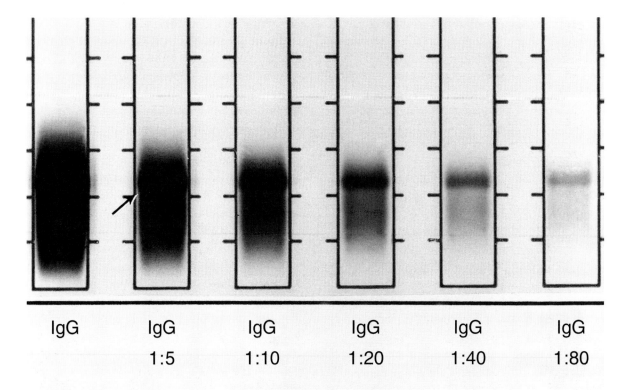

IgG	IgG	IgG	IgG	IgG	IgG
	1:5	1:10	1:20	1:40	1:80

Figure 3.15 Immunofixation of serum from patient with an IgG κ monoclonal gammopathy. The concentration of the IgG was 1740 mg/dl. The monoclonal band is readily seen at the 1:20, 1:40, and 1:80 dilutions. At 1:10 the band is visible, but is considerably obscured by the amount of polyclonal IgG present. At 1:5, there is a slight indentation at the junction of the band with the cathodal end of the polyclonal IgG (arrow). However, I believe that this would have been missed. With the IgG undiluted (left lane), the band is not detectable. (Paragon system stained with Paragon Violet; anode at the top.)

the correct interpretation difficult. When in doubt, one should always rerun the sample at other dilutions or with other antisera.

In the past few years, manufacturers have recognized the need to adjust the concentration of serum to account for extremely high and low levels of antigen. For example, the Sebia Hydragel system (Sebia, Issy-les-Moulineaux, France) offers several levels of dilutions depending on the total immunoglobulin concentration (Table 3.1). In our laboratory, strict adherence to these instructions has provided excellent results with this kit.

When using 'home brew' IFE, or kits where no suggestion is provided for diluting the serum, adjustment for concentration of a monoclonal protein may be accomplished in one of several ways. One method involves evaluating serum

samples for monoclonal gammopathies by first performing protein electrophoresis and quantification of IgG, IgA, IgM, κ and λ by standard immunological methods (nephelometry, turbidimetry, or radial immunodiffusion). Using this method, one can characterize large M-proteins without resorting to either IEP or IFE (see Chapter 8). In those cases where IFE is necessary (identifying small bands, abnormal κ:λ ratio, clinical picture compatible with monoclonal gammopathy process despite normal protein electrophoresis and quantifications), one already knows the immunoglobulin concentrations and this allows one to estimate readily the dilution to obtain optimal results. For example, with the immunoglobulin values shown in Table 3.2, results are expressed as mg/dl. To determine the appropriate dilution to use for a given sample, divide the

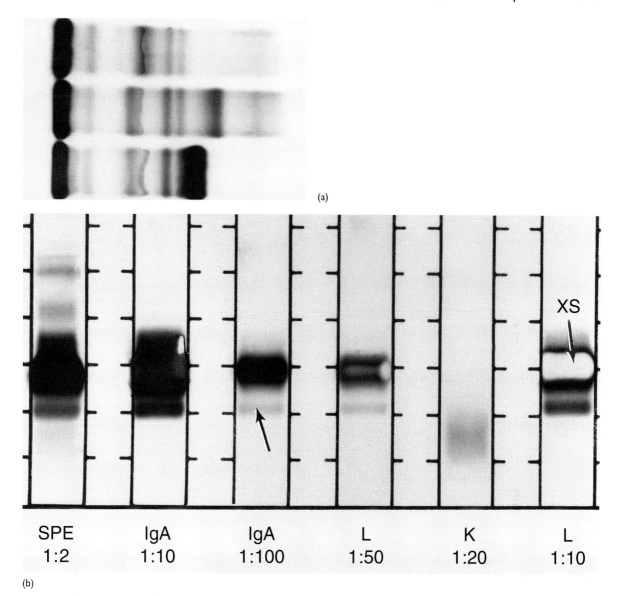

Figure 3.16 (a) Serum protein electrophoresis gel where the bottom lane contains a massive monoclonal gammopathy in the slow β-region. The serum immediately above this sample has a small monoclonal band in the β–γ and another tiny restriction in the slow γ-region. Since the monoclonal band in the bottom lane stains as dense as the albumin bands on this gel and is considerably broader, one can estimate that it will have about three times as much protein as albumin (about 12 g). (Paragon SPE2 system stained with Paragon Violet.) (b) Immunofixation of the sample from the bottom lane of (a) using the indicated dilutions of patient's serum. Note that the serum protein electrophoresis (SPE) lane did not fix the albumin band. Therefore, one cannot use it to estimate the protein concentration of the monoclonal band. A large IgA restriction is easily seen in the 1:10 and 1:100 dilution, but a second IgA band is somewhat obscured by the adjacent large IgA band at 1:10, but stands out nicely (arrow) at the 1:100 dilution. The center of the λ precipitate has washed away at the 1:10 (XS arrow) dilution and even to some extent at the 1:50 dilution. A diffuse κ band is seen using the 1:20 dilution. It indicates the presence of this patient's normal κ. (Paragon system stained with Paragon Violet; anode at the top.)

Table 3.1 Manufacturer's suggested dilutions related to total immunoglobulin concentration[a]

Track	Ratio of serum to diluent
IgG (total immunoglobulin > 0.5 < 2.0 g/dl)	1:5
IgA, IgM, κ, λ (total immunoglobulin > 0.5 < 2.0 g/dl)	1:2
IgG (total immunoglobulin > 2.0 g/dl)	1:10
IgA, IgM, κ, λ (total immunoglobulin > 2.0 g/dl)	1:4
IgG (total immunoglobulin < 0.5 g/dl)	1:2.5
IgA, IgM, κ, λ (total immunoglobulin < 0.5 g/dl)	1:1

[a]Sebia Hydragel instructions.

quantity of the immunoglobulin by 100. If the number is less than 1, in theory the serum should be undiluted (neat). However, we usually dilute samples at least 1:2 because when serum is used neat, it is more likely to give us problems with artifact bands at the origin. When very tiny amounts are present < 50 mg/dl we will apply the patient's serum neat. Note that the optimal dilution will vary somewhat between different antisera. However, diluting the immunoglobulin to be assayed to a concentration of 100 mg/dl is often a reasonable approximation for most antisera that I have used. This might not be ideal for all antisera and controls should be studied with each new lot of antisera to ensure appropriate reactivity.

Dilution of serum for IgX = [IgX]/100

If the laboratory does not perform immunoglobulin measurements as part of the evaluation of serum for monoclonal gammopathies, one may use the densitometric scan value of the γ-region as a crude estimate of IgG. This is similar to the Sebia scheme above, since, in most sera, IgG makes up most of the total immunoglobulin. With most γ-migrating IgG monoclonal gammopathies, there is a reasonable correlation between the densitometric scan of the γ-region and the IgG determined by immunochemical methods.[42] Although these estimates can be used as approximations of the IgG concentration, they are far from perfect.[42–44] This method only approximates the IgG values, and one must estimate to the concentrations for the other chains. Usually, however, two-thirds of the IgG is κ and one third is λ. Also, IgA and IgM tend to be present in relatively low concentrations (less than 300 mg/dl for most individuals). When using this method to estimate dilution, I prefer to start with a 1:2 dilution for IgA and IgM. It is unusual that greater than a 1:4 dilution is required to avoid extreme antigen excess for IgA and IgM under normal circumstances. Since IFE is performed after serum protein electrophoresis, the presence of a suspicious band, and its size, will alert us to the occasional

Table 3.2 Calculation of dilution of serum to use with home brew immunofixation

Immunoglobulin	Concentration (mg/dl)	Dilution[a]
IgG	1434	1:14
IgA	241	1:2
IgM	126	1:2[b]
κ-Containing	1047	1:10
λ-Containing	683	1:7[c]

[a]In this example, dilution is based on closest approximation to 100 mg/dl. If antigen excess effect is seen, a greater dilution of serum will be needed for your system. If no precipitate is seen, a more concentrated patient sample may be needed. Adjustment needs to be made against controls with each new lot of antisera.
[b]Dilute at least 1:2 to minimize the origin artifact (see text).
[c]Antisera against λ-containing immunoglobulins tend to give weak precipitates; adjustment of the dilution of patient's sample may be necessary.

need to increase the dilution. When there is a gross discrepancy between nephelometry and densitometry, we usually favor the densitometric value because, unlike nephelometry, it is not influenced by the vagaries of the different commercial antisera and antigen excess effects.[45,46]

One may estimate the γ-globulin region of the serum protein electrophoresis without densitometry. Because serum protein electrophoresis is always performed before IFE in our laboratory, our technologists become adept in estimating the concentration of the γ-region by examining the stained gel itself. They base their dilutions on their experience examining the gels. This method has been adequate to estimate the dilution that should be used to perform IFE and avoid the problems associated with the extreme antigen excess effect. Whether one uses a standard dilution or one of the above methods for performing dilutions of the patient's serum for IFE, it is recommend that a serum protein electrophoresis be performed at the same time and compared with the IFE. Many of the commercial products provide a serum protein electrophoresis lane that can be used to detect the M-protein for comparison purposes. Any band that is not explained by the IFE should result in a repeat IFE with other dilutions, or occasionally other antisera.

Performance of immunofixation

When performing IFE, a separate sample of serum, appropriately diluted, is applied to the gel for each immunoglobulin or other antigen to be assayed (Fig. 3.17). For example, although the typical IFE uses antisera against IgG, IgA, IgM, κ, and λ, antisera against other antigens such as IgE, IgD, fibrinogen, or C-reactive protein may be useful for identifying unusual bands in specific samples (see Chapter 6). A sample of the patient's serum should also be placed in one lane for a comparison protein electrophoresis. Some manufacturers have come up with ways to streamline the initial screening. By using a pentavalent (Penta) antisera (reactive with IgG, IgA, IgM, κ

and λ), an M-protein can be readily ruled out (Fig. 3.18).

Following electrophoresis, the lane for comparison protein electrophoresis is fixed with acid. For the pentavalent or specific immunoglobulin lanes, the patient's sample is overlaid with specific antiserum. In some systems, commercial antiserum against a specific isotype is coated onto a strip of cellulose acetate. This strip is placed directly on top of the electrophoresed sample. In other systems, using a template, the commercial antiserum is directly layered onto the gel. Diffusion of the antisera directly into the thin gel beneath is quite rapid, requiring from as little as 5–30 min depending on the system used.

Following removal of the antisera, gels are washed and stained with Coomassie Brilliant Blue, Amido Black or a specific commercial protein dye provided by the manufacturer's kit. Some workers advocate silver staining for enhanced sensitivity. I find silver stains clumsy and messy to use. In addition, the background that may result makes the interpretation difficult. Sensitivity is not usually a problem with IFE. Indeed, one of the more common complaints about the technique is that with the standard dyes, IFE shows too many bands, not too few.

INTERPRETATION OF IMMUNOFIXATION

Before looking at the IFE results, one should review the comparison protein electrophoresis gel on the sample. Any suspicious band located anywhere from the α- to the γ-region should be noted and the IFE must resolve its identity. If the antisera used do not identify a band seen in the comparison protein electrophoresis lane, one must consider additional steps: altered dilution of the antisera (addressing the possibility of antigen excess) or use of antisera against other proteins. Other reagents used include antisera against IgE or IgD, fibrinogen, C3, C-reactive protein depending on the position of the restriction and the clinical situation. Another approach that is sometimes used involves pretreatment of the sample with 2-mercaptoethanol. Before IFE became so well standardized, I would occasion-

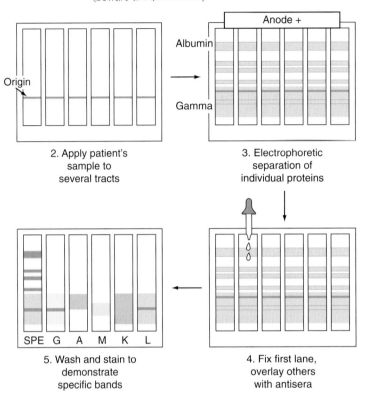

1. Select the appropriate dilution of patient's serum:
    A. Quantify IgG, IgA, IgM, Kappa, Lambda
    B. Standard dilution of manufacturer
    C. Estimate by densitometry
    D. Estimate by experience reviewing the HRE gel
        (beware antigen excess)

2. Apply patient's sample to several tracts

3. Electrophoretic separation of individual proteins

5. Wash and stain to demonstrate specific bands

4. Fix first lane, overlay others with antisera

Figure 3.17 Schematic overview of immunofixation electrophoresis. The dilution of the patient's serum for each lane must be selected (Step 1) by using one of four common methods for doing this – listed A–D and discussed in the text. After applying the serum to the origin (Step 2), the proteins are separated by electrophoresis (Step 3) (in this example, the anode is at the top) and then (Step 4), acid (SPE lane), or specific antisera (anti-G, A, M, κ or λ) are placed on each lane to precipitate the protein. In Step 5, the gels are washed and stained to reveal, in this case, an IgG λ monoclonal protein. Note that ideally, one should see diffusely staining bands at the appropriate locations for the other analytes. This is achieved by appropriate dilutions of the patient's serum. If a lane is empty, one would not be certain if the correct antisera (or any antisera) was added, if the patient's sample was added, or if there is just too little protein because too large a dilution of the patient's serum was used. HRE, high-resolution electrophoresis.

ally purify IgG and IgM with charge or sizing columns. With currently available IFE and immunosubtraction (ISUB; see below) techniques, I have not needed to use column purification for several years.

The M-protein must have the same electrophoretic mobility in the IFE gel as it does in the protein electrophoresis lane. Compare the migration across the gel as this identifies the position of the suspected monoclonal protein. It is important

that the monoclonal protein lines up with the suspected band because other proteins, such as fibrinogen, genetic variants (C3, transferrin) or such as C-reactive protein can produce a suspicious band in protein electrophoresis.[47] When two or more *small* bands are detected in the γ-region, one is usually dealing with an oligoclonal expansion in a patient with an infectious disease, an autoimmune disease, or some lymphoproliferative processes (see Chapters 4 and 6) with polyclonal

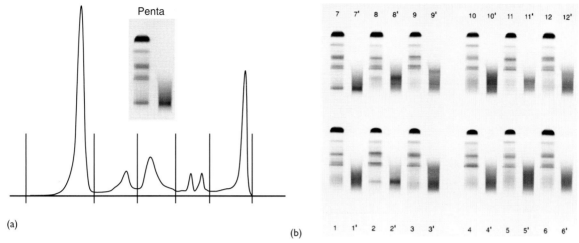

Penta

(a)

(b)

Figure 3.18 (a) This capillary zone electrophoresis pattern contains a large γ-region spike. The insert with the result of a Penta (pentavalent) immunofixation is shown. On the left is the patient's serum fixed with acid. On the right is the result of the immunofixation with antipentavalent serum. This immunofixation demonstrates that the spike is caused by an immunoglobulin. To characterize the immunoglobulin, further studies need to be performed. With such a large γ-spike, I usually go straight to immunosubtraction or immunofixation. However, with subtle β-region restrictions, the confirmation that it is due to an immunoglobulin can save us from performing immunofixation on a radiocontrast dye. (b) Overview of an entire Penta Gel. Each number represents the serum protein electrophoresis pattern and the number prime to its right is the Penta immunofixation on that serum. For this gel, M-proteins that require a more complete characterization are found in cases 2, 7, 8, 9 and 11. Case number 3 is arguable, but I would also perform a characterization of that one. (Sebia Penta gel.)

expansion of immunoglobulins (Fig. 3.19). Small bands on IFE can occur as artifacts. Cryoglobulins can produce small, artifactual bands at the origin (see Chapter 6). Some commercial antibodies against human immunoglobulins contain antibodies that react with normal β- or γ-region components. We have seen reactivity against several such bands masquerading as monoclonal proteins (Table 3.3).[47]

Biclonal gammopathies occur uncommonly and can be readily diagnosed by IFE (Figs. 3.20 and 3.21). They are usually easily distinguished from the polyclonal process, which has tiny oligoclonal bands shown in Fig. 3.19. The monoclonal proteins resulting from myeloma or B-cell lymphoproliferative disorders are usually present in a greater concentration than oligoclonal bands found in polyclonal processes. Furthermore there is often an accompanying decrease in concentration of the polyclonal immunoglobulins in myeloma

and B-cell neoplasms as opposed to the diffuse increase of immunoglobulins that accompanies polyclonal processes in many infections and autoimmune diseases.

Table 3.3 Reactivities of anti-immunoglobulin reagents that can produce discrete bands resembling monoclonal gammopathies

Additional reactant[a]
Fibrinogen
Transferrin
C3
C4

[a]Reactivities of commercial antisera with specificity for an immunoglobulin.

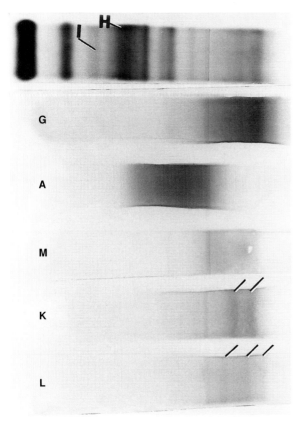

Figure 3.19 Immunofixation of serum from a patient with pneumonia who had a few (oligoclonal) bands in the γ-region. The serum protein electrophoresis lane at the top shows increased α_1-antitrypsin band, increased α_1–α_2 interzone (I), hemoglobin-haptoglobin complex (H) due to hemolysis during sample preparation and several tiny γ-region bands. Immunofixation shows that the bands are both κ (K) and λ (L), therefore polyclonal. Some of the small, round, clear areas seen best in the anti-IgG and anti-IgM reaction are caused by air bubbles that prevent the precipitin reaction from occurring. Dilutions used: IgG, 1:15; IgA, 1:3; IgM, 1:2; κ, 1:10; λ, 1:4. (Panagel system stained with Coomassie Blue; anode at the left.)

Limitations of immunofixation

As mentioned above, a key problem to avoid is the antigen excess effect. When a large amount of monoclonal protein is present, an antigen excess effect can be seen (Fig. 3.22). In antigen excess, the immune complexes formed are small and wash away during the washing steps.

The lack of within-gel internal controls is a potential source of error. Immunoelectrophoresis

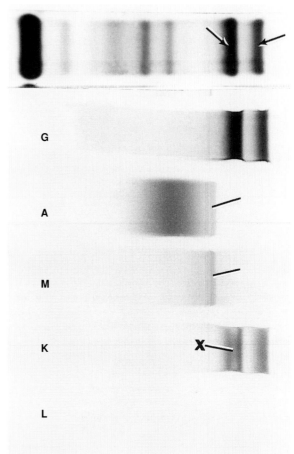

Figure 3.20 Serum protein electrophoresis (top lane) of this sample shows lightly staining α_1- and α_2-regions indicating that an inflammatory response is not likely. Two distinct γ-bands (arrows) are identified as IgG κ by immunofixation electrophoresis. The two bands may reflect a monoclonal protein that forms monomers and dimers, post-translational modification of a monoclonal protein or may be a true double (biclonal) gammopathy. In this case, subclass determination showed that the two bands were of different subclasses of IgG (a double gammopathy). Unlike Fig. 3.19, which shows a diffuse increase in the IgG proteins caused by an infectious process, there is a relative hypogammaglobulinemia in this IgG lane. The origin artifact (indicated) in the IgA and IgM reactions is seen in some monoclonal proteins, which tend to self-aggregate, but may also be seen with cryoglobulins or in normal samples with a polyclonal increase in γ-globulins. The origin artifact is most obvious in the IgA and IgM reactions here because the serum samples were applied undiluted to the gel (since the IgA and IgM concentrations were low). A slight antigen excess effect (x) is seen in one of the κ bands. Note also that no precipitate was seen with the anti-λ reagent, likely due to too large a dilution of the patient's serum being used in this lane. Dilutions used: IgG, 1:20; IgA, neat; IgM, neat; κ (K), 1:12; λ (L), 1:10. (Panagel system stained with Coomassie Blue; anode at the left.)

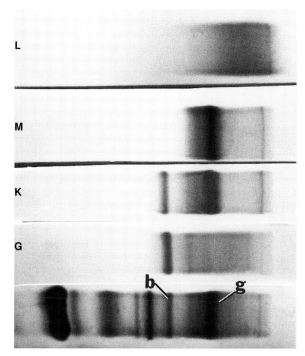

Figure 3.21 A double gammopathy. In the bottom lane, the serum protein electrophoresis lane shows a discrete abnormal band in the β-region (b), and a broader band in the γ-region (g). Immunofixation discloses an IgG κ and an IgM κ double gammopathy. (Cellogel high-resolution acetate system; anode to the left.) This photograph was contributed by Francesco Aguzzi.

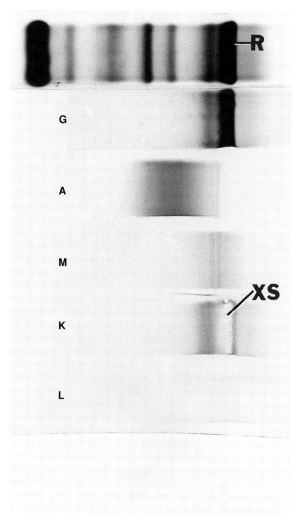

Figure 3.22 The serum protein electrophoresis lane (top) shows a large γ-restriction (R–). It is obvious from the immunofixation that this is IgG. However, the light chain is not as readily seen, because the wrong dilution of serum was placed in the κ reaction. Too much κ-containing immunoglobulin was present for the amount of anti-κ reactivity in the antiserum used. This created an antigen-excess situation (XS). The small complexes formed were removed during the wash step, leaving the clear area indicated. Note that it is surrounded by κ, which was present at a lower concentration. Once again, the anti-λ reaction is barely visible. Dilutions used: IgG, 1:25; IgA, 1:2; IgM, neat; κ (K), 1:15; λ (L), 1:5. (Panagel system stained with Coomassie Blue; anode at the left.)

has a control serum sample alternating with the patient's sample, allowing for comparison across the specific antibody trough. IFE does not routinely employ a control serum on each sample gel. If the appropriate dilution of the serum is used, the normal immunoglobulins for each class should be precipitated along with the monoclonal protein. A normal serum should have a precipitate for each immunoglobulin class, unless its concentration is relatively low, such as may be true for IgA and IgM (Fig. 3.23). In a serum with normal immunoglobulins, there should always be a diffuse precipitate in the IgG, κ, and λ lanes. This precipitate represents the normal immunoglobulin of the isotype for which one is testing. However, when examining urine for MFLC, one will often find no reaction at all in wells for the non-MFLC light chain isotype. This is because, under normal circumstances, only trivial quantities of light

chains are found in urine. When the protein electrophoresis pattern and IFE pattern match well (monoclonal spike in protein electrophoresis of urine and one light chain type by IFE that corresponds to the spike in migration), the diagnosis is usually clear-cut (Fig. 3.24).

Instances where the protein electrophoresis shows γ-globulins, but on immunofixation a polyclonal pattern is seen with only one light chain type while the other light chain shows no reaction should give the interpreter pause. Since there is no built-in control, the interpreter may suspect that one of the anti-light chain antisera had not been added. In that case, the sample should be run again.

The position of the precipitin bands is useful for identifying the antisera used. For example, note that the IgA in Fig. 3.23 gives a diffuse staining in the β-region, because IgA is mainly a β-migrating

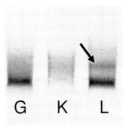

Figure 3.24 Urine immunofixation from a patient with a known IgG λ monoclonal gammopathy in serum. This immunofixation demonstrates a slow γ IgG band, no bands in the κ (K) lane and two bands in the λ (L) lane. The slower λ band migrates the same as the IgG and reflects the λ bound to the intact monoclonal protein. The smaller and faster-migrating λ-band (arrow) is the monoclonal free light chain (MFLC).

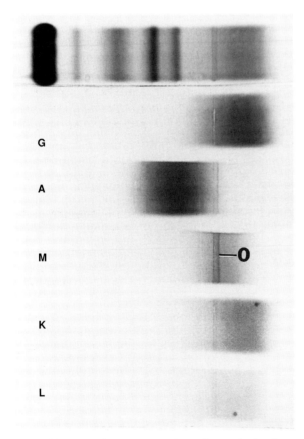

Figure 3.23 Normal serum protein electrophoresis (top) and immunofixation. Note origin artifact (O–) with undiluted IgM. Dilutions used: IgG, 1:10; IgA, 1:2; IgM, neat; κ (K), 1:6; λ (L), 1:3. (Panagel system stained with Coomassie Blue; anode at the left.).

globulin. Similarly, IgM stays near the origin and IgG is mainly a γ-migrating globulin. If the IgA lane has a broad precipitate in the γ-region, the wrong antiserum (likely anti-IgG or one of the anti-light chain antisera) was used and the sample should be run again.

The sensitivity of IFE requires one to deal with the small oligoclonal bands that may be seen in patients with infections, autoimmune diseases, and occasionally patients with lymphoproliferative disorders. Occasionally, one of these bands may be relatively prominent. With difficult cases, I always suggest that the clinician send urine to rule out an MFLC. In addition, I recommend repeating the IFE in 3–6 months. If the process is reactive, it may resolve. If it reflects a true M-protein, the band will persist and may increase.

As mentioned above, the sensitivity and resolution of IFE may result in the detection of minor antibody reactivities in reagent antisera, which can create confusing patterns.[47] Most commercial antisera are monospecific reagents for the stated immunoglobulin. Some reagent antisera against immunoglobulins have minor cross-reactivities with other serum proteins. This can lead to a false positive interpretation for heavy or light chain disease. The most common problematic reactivities have been seen with β-region-migrating proteins: fibrinogen, C3, C4, and transferrin. I have found this problem most frequently with anti-IgM and

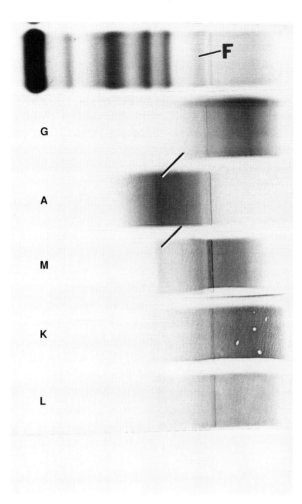

Figure 3.25 Serum protein electrophoresis lane (top) shows a faint fibrinogen band (F–) just anodal to the origin. The anti-IgA and anti-IgM reactions show a faint but distinct band (indicated) due to reactivity of these commercial antisera with a β-region migrating protein (possibly C4 or a complement breakdown product). It was not identified. Minor reactivities such as these usually were too small to be noticed with immunoelectrophoresis. They can be controlled for by testing reagents with a normal serum prior to use with patient samples. Dilutions used: IgG, 1:10; IgA, 1:2; IgM, neat; κ (K), 1:6; λ (K), 1:3. (Panagel system stained with Coomassie Blue; anode at the left.)

sample has to be used in an immunofixation, control plasma must be run to check all reagents for antifibrinogen reactivity that could give a false positive band.

For laboratories with larger volumes, semi-automated IFE procedures are available that may perform as many as nine IFE simultaneously. As shown in Fig. 3.26, nine monoclonal gammopathies were easily characterized on this one gel. The instrument performs the application of the samples, washing and staining while the operator places the samples in wells and determines which antisera to add.

IMMUNOSUBTRACTION

Immunosubtraction was first used as an adjunct to gel-based protein electrophoresis. In 1977, Aguzzi and Poggi[48] reported a method to precipitate specific individual proteins by the use of monospecific antisera prior to performing protein electrophoresis on cellulose acetate. They noted that by removing specific bands or zones (immunosubtracting them) from the electrophoretic pattern it improved their ability to interpret the final product. A few years later Merlini et al.[49] passed plasma samples through a layer of monospecific

anti-IgA reagents (Fig. 3.25). This reaction is usually seen when the sample is run undiluted. The lack of reactivity of the same area with either of the two light chain antisera should cause the interpreter to suspect this cross-reactivity. Quality control of such reagents should include use of a normal serum run undiluted for IFE. If a plasma

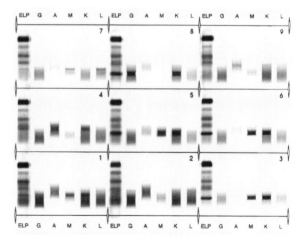

Figure 3.26 Semi-automated immunofixation gel from Sebia that demonstrates nine monoclonal proteins with minimal technologist time. The application, wash and staining steps are performed by the instrument.

antisera to subtract out certain proteins. They applied this technique to prealbumin, albumin, α-lipoprotein, α_1-antitrypsin, Gc-globulins, α_2-macroglobulin, haptoglobin, fibronectin, transferrin, β-lipoprotein, C3 and fibrinogen to demonstrate protein polymorphisms. The technique was used to detect M-proteins by White and Attwood[50] who found that it may be more sensitive than IEP.

In the past decade, the high-quality resolution of capillary zone immunosubtraction together with its ability for automation has found its way into several high-volume clinical laboratories. Currently, it is available on the Paragon CZE 2000 technique as an automated module. For this, the patient's serum containing an M-protein is pre-incubated in individual wells containing Sepharose beads coated with antibodies against one of the following: anti-IgG, anti-IgA, anti-IgM, anti-κ or anti-λ. After a short incubation period, the beads settle and the supernatant of each well is sampled for electrophoresis.[51] As shown in Fig. 3.27, the M-protein is removed by the antisera against its heavy and light chain components. As currently performed on the Paragon CZE 2000, it is an entirely hands-free operation. A prepackaged container with the antisera is placed on the instrument along with the patient sample to be tested. The instrument mixes the serum with the reagents and following an incubation period aspirates the sample into the capillary for protein electrophoresis. Removal of the M-protein peak identifies its immunoglobulin composition[52] (Fig. 3.28). Because the analysis is performed in all six capillaries simultaneously, the entire procedure takes under 10 min.

How well does ISUB compare with IFE for accuracy and sensitivity? ISUB is convenient and when the M-protein is readily identifiable in the CZE, ISUB should demonstrate the same heavy and light chain types as IFE. However, there are some problems. Immunofixation electrophoresis is slightly more sensitive than ISUB because ISUB can only deal with a band that one can remove. Henskens et al.[53] found that classification of four IgG monoclonal components was more readily

Immunosubtraction

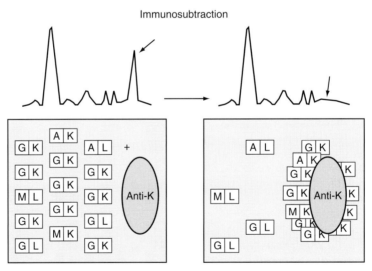

Figure 3.27 Schematic representation of one immunofixation lane. On the top, an arrow points to the prominent γ-region spike in the capillary zone electrophoresis pattern. Directly below is a diagram using block figures for the immunoglobulin molecules in this patient's serum. Beads are added to this serum that are coated with specific antisera. In this case the bead is coated with anti-κ. On the top right, after immunosubtraction with anti-κ, the spike is gone. Below, the κ-containing molecules have all bound to the beads (one shown), leaving on the λ-containing immunoglobulins in the serum. Note that the beads bind not only the monoclonal κ immunoglobulins, but all κ-containing immunoglobulins.

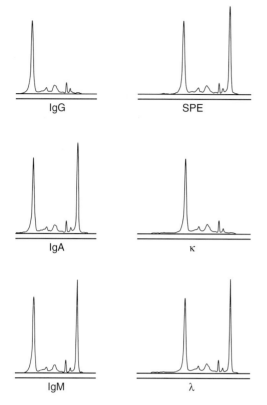

Figure 3.28 Immunofixation pattern from a patient with an IgG κ monoclonal gammopathy. The top right segment (serum protein electrophoresis, SPE) is the capillary zone electrophoresis (CZE) pattern on this patient. A large γ-spike is evident. The top left segment displays the CZE pattern of this serum after it reacted with beads coated with anti-IgG. The spike has been removed by this treatment. Below this, reaction with beads coated with anti-IgA had no effect on the spike, neither did the beads coated with anti-IgM or anti-λ. However, the beads coated with anti-κ also removed the spike.

performed by IFE than by ISUB, although one IgM M-protein was detected more readily by ISUB. There are occasional exceptions such as the report by Oda et al.[54] of a case with a small M-protein. In that patient, there was also a marked polyclonal increase in IgG that obscured detection of the M-protein by IFE, whereas ISUB was able to detect it.

In a multicenter report by Bienvenu et al.,[51] of M-proteins with > 10 g/l (1 g/dl), the M-protein was completely characterized by ISUB in 115 of 118 cases. The other three cases were detected, but the light chain could not be characterized. With M-proteins < 5 g/l (0.5 g/dl), either they could not identify both the heavy and light chain type, or the ISUB missed the M-protein in 33 per cent of the sera.[51] Similarly, Litwin et al.[55] compared the ability of four readers to interpret ISUB on 48 serum samples with an M-protein confirmed with IFE. They found that only 60–75 per cent were correctly typed by ISUB when the interpreters were blinded to the IFE results. They recorded several concerns with ISUB patterns. Sometimes they found normal-appearing CZE electropherograms but abnormal ISUB patterns that two of their interpreters found confusing.[55] This produced a result of false positives in 3 per cent and 7 per cent of negative controls by those individuals. Further, they reported only 30–40 per cent of IgM M-proteins correctly identified by ISUB.[55] Biclonal gammopathies, especially of the IgM class were difficult to interpret.

On the other hand, Katzmann et al. found that of 208 samples studied, 16 with small M-proteins were initially thought to be polyclonal when evaluated by IFE using their IFE assay.[56] On dilution, these were found all to contain a monoclonal protein. They noted that complicated electrophoretic patterns, biclonal gammopathies, light chain disease and hypogammaglobulinemia with no M-protein spike may still need IFE.[57] An important conclusion of both the Katzmann et al. and Litwin et al. studies was that with experience, the correct interpretations and better understanding of the limitations of ISUB would improve.[55,56]

Bossuyt et al.[57] evaluated 58 selected specimens with M-proteins identified by either IEP or IFE by the same Paragon CZE 2000 system as Katzmann et al.[56] and Litwin et al.[55] It was found that ISUB was able to correctly identify 91 per cent of these cases. The few that were not detected by ISUB were also not visible on their agarose and cellulose acetate electrophoresis.[57]

In my laboratory when an M-protein peak can be seen on CZE, ISUB will almost always characterize it correctly and quickly. However, when the peak is quite small, the ISUB may miss it and IFE is superior. Some clinically significant M-proteins are present in small quantities that may not be detectable by either CZE or gel-based electrophoresis, yet would be detectable by IFE.[58] Recall that in IFE, following electrophoresis, specific antiserum is applied to each lane. This precipitates the protein of interest and all other proteins are washed away. Therefore, an M-protein that may be obscured by transferrin or C3 in the β-region will be seen detected after these proteins are washed away in an IFE reaction. However, ISUB does not remove the large amount of normal proteins and usually will not improve on the sensitivity of CZE to detect small M-proteins. This would be relevant in patients with small monoclonal gammopathies associated with neuropathies where the M-proteins often require IFE for detection.[59]

Immunosubtraction on the Paragon CZE 2000 is set up to look at the major immunoglobulin classes, but does not (at the time of writing) have kits for IgD and IgE. Whereas IgE is exceedingly rare, IgD is seen in about 1–2 per cent of M-proteins in our laboratory and is easy to test for by IFE. In cases where only a light chain is identified, then, laboratories performing ISUB would either have to perform their own in-house ISUB, maintain a redundant IFE procedure to deal with IgD and IgE or send the sample to a reference laboratory. Since light chain myeloma represents about 15–20 per cent of cases we see, this part of the evaluation is not rare. Therefore, laboratories that wish to take advantage of ISUB in the automated system need to be capable of also performing IFE, or setting up an ISUB with a 'home-brew' method to deal with this issue.

Overall, ISUB is a useful technique for relatively large laboratories (processing about 40 or more samples for serum protein electrophoresis daily). It is recommended that laboratories using it maintain an IFE technique to deal with ruling out IgD and rare IgE in the many cases of light chain disease that they will detect. In occasional difficult cases, the wide availability of reagents that can be used in IFE provides a further advantage.

IMMUNOSELECTION

Immunoselection techniques were technically an early version of immunosubtraction. It is now used mainly in the detection of heavy chain disease. In detecting free heavy chains, one wishes to distinguish between intact immunoglobulin molecules and those that contain only heavy chains of a specific isotype (γ, α, or μ). One may do this by purifying the heavy chain fractions or separating them in a manner that allows their identification by both molecular size and immunochemical characteristics. However, such purification methods are laborious.

Early studies by Seligmann et al.[60] and by Radl[61] used IEP with agarose that contained antisera against light chains in the gel itself. By mixing highly selective antisera into agarose used for IEP, early workers were able to selectively remove light chain determinants bound to intact immunoglobulins.

In the original technique, antisera against κ and λ light chains were mixed into the agarose when it was in liquid form, but below 50°C. This was then poured into the appropriate mold for standard IEP (see above) and allowed to cool. When IEP was performed on the patient's serum or urine, intact molecules of immunoglobulin which all contain light chains would precipitate around the well of origin. Only molecules of free heavy chain were able to migrate from the well of origin. Following electrophoresis, antisera against the heavy chain of interest was placed in the trough and allowed to diffuse overnight toward the electrophoresed serum. The presence of a precipitin arc identified the free heavy chain. The α heavy chains usually migrate toward the β-region and will precipitate in an arc with antisera against IgA.[62,63] Other isotypes of heavy chain disease also can be detected by this technique.

Sun et al.[64] report a modified ISEL procedure that can be used in an IFE format to identify the heavy chain. For this procedure, anti-κ and anti-λ reagent is placed on an IFE gel in the region of the application (Fig. 3.29). After allowing the antisera to diffuse into the gel, specific antisera against the

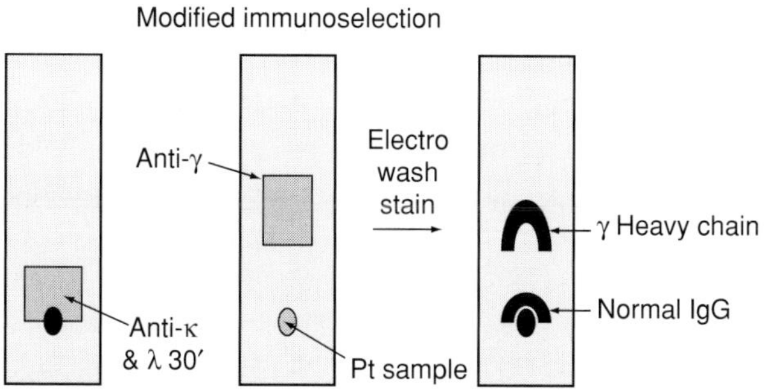

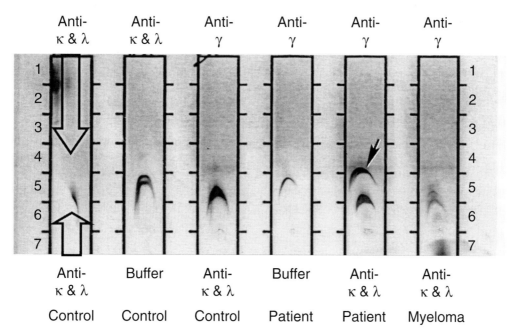

Modified immunoselection

Figure 3.29 A modified immunoselection technique requires that specific antisera against both κ and λ light chain (anti-κ & λ) are applied to the agarose just anodal to the origin 30 min prior to application of the patient's serum (left). Just before application of the patient's serum, antiserum to the specific heavy chain is placed more anodal on the gel (center). After electrophoresis, any intact immunoglobulins will precipitate near the well because it will react with the anti-κ & λ present in the gel (Normal IgG). Free heavy chains can diffuse through this area containing anti-κ & λ and eventually precipitates with the anti-IgG (γ Heavy Chain).

heavy chain of interest is applied to a more anodal portion of the gel. The patient's sample is applied in a well at the origin and the serum is electrophoresed. As with ISEL in the IEP format, any

immunoglobulin containing either κ or λ chains would precipitate near the origin. Any free heavy chains are able to migrate toward the anode where they precipitate as a second arc (Fig. 3.30).

Figure 3.30 Immunoselection on a case of γ-heavy chain disease. On the bottom are listed the samples placed in each well (control serum, patient with γ-heavy chain disease and a myeloma patient that had an IgG κ monoclonal gammopathy). Just above the contents of the well are listed the reagents applied near the origin (anti-κ & λ, or buffer). At the top of the gel are listed the reagents applied at the more anodal end of the gel (anti-κ & λ, or anti-γ). Note only the lane that had the patient's serum treated around the origin with anti-κ & λ and treated toward the anode with anti-γ has the band characteristic of γ-heavy chain disease (arrow).

Both the classic and modified forms of ISEL are complex and require considerable technical training and expertise in clinical immunology for proper interpretation and to avoid pitfalls of antigen excess situations.

IMMUNOBLOTTING

Immunoblotting (Western blotting) techniques improve the sensitivity for detection of both small monoclonal bands and oligoclonal bands.[65] Radl et al. have found these techniques especially useful in detecting monoclonal gammopathies that are associated with dysregulation of the immune system that may occur as a result of congenital, acquired or iatrogenic conditions.[66–69] This technique may be able to detect monoclonal proteins (homogeneous immunoglobulins) in concentrations as low as 0.5 mg/l.[65]

For immunoblotting, sera are diluted up to 1000-fold in the buffer system. The dilution depends upon the concentration of the analyte sought.[65,70] After electrophoresis (which may be performed on agarose or by isoelectric focusing), the separated proteins are blotted onto nitrocellulose paper. This paper is cut into strips that are incubated with specific antisera to the antigens of interest. The reaction is often enhanced by having a peroxidase-conjugated reagent as the detecting antibody. Its most common use in our laboratory is for the Western blot confirmation for human immunodeficiency virus (HIV) infection.

Norden et al.[70] reported that this method was able to examine 10 sera for the presence of monoclonal gammopathies in only 1 h. Further, its performance matched that of immunofixation in a study of 121 serum specimens. In addition to the study of individuals with a variety of immunodeficiency conditions, this method is sensitive for detecting the small amounts of M-proteins present in B-cell chronic lymphoproliferative processes, MFLC in urine, and typing subclasses of M-proteins.[71–75]

REFERENCES

1. Davies C. Introduction to immunoassay principles. In: Wild D, ed. *The immunoassay handbook*. London: Nature Publishing Group, 2001.
2. Davies DR, Metzer H. Structural basis of antibody function. *Ann Rev Immunol* 1983;1:87.
3. Harrison HH. The 'ladder light chain' or 'pseudo-oligoclonal' pattern in urinary immunofixation electrophoresis (IFE) studies: a distinctive IFE pattern and an explanatory hypothesis relating it to free polyclonal light chains. *Clin Chem* 1991;37:1559–1564.
4. Ouchterlony O. The antigenic pattern of immunoglobulins. *G Mal Infett Parassit* 1966;18(Suppl):942–948.
5. Penn G, Batya J. *Interpretation of immunoelectrophoretic patterns*. Chicago: ASCP Press, 1978.
6. Chowdhury MM, Serhat Inaloz H, Motley RJ, Knight AG. Erythema elevatum diutinum and IgA paraproteinaemia: 'a preclinical iceberg'. *Int J Dermatol* 2002;41:368–370.
7. Ritchie RF, Smith R. Immunofixation. III. Application to the study of monoclonal proteins. *Clin Chem* 1976;22:1982–1985.
8. Pudek MR. Investigation of monoclonal gammapathies by immunoelectrophoresis and immunofixation. *Clin Chem* 1982;28:1231–1232.
9. Whicher JT, Chambers RE. Immunofixation can replace immunoelectrophoresis. *Clin Chem* 1984;30:1112–1113.
10. Aguzzi F, Kohn J, Merlini G, Riches PG. More on immunofixation vs. immunoelectrophoresis. *Clin Chem* 1984;30:1113.
11. Reichert CM, Everett DF Jr, Nadler PI, Papadopoulos NM. High-resolution zone electrophoresis, combined with immunofixation, in the detection of an occult myeloma paraprotein. *Clin Chem* 1982;28:2312–2313.
12. Smith AM, Thompson RA, Haeney MR. Detection of monoclonal immunoglobulins by immunoelectrophoresis: a possible source of error. *J Clin Pathol* 1980;33:500–504.

13. Sun T, Lien YY, Degnan T. Study of gammopathies with immunofixation electrophoresis. *Am J Clin Pathol* 1979;**72**:5–11.

14. Kahn SN, Bina M. Sensitivity of immunofixation electrophoresis for detecting IgM paraproteins in serum. *Clin Chem* 1988;**34**:1633–1635.

15. Normansell DE. Comparison of five methods for the analysis of the light chain type of monoclonal serum IgM proteins. *Am J Clin Pathol* 1985;**84**:469–475.

16. Lane JR, Bowles KJ, Normansell DE. Detection of IgM monoclonal proteins in serum enhanced by removal of IgG. *Lab Med* 1985;**16**:676–678.

17. Prokesova L. Study of properties of structural subunits of IgM immunoglobulin obtained by reduction with 2-mercaptoethanol or by oxidative sulphitolysis. *Folia Microbiol* 1969;**14**:82–88.

18. Herrlinger JD, Kriegel W. Effect of D-penicillamine and 2-mercaptoethanol on human IgM in normal serum. *Z Rheumatol* 1976;**35**:108–112.

19. Orr KB. Use of 2-mercaptoethanol to facilitate detection and classification of IgM abnormalities by immunoelectrophoresis. *J Immunol Methods* 1979;**30**:339–347.

20. Capel PJ, Gerlag PG, Hagemann JF, Koene RA. The effect of 2-mercaptoethanol on IgM and IgG antibody activity. *J Immunol Methods* 1980;**36**:77–80.

21. Sorensen S. Monoclonal IgM kappa with rheumatoid factor activity and cryoprecipitability identified only by immunofixation electrophoresis after 2-mercaptoethanol treatment. *Clin Chim Acta* 1988;**173**:217–224.

22. Kyle RA. 'Benign' monoclonal gammopathy – after 20–35 years of follow-up. *Mayo Clin Proc* 1993;**68**:26–36.

23. Kyle RA, Therneau TM, Rajkumar SV, et al. A long-term study of prognosis in monoclonal gammopathy of undetermined significance. *N Engl J Med* 2002;**346**:564–569.

24. Kyle RA, Rajkumar SV. Monoclonal gammopathies of undetermined significance. *Hematol Oncol Clin North Am.* 1999;**13**:1181–1202.

25. Bajetta E, Gasparini G, Facchetti G, Ferrari L, Giardini R, Delia D. Monoclonal gammopathy (IgM-κ) in a patient with Burkitt's type lymphoblastic lymphoma. *Tumori* 1984;**70**:403–407.

26. Braunstein AH, Keren DF. Monoclonal gammopathy (IgM-kappa) in a patient with Burkitt's lymphoma. Case report and literature review. *Arch Pathol Lab Med* 1983;**107**:235–238.

27. Cesana C, Klersy C, Barbarano L, et al. Prognostic factors for malignant transformation in monoclonal gammopathy of undetermined significance and smoldering multiple myeloma. *J Clin Oncol* 2002;**20**:1625–1634.

28. Steck AJ, Murray N, Meier C, Page N, Perruisseau G. Demyelinating neuropathy and monoclonal IgM antibody to myelin-associated glycoprotein. *Neurology* 1983;**33**:19–23.

29. Dalakas MC, Engel WK. Polyneuropathy with monoclonal gammopathy: studies of 11 patients. *Ann Neurol* 1981;**10**:45–52.

30. Driedger H, Pruzanski W. Plasma cell neoplasia with peripheral polyneuropathy. A study of five cases and a review of the literature. *Medicine (Baltimore)* 1980;**59**:301–310.

31. Lee YC, Came N, Schwarer A, Day B. Autologous peripheral blood stem cell transplantation for peripheral neuropathy secondary to monoclonal gammopathy of unknown significance. *Bone Marrow Transplant* 2002;**30**:53–56.

32. Gorson KC, Ropper AH, Weinberg DH, Weinstein R. Efficacy of intravenous immunoglobulin in patients with IgG monoclonal gammopathy and polyneuropathy. *Arch Neurol* 2002;**59**:766–772.

33. Kvarnstrom M, Sidorova E, Nilsson J, et al. Myelin protein P0-specific IgM producing monoclonal B cell lines were established from polyneuropathy patients with monoclonal gammopathy of undetermined significance (MGUS). *Clin Exp Immunol* 2002;**127**:255–262.

34. Fisher MA, Wilson JR. Characterizing neuropathies associated with monoclonal gammopathy of undetermined significance (MGUS): a framework consistent with classifying injuries according to fiber size. *Neurol Clin Neurophysiol* 2002;3:2–7.

35. Janik B. Identification of monoclonal proteins by immunofixation. *Electrophor Today* 1981;2:1–4.

36. Cawley LP, Minard BJ, Tourtellotte WW, Ma BI, Chelle C. Immunofixation electrophoretic techniques applied to identification of proteins in serum and cerebrospinal fluid. *Clin Chem* 1976;22:1262–1268.

37. Ritchie RF, Smith R. Immunofixation. I. General principles and application to agarose gel electrophoresis. *Clin Chem* 1976;22: 497–499.

38. Ritchie RF, Smith R. Immunofixation. II. Application to typing of alpha1-antitrypsin at acid pH. *Clin Chem* 1976;22:1735–1737.

39. Richard S, Miossec V, Moreau JF, Taupin JL. Detection of oligoclonal immunoglobulins in cerebrospinal fluid by an immunofixation-peroxidase method. *Clin Chem* 2002;48: 167–173.

40. Monos DS, Bina M, Kahn SN. Evaluation and optimization of variables in immunofixation electrophoresis for the detection of IgG paraproteins. *Clin Biochem* 1989;22: 369–371.

41. Hadler MB, Kimura EY, Leser PG, Kerbauy J. Standardization of immunofixation technique for the detection and identification of serum paraproteins. *Rev Assoc Med Bras* 1995;41:119–124.

42. Keren DF, Di Sante AC, Bordine SL. Densitometric scanning of high-resolution electrophoresis of serum: methodology and clinical application. *Am J Clin Pathol* 1986;85:348–352.

43. Uriel L. Interpretation quantitative des resultats après electrophorese en gelose, 1. Considerations générales, application a l'étude de constituants proteiques isolés. *Clin Chim Acta* 1958;3:234–238.

44. Schreiber WE, Chiang E, Tse SS. Electrophoresis underestimates the concentration of polyclonal immunoglobulins in serum. *Am J Clin Pathol* 1992;97:610–613.

45. Bush D, Keren DF. Over- and underestimation of monoclonal gammopathies by quantification of kappa- and lambda-containing immunoglobulins in serum. *Clin Chem* 1992;38:315–316.

46. Su L, Keren DF, Warren JS. Failure of anti-lambda immunofixation reagent mimics alpha heavy-chain disease. *Clin Chem* 1995;41: 121–123.

47. Register LJ, Keren DF. Hazard of commercial antiserum cross-reactivity in monoclonal gammopathy evaluation. *Clin Chem* 1989;35:2016–2017.

48. Aguzzi F, Poggi N. 'Immunosubtraction' electrophoresis: a simple method for identifying specific proteins producing the cellulose acetate electrophoretogram. *Boll Ist Sieroter Milan* 1977;56:212–216.

49. Merlini G, Pavesi F, Carini A, Zorzoli I, Valentini O, Aguzzi F. Identification of specific plasma proteins determining the agarose gel electrophoresis by the immunosubtraction technique. *J Clin Chem Clin Biochem* 1983;21:841–844.

50. White WA, Attwood EC. Immunofixation and immunosubtraction on agarose gel: an aid in the typing of monoclonal gammopathies. *Ann Clin BioChem* 1984;21:467–470.

51. Bienvenu J, Graziani MS, Arpin F, et al. Multicenter evaluation of the Paragon CZE 2000 capillary zone electrophoresis system for serum protein electrophoresis and monoclonal component typing. *Clin Chem* 1998;44:599–605.

52. Klein GL, Jolliff CR. Chapter 16. Capillary electrophoresis for the routine clinical laboratory. In: Landers JP, ed. *Handbook of capillary electrophoresis*. Boca Raton: CRC Press, 1993.

53. Henskens Y, de Winter J, Pekelharing M, Ponjee G. Detection and identification of monoclonal gammopathies by capillary electrophoresis. *Clin Chem* 1998;44:1184–1190.

54. Oda RP, Clark R, Katzmann JA, Landers JP. Capillary electrophoresis as a clinical tool for the analysis of protein in serum and other body fluids. *Electrophoresis* 1997;**18**: 1715–1723.

55. Litwin CM, Anderson SK, Philipps G, Martins TB, Jaskowski TD, Hill HR. Comparison of capillary zone and immunosubtraction with agarose gel and immunofixation electrophoresis for detecting and identifying monoclonal gammopathies. *Am J Clin Pathol* 1999;**112**:411–417.

56. Katzmann JA, Clark R, Sanders E, Landers JP, Kyle RA. Prospective study of serum protein capillary zone electrophoresis and immunotyping of monoclonal proteins by immunosubtraction. *Am J Clin Pathol* 1998;**110**:503–509.

57. Bossuyt X, Bogaerts A, Schiettekatte G, Blanckaert N. Detection and classification of paraproteins by capillary immunofixation/subtraction. *Clin Chem* 1998;**44**:760–764.

58. Keren DF. Detection and characterization of monoclonal components in serum and urine. *Clin Chem* 1998;**44**:1143–1145.

59. Vrethem M, Larsson B, von Schenck H, Ernerudh J. Immunofixation superior to plasma agarose electrophoresis in detecting small M-components in patients with polyneuropathy. *J Neurol Sci* 1993;**120**:93–98.

60. Seligmann M, Mihaesco E, Hurez D, Mihaesco C, Preud'homme JL, Rambaud JC. Immunochemical studies in four cases of alpha chain disease. *J Clin Invest* 1969;**48**: 2374–2389.

61. Radl J. Light chain typing of immunoglobulins in small samples of biological material. *Immunology* 1970;**19**:137–149.

62. Al-Saleem TI, Qadiry WA, Issa FS, King J. The immunoselection technic in laboratory diagnosis of alpha heavy-chain disease. *Am J Clin Pathol* 1979;**72**:132–133.

63. Seligmann M, Mihaesco E, Preud'homme JL, Danon F, Brouet JC. Heavy chain diseases: current findings and concepts. *Immunol Rev* 1979;**48**:145–167.

64. Sun T, Peng S, Narurkar L. Modified immunoselection technique for definitive diagnosis of heavy-chain disease. *Clin Chem* 1994;**40**:664.

65. Radl J, Wels J, Hoogeveen CM. Immunoblotting with (sub)class-specific antibodies reveals a high frequency of monoclonal gammopathies in persons thought to be immunodeficient. *Clin Chem* 1988;**34**:1839–1842.

66. Gerritsen E, Vossen J, van Tol M, Jol-van der Zijde C, van der Weijden-Ragas R, Radl J. Monoclonal gammopathies in children. *J Clin Immunol* 1989;**9**:296–305.

67. Radl J. Monoclonal gammapathies. An attempt at a new classification. *Neth J Med* 1985;**28**:134–137.

68. Radl J, Liu M, Hoogeveen CM, et al. Monoclonal gammapathies in long-term surviving rhesus monkeys after lethal irradiation and bone marrow transplantation. *Clin Immunol Immunopathol* 1991;**60**:305–309.

69. Radl J, Valentijn RM, Haaijman JJ, Paul LC. Monoclonal gammapathies in patients undergoing immunosuppressive treatment after renal transplantation. *Clin Immunol Immunopathol* 1985;**37**:98–102.

70. Norden AG, Fulcher LM, Heys AD. Rapid typing of serum paraproteins by immunoblotting without antigen-excess artifacts. *Clin Chem* 1987;**33**:1433–1436.

71. Nooij FJ, van der Sluijs-Gelling AJ, Jol-Van der Zijde CM, van Tol MJ, Haas H, Radl J. Immunoblotting techniques for the detection of low level homogeneous immunoglobulin components in serum. *J Immunol Methods* 1990;**134**:273–281.

72. Withold W, Reinauer H. An immunoblotting procedure following agarose gel electrophoresis for detection of Bence Jones proteinuria compared with immunofixation and quantitative light chain determination. *Eur J Clin Chem Clin Biochem* 1995;**33**:135–138.

73. Withold W, Rick W. An immunoblotting procedure following agarose gel electrophoresis for subclass typing of IgG paraproteins in human sera. *Eur J Clin Chem Clin Biochem* 1993;**31**:17–21.

74. Beaume A, Brizard A, Dreyfus B, Preud'homme JL. High incidence of serum monoclonal Igs detected by a sensitive immunoblotting technique in B-cell chronic lymphocytic leukemia. *Blood* 1994;84:1216–1219.

75. Musset L, Diemert MC, Taibi F, et al. Characterization of cryoglobulins by immunoblotting. *Clin Chem* 1992;38:798–802.

4

Proteins identified by serum protein electrophoresis

The protein bands identified by serum protein electrophoresis may be divided into major and minor protein bands (Tables 4.1 and 4.2). The major bands constitute those that are virtually always seen in normal serum (fibrinogen is included because it is visualized in various circumstances). Minor bands are those that stain weakly or not at all in normal serum but may affect the electrophoretic pattern in a variety of clinical situations. The exact position of some of the bands will vary slightly with the methodology employed.

TRANSTHYRETIN (PREALBUMIN)

Transthyretin, a 55 kDa protein, is the first band encountered from the anodal side of the

Table 4.1 Major proteins visible on high-resolution electrophoresis

Protein	Concentration[a]	Function
Albumin	3.5–5.0	Transport/oncotic
α_1-Antitrypsin	90–200	Protease inhibitor
Haptoglobin	38–227	Binds hemoglobin
α_2-Macroglobulin	130–300	Protease inhibitor
Transferrin	240–480	Binds iron
C3	90–180	Host defense
Fibrinogen	200–400[b]	Thrombosis
Immunoglobulins		Host defense
IgA β-region	70–400	
IgM origin	40–230	
IgG γ-region	700–1600	

[a]Concentrations are for adults. They are expressed as g/dl for albumin and in mg/dl for all other proteins.
[b]Concentration in plasma.

Table 4.2 Minor proteins occasionally seen on high-resolution electrophoresis

Protein	Concentration (mg/dl)	Function
Transthyretin (prealbumin)	20–40	Transport
α_1-Lipoprotein	27–96	High-density lipoprotein
α_1-Acid glycoprotein	50–150	Acute phase
Inter-α-trypsin inhibitor	20–70	Protease inhibitor
Group-specific component	20–50	Carrier protein
α_1-Antichymotrypsin	30–60	Protease inhibitor
Pregnancy zone protein	0.5–2.0	Protease inhibitor
Ceruloplasmin	14–46	Copper binding
Fibronectin	19–35	Wound healing
β-Lipoprotein	57–206	Low-density lipoprotein
C4	10–40	Host defense
C-reactive protein	< 2	Inflammation

electrophoresis strip.[1] Its concentration in serum is 20–40 mg/dl. Structurally, transthyretin is a symmetrical tetramer composed of four monomers each 13.7 kDa and 127 amino acids long.[2] It is synthesized mainly in the liver and it provides transport for about 20 per cent of serum thyroxin (T4) (each molecule of transthyretin combines with one molecule of T4).[3] Together with retinol-binding protein, transthyretin also acts as a carrier for vitamin A.[2] Its older name, prealbumin, merely referred to its position just anodal to albumin. Transthyretin is the new name for this protein that reflects its roles *trans*porting *thy*roxin and *retin*ol-binding protein.[4]

By most electrophoretic procedures on serum, transthyretin produces only a weak, diffuse band just proximal to albumin (Fig. 4.1). The best resolution of transthyretin in serum from currently available techniques is that provided by capillary zone electrophoresis (CZE) (Fig. 4.2). However, even with CZE, the transthyretin band is too small for useful monitoring of any clinical condition. Although it is present in lower concentration in CSF than it is in serum (Fig. 4.1), transthyretin is a more substantial fraction in CSF than serum

because it is synthesized locally in the choroid plexus.[5]

Transthyretin has become a mainstay in assessing the nutritional status of patients. When using transthyretin levels to follow nutritional status, quantification should be performed by nephelometry or radial immunodiffusion, not electrophoresis. It has advantages over albumin as a monitor for protein-calorie malnutrition because albumin has a relatively long half-life (3 weeks, compared with the 2-day half-life of transthyretin). Therefore, transthyretin is a more sensitive indicator of change in protein-calorie status.[6,7] It is not, however, a perfect test for malnutrition because transthyretin levels are low in patients with severe liver disease.

Rarely, one can note a marked increase in the transthyretin band. This has been reported in patients with inflammatory bowel disease. In an extreme case of a patient with a long history of diarrhea complicated with hepatitis B, Jolliff[8] noted that it could be confused with bisalbuminemia (a matter that can be cleared up by immunofixation) (Fig. 4.3).

Structurally, transthyretin has a high content of

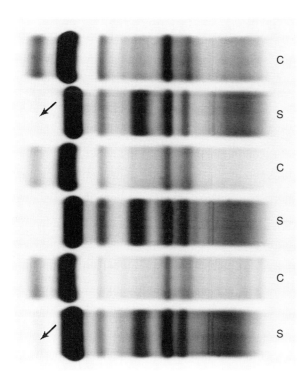

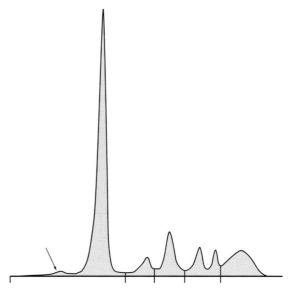

Figure 4.2 Capillary zone electropherogram of normal serum. The small, but distinct, transthyretin band is indicated by the arrow. (Sebia Capillarys.)

Figure 4.1 Cerebrospinal fluid (C) and serum (S) samples are alternated on this gel. The cerebrospinal fluid has been concentrated 80-fold, whereas the serum has been diluted 1:3. Note the prominence of the transthyretin band in the cerebrospinal fluid samples. The transthyretin band in the serum is barely visible in the top and bottom sera (arrows), but is not visible (although it is present) in the middle serum. (Paragon SPE2 system stained with Paragon Violet).

try, these variants may be characterized in cerebrospinal fluid.[12]

ALBUMIN

Albumin is a 69 kDa protein with a concentration of 3.5–5.0 g/dl (35–50 g/l) in adults 18–60 years old. It is the most prominent protein in normal serum. Similar to transthyretin, albumin is synthesized in the liver and functions as a transport protein. Albumin accounts for much of the osmotic effect of serum proteins and transports a variety of endogenous and exogenous molecules, including bilirubin, enzymes, hormones, lipid, metallic ions, and drugs. Many of these molecules are poorly soluble in aqueous solution alone. The breadth of the normal albumin band is partly due to its great serum concentration and partly due to the microheterogeneity resulting from the charges and size of various molecules transported by albumin.[13] The tendency for albumin to transport a variety of substances accounts for some of the abnormal patterns of electrophoresis in patients receiving albumin-binding drugs (such as antibiotics, especially

β-sheets with only one small α-helix by X-ray crystallography.[1] The predominance of β-sheet structure may be related to the propensity of familial amyloidosis to occur with minor structural changes to transthyretin. Transthyretin variants are associated with two types of amyloidosis: senile systemic amyloidosis (SSA) and familial amyloidotic polyneuropathy (FAP).[9,10] Familial amyloidotic polyneuropathy is the most common form of inherited amyloidosis with an incidence of 1 in 100 000.[10] it may also be caused by a genetic variant of apolipoprotein A1.[11] Genetic variants of transthyretin associated with FAP cannot be identified by serum protein electrophoresis. However, using mass spectrome-

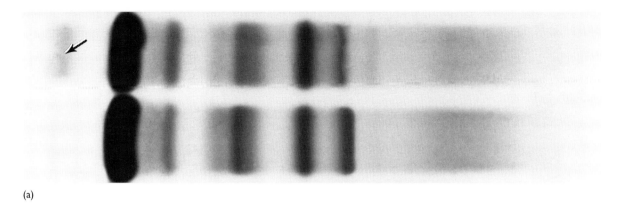

(a)

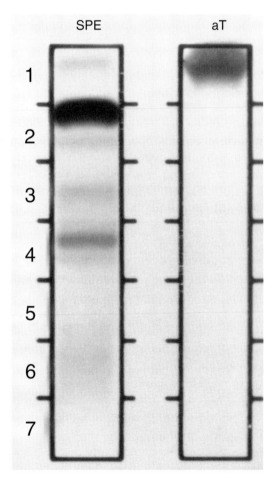

(b)

Figure 4.3 (a) Serum from patient with prominent transthyretin band (arrow) is on top and a normal control is on the bottom. (b) Serum protein electrophoresis (SPE) and immunofixation of serum from patient in (a) demonstrates that the fast-moving band reacts with anti-transthyretin (aT). Anode at the top. Figures contributed by Carl R. Jolliff.

penicillin) or in patients with hyperbilirubinemia (Fig. 4.4).[14–18]

Decreased albumin

Decreased concentration of serum albumin indicates significant pathology either in the production of albumin by the liver or its leakage through a damaged surface (glomerular disease, gastrointestinal loss, or thermal injury) (Table 4.3). In Western countries, a decrease in the production of albumin most commonly reflects severe liver injury. Because of the large reserve capacity of the liver, hypoalbuminemia resulting from liver damage occurs after most of the hepatocytes have been damaged or destroyed. Such a decrease may be accompanied by clotting abnormalities and decreased synthesis of other hepatocyte products, including haptoglobin and transferrin.[19] Mendler et al.[20] recommend that serum albumin concentrations should be performed by nephelometric techniques rather than serum protein electrophoresis in these patients.

In underdeveloped countries, severe protein malnutrition (kwashiorkor) is the leading cause of decreased synthesis of albumin. Patients with neoplasia or other chronic diseases also develop a nutritionally related hypoalbuminemia. Despite the decreased serum albumin band on serum protein electrophoresis in most cases of protein-

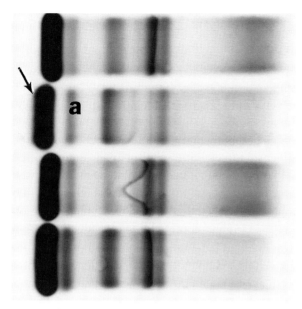

Figure 4.4 The second serum from the top has an anodal migration of albumin due to penicillin treatment. This can be seen by comparing the anodal edge of that albumin band (arrow) with the other albumin bands on the gel. Also, the distance between the cathodal edge of the albumin band in the second sample is further from the α_1-antitrypsin band (a) than for the other three samples on the gel. (Paragon SPE2 system stained with Paragon Violet.)

Table 4.3 Alterations of serum albumin

Alteration	Pathophysiology
Analbuminemia	Congenital
Bisalbuminemia	Congenital
Hypoalbuminemia	A. Decreased production
	1. Liver disease
	2. Protein malnutrition
	3. Acute inflammation
	B. Increased loss
	1. Kidney disease
	2. Protein-losing enteropathy
	3. Thermal injury
Anodal smearing	Binding with anionic molecules
	1. Bilirubin
	2. Antibiotics (Penicillin, etc.)

calorie malnutrition, quantification of albumin and transthyretin (prealbumin) by immunoassays such as enzyme immunoassay or nephelometry is the preferred method to follow these individuals.[7]

A decreased concentration of serum albumin may also result from excessive loss through injury to the kidneys, gastrointestinal tract, thermal injury to the skin or severe eczema, and in hypercatabolic states. When renal damage is severe enough to allow albumin to pass in large amounts into the urine, there is a corresponding loss of other serum proteins including γ-globulins.[21-23] Some of the largest serum proteins, such as α_2-macroglobulin with a molecular mass of 720 kDa, remain in the serum and are synthesized at an increased rate. These probably constitute the body's attempt to stabilize oncotic pressure.

Gastrointestinal loss in various protein-losing enteropathies is also associated with hypoalbuminemia and a decrease in concentration of other serum proteins.[24] In studies of protein loss in otherwise healthy adults, although albumin was often found in stool, α_2-macroglobulin was not.[25] In protein-losing enteropathies, a normal to increased level of α_2-macroglobulin is seen in the serum along with decreased albumin. Serum protein electrophoresis cannot distinguish between gastrointestinal loss of protein versus renal loss. When in doubt as to the interpretation of a pattern, contacting the clinician often results in a useful exchange whereby one learns, for example, that the patient has had chronic diarrhea and the clinician learns that the protein loss is severe enough to create a marked hypoalbuminemia. Such information can be of further use in helping the clinician explain a corresponding lymphocytopenia that may accompany protein loss in the gut.[23]

Albumin is also often decreased in patients with acute inflammation and during chronic infections and hemodialysis[26]. This may be related to the production of interleukin-6 that increases the synthesis of acute-phase reactants by the hepatocytes while decreasing the production of albumin.[27] Other work suggests that tumor necrosis factor (TNF) also plays a significant role in this process.[28]

Analbuminemia

Analbuminemia (markedly decreased or absent synthesis of albumin due to inheritance of recessive genes) is extremely rare (Fig. 4.5). Surprisingly, some of the patients (30 cases are available for review at the on-line Register of Analbuminemia cases)[29] who have analbuminemia are clinically well, presumably owing to the maintenance of oncotic pressure and transport function by other serum proteins.[30,31] However, some patients have required diuretics to control mild edema.[32] Laboratory investigations have occasionally disclosed elevated cholesterol,

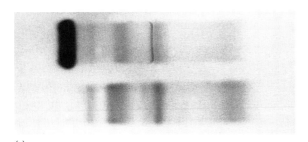

(a)

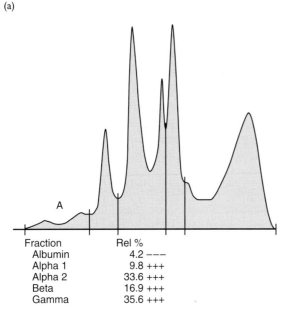

Fraction	Rel %	
Albumin	4.2	———
Alpha 1	9.8	+++
Alpha 2	33.6	+++
Beta	16.9	+++
Gamma	35.6	+++

A/G: 0.04

(b)

Figure 4.5 (a) Serum from a normal individual (top) is compared with serum from a patient with analbuminemia (bottom). (b) Densitometric scan from analbuminemic patient. Only minor background distortions are seen in the albumin area (a). Figures contributed by Carl R. Jolliff.

β-lipoproteins, acute-phase proteins and immuno-globulins.[33] A follow-up 38 years after the report of the first two cases revealed that the female patient required occasional replacement therapy with human serum albumin for edema, developed lipody-strophy by her fourth decade and died at age 69 years of a granulosa cell tumor.[34] Her brother did not require albumin therapy and died of colon cancer at 59 years of age.[34] Both patients had osteoporosis.[34] (If you come across one of these rare cases, please report it to the Analbuminemia Register: Dr Theodore Peters, Jr at (+1) 607 5473673 or Dr Roberta G. Reed at (+1) 607 5473676). Information about patients with analbuminemia is available on the internet at www.albumin.org.)

Bisalbuminemia or alloalbuminemia

Another inherited abnormality of albumin is bis-albuminemia, in which two types of albumin that have slightly different electrophoretic mobility are produced; this results in two distinct and equal peaks if the variant produces albumin at the same rate as the normal gene. Occasionally, two distinct peaks of different heights are seen when the variant gene produces less albumin[35] (Fig. 4.6). Variant albumins may also alter binding of drugs such as warfarin.[36] Capillary zone electrophoresis is supe-rior to gel-based techniques in demonstrating this finding (Fig. 4.7).[37] Some variants of albumin will migrate anodally to the normal position and others will migrate cathodally; the albumin variant depicted in Fig. 4.6 migrates cathodally to the nor-mal albumin. Clinically, bisalbuminemia (alloalbu-minemia) has no pathological consequence.[38] The incidence varies depending on the population with a range of 1:1000–1:10 000 in Caucasians and Japanese, but with a frequency as high as about 1 per cent in some American Indian tribes.[39,40]

Increased albumin

An elevated albumin indicates acute dehydration and is accompanied by an increase in the other serum proteins. Albumin has been used for many

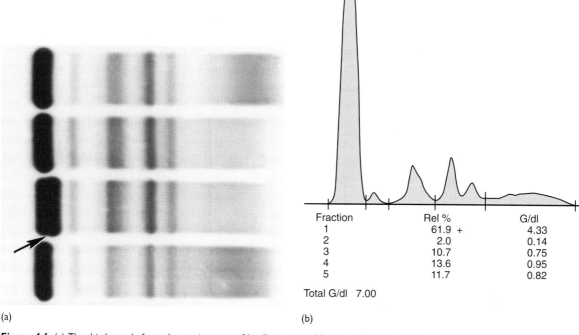

Fraction	Rel %	G/dl
1	61.9 +	4.33
2	2.0	0.14
3	10.7	0.75
4	13.6	0.95
5	11.7	0.82

Total G/dl   7.00

(a)   (b)

Figure 4.6 (a) The third sample from the top is a case of bisalbuminemia. Note that the two bands did not separate completely on this gel. The fact that there are two bands present can be noted by the indentation (arrow) between the two bands. (Paragon SPE2 system stained with Paragon Violet.) (b) Densitometric scan of the case of bisalbuminemia from (a). Note the two peaks atop the broadened albumin band.

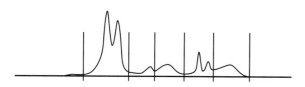

Figure 4.7 Clear separation of the bisalbuminemia bands on this capillary zone electropherogram. (Paragon CZE 2000.)

years as a safe way to restore oncotic pressure in individuals that have received plasmapheresis.[41] Consequently, patients with normal or elevated albumin and a decrease in the other serum proteins may be the result of plasmapheresis for either therapeutic reasons or donation (Fig. 4.8).

α-REGION

α₁-Lipoprotein

With gel-based electrophoresis and some capillary zone methods, α_1-lipoprotein forms a broad,

usually faint band that extends from albumin into the α_1 zone (Fig. 4.9). This band consists of high-density lipoproteins (HDLs) that may vary considerably in concentration in normal individuals as the result of diet, gender, and genetic differences. The range for adults is 27–75 mg/dl (0.27–0.75 g/l) for men and 33–96 mg/dl (0.33–0.96 g/l) for women.[42] α_1-Lipoproteins are absent in Tangier disease, an inherited HDL deficiency caused by a mutation in a cell-membrane protein called ABCA1.[43] α_1-Lipoprotein can be decreased in a variety of acquired conditions including chronic liver disease, renal disease, and with acute inflammation.[44,45] Elevated α_1-lipoprotein is often seen in pregnancy, with use of oral contraceptives, in postmenopausal estrogen supplementation, and among patients receiving statin therapy.[46,47] The effect of estrogens to increase α_1-lipoprotein has been associated with its ability to bind to the high-density lipoprotein receptor.[46]

Owing to the variability of α_1-lipoprotein in

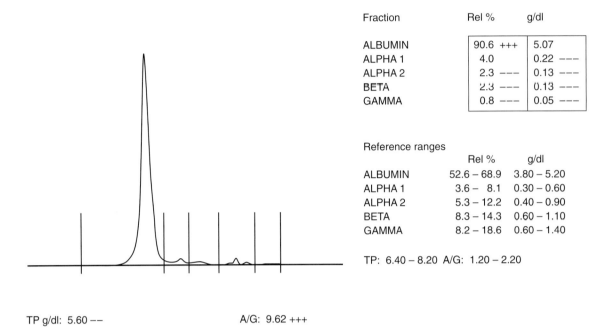

Fraction	Rel %	g/dl
ALBUMIN	90.6 +++	5.07
ALPHA 1	4.0	0.22 ---
ALPHA 2	2.3 ---	0.13 ---
BETA	2.3 ---	0.13 ---
GAMMA	0.8 ---	0.05 ---

Reference ranges

	Rel %	g/dl
ALBUMIN	52.6 – 68.9	3.80 – 5.20
ALPHA 1	3.6 – 8.1	0.30 – 0.60
ALPHA 2	5.3 – 12.2	0.40 – 0.90
BETA	8.3 – 14.3	0.60 – 1.10
GAMMA	8.2 – 18.6	0.60 – 1.40

TP: 6.40 – 8.20 A/G: 1.20 – 2.20

TP g/dl: 5.60 --   A/G: 9.62 +++

Figure 4.8 Normal albumin (but increased relative per cent of albumin) with decrease in all other bands. This serum is from a patient that received plasmapheresis for a neuropathy and had albumin replacement. (Paragon CZE 2000.)

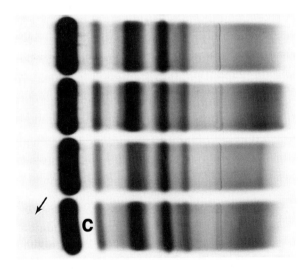

Figure 4.9 Darkly-stained gel emphasizes the α_1- lipoprotein region. Note the relatively clear area (C) between albumin and α_1-antitrypsin in the bottom serum. This patient was receiving heparin and the α_1-lipoprotein migrates anodally to albumin as a diffuse hazy band which extends out to and combines with the transthyretin area (arrow). The top three samples have normal-appearing α_1-lipoprotein regions. (Paragon SPE2 system stained with Paragon Violet.)

normal subjects and the diffuse nature of the band itself, the examination of this area alone is not particularly useful for clinical diagnosis. In trying to assess the ramifications of this fraction for the patient's lipid status, it is better to quantify the HDL or apolipoprotein A (which reflects the HDL concentration) than to comment on the protein electrophoresis findings. With the two currently available CZE methods, α_1-lipoprotein (HDL) may migrate as part of the albumin band (Sebia Capillarys; Sebia, Issy-les-Moulineaux, France) where it is not seen at all, or show up as a diffuse increase in the baseline between albumin and α_1-acid glycoprotein (orosomucoid) (Paragon CZE 2000; Beckman Coulter, Fullerton, CA, USA) (Fig. 4.10). In neither system does it interfere with visualization of the α_1-antitrypsin band.

On some of the gel-based systems, α_1-lipoprotein may present such a dense band that it obscures the α_1-antitrypsin band. This can be a serious problem. When the α_1-lipoprotein band is dense and abuts on the α_1-antitrypsin band, one cannot confidently rule out an α_1-antitrypsin deficiency. In those cases, one must use an alternative electrophoretic technique (see below), or at least measure the amount of α_1-antitrypsin by immunoassay to detect any deficiency. Even with quantification, however, an α_1-antitrypsin deficiency may be missed. Some patients with α_1-antitrypsin deficiency can have borderline low normal amounts of α_1-antitrypsin during an acute-phase reaction. Therefore, a C-reactive protein assay should be run simultaneously to rule out that possibility.

The laboratory can remove the effect of α_1-lipoprotein by taking a lesson from the heparin effect. Heparin alters the migration of the lipoprotein bands, but its effect depends on whether the heparin was administered to the patient, or if the tube to obtain the blood sample was heparinized. When the patient is receiving heparin, it affects the migration of both α_1-lipoprotein and β_1-lipoprotein (Figs 4.9 and 4.11),[48] whereas, when heparin is added to the blood sample from a non-heparinized patient, only the migration of β_1-lipoprotein is affected.[49]

These different effects of heparin reflect two processes. When heparin is added to a sample *in vitro*, it binds directly to β_1-lipoprotein, resulting in a more diffuse and anodal migration.[49] However, *in vivo*, heparin activates lipoprotein lipase in endothelial cells, resulting in the liberation of free

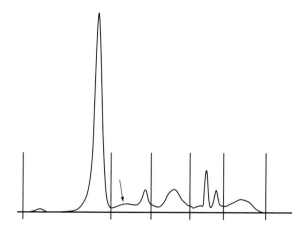

Figure 4.10 Prominent α_1-lipoprotein is present in this capillary zone electropherogram. (Paragon CZE 2000.)

fatty acids from triglycerides.[48] The free fatty acids bind to α_1-lipoprotein and result in the anodal migration of this band. The clinical laboratory can take advantage of the effect of free fatty acids to alter the migration of α_1-lipoprotein in cases where the density of this region obscures the view of α_1-antitrypsin. For this method, my laboratory prepares a 17.2 mmol/l solution of lauric acid in the standard electrophoresis buffer.[50] The serum is diluted 1:4 in this lauric acid solution and routine electrophoresis is performed on the sample. This displaces the α_1-lipoprotein to the prealbumin region thus allowing a clear view of the α_1-antitrypsin band.

α_1-Antitrypsin

Second in importance only to detection of monoclonal gammopathies is the detection of decreased and/or abnormal α_1-antitrypsin. α_1-Antitrypsin is the major protein accounting for the discrete band in the cathodal end of the α_1-region. Although it is represented by a relatively small band, in this text it is classified as a major protein band because of the clinical importance of abnormalities associated with this band. This 52 kDa, single-chain glycoprotein has a normal concentration of 90–200 mg/dl (0.9–2.0 g/l) in adults. It is produced in the liver and is classified as a member of a family of protease inhibitors called serpins (*ser*ine *pro*tease *in*hibitors) that react with proteolytic enzymes containing a serine group at their active site.[51]

Although its name implies that its functions are limited to inhibiting the activity of trypsin, α_1-antitrypsin interferes with the enzymatic activity of a variety of enzymes including trypsin, chymotrypsin, pancreatic elastase, skin collagenase, renin, urokinase, Hageman-factor cofactor, and leukocyte neutral proteases.[52] It is by far the most significant protease inhibitor in serum, accounting for the vast majority of the trypsin-inhibiting capacity of human serum.[53] α_1-Antitrypsin works like a mousetrap to inhibit proteases.[54] It docks with the target protease, which cleaves the reactive

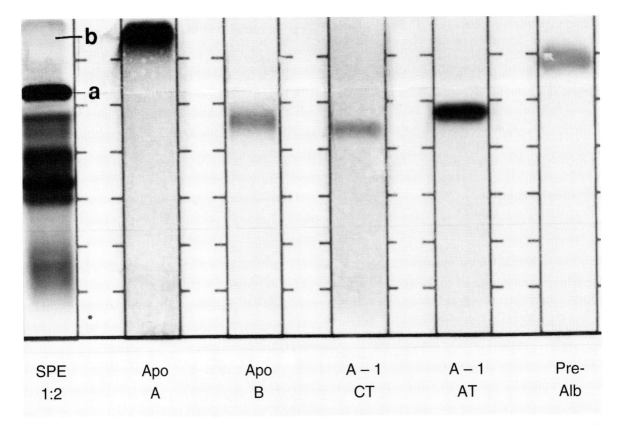

Figure 4.11 Immunofixation of serum from a patient receiving heparin. The anode is at the top of the gel. There is a faint diffuse band (b) anodal to albumin (a). The extreme anodal edge of this band reacts with antisera against apolipoprotein A (α_1-lipoprotein component) (Apo A). Transthyretin (prealbumin) is more cathodal to this band (Pre-Alb). For comparison, antisera against apolipoprotein B (Apo B) which reacts with a β_1-lipoprotein component, α_1-antichymotrypsin (A-I CT) and α_1-antitrypsin (A-I AT) are shown. Anode at the top. (Paragon Immunofixation stained with Paragon Violet.)

center of α_1-antitrypsin. This shifts the protease to the other side of α_1-antitrypsin which destroys the protease.[52] Mutations at the active site such as the Z mutation (discussed below) inactivate the active site of α_1-antitrypsin.[52]

In the lung, it binds to and inhibits the proteolytic activity of neutrophil elastase. The protease–antiprotease theory of emphysema predicts that an excess of proteases or a deficiency in antiprotease leads to destruction of alveolar walls.[55] Therefore, patients with α_1-antitrypsin deficiency are vulnerable to this condition, and examination of the α_1-antitrypsin band by serum protein electrophoresis is an important mechanism to detect individuals or families of individuals who may be at risk.

DECREASED α_1-ANTITRYPSIN

By serum protein electrophoresis, decrease, altered migration or absence of this band strongly suggests the presence of an α_1-antitrypsin variant. The examiner must inspect each sample on the gel for this possibility, as subtle decreases may be missed. Although the results of a densitometric scan may be helpful, it may provide a false sense of security. A borderline normal densitometric scan (one with values in the low, but normal range) does not rule out an α_1-antitrypsin deficiency. Since the densitometric scan of the α_1-region usually includes α_1-lipoprotein, a relatively large α_1-lipoprotein band could obscure a small or absent α_1-antitrypsin band. Further, densitometers are inherently inaccurate when trying to measure such small bands. We reported a coefficient variation of 10 per cent in this region.[56]

Capillary zone electrophoresis offers excellent resolution in the α_1-region. Because it detects peptide bonds, it is superior to gel-based techniques in detecting α_1-acid glycoprotein (a heavily glycated molecule that does not stain well with the dyes used in gel-based methods). In addition, α_1-lipoprotein and, in some CZE systems, triglycerides are also included in the α_1-region. Because of this, Gonzalez-Sagrado et al.[57] reported that neither CZE nor gel-based methods provide an adequate method to rule out deficiencies of α_1-antitrypsin. They recommended performing nephelometry and phenotyping when the total α_1-region (which includes all the proteins in this region) on CZE is less than 400 mg/dl (4 g/l).

Detection of a decreased α_1-antitrypsin band, or one with altered migration, must be followed up by genetic studies to provide prognostic information to the family.[58,59] Because of interference with nephelometric quantification of α_1-antitrypsin levels by anticoagulants, serum should be used to measure the functional activity or the concentration by immune precipitation assays. However, either serum or plasma is suitable for phenotypic analysis by reference electrophoretic methods (isoelectric focus).[60,61] Molecular techniques to detect both the Z and S mutations are now widely available.[62–66]

A complete discussion of the genetic polymorphism (involving about 75 possible alleles) of α_1-antitrypsin is beyond the scope of this volume, but some details are relevant to interpreting and understanding the serum protein electrophoresis patterns that one will come across in patients with α_1-antitrypsin deficiency. Expression of the two alleles of any individual is codominant, that is, each allele controls production of a specific α_1-antitrypsin molecule unaffected by the other allele. The alleles are referred to as Pi (protease inhibitor) followed by a third letter characterizing the particular allele. For example, the α_1-antitrypsin from the allele with the lowest concentration (PiZ) has the slowest electrophoretic mobility. The most common allele is PiM (Table 4.4). There are several variants of this gene, M1, M2, and M3, which produce 'normal' levels of α_1-antitrypsin.[67] Combined, the M alleles have a gene frequency of about 0.94. The most cathodal molecule denoted by the letter Z (it is the 'zlowest') is the most common variant associated with severe deficiency. It results in a change from Glu to Lys at position 342 of the M allele.[68] About 3 per cent of the population in the USA are phenotype PiMZ, which usually does not result in clinical deficiency (because of the presence of the single PiM allele). Fortunately, only one in 3630 individuals has the severe deficient PiZZ phenotype.[69,70]

The second most common deficient phenotype, referred to as S, migrates between M and Z on serum protein electrophoresis. The S gene has an allelic frequency of 0.02–0.04. Individuals with PiSZ also have clinically significant deficiency; this occurs in about one in 500 individuals in the USA.[71] The S mutation is a single base-pair substitution from Glu to Val at codon 264.[72] Several other deficient genes have been described. However, these are very rare and often difficult to detect by serum protein electrophoresis. Some have extraordinary symptoms such as in the case of antitrypsin Pittsburgh (single base-pair substitution from Met to Arg at codon 358), wherein the mutated protein serves as an inhibitor of thrombin (Fig. 4.12).[73,74] One of the two cases of antitrypsin Pittsburgh was associated with severe hemorrhage. Both cases were associated with a minor slow

Table 4.4 Common α_1-antitrypsin phenotypes

Genotype	Electrophoretic migration
MM[a]	Mid-α_1
MS	Mid- and slow α_1
S S	Slow α_1
MZ	Mid- and slowest α_1
SZ	Slow and slowest α_1
ZZ	'z'lowest α_1

[a]Includes M1, M2, M3, M4 (types of M with normal activity).

albumin component (called proalbumin), which migrated more toward the cathode than normal (Fig. 4.12).

Genetic deficiency of α_1-antitrypsin is associated with severe lung and, less commonly, liver disease. Some affected individuals develop neonatal hepatitis (those who do are highly likely to develop serious liver disease), others develop liver disease later as children, or in adult life they develop a form of cirrhosis (about 20 per cent of patients), which is characterized by large globules of amorphous periodic acid–Schiff-positive material that occurs within the cytoplasm of hepatocytes in the periportal areas (Fig. 4.13).[75-77] These globules are distinguished from glycogen because they are not digested by treatment with diastase. Immunohistochemical studies have shown this material to be an α_1-antitrypsin precursor that is not excreted.[52,78] They represent aggregation of the

PiZZ product. The variant molecule of PiZZ is predisposed to polymerize resulting in disease due to conformational change of the molecule.[52] The PiZZ and PiSZ genotypes may have hepatic dysfunction as early as the first three months of life.[79] However, the presentation even in individuals with PiZZ is quite variable. Some cases of PiZZ do not present until their seventh decade (at that time with both cirrhosis and emphysema).[80] Others have noted cases with no clinical symptoms until their eighth decade.[81] As discussed above, this variability implies that factors other than the genotype are involved. Personal and environmental factors such as smoking or living in an area with considerable air pollution have a negative effect on the lung process.[82] However, since only 10–15 per cent of individuals with PiZZ develop clinically significant liver disease, a second factor relating to the ability of an individual to degrade these aggregates in

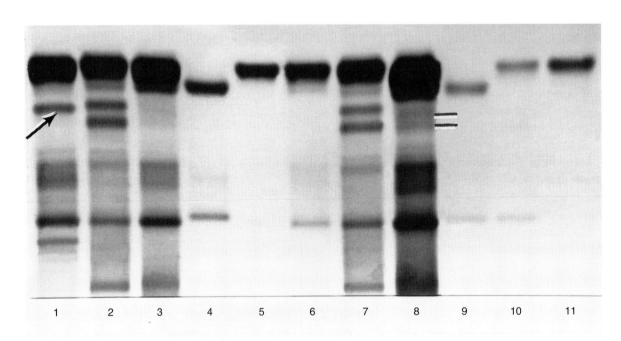

Figure 4.12 Agarose gel with the anode at the top. In lane 1, 2.5 μl of normal serum is shown to compare migration of albumin (note location of the cathodal end) and α_1-antitrypsin band (arrow). In lanes 2 and 7, 2.5 μl of plasma is shown from the original case of α_1-antitrypsin (Pittsburgh). Note the broader cathodal migration of the albumin band and the double α_1-antitrypsin bands. In lanes 3, 6, and 8 are 2.5, 0.6, and 5.0 μl, respectively, of plasma samples from a second case of α_1-antitrypsin (Pittsburgh). The cathodal migration of albumin is apparent on all three, but the double α_1-antitrypsin band is only seen to advantage in lane 8 (indicated). The abnormal albumin was purified by diethylaminoethyl Sephadex column (lane 9) and trypsin treatment restored the normal migration of this band (lane 10). In lane 11, a normal albumin is shown for comparison. (Figure contributed by Stephen O. Brennan.)[74]

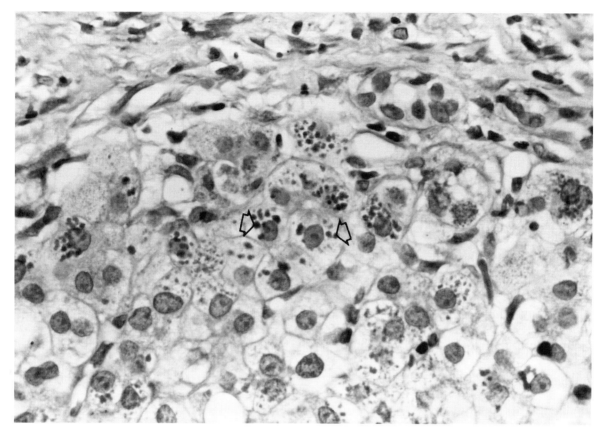

Figure 4.13 Liver from patient with α₁-antitrypsin deficiency (PiZZ) is depicted. The periodic acid–Schiff (PAS), diastase stain, and digestion disclose the large granules (arrows), which are especially prominent in the periportal hepatocytes.

hepatocytes may determine which patients develop cirrhosis.[68] In addition to the more obvious association of PiZZ with cirrhosis, occasional cases of chronic liver failure have been reported in patients that are PiMZ or PiMS.[83,84]

The lung injury, resulting in early emphysema, has been hypothesized to result from the unchecked endogenous activity of a variety of proteolytic enzymes that are liberated by minor inflammatory events in these tissues.[85–87] Damage is most severe at the base of the lung and is compounded by environmental factors, especially smoking.[68] However, even in non-smoking individuals homozygous for PiZZ, lung function will decline, especially after 50 years of age.[82,88] Although heterozygotes are usually clinically well, Dahl et al.[89] reported that individuals with clinically established chronic obstructive pulmonary

disease (COPD) who also have PiMZ have a decreased forced expiratory volume (FEV1) compared with PiMM individuals with COPD.

In examining electrophoretic patterns, the most common genetic variant seen in the laboratory is the PiMS banding (Figs 4.14 and 4.15). As with the PiMZ heterozygotes, these patients have no clinical disease, although family studies should be recommended because siblings or children may carry two defective genes. Another genetic variant likely to be detected by examination of serum is reflected by the protein product that migrates anodal to PiM and is termed PiF (fast). This genotype is not known to be associated with antitrypsin deficiency.

While serum protein electrophoresis cannot detect some of the more unusual variants, it can be helpful in detecting the more common ones. When examining the electrophoretic pattern, one should

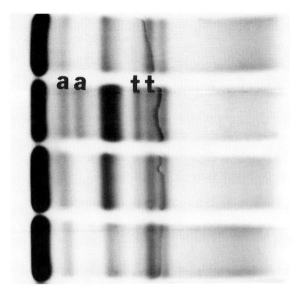

Figure 4.14 A PiMS pattern with two bands of identical intensity (a a) in the α_1-region is shown in the second serum from the top. Although one can clearly see the two bands with high-resolution electrophoresis (HRE) technique, the actual phenotype a PiMS was performed by isoelectric focusing, not by HRE. This patient had an acute-phase reaction which is why the α_1-antitrypsin bands stain more strongly than an MS pattern usually would. This serum was particularly interesting because the individual also was a heterozygote for the transferrin band. Two bands of equal intensity are seen in the β_1 region (t t). (Paragon SPE2 system stained with Paragon Violet.)

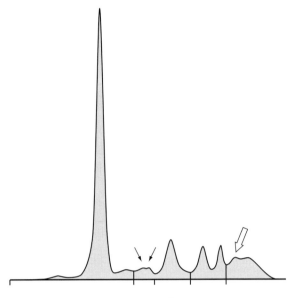

Figure 4.15 A PiMS pattern on a capillary zone electropherogram. The two bands are of equal intensity (small arrows). Interestingly, there is a small monoclonal gammopathy in the fast γ-region (block arrow). (Sebia Capillarys).

be especially aware of samples with decreased α_1-antitrypsin band and/or altered electrophoretic migration. By noting that the weak α_1-antitrypsin band is located in the slow α_1-region, it suggests the possibility of either PiZZ, PiSS or PiSZ phenotypes (Figs 4.16 and 4.17).

Some clinically significant variants such as PiM_{Malton} and $PiM_{Heerlen}$ require molecular techniques to detect because they have a normal electrophoretic migration, but have a substantial decrease in the α_1-antitrypsin level and can cause liver and lung disease.[90,91] Some rare individuals are unable to produce serum α_1-antitrypsin because of deletions and stop codons of α_1-antitrypsin coding null variants.[92] These patients are at high risk for the early development of emphysema even if they do not smoke.[93] Therefore, when there is altered electrophoretic mobility, or decrease or absence

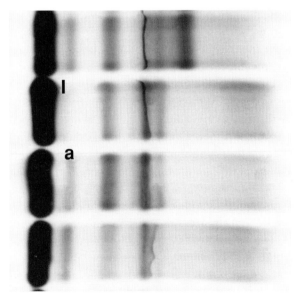

Figure 4.16 The second sample from the top is from a patient with α_1-antitrypsin deficiency (PiZZ). No band can be seen in the α_1-antitrypsin area. A very faint α_1-lipoprotein band (l) extends from the cathodal end of the albumin band to the area where the normal α_1-antitrypsin band would normally be seen. For comparison, the α_1-antitrypsin PiMM band in the third sample is labeled 'a'. (SPE2 system stained with Paragon Violet.)

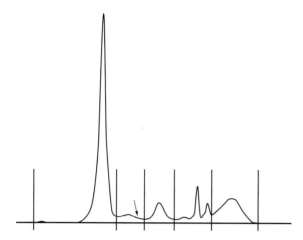

Figure 4.17 An α_1-antitrypsin deficiency (PiZZ) deficiency on capillary zone electrophoresis. The scaphoid region (arrow) at the end of the α_1-region denotes the absence of the α_1-antitrypsin band. (Paragon CZE 2000.)

of the α_1-antitrypsin band, determination of α_1-antitrypsin levels and phenotype studies are recommended.

α_1-Antitrypsin augmentation therapy is available that may decrease the rate of decline of the FEV1 in patients with severe deficiencies.[94–96] Perhaps because of this available therapy, an expert group from the World Health Organization (WHO) has recommended that α_1-antitrypsin measurement be included in neonatal screening programs.[97,98] The mean annual cost of this therapy in the USA was estimated at \$40 123 in 2001.[99] A study of the cost-effectiveness noted that although the cost of the current pooled human α_1-antitrypsin therapy is high (their estimate was \$52 000/year), the incremental cost per year of life saved varies considerably depending on the reduction in mortality.[100] Unfortunately, because of the demonstration of a deceleration in the rate of FEV1 decline in non-randomized studies, it may not be ethically possible to conduct placebo-controlled randomized trials.[100]

Increased α_1-antitrypsin

Increased levels of α_1-antitrypsin occur in a variety of conditions and can be useful in pattern interpre-

tations (see Chapter 5). α_1-Antitrypsin is increased as part of the acute-phase response, in patients with hyperestrogenemia caused by pregnancy, oral contraceptives, tamoxifen, tumors or with liver disease.[101–104] An increase in α_1-antitrypsin levels may help to predict the onset of labor.[105]

Note that α_1-antitrypsin levels increase during an acute-phase reaction even in patients with genetic deficiency of this protein (except those with null variants). Therefore, in a patient suspected of having α_1-antitrypsin deficiency, a normal level of this protein does not in itself rule out a deficiency. When measuring α_1-antitrypsin in patients suspected of having a deficiency, one should also measure the C-reactive protein levels. When the C-reactive protein is elevated, a 'normal' level of α_1-antitrypsin does not rule out the possibility of deficiency. In such cases, more definitive testing of α_1-antitrypsin such as isoelectric focusing or molecular studies should be performed.

α_1-Fetoprotein

α_1-Fetoprotein is present in neonatal serum in concentrations up to 50 mg/dl (500 mg/l), whereas normal adults have less than 20 µg/dl (200 µg/l), except for pregnant females who may achieve levels of 1 mg/dl (10 mg/l) at 20 weeks gestation.[106] Although this tiny amount is never seen on routine serum protein electrophoresis, some patients with hepatocellular carcinoma, yolk sac tumors, gastric cancer and chronic hepatitis have markedly elevated levels of this protein produced by the liver.[107] These determinations require immunoassay studies for accurate measurement.

α_1-Acid glycoprotein

α_1-Acid glycoprotein, also known as orosomucoid, is a minor band that occurs just anodal to α_1-antitrypsin. Its serum concentration for ages > 1 year ranges from 50 to 150 mg/dl (0.5–1.50 g/l).[101] However, α_1-acid glycoprotein is not usually visible on gel-based electrophoresis methods unless greatly elevated because its high sialic acid content interferes with binding of most stains such as Amido Black or Coomassie Brilliant Blue. With gel-based techniques, when α_1-acid glycoprotein is increased to greater than 200 mg/dl it may show up with a fuzzy appearance on the anodal side of α_1-antitrypsin.[104] However, since CZE measures peptide bond absorbance at from 200 to 214 nm, α_1-acid glycoprotein more commonly is seen as a prominent anodal shoulder to α_1-antitrypsin (Fig. 4.18).

α_1-Acid glycoprotein is commonly seen with the acute-phase reaction pattern and in uremic patients receiving hemodialysis (this protein is normally lost through the glomerulus).[108] Quantification of this protein may be useful in detecting neonatal infections because the serum levels of α_1-acid glycoprotein are much lower at birth than by 1 year of age. Because of this, levels of 60–80 mg/dl (6–8 g/l) have been associated with neonatal sepsis.[109] These levels are too low, however, to be measured reliably by either CZE or gel-based techniques and should be determined by immunochemical methods.[109] Similar to C-reactive

protein (see below), increased α_1-acid glycoprotein levels are associated with an increased risk for myocardial infarction.[110]

α_1-Antichymotrypsin

α_1-Antichymotrypsin is a tiny band that may be found in the α_1–α_2 interregion (Fig. 4.19). It is about 1/10 as prominent as the α_1-antitrypsin band with a serum concentration of 30–60 mg/dl, and a molecular mass of 69 kDa. It is a serine protease inhibitor genetically linked with α_1-antitrypsin.[111] α_1-Antichymotrypsin reacts with the potent neutrophil protease cathepsin G, mast cell chymase, and prostate specific antigen.[112,113] Recent data for the α_1-antichymotrypsin–prostate-specific antigen (PSA) complexes have been used to evaluate modestly elevated PSA levels.[114] However, this does not affect the pattern visible on protein electrophoresis. While its name implies a role in chymotrypsin inhibition, it provides less significant inhibition for chymotrypsin than does α_1-antitrypsin.[115]

In serum, its concentration increases rapidly after acute injury, perhaps acting to inhibit some of the enzymes liberated during this process, and it may be seen in hepatocytes of individuals with hepatitis C.[116] It is increased in the serum and cerebrospinal

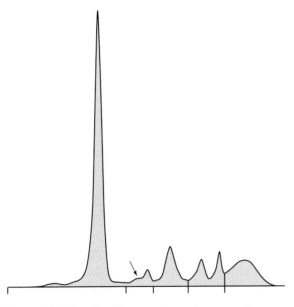

Figure 4.18 Normal capillary zone electropherogram with a relatively prominent α_1-acid glycoprotein (orosomucoid) shoulder (arrow) just anodal to α_1-antitrypsin. (Sebia Capillarys.)

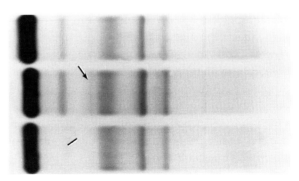

Figure 4.19 Small bands are often seen in the inter-α_1–α_2 region. In this photograph, they are best seen in the middle sample (arrow). The serum in the bottom lane contains a slow-migrating, weakly staining α_1-antitrypsin band (indicated) which should be further evaluated as a possible PiSS (see text). (Panagel System stained with Amido Black.)

fluid of patients with Alzheimer's disease, and the levels may be related to its heightened secretion by astrocytes in those patients.[117] By serum protein electrophoresis, one only sees a variable, deeper staining in the α_1–α_2 interregion. Obviously, a decreased α_1-antichymotrypsin concentration would not be detectable by routine electrophoretic techniques. Congenital deficiencies of this inhibitor can be detected by immunoassay methods and may predispose individuals to liver and lung disease.[113]

α_2-Macroglobulin

α_2-Macroglobulin is one of the two major proteins in the α_2-region. Synthesized mainly by the liver, it is a tetramer composed of four identical subunits. The subunits are held together as dimers by disulfide bonds. Each α_2-macroglobulin molecule consists of two sets of these dimers.[118] With a serum concentration of 130–300 mg/dl (1.3–3.0 g/l) in adults, it accounts for about 3 per cent of the total protein in serum.[106] Because of the variable migration of the haptoglobin types (see below), α_2-macroglobulin is often adjacent to, or comigrating with, haptoglobin and is therefore not seen as a discrete band. An important exception is in the neonate where little haptoglobin is present and α_2-macroglobulin levels are much higher than in adults. α_2-Macroglobulin is an enormous molecule (the tetramer has a molecular mass of 720 kDa) that functions as a protease inhibitor.

It is part of the family called thiol ester plasma proteins. This name refers to the presence of a unique cyclic thiol ester bond that is involved with the covalent binding of proteases.[118] This family of proteins includes α_2-macroglobulin, pregnancy zone protein, C3, C4, and C5.[118] While it is particularly effective at inhibiting plasmin activity, α_2-macroglobulin is also able to inhibit the enzymes trypsin, chymotrypsin, thrombin, and elastase.[119] α_2-Macroglobulin binds to β_2-microglobulin, suggesting that it may modify the metabolism of the latter.[120] It is also the major serum collagenase inhibitor, which may be important in wound healing.[121] In addition, α_2-macroglobulin has been shown to bind to various cytokines, including interleukin-2, interleukin-8, and heat shock protein receptor CD91, indicating the diverse repertoire of this molecule.[122–124]

α_2-Macroglobulin has a unique mechanism for trapping the proteinase in the 'bait' region of the disulfide-linked dimers. After the proteinase has cleaved a peptide bond in the middle of each of the four α_2-macroglobulin subunits (bait region), α_2-macroglobulin rearranges itself to trap the proteinase within the α_2-macroglobulin molecule.[125,126] The α_2-macroglobulin bound to the proteinase creates a conformational change that results in a more compact α_2-macroglobulin molecule that migrates faster during electrophoresis.[125,126] Despite its effects on proteinases, α_2-macroglobulin is not an acute-phase reactant and does not increase with inflammation.

The function of α_2-macroglobulin as a pan-proteinase inhibitor, together with its ability to bind with β-amyloid have raised the suggestion that the α_2-macroglobulin gene is a candidate gene in relation to Alzheimer's disease.[127–129] However, this area of investigation is controversial because recent investigations have not confirmed this association.[129–131]

INCREASED α_2-MACROGLOBULIN

α_2-Macroglobulin is elevated in neonates, the elderly, patients with elevated estrogen levels, and especially as part of the nephrotic pattern in patients with selective glomerular leakage. In nephrotic syndrome, there is a compensatory increase in the synthesis of α_2-macroglobulin. Further, its large size prevents its passage into the urine.

Serum levels of α_2-macroglobulin are moderately higher in cord blood and during the first week of life than in adults, averaging about 300 mg/dl.[118,132] Elevated levels of α_2-macroglobulin are commonly seen in diabetics, especially in longstanding cases.[133] In patients with α_1-antitrypsin deficiency and emphysema an increased concentration of α_2-macroglobulin is often also present.[134]

On serum protein electrophoresis, elevated α_2-macroglobulin may be seen as a sharp band in the

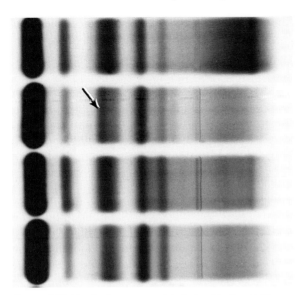

Figure 4.20 Usually the α_2-macroglobulin band is obscured by haptoglobin. However, in the second serum from the top, the haptoglobin stains relatively weakly, and the α_2-macroglobulin band shows to advantage in the anodal side of the α_2 region band (arrow). (Paragon SPE2 system stained with Paragon Violet.)

anodal end of the α_2-globulin region (Fig. 4.20), but usually it is masked by haptoglobin.

DECREASED α_2-MACROGLOBULIN

A decrease in α_2-macroglobulin is extremely difficult to detect using serum protein electrophoresis. Haptoglobin will almost always hinder such an interpretation. Therefore, although a decreased α_2-macroglobulin has occasionally been reported in a variety of diseases, electrophoresis is not the way to detect it.

Haptoglobin

Haptoglobin is a glycoprotein of variable molecular weight and migration (depending on its expressed alleles – see below), synthesized by the liver with a serum concentration (in adults) of 38–227 mg/dl (0.38–2.27 g/l).[135] It is an α_2-migrating protein that binds free hemoglobin liberated during intravascular hemolysis, thereby preserving iron and preventing renal damage.[135–137] During this process, each molecule of haptoglobin can bind

with two hemoglobin molecules via the globin portion of the hemoglobin molecule; this prevents loss of the iron in free hemoglobin that would readily pass through the glomeruli if not bound to the bulky haptoglobin molecule.[138] The resulting hemoglobin–haptoglobin complex is rapidly taken-up by the reticuloendothelial system, where the iron is liberated for reuse. This hemoglobin–haptoglobin complex is removed at the rate of 15 mg/100 ml.hour. Therefore, intravascular hemolysis usually is associated with a decreased haptoglobin rather than free hemoglobin, unless there has been recent massive hemolysis.[139,140]

On serum protein electrophoresis, haptoglobin has a relatively complex pattern that often causes confusion about what is happening in the α_2-region. This complexity results from genetic polymorphism and alteration of electrophoretic mobility by the hemoglobin–haptoglobin complex; the latter is most commonly a result of improper handling of blood specimens, but may be seen with recent massive hemolysis.

There are three major phenotypes of haptoglobin resulting from the codominant expression of two alleles, Hp[1] and Hp[2], which can be discerned by their serum protein electrophoretic pattern (Table 4.5).[135,138,141] The polymorphisms of haptoglobin were first demonstrated with starch gel electrophoresis.[142] Haptoglobin 1-1 is found in individuals that are homozygous for the Hp[1] allele.[138,141] Its name reflects the fact that it has the fastest electrophoretic mobility, migrating anodal to α_2-macroglobulin and obscuring the minor inter-α-trypsin inhibitor region bands (Fig. 4.21). The cathodal end of this haptoglobin type merges with α_2-macroglobulin, preventing complete separation of these two proteins by most serum protein electrophoresis techniques. The presence of haptoglobin 1-1 phenotype has been associated with a significantly lower prevalence of diabetic retinopathy than in diabetics with the other two major phenotypes.[143] The superior antioxidant capacity of this phenotype may prevent the damage.

Individuals that are homozygous for Hp[2] produce a pattern of several slower-moving bands on starch gel electrophoresis. However, by serum protein

Table 4.5 Characteristic of the major haptoglobin phenotypes[a]

Genotype	Concentration (mg/dl)	Mass (kDa)	HRE[b] migration
Hp type 1–1	57–227[b]	86	Mid-α_{1-2}
Hp type 2–1	44–183	86–300	Mid-α_2
Hp type 2–2	38–150	170–900	Slow α_2

[a]Data adapted in part from Langlois and Delanghe,[135] Janik,[137] and in part from van Lente.[138]
[b]High-resolution electrophoresis.

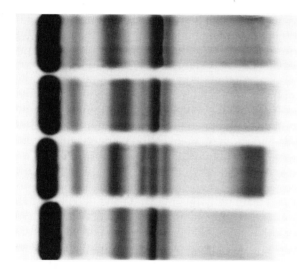

Figure 4.21 The top sample shows a haptoglobin 1-1 variant which migrates anodally from α_2-macroglobulin. The second and fourth sera contain a haptoglobin 2-2 variant that migrates cathodally to α_2-macroglobulin (from which it cannot be distinguished on this gel). The third serum contains a haptoglobin 1-2 variant which migrates directly over α_2-macroglobulin. The slow γ-region of the third serum also contains two dark-staining bands which may represent a double gammopathy (see Chapter 7). (Paragon SPE2 system stained with Paragon Violet.)

electrophoresis agarose gels and CZE, Hp^2 homozygotes produce a broad band at the cathodal end of α_2-macroglobulin (Fig. 4.21). Heterozygotes, Hp^1/Hp^2, on starch gels produce one weak band in the 1-1 position and several slower-migrating bands. By routine serum protein electrophoretic techniques, these appear as a broad band just anodal to and directly over α_2-macroglobulin. Molecular studies on the genotype distribution of three phenotypes demonstrated the presence of 14.5 per cent Hp1-1, 48.2 per cent Hp2-1, and 37.3 per cent Hp2-2.[144]

DECREASED HAPTOGLOBIN

There are several important causes of decreased haptoglobin levels that can be suggested by the serum protein electrophoresis pattern (Table 4.6); the most clinically significant causes relate to hemolysis. For the clinician studying a patient with a hemolytic process, the laboratory interpretation of 'marked decrease in haptoglobin consistent with intravascular hemolysis' is useful both in suggesting the process (extravascular versus intravascular) and in following the response of the patient to therapy. As usual, there are some caveats. An uncommon phenotypic variant of haptoglobin, referred to as type 0-0 (seen most frequently in black patients), produces no or very low levels of haptoglobin.[145] Further, infants have little haptoglobin; the concentration goes from virtually zero in the cord blood to adult levels by around the age of 4 months.[139] Ineffective erythropoiesis such as occurs during folate and vitamin B_{12} deficiency may produce a decrease in the haptoglobin region on serum protein electrophoresis.

When hemoglobin binds to haptoglobin, the complex has a much slower migration than that of haptoglobin alone (Figs 4.22 and 4.23). The patterns shown in these figures usually indicate that the sample has been handled poorly, with *in vitro* hemolysis producing the haptoglobin–hemoglobin band indicated. However, in cases of recent

Table 4.6 Decreased haptoglobin

Decreased haptoglobin	
Hemolysis	
In vivo:	Intravascular hemolysis
	Extravascular hemolysis
In vitro:	Poor blood sampling technique; (hemoglobin–haptoglobin migrates between α_2-macroglobulin and transferrin)
Ineffective erythropoiesis	Vitamin B_{12} deficiency
	Folate deficiency
Congenital	Hp O phenotype
Neonates	Normally have low haptoglobin

incompatible blood transfusion or recent massive hemolysis, as in malarial or clostridial infections, this may reflect an urgent clinical situation. Haptoglobin may also be decreased in severe liver disease. The presence of a decreased haptoglobin in a patient with hypoalbuminemia, polyclonal hypergammaglobulinemia and a low transferrin often has a poor prognosis.[146]

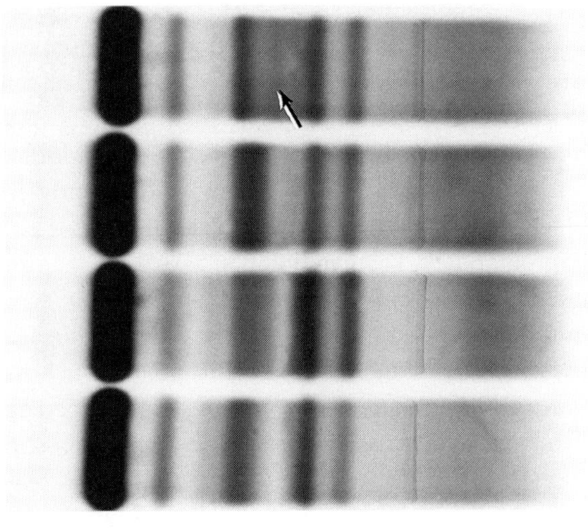

INCREASED HAPTOGLOBIN

An increased synthesis of haptoglobin occurs in patients with an acute inflammatory response and in patients with increased corticosteroid and estrogen stimulation (pregnancy, contraceptive drugs, estrogen-secreting neoplasms, and cirrhosis).[27,147,148] A key factor for the increased synthesis of haptoglobin and other acute-phase reactants is interleukin-6.[27] A patient may have simultaneous processes that result in a normal haptoglobin level when an elevation or decrease may have been predicted by the available clinical information. For example, if a patient has an autoimmune disease with intravascular hemolysis, the haptoglobin should be decreased. However, if the autoimmune disease is clinically active, with acute inflammation, there will be a stimulation of haptoglobin production by the liver. In this case, the haptoglobin band on electrophoresis will depend on which process predominates at the time the sample was taken.[149]

Patients with renal disease may have an increase or a decrease in haptoglobin depending on the phenotype and on the degree of renal damage. During renal injury, there may be inflammation resulting in an increase in haptoglobin. However,

Figure 4.22 The top sample shows a prominent α_2-macroglobulin band followed by a broad, moderately dense staining band that extends to the transferrin band. This is hemoglobin–haptoglobin complex. A wavy band due to β_1-lipoprotein can be seen just anodal to transferrin. A thin, straight band in the middle of the hemoglobin–haptoglobin complex was not identified (arrow). (Paragon SPE2 system stained with Paragon Violet.)

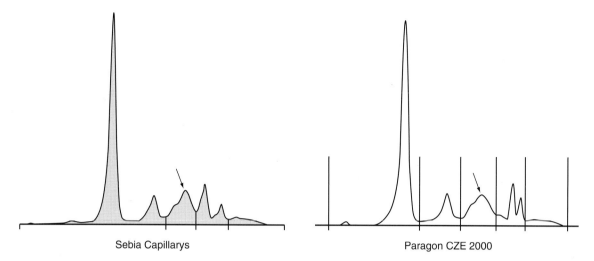

Sebia Capillarys

Paragon CZE 2000

Figure 4.23 Capillary zone electropherograms display a prominent slow β-region band (arrow) due to hemoglobin–haptoglobin complex. Note, more recent buffer solution on the Sebia Capillarys may result in a different migration of the hemoglobin–haptoglobin complex.

since the Hp 1-1 phenotype is a relatively smaller molecule than Hp 2-1 or 2-2, it will more readily pass into the urine, so the first phenotype haptoglobin may be decreased, whereas the last two would be increased.[135]

Ceruloplasmin

Ceruloplasmin is an important copper-binding transport protein produced by the liver. However, its low concentration in normal serum (14–46 mg/dl)[150] precludes its detection on the routine serum protein electrophoresis. Even during the acute-phase response, when its concentration in serum often increases, it is difficult to see because of the increase in the other α_2-globulins. It is elevated during an acute-phase response, in steroid therapy, and in cases of biliary tract obstruction.

The most significant feature of ceruloplasmin is its marked decrease in patients with Wilson's disease. The lack of this transport protein is thought to play the major role in hepatolenticular degeneration. However, serum protein electrophoresis will not aid in the detection of a decreased ceruloplasmin, and specific assays must be performed.

Pre-β_1-lipoprotein

Serum protein electrophoresis is not used to screen for lipid abnormalities. However, some patients will have unusual electrophoretic patterns due to hyperlipidemia. In the Preβ area (very low density lipoproteins, VLDL), a broad elevation similar to that seen with hemolysis has also been seen in patients with elevated preβ₁-lipoprotein. Therefore, when there is an elevation in this area, look at the serum sample to see if it is hemolysed. If it is not, a lipoprotein evaluation may be appropriate.

Fibronectin

Fibronectin (cold insoluble globulin) is a faint band regularly seen between the α_2 and β_1 region. It is a large, 440 kDa protein that derived its original name from the observation that it precipitates in the cold and with heparin. The band becomes more prominent during pregnancy or with cholestasis, when the protein increases beyond its usual serum range of from 19 mg/dl to 35 mg/dl.[151] Fibronectin acts in the wound-healing process by interaction with fibrinogen and by mediating the adherence of fibroblasts and monocytes at sites of tissue damage.[152,153] It is increased in pregnancy, and is further increased in patients with pre-eclampsia caused by vascular injury, increased production or enzymatic degradation.[154] It is also increased in the vessel walls in active central nervous system plaques from patients with multiple sclerosis, possibly enhancing myelin phagocytosis in these lesions.[155] During the first 48 h of the acute inflammatory response, fibronectin is rapidly deposited in the damaged tissue. Simultaneously, its concentration in the serum decreases.[152]

β-REGION

Transferrin

Transferrin is the major band at the anodal end of the β_1-region. Transferrin is a single polypeptide glycoprotein with a molecular mass of 76.5 kDa that functions to transport non-heme ferric iron from the gastrointestinal tract and from the breakdown of hemoglobin to the bone marrow.[156,157] Each transferrin molecule can bind two molecules of free iron, but normally only about one-third of the transferrin molecules are saturated with iron. The total iron-binding capacity of serum is a reflection of the amount of transferrin present; its normal concentration is 240–480 mg/dl.[158] In patients with iron-deficiency anemia, the levels of transferrin are considerably increased. Determinations of the transferrin levels are useful in distinguishing between iron deficiency anemia (inadequate intake or chronic hemorrhage with loss of iron stores) from iron-refractory anemias. In iron deficiency, the concentration of serum transferrin goes up, but as less iron is available for transport, the saturation of the transferrin falls, often to less than 15 per cent compared with 33 per cent normally. Transferrin levels are also increased in patients that are pregnant or in patients receiving estrogens.[159]

TRANSFERRIN VARIANTS

Serum protein electrophoresis using either CZE or gels that provide a crisp separation of transferrin from C3 can demonstrate relatively common genetic variants of transferrin (Fig. 4.24). Both electrophoretically slow and fast variants of transferrin have been identified. The major factors that determine the migration of transferrin on serum protein electrophoresis are the amino acid sequence and its sialic acid content (see below). The migration of transferrin in the vast majority of people (98 per cent) is due to the common type of transferrin TfC.[160,161] Variants that move toward the anode are TfB and those which move toward the cathode are TfD.[156] Black patients have a gene frequency of about 10 per cent for TfCD.[157,162] Homozygotes of fast or slow variants are rarely detected, perhaps because they would be difficult to recognize by commonly used clinical laboratory electrophoretic techniques. However, heterozygotes are easily seen as two equal staining bands in the β_1 region (Fig. 4.24). These represent the codominant expression of two alleles, one normal and one variant.[13] Despite the characteristic appearance, I recommend performing an immunofixation, usually a Penta (pentavalent, see Chapter 3), to rule out a small monoclonal gammopathy. While earlier studies suggested that these variants had no functional effect on patients, recent studies by Kasvosve et al.[157] demonstrate that the amount of iron bound by transferrin is lower in TFCD individuals than with the other common variants. This may have functional significance to partly protect these individuals from increased iron accumulation in iron overload situations.[157] While one cannot subtype the variants by serum protein electrophoresis, it is important to understand the possible double bands associated with the variants. These may be mistaken for small M-proteins. This is one reason why bands suspected of being M-proteins must always be characterized to prove that they are immunoglobulins.

Serum transferrin is normally glycosylated by two complex carbohydrate chains that contain a charged component, sialic acid. The number of the sialic acids attached varies from none to six. Each sialic acid molecule present has been estimated to alter the pI of transferrin by 0.1 pH unit.[163–165] The

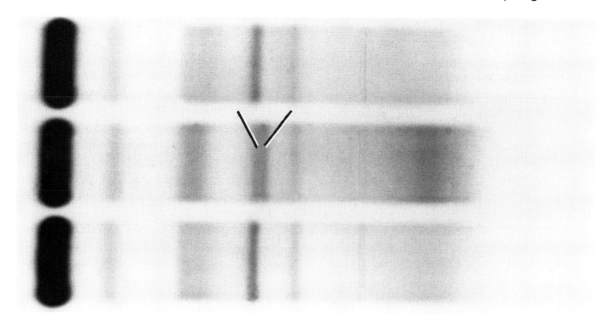

Figure 4.24 The middle serum is from a patient with genetic variant of transferrin. There are two discrete bands (indicated) of equivalent intensity. (Panagel System stained with Amido Black.)

most common isoform of transferrin has two molecules of sialic acid attached.[163] Patients that consume large quantities of alcohol contain serum transferrin with few sialic acid molecules attached (none, monosialated and disialated transferrin) with a correspondingly higher pI than normal.[166,167] This is referred to as carbohydrate-deficient transferrin. Although, using iron specific stains, a difference in migration on serum protein electrophoresis has been reported, it is not possible to distinguish reliably this difference in migration using routine protein stains.[168] A variety of methods, including CZE have been used to measure carbohydrate-deficient transferrin.[166] However, those capillary zone studies were performed on a P/ACE 5000 electropherograph (Beckman), not the currently available clinical equipment (Paragon CZE 2000 or Sebia Capillarys).[169]

We can detect cerebrospinal fluid leakage in the ear or nose following head trauma or secondary to a neoplasm by taking advantage of the fact that serum transferrin usually contains sialic acid molecules.[170-173] Cerebrospinal fluid normally contains both the β1-migrating glycated transferrin seen in serum and a substantial transferrin band in the β2-region (τ fraction of transferrin) that lacks the sialic acid residues. This form of transferrin is also present in human aqueous humor.[174] Immunofixation of cerebrospinal fluid with antibodies against transferrin will demonstrate two transferrin bands (discussed in Chapter 8): one due to the normal sialated transferrin seen in the β1-region of serum and the other due to the τ fraction, (a serum control from the same patient must be examined in tract next to the nasal or ear fluid to be sure the patient does not normally have asialated transferrin in his or her serum).[172] Altered migration of genetic variants and changes of sialic acid in alcoholics could result in false positives if a control serum was not studied.[172,175]

INCREASED TRANSFERRIN

The transferrin band is increased in patients with iron deficiency anemia.[176] Occasionally, the transferrin band may be obscured by the presence of a large amount of preβ or β1-lipoprotein (see below). Also, when using CZE, some radiocontrast dyes will migrate in the β-region and may result in a falsely elevated transferrin band (see Chapter 2).[177]

In addition to radiocontrast dyes, the antibiotic piperacillin-tazobactam (Tazocin; Wyeth Lederle, Collegeville, PA, USA) was reported to produce a small peak just anodal to the transferrin band on the Paragon CZE 2000, using software version 2.21).[178] Such a band could be mistaken either for a transferrin variant or small M-protein. It is recommended that all suspicious unidentified bands be evaluated by performing an immunofixation. A Penta screen may be the most efficient manner in which to do this (see Chapter 3).

Lower resolution electrophoretic methods that do not provide a crisp separation of the β_1 and β_2 bands do not offer a clear view of the transferrin band. This may make it more difficult to detect the occasional M-protein that causes a distortion of the transferrin band. When β_1-lipoprotein obscures this region, electrophoresis could be repeated with heparin added to the sample. This pulls the β_1-lipoprotein toward the anode and gives the interpreter a better view of the β_1-region (Fig. 4.25).[49,50] Another strategy to rule out a monoclonal gammopathy is to perform an immunofixation (such as a Penta evaluation) on all suspicious bands (see Chapter 3).

DECREASED TRANSFERRIN

Transferrin is usually decreased in alcoholic cirrhosis.[179] During acute inflammation, the synthesis of transferrin by the liver is largely shut down, resulting in a faint β_1-region band. Transferrin is also decreased during renal disease and thermal injuries because of loss through the glomeruli and damaged skin, respectively.[180] Congenital atransferrinemia has been reported, but is quite rare. These individuals suffer from a microcytic, hypochromic anemia, despite the presence of normal serum iron levels; they have a very low iron-binding capacity. Their poor ability to transport iron also predisposes them to develop hemochromatosis unless detected and treated early.[181,182] Therefore, the finding of an undetectable β_1-region band on serum protein

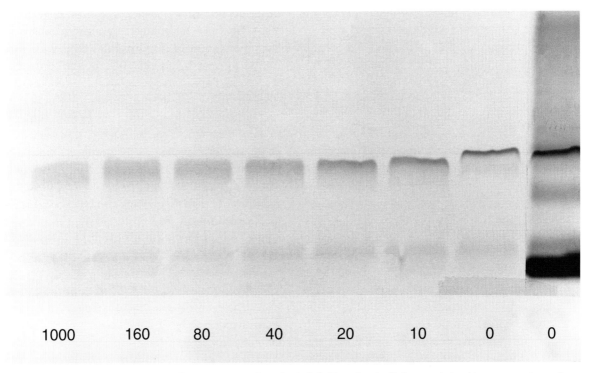

Figure 4.25 Serum treated with heparin (concentrations indicated units/ml). Note that the β_1-lipoprotein band becomes weaker and moves toward the anode, making the transferrin region easier to inspect in difficult cases. (Photograph from work by Jeffrey Pearson).

electrophoresis has considerable importance both to the individual patient and for their relatives.

β-Migrating monoclonal gammopathies (M-proteins) and complement activation products

The major proteins that need to be distinguished from the double transferrin band (heterozygous variants), are β-migrating monoclonal gammopathies and complement activation products. It is often erroneously believed that immunoglobulin molecules only migrate in the γ-region. Although IgG usually resides in the γ-region, IgA is mainly a β-migrating molecule while IgM tends to migrate between these two (Fig. 4.26). Further, the rare IgE, uncommon IgD and the more common monoclonal free light chains (Bence Jones protein) will frequently occur in the β- or even in the α-globulin regions. Monoclonal gammopathies involving any of these β-migrating immunoglobulins may only produce a subtle distortion of the transferrin or C3 band.

The presence of a second β1-region band (other than transferrin) that is due to an M-protein of the IgA class usually will be denser and more diffuse than the equal transferrin lines seen in the heterozygous variants. Performing an immunofixation to rule out the possibility of an M-protein is recommended when there is distortion of the transferrin or C3 band. Occasionally, monoclonal bands lie

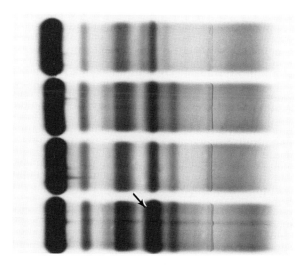

Figure 4.27 The bottom sample has a very prominent β1-region (transferrin region) band (arrow) because of an IgA κ monoclonal gammopathy which migrated on top of the usual transferrin location. Even though the migration is correct for transferrin, the band is far too dense in staining to be caused by iron deficiency. This sample, and those above it, also suffer from a slight distortion, possibly owing to insufficient blotting. (Paragon SPE2 system stained with Paragon Violet.)

directly over transferrin or C3 (see later) (Fig. 4.27). Therefore, when there is a large transferrin band, but no clinical history of an iron-deficient anemia, one should rule out the possibility of a monoclonal gammopathy by an immunofixation (such as the studies discussed in Chapter 3).

The other major protein band of clinical significance that is often confused with the transferrin variant is the C3c product of complement (see later). This may be an *in vitro* activation due to poor specimen handling or may reflect *in vivo* activation caused by autoimmune disease or ongoing inflammation. As discussed later, when the band anodal to transferrin is C3c, the usual C3 band (β2-region) is decreased or absent. To be certain, however, performing an immunofixation is recommended to rule out an M-protein.

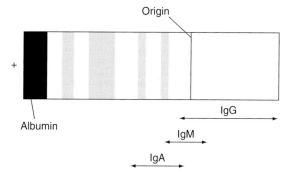

Figure 4.26 The usual locations of IgG, IgA and IgM molecules are indicated by lines below the schematic representation of the gel.

β1-lipoprotein

β1-lipoprotein (low density lipoprotein) is an unusual molecule in two respects: its position on gel electrophoresis varies considerably depending on its concentration and it has an irregular anodal front (Fig. 4.28). The molecular mass is enormous (2750 kDa) and its concentration spans a wide range between 57 mg/dl and 206 mg/dl in adults.[183] β_1-Lipoprotein transports lipids, cholesterol and hormones. Because the migration of β_1-lipoprotein on gels decreases with increasing concentration, it may be found anodal to the transferrin band or cathodal to the C3 band. The band often has an irregular anodal front on gel-based systems. This irregularity may also occur with a few IgM and IgA monoclonal gammopathies. It is related to the tendency of these molecules to aggregate when present in high concentration. At the anodal edge, the movement is faster and the concentration of the molecules is lower than at the center of the band, where aggregation tends to occur. The aggregation of these large molecules interferes with the electri-

cal field and endosmotic flow (see Chapter 1) in the immediate vicinity of the band. The partial interruption of the regular electrical field and endosmotic flow results in an irregular band or small parallel bands (striae) with a crescent-type shape (the middle lagging toward the cathode).[184]

Such irregularity is especially pronounced in gels that have narrower pores with a greater molecular sieve capability. While these gels often give excellent resolution of γ-globulins, the β_1-lipoprotein band can obscure other important beta proteins. This tendency will vary from one manufacturer to another. Therefore, in some systems, β_1-lipoprotein will produce a diffuse band that does not significantly interfere with the interpretation of the major β-region bands or β-migrating monoclonal gammopathies; other systems can give irregular, dark-staining bands that may obscure transferrin variants or small monoclonal gammopathies. In CZE systems, β_1-lipoprotein may migrate in the slow α_2-region (Fig. 4.29). Because of a relative constancy of position of β_1-lipoprotein on the available CZE systems, I have found less confusion with an M-protein than in gel-based systems.

Elevated levels of β_1-lipoprotein are seen in conditions with increased cholesterol, such as the nephrotic syndrome and Type II hypercholesterolemia. Preincubating a patient's serum with 40 units of heparin/ml serum will cause the β_1-lipoprotein band to migrate more anodally as a faint diffuse band and may provide the interpreter with a better view of the β_1-region [49,50] (Fig. 4.25).

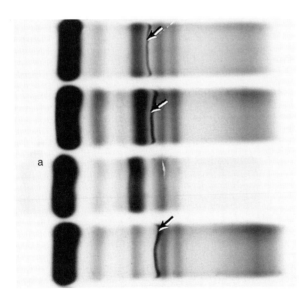

Figure 4.28 The irregular β_1-lipoprotein bands migrate slightly anodal to transferrin in the top two samples (arrows) and almost abuts on the anodal edge of the transferrin band in the bottom sample (arrow). The β_1-lipoprotein band is not visible in the third sample. This serum was from a patient receiving heparin. The α_1-lipoprotein and β_1-lipoprotein become more diffuse and migrate more anodally in patients receiving heparin. A faint anodal slurring (a) is barely seen just anodal to albumin, and the β_1-lipoprotein band is now merged into the haptoglobin band and, consequently, is not seen. (Paragon SPE2 system stained with Paragon Violet.)

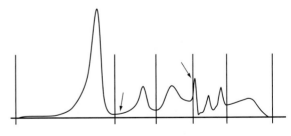

Figure 4.29 The arrow in the α_1-region indicates the usual position of α_1-lipoprotein and the other arrow points to the prominent β_1-lipoprotein restriction at the junction of the α_2- and β-regions in this electropherogram of serum that contains increased β_1-lipoprotein. (Paragon CZE 2000.)

C3

After transferrin, C3, the third component of complement, is the next major protein band seen on both gel-based and CZE. C3 is the only component of complement present in sufficient concentration to allow its recognition by serum protein electrophoresis. It is normally present in serum at a concentration of 90–180 mg/dl (0.9–1.8 g/l) and consists of a 110 kDa α-chain and a 75 kDa β-chain that are linked by disulfide bonds.[106,185,186] This band's density and position can vary depending on genetic differences, active inflammatory disease, or poor specimen processing.

The use of calcium lactate in the buffer systems for some gels improves the resolution of this band. Occasionally, the C3 band may appear as a double line (Fig. 4.30). This is because of the genetic polymorphism of C3.[106] C3 variants are uncommon. They represent the codominant production of variant genes. Rarely, C3 variants have been associated with a decrease in hemolytic complement.[187] However, most C3 variants have no known significance but need to be distinguished from small M-proteins.[188] The two C3 bands are of identical intensity when they result from a genetic variant. With a monoclonal gammopathy, one almost always sees variation (darker or lighter) of the band. When in doubt, I perform the immunofixation to be certain.

Complement is a complex group of serum proteins that may be activated by the interactions of antibody with antigen or by a variety of other stimuli (Table 4.7). C3 may be decreased when either the classical or alternative pathway is activated. Either pathway ends up with the formation of C3 convertase that cleaves native C3 into the small, active molecule C3a and a larger C3b fragment. C3a is a particularly active 9 kDa polypeptide. It can induce vasodilatation and increase vascular permeability, with resulting edema. The presence of C3b on the initiating

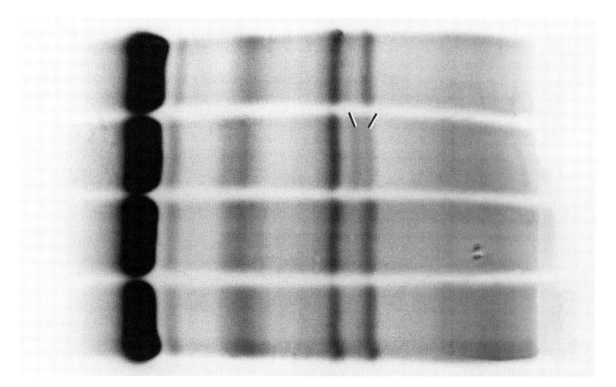

Figure 4.30 The second sample from the top is a C3 heterozygote with two identical staining bands (indicated) in the β₂-region. (Cellogel high-resolution acetate system.) This photograph was contributed by Francesco Aguzzi.

complex has functional significance. Most phago-cytic cells have receptors for C3b on their surface and, therefore, the C3b on the surface of the initi-ating complex will be more readily attached to neutrophils, monocytes and eosinophils.

During inactivation of C3b, Factor I first cleaves the α-chain into the inactive C3bi. Next, in a kinet-ically slower reaction, Factor I cleaves the terminal end of the α-chain, resulting in the formation of a small C3d fragment (30 000) and a larger C3c fragment (160 000) that is often seen in serum protein electrophoresis of poorly processed blood samples or aged specimens. In the latter reaction, Factor I collaborates with serum proteases. Usually, the presence of the C3c band indicates that the specimen has been processed poorly, stored at room temperature, lyophilized (most lyophilized commercial standard serum prepara-tions will show this C3c band), or stored for several days at 4°C. Owing to the short biological half-life of C3c, it is unlikely that the C3c seen by serum protein electrophoresis relates to active inflammation *in vivo*.

When C3 breaks down, the C3 (β_2-band) band declines and smaller, smudgy bands may appear anodal to transferrin, cathodal to transferrin or even in the γ-region, depending on which electro-phoretic system is used. On most gel-based systems and on the Paragon CZE 2000, using Version 1.5, the C3c breakdown product appears anodal to transferrin (Figs. 4.31 and 4.32).

The concentration of C3 is elevated late during the acute phase response. It is upregulated by inter-leukin-1 (IL-1), interleukin-6 (IL-6) and TNF.[189-192] Because C3 and haptoglobin are often elevated after other acute phase reactants, α_1-acid glycopro-tein (orosomucoid), α_1-antitrypsin and C-reactive protein have declined to the normal range, I refer to the combined elevation of α_2-region and C3 (β_2-band) together with a decrease in transferrin as a subacute reaction.

The use of a CZE system may be more sensitive than gel-based methods for detecting C3 even on specimens that have been stored for up to 30 days at 4°C.[193] In addition to providing crisp separa-tion of the β_1- and β_2-regions, it has been sug-gested that CZE may provide useful estimates of both C3 and transferrin that did not require further testing.[194] However, I recommend specific nephelometric-determined concentrations for best accuracy.

C4

C4 is a large, 206 kDa glycoprotein with an adult serum concentration of from 10 to 40 mg/dl (0.1–0.4 g/l).[106] It is visible on serum protein electrophoresis only when present in the higher normal range or when increased such as in patients with an acute inflammatory reaction. It may be seen just anodal to C3 on gel-based

Table 4.7 Factors that activate complement

Classical pathway	Alternative pathway
Immune complexes	Aggregated immunoglobulins
Staphylococcal protein A-IgG	Complex polysaccharides:
Polyanions:	Bacterial dextrans
DNA	Inulin
Dextran sulfate	Cobra venom factor
Some RNA viruses	Erythrocyte stroma
C-reactive protein	Nephritic factor

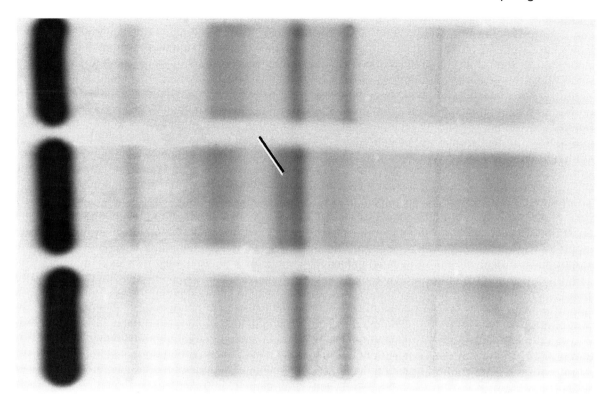

Figure 4.31 C3 activation is seen in the middle sample. A C3c band is now present (indicated), while the usual position of intact C3 in the β₂-region shows only a faint slur toward the anode. (Panagel system stained with Amido Black.)

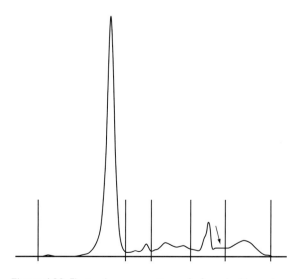

Figure 4.32 Electropherogram pattern of a 2-week-old sample sent to the laboratory. The arrow notes the lack of C3. In addition, an anodal shoulder to the transferrin band indicates the position of some C3 breakdown products. (Paragon CZE 2000.)

systems that provide crisp separation of the β₁- and β₂-bands. On the Paragon CZE 2000 Version 1.6 CZE system, C4 may give a tiny deflection to the C3 band. Knowing the position of this band is relevant primarily to the possibility of a β-migrating monoclonal protein. Nonetheless, check distortion in this region by performing an immunofixation. Genetically heterogeneous C4 deficiencies have been reported in association with systemic lupus erythematosus.[195,196] However, serum protein electrophoresis is too insensitive a tool to use to detect such deficiencies.

Fibrinogen

Fibrinogen is a 340 kDa protein that is present in plasma in concentrations of 200–400 mg/dl in adults.[197] Although fibrinogen is not present in properly processed normal serum, a small fibrinogen band may be seen in serum protein electrophoresis due to insufficient clotting or failure to remove the serum from the clot with the release of fibrinogen breakdown products (Fig. 4.33). Fibrinogen may also be seen in the 'serum' of patients on heparin therapy. It is strongly recommended that plasma not be used for serum protein electrophoresis because the fibrinogen band obscures detection of monoclonal gammopathies in an important part of the β–γ region.

One may suspect that this band is fibrinogen because of its position on gel electrophoresis or CZE, and on gel-based systems fibrinogen may have an irregular, curved or non-symmetrical pattern. However, because monoclonal gammopathies may also migrate as irregular bands if they self-aggregate or occur as cryoglobulins, one may perform an immunofixation to identify the fibrinogen (Fig. 4.34). The fibrinogen band has led to false positive reports of monoclonal gammopathies.[198] However, currently, I prefer the use of an immunofixation to rule out that the band is an immunoglobulin (see Chapter 3). If one wishes to avoid the expense of the immunofixation, one may wait 24 h and repeat the serum protein electrophoresis. The fibrinogen band may disappear after allowing the sample to clot further. However, the time and the

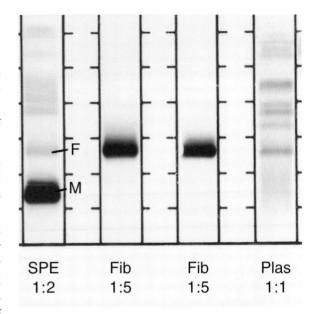

SPE 1:2   Fib 1:5   Fib 1:5   Plas 1:1

Figure 4.34 Immunofixation, with the anode at the top, comparing a serum (serum protein electrophoresis, SPE) containing a prominent monoclonal gammopathy (M) and a second band which migrates in the fibrinogen region (F) with plasma (Plas). The lane to the right of the SPE is the serum diluted 1:5 and reacted with anti-fibrinogen (Fib). The lane to the left of the Plas is the plasma diluted 1:5 and reacted with anti-fibrinogen (Fib). The fibrinogen bands from both have the same migration. This identifies the second band in the serum as fibrinogen. (Paragon Immunofixation stained with Paragon Violet).

Figure 4.33 The bottom sample has a prominent fibrinogen band (arrow). This dense band indicates that the sample was likely plasma and not serum. This band must be distinguished from a monoclonal gammopathy by immunological studies. When it is this prominent, it may mask a monoclonal gammopathy in this region. The slight irregularity to all the bands on this gel indicates there has been inadequate blotting. (SPE2 system stained with Paragon Violet.)

additional electrophoresis are just about as costly as adding one sample to immunofixation gel (see Chapter 3).

Occasionally, one may be misled by immunofixation studies because of the non-specificity of some immunological reagents that may not be caught by the laboratory's quality control measures. Some reagent antisera against immunoglobulins may contain minor crossreactivities against fibrinogen.[199] This seems to be a problem particularly with reagents directed against IgA and IgM. Laboratories typically test new lots of reagents against serum and known monoclonal proteins. However, they should also test the lots against plasma because cross-reactivity against fibrinogen cannot be detected using serum controls.

γ-REGION

C-reactive protein

C-reactive protein is a hepatocyte-derived, 135 kDa non-immunoglobulin protein that migrates in the γ-region. C-reactive protein derives its name from the fact that it reacts with the capsular polysaccharide of *Streptococcus pneumoniae*.[200] It is one of the most reliable, objective measures of the acute inflammatory response, and has been recommended as a particularly strong indicator of bacterial infections.[201-208] C-reactive protein may be superior to cytokine assays of IL-6, IL-1β and TNF for detecting inflammation in serum from patients receiving intensive care.[209] In the past few years, the use of high-sensitive (hs) C-reactive protein assays has proven useful to detect individuals at increased risk of stroke or coronary artery disease.[210-212]

As with most of the bands discussed above, its exact position will vary somewhat depending upon the particular electrophoretic system being used. Although its normal concentration of < 2 mg/dl is far too low to be noticed, C-reactive protein levels may increase considerably during acute inflammatory responses to concentrations greater than 100 mg/dl. In the latter case, it creates a distinctive mid-γ-band that may be mistaken for a small monoclonal protein on gel-based systems (Fig. 4.35).[213] Interestingly, using nephelometric assays, the opposite has also occurred. Crossreactivity between C-reactive protein and M-proteins has resulted in falsely elevated levels of C-reactive protein with some immunochemical quantitative methods.[214] Since the elevation of C-reactive protein occurs during an acute inflammatory response, the presence of the typical acute phase pattern should alert the reader to include C-reactive protein in the differential. Quantification of the C-reactive protein level or immunofixation for C-reactive protein (Chapter 3) may be used to achieve correct identification of the restricted area.

Immunoglobulins

The γ-region consists overwhelmingly of immunoglobulin molecules. Because of the relatively slow migration of most of these molecules, the term 'gamma globulin' was at one time synonymous with immunoglobulin. It is now clear that antibody molecules migrate anywhere from the α- to the slow γ-region. The term γ-globulin reflected the fact that most of the serum antibodies are of the IgG class that primarily migrates in a γ location.

To understand the unusually broad electrophoretic migration of this important group of molecules, it is worth reviewing the structure of immunoglobulin molecules. More details can be found in the review by Frazer and Capra.[215] The basic monomeric unit of an immunoglobulin molecule consists of two identical heavy polypeptide chains and two identical light polypeptide chains (Fig. 4.36). One reason for the vast heterogeneity of pI values among immunoglobulins is that there are five different major classes of immunoglobulin in heavy chains and two different classes of immunoglobulin light chains, each with differing electrophoretic mobilities. In addition, there are millions of possible antigen-combining sites, creating an enormous diversity of charge.

The characteristics of the heavy chain bands are listed in Table 4.8. Immunoglobulin molecules are named for their heavy chain class, which is of major importance in the biologic capabilities of the molecule (Table 4.8). The five major classes of immunoglobulins are IgG, IgA, IgM, IgD, and IgE. Some of the heavy chains have been further divided into subclasses, such as IgG1, 2, 3 and 4. Any heavy chain may be combined with either of the two light chain types, κ or λ. Only one type of light chain is produced by a clone of B cells or in a plasma cell tumor. The normal κ/λ ratio of intact immunoglobulins in serum is 2:1. Interestingly, however, plasma cells all produce excess amounts of free κ and λ light chains. Since free polyclonal κ light chains occur mainly as monomers, they have a more rapid glomerular filtration rate than free λ polyclonal light chains that occur as dimers.[216] As a

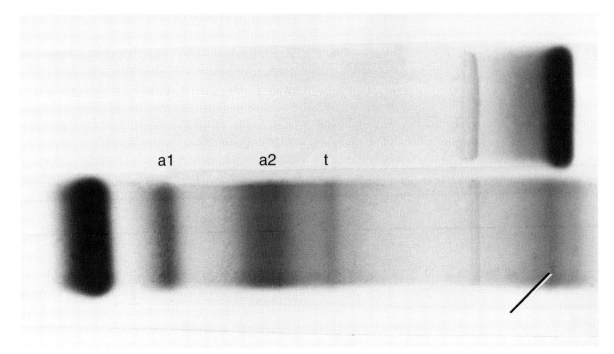

Figure 4.35 The top lane is an immunofixation with anti-C-reactive protein. It indicates the location of this prominent protein. The second lane is the serum protein electrophoresis from the same case. This serum was from a patient with an acute phase reaction and shows increased α_1 (a1), α_2 (a2), and decreased transferrin (t). The markedly elevated C-reactive protein is demonstrated by the discrete band (indicated) in the mid-γ-region. The C3 band is missing because the sample had been stored for several days before the immunofixation was performed for this demonstration. (Panagel stained with Coomassie Blue.)

Table 4.8 Characteristics of human immunoglobulins

Characteristics	IgG	IgA	IgM	IgD	IgE
Concentration (mg/dl)[106]	700–1600	70–400	40–230	0–8	–
Mass (kDa)[106]	160	160 (serum)	900	184	188
		380 (mucosal)			
Subclasses[a]	G1, G2, G3, G4	A1, A2	None	None	None
Serum half life[a]	23[b]	6.5	7	0.4	0.016
Complement activation	Yes	No	Yes	No	No
Opsoniztion	Yes	Yes	No	No	No
Secretory	No	Yes	Yes	No	No
Cross placenta	Yes	No	No	No	No
Binds Mast cells	No	No	No	No	Yes
Surface membrane of early B-cells[c]	No	No	Yes	Yes	No

[a]Subclasses, serum half life (days).[215]
[b]Half-life varies with different subclasses.
[c]All of the immunoglobulins appear on the surface of some B cells, however, only IgD and IgM are present on early B-lymphocytes.

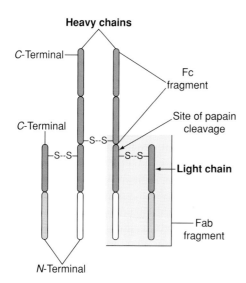

Heavy chains

C-Terminal

Fc
fragment

C-Terminal

Site of papain
cleavage

–S--S–

–S--S–

–S--S–

Light chain

Fab
fragment

N-Terminal

Figure 4.36 Schematic representation of the immunoglobulin molecules. The constant regions are indicated by dark gray shading; the variable regions are indicated by mid gray shading. Note that the variable regions for both light chains are identical and the variable regions from both heavy chains are identical. The fragment with antigen binding (Fab) capability is identified by light shading. The fragment which crystallizes (Fc) upon papain digestion is indicated. (Note, the designations of Fab and Fc were derived from papain digestion studies on rabbit immunoglobulin.)

result, when the new specific assay for free serum light chains is performed, the free κ/λ ratio range is 0.26–1.65.[217] So when reading about the κ/λ ratio, be sure to check whether the authors mean free or bound κ/λ.

During early studies of the basic molecular structure of immunoglobulins, it was found that the enzyme papain cleaved the molecule into two major fragments. One fragment was called Fc (the fragment that crystallized), while the other fragment was called Fab (the fragment with antigen binding activity) (Fig. 4.36). Fc is the C-terminal half of the two heavy chains, and Fab contains an intact light chain and the N-terminal half of one heavy chain. Further investigation into amino acid sequences of the immunoglobulins took advantage of the fact that myeloma proteins provide extremely large amounts of a single, specific immunoglobulin molecule. These studies disclosed that the C-terminal half of all κ light chains is

remarkably homogeneous (constant region) while the N-terminal half varies considerably from one κ light chain protein to another (variable region). Similarly, the N-terminal quarter of a γ heavy chain (variable region) varies in amino acid composition from one myeloma protein to another, while the C-terminal three-quarters (constant region) are almost identical.[215]

There is occasional confusion about nomenclature of immunoglobulins. Isotype refers to the heavy chain class. Therefore, there are five major isotypes; IgG, IgA, IgM, IgD, and IgE. The term idiotype refers to the specific antigenic makeup of the variable portion of the Fab region; this is unique to an antibody molecule of a certain specificity. I like to oversimplify and think of the 'idiotype' as referring to the specific binding site of the antibody molecule. Allotype refers to certain genetic alleles that may occur within the population such as Gml or Gm2.[215]

The Fc portion of the immunoglobulin molecule determines the biological capabilities of the antibody. For example, IgG1, IgG3, and IgG4 have Fc regions that are capable of activating complement through the classical pathway following an antibody–antigen interaction. There is practical importance in being aware of the different IgG subclasses, as this will help one to understand some unusual monoclonal and polyclonal patterns. For example, although most serum with polyclonal increases in IgG due to an infectious process shows a broad increase in γ-globulin, patients with a marked polyclonal increase in IgG4 subclass may show an atypical electrophoretic pattern with a relatively broad peak in the β–γ region.[218,219] Such atypical peaks have been mistaken for monoclonal gammopathies. IgA and IgG2 are relatively ineffective at complement activation because of the structure of their Fc regions.

IMMUNOGLOBULIN G

Immunoglobulin G, the major immunoglobulin in the γ-region, is responsible for the broad, light-staining characteristic of this region in normal individuals. In adults, IgG occurs in a concentration of 700–1600 mg/dl, whereas in children the concentration is variable. Neonates will have almost normal adult levels, because this molecule is transported across the placenta through specific Fc-receptors on the placenta. However, the newborn child does not begin to synthesize IgG for several months. The maternal IgG that crossed the placenta is gradually catabolized so that at about 6 months a nadir is reached where extremely low levels of IgG are present in the serum. Children with congenital immunodeficiencies are asymptomatic until after 6 months of age because they are protected by the maternal IgG that crossed the placenta.[220]

Many abnormalities of IgG occur. Elevations with infections, autoimmune diseases, liver disease and multiple myeloma are commonly seen. It is the immunoglobulin most often found in multiple myeloma.[221] Two or three discrete bands (oligoclonal bands) occur in the γ-region of serum from patients with circulating immune complexes, acquired immune deficiency syndrome, hepatitis or polyclonal increases in response to other infections.[222] The bands may be especially prominent during infectious diseases.[223]

An isolated decrease in IgG demonstrated by finding faint staining of the γ-region with other fractions being normal is a highly significant pattern that may indicate a variety of conditions including: humoral immunodeficiency, B-cell lymphoproliferative disorders, light chain disease, nonsecretory myeloma, or amyloid AL (amyloid associated with immunoglobulin light chain; discussed in Chapter 6).

IMMUNOGLOBULIN M

Immunoglobulin M usually occurs as a pentamer in the serum. As such, it has a molecular mass of about 900 kDa. The normal concentration is 40–230 mg/dl.[106] It is the most effective of all isotypes in activating complement. Whereas IgG may require 100 or more molecules in an antibody–antigen complex to effectively activate complement, a single IgM molecule will suffice for complete activation through the terminal lytic sequence on a cell surface.[215] This property probably relates not only to the amino acid makeup of

the Fc portion, but also to the physical proximity of the five Fc groups of the pentamer. The pI and bulk of IgM are such that it does not stray far from the origin on gel-based electrophoresis. Depending on the character of the support medium for the particular electrophoretic system, it may migrate slightly cathodally or slightly anodally. By CZE, IgM migrates mainly in the β–γ region. Marked elevations may occur with infections or malignancy and are reviewed further in Chapter 7.

IMMUNOGLOBULIN A

Immunoglobulin A is the other major isotype of immunoglobulin in the serum. In the serum, IgA occurs mainly as a 160 kDa 7S monomer that has a concentration of 70–400 mg/dl in adults. It is thought to function mainly along mucosal surfaces where it is the most prominent immunoglobulin isotype.[215] Along the mucosa of the gastrointestinal tract, respiratory tract, mammary or other glands, IgA is secreted from plasma cells as a dimer attached by a small 12 kDa polypeptide called joining (J) chain. It then migrates toward the epithelium where it combines with a 60 kDa glycoprotein called secretory component, which is made by the surface epithelium. This is secreted into the mucosal lumen where it prevents pathogenic microorganisms and their toxic products from entering and attaching to the surface epithelium. It is a poor activator of complement and does not opsonize for phagocytosis. Its role in the serum is unclear, although some studies suggest it may be able to collaborate with killer cells in cytotoxic reactions.[224,225]

Isolated IgA deficiency is the most common of all congenital immunodeficiencies, occurring in 1 of 133 blood donors.[220] Although there is usually a compensatory increase of IgM along the mucosal surfaces, these patients have a significantly increased incidence of autoimmune disease and allergy. They may also be subject to severe life-threatening anaphylactic reactions to blood products if they have developed a hypersensitivity to the IgA from previous transfusions. When these patients require intravenous immunoglobulin therapy, preparations of IgA-depleted immunoglobulin have proved to be safer than standard γ-globulin preparations.[226] In normal concentration, IgA is too diffuse to be seen by serum protein electrophoresis. However, in patients with IgA myeloma, or with cirrhosis, which results in a polyclonal increase in IgA (Chapter 5), a discrete increase in the β- and β–γ zone (β–γ bridging), respectively, is seen.

IMMUNOGLOBULIN E

Except in the extremely rare case of an IgE myeloma, IgE is normally present in insufficient quantity to be seen by serum protein electrophoresis. It is the immunoglobulin class that is elevated in atopic conditions and in parasitic infestation. In allergies, the Cε3 portion of antigen-specific IgE attaches to two specific receptors on mast cells in the mucosa and skin.[227–229] When the specific allergen attaches to these cells, it causes an immediate release of the mast cell products that are responsible for the acute symptoms seen. Reagenic antibodies such as IgE may play an important role in host defense against parasitic infections.[230–233]

IMMUNOGLOBULIN D

Similar to IgE, IgD is normally present in quantities too small to be seen by serum protein electrophoresis. While having only a minor defined role in the serum (possibly related to immunological memory), IgD is one of the major immunoglobulin classes found on the surface membrane of B lymphocytes (Chapter 6).[215] One interesting feature of IgD myeloma is that whereas most cases of IgG and IgA myeloma contain κ light chain, IgD myelomas have a clear predominance of λ with a κ/λ ratio reported in studies variously from 1:2 to 1:9.[234] The concentration of IgD is considerably increased in patients with IgD myeloma, yet the M-protein spike typically is smaller than seen with IgG or IgA myeloma.[235]

A polyclonal increase in IgD occurs as part of the hyper-IgD syndrome. In this condition, patients (usually children) suffer from recurrent febrile episodes, lymph node enlargement and occasionally arthritis.[236–239] These patients have IgD levels above 150 IU/ml. Among these patients, the κ/λ ratio is usually elevated (often > 5.0), along with a polyclonal increase of IgG and decreased IgM.[240]

REFERENCES

1. Hamilton JA, Benson MD. Transthyretin: a review from a structural perspective. *Cell Mol Life Sci* 2001;58:1491–1521.
2. Kanda Y, Goodman DS, Canfield RE, Morgan FJ. The amino acid sequence of human plasma prealbumin. *J Biol Chem* 1974;**249**:6796–6805.
3. Ferguson RN, Edelhoch H, Saroff HA, Robbins J, Cahnmann HJ. Negative cooperativity in the binding of thyroxine to human serum prealbumin. Preparation of tritium-labeled 8-anilino-1- naphthalenesulfonic acid. *Biochemistry* 1975;14:282–289.
4. (NC-IUB) NCoI. IUB-IU-PAC Joint Commission on Biochemical Nomenclature. *J Biol Chem* 1981;**256**:12–14.
5. Dickson PW, Aldred AR, Marley PD, Tu GF, Howlett GJ, Schreiber G. High prealbumin and transferrin mRNA levels in the choroid plexus of rat brain. *Biochem Biophys Res Commun* 1985;**127**:890–895.
6. Ingenbleek Y, Young V. Transthyretin (prealbumin) in health and disease: nutritional implications. *Annu Rev Nutr.* 1994;**14**:495–533.
7. Bernstein LH, Leukhardt-Fairfield CJ, Pleban W, Rudolph R. Usefulness of data on albumin and prealbumin concentrations in determining effectiveness of nutritional support. *Clin Chem* 1989;**35**:271–274.
8. Jolliff C. Case studies in electrophoresis. *Beckman Puzzler* 1991;**6**:11.
9. Cornwell GG 3rd, Sletten K, Olofsson BO, Johansson B, Westermark P. Prealbumin: its association with amyloid. *J Clin Pathol* 1987;**40**:226–231.
10. Benson MD. Amyloidosis. In: Scriver CR, Beaudet AL, Sly WS, et al., eds. *The metabolic basis of inherited disease.* Columbus: McGraw-Hill; 2001.
11. Nichols WC, Dwulet FE, Liepnieks J, Benson MD. Variant apolipoprotein AI as a major constituent of a human hereditary amyloid. *Biochem Biophys Res Commun* 1988;**156**:762–768.

12. Bergquist J, Andersen O, Westman A. Rapid method to characterize mutations in transthyretin in cerebrospinal fluid from familial amyloidotic polyneuropathy patients by use of matrix-assisted laser desorption/ionization time-of-flight mass spectrometry. *Clin Chem* 2000;**46**:1293–1300.

13. Laurell CB. Composition and variation of the gel electrophoretic fractions of plasma, cerebrosinal fluid and urine. *Scand J Clin Lab Invest Suppl* 1972;**124**:71–82.

14. Yeung CY, Fung YS, Sun DX. Capillary electrophoresis for the determination of albumin binding capacity and free bilirubin in jaundiced neonates. *Semin Perinatol* 2001;**25**:50–54.

15. Fung YS, Sun DX, Yeung CY. Capillary electrophoresis for determination of free and albumin-bound bilirubin and the investigation of drug interaction with bilirubin-bound albumin. *Electrophoresis* 2000;**21**:403–410.

16. Zhang B, Fung Y, Lau K, Lin B. Bilirubin–human serum albumin interaction monitored by capillary zone electrophoresis. *Biomed Chromatogr* 1999;**13**:267–271.

17. Tarnoky AL. Genetic and drug-induced variation in serum albumin. *Adv Clin Chem* 1980;**21**:101–146.

18. Arvan DA, Blumberg BS, Melartin L. Transient 'bisalmuminemia' induced by drugs. *Clin Chim Acta* 1968;**22**:211–218.

19. Daniels JC. Carrier protein abnormalities. In: Ritzmann SE, Daniels JC, eds. *Serum protein abnormalities, diagnostic and clinical aspects.* Boston: Little, Brown & Co., 1975.

20. Mendler MH, Corbinais S, Sapey T, et al. In patients with cirrhosis, serum albumin determination should be carried out by immunonephelometry rather than by protein electrophoresis. *Eur J Gastroenterol Hepatol* 1999;**11**:1405–1411.

21. Chew ST, Fitzwilliam J, Indridason OS, Kovalik EC. Role of urine and serum protein electrophoresis in evaluation of nephrotic-range proteinuria. *Am J Kidney Dis* 1999;**34**:135–139.

22. Levinson SS. Urine protein electrophoresis and immunofixation electrophoresis supplement one another in characterizing proteinuria. *Ann Clin Lab Sci* 2000;**30**:79–84.

23. Lessard F, Bannon P, Lepage R, Joly JG. Light chain disease and massive proteinuria. *Clin Chem* 1985;**31**:475–477.

24. Cacoub P, Sbai A, Toan SV, Bellanger J, Hoang C, Godeau P, Piette JC. Collagenous colitis. A study of 11 cases. *Ann Med Interne (Paris)* 2001;**152**:299–303.

25. Schmidt PN, Blirup-Jensen S, Svendsen PJ, Wandall JH. Characterization and quantification of plasma proteins excreted in faeces from healthy humans. *Scand J Clin Lab Invest* 1995;**55**:35–45.

26. Bologa RM, Levine DM, Parker TS, Cheigh JS, Serur D, Stenzel KH, Rubin AL. Interleukin-6 predicts hypoalbuminemia, hypocholesterolemia, and mortality in hemodialysis patients. *Am J Kidney Dis* 1998;**32**:107–114.

27. Mackiewicz A, Schooltink H, Heinrich PC, Rose-John S. Complex of soluble human IL-6-receptor/IL-6 up-regulates expression of acute-phase proteins. *J Immunol* 1992;**149**:2021–2027.

28. Odamaki M, Kato A, Takita T, et al. Role of soluble receptors for tumor necrosis factor alpha in the development of hypoalbuminemia in hemodialysis patients. *Am J Nephrol* 2002;**22**:73–80.

29. Holowachuk EW. Register of Analbuminemia Cases 2001. www.albumin.org

30. Gossi B, Kleinert D, Gossi U. A further case of analbuminemia. *Schweiz Med Wochenschr* 2000;**130**:583–589.

31. Benhold H, Klaus D, Schluren PG. Volumen regulation and renal function in analbuminemia. *Lancet* 1960;**ii**:1169–1170.

32. Platt HS, Barron N, Giles AF, Midgley JE, Wilkins TA. Thyroid-function indices in an analbuminemic subject being treated with thyroxin for hypothyroidism. *Clin Chem* 1985;**31**:341–342.

33. Dammacco F, Miglietta A, D'Addabbo A, Fratello A, Moschetta R, Bonomo L.

Analbuminemia: report of a case and review of the literature. *Vox Sang* 1980;**39**:153–161.

34. Kallee E. Bennhold's analbuminemia: a follow-up study of the first two cases (1953–1992). *J Lab Clin Med* 1996;**127**:470–480.

35. Minchiotti L, Campagnoli M, Rossi A, et al. A nucleotide insertion and frameshift cause albumin Kenitra, an extended and *O*-glycosylated mutant of human serum albumin with two additional disulfide bridges. *Eur J Biochem* 2001;**268**:344–352.

36. Petersen CE, Ha CE, Harohalli K, Park DS, Bhagavan NV. Familial dysalbuminemic hyperthyroxinemia may result in altered warfarin pharmacokinetics. *Chem Biol Interact* 2000;**124**:161–172.

37. Jaeggi-Groisman SE, Byland C, Gerber H. Improved sensitivity of capillary electrophoresis for detection of bisalbuminemia. *Clin Chem* 2000;**46**:882–883.

38. Huss K, Putnam FW, Takahashi N, Takahashi Y, Weaver GA, Peters T Jr. Albumin Cooperstown: a serum albumin variant with the same (313 Lys–Asn) mutation found in albumins in Italy and New Zealand. *Clin Chem* 1988;**34**:183–187.

39. Madison J, Arai K, Sakamoto Y, et al. Genetic variants of serum albumin in Americans and Japanese. *Proc Natl Acad Sci USA* 1991;**88**:9853–9857.

40. Takahashi N, Takahashi Y, Isobe T, et al. Amino acid substitutions in inherited albumin variants from Amerindian and Japanese populations. *Proc Natl Acad Sci USA* 1987;**84**:8001–8005.

41. Wolf M, Kronenberg H, Dodds A, et al. A safety study of Albumex 5, a human albumin solution produced by ion exchange chromatography. *Vox Sang* 1996;**70**:198–202.

42. Rifai N, Bachorik PS, Albers JJ. Lipids, lipoproteins, and apolipoproteins. In: Burtis CA, Ashwood ER, eds. *Tietz textbook of clinical chemistry*. Philadelphia: WB Saunders, 1999.

43. Oram JF. Molecular basis of cholesterol homeostasis: lessons from Tangier disease and ABCA1. *Trends Mol Med* 2002;**8**:168–173.

44. Lampreave F, Gonzalez-Ramon N, Martinez-Ayensa S, Hernandez MA, Lorenzo HK, Garcia-Gil A, Pineiro A. Characterization of the acute phase serum protein response in pigs. *Electrophoresis* 1994;**15**:672–676.

45. Feher J, Romics L, Jakab L, Feher E, Szilvasi I, Papp G. Serum lipids and lipoproteins in chronic liver disease. *Acta Med Acad Sci Hung* 1976;**33**:217–223.

46. Lopez D, Sanchez MD, Shea-Eaton W, McLean MP. Estrogen activates the high-density lipoprotein receptor gene via binding to estrogen response elements and interaction with sterol regulatory element binding protein–1A. *Endocrinology* 2002;**143**:2155–2168.

47. Jula A, Marniemi J, Huupponen R, Virtanen A, Rastas M, Ronnemaa T. Effects of diet and simvastatin on serum lipids, insulin, and antioxidants in hypercholesterolemic men: a randomized controlled trial. *JAMA* 2002;**287**:598–605.

48. Muckle TJ. The 'post-heparin pattern' in routine serum protein immunoelectrophoresis. *Am J Clin Pathol* 1977;**67**:450–454.

49. Pearson JP, Keren DF. The effects of heparin on lipoproteins in high-resolution electrophoresis of serum. *Am J Clin Pathol* 1995;**104**:468–471.

50. Su LD, Keren DF. The effects of exogenous free fatty acids on lipoprotein migration in serum high-resolution electrophoresis: addition of free fatty acids improves visualization of normal and abnormal alpha1-antitrypsin. *Am J Clin Pathol* 1998;**109**:262–267.

51. Khan H, Salman KA, Ahmed S. Alpha-1 antitrypsin deficiency in emphysema. *J Assoc Physicians India* 2002;**50**:579–582.

52. Carrell RW, Lomas DA. Alpha$_1$-antitrypsin deficiency – a model for conformational diseases. *N Engl J Med* 2002;**346**:45–53.

53. Hibbetts K, Hines B, Williams D. An overview of proteinase inhibitors. *J Vet Intern Med* 1999;**13**:302–308.

54. Lomas DA, Lourbakos A, Cumming SA, Belorgey D. Hypersensitive mousetraps, alpha$_1$-antitrypsin deficiency and dementia. *Biochem Soc Trans* 2002;**30**:89–92.

55. Weinberger SE. Recent advances in pulmonary medicine (1). *N Engl J Med* 1993;**328**:1389–1397.

56. Keren DF, Di Sante AC, Bordine SL. Densitometric scanning of high-resolution electrophoresis of serum: methodology and clinical application. *Am J Clin Pathol* 1986;85:348–352.

57. Gonzalez-Sagrado M, Lopez-Hernandez S, Martin-Gil FJ, Tasende J, Banuelos MC, Fernandez-Garcia N, Arranz-Pena ML. Alpha$_1$-antitrypsin deficiencies masked by a clinical capillary electrophoresis system (CZE 2000). *Clin Biochem* 2000;33:79–80.

58. Malfait R, Gorus F, Sevens C. Electrophoresis of serum protein to detect alpha$_1$-antitrypsin deficiency: five illustrative cases. *Clin Chem* 1985;31:1397–1399.

59. Mullins RE, Miller RL, Hunter RL, Bennett B. Standardized automated assay for functional alpha 1-antitrypsin. *Clin Chem* 1984;30:1857–1860.

60. Jeppsson JO, Einarsson R. Genetic variants of alpha 1-antitrypsin and hemoglobin typed by isoelectric focusing in preselected narrow pH gradients with PhastSystem. *Clin Chem* 1992;38:577–580.

61. Berninger RW, Teixeira MF. Alpha 1-antitrypsin: the effect of anticoagulants on the trypsin inhibitory capacity, concentration and phenotype. *J Clin Chem Clin Biochem* 1985;23:277–281.

62. von Ahsen N, Oellerich M, Schutz E. Use of two reporter dyes without interference in a single-tube rapid-cycle PCR: alpha(1)-antitrypsin genotyping by multiplex real-time fluorescence PCR with the LightCycler. *Clin Chem* 2000;46:156–161.

63. Ortiz-Pallardo ME, Zhou H, Fischer HP, Neuhaus T, Sachinidis A, Vetter H, Bruning T, Ko Y. Rapid analysis of alpha1-antitrypsin PiZ genotype by a real-time PCR approach. *J Mol Med* 2000;78:212–216.

64. Rieger S, Riemer H, Mannhalter C. Multiplex PCR assay for the detection of genetic variants of alpha1-antitrypsin. *Clin Chem* 1999;45: 688–690.

65. Aslanidis C, Nauck M, Schmitz G. High-speed detection of the two common alpha(1)-

antitrypsin deficiency alleles Pi*Z and Pi*S by real-time fluorescence PCR and melting curves. *Clin Chem* 1999;45:1872–1875.

66. Hammerberg G, Keren DF. Polymerase chain reaction-mediated site directed mutagenesis detection of Z and S alpha-1-antitrypsin alleles in family members. *J Clin Lab Anal* 1996;10: 384–388.

67. Pierce JA, Eradio B, Dew TA. Antitrypsin phenotypes in St Louis. *JAMA* 1975;231: 609–612.

68. Teckman JH, Burrows J, Hidvegi T, Schmidt B, Hale PD, Perlmutter DH. The proteasome participates in degradation of mutant alpha-1-antitrypsin Z in the endoplasmic reticulum of hepatoma-derived hepatocytes. *J Biol Chem* 2001;276:44865–44872.

69. Morse JO. Alpha1-antitrypsin deficiency (first of two parts). *N Engl J Med* 1978;299: 1045–1048.

70. Morse JO. Alpha1-antitrypsin deficiency (second of two parts). *N Engl J Med* 1978;299: 1099–1105.

71. Morse JO, Lebowitz MD, Knudson RJ, Burrows B. Relation of protease inhibitor phenotypes to obstructive lung diseases in a community. *N Engl J Med* 1977;296:1190–1194.

72. Elliott PR, Stein PE, Bilton D, Carrell RW, Lomas DA. Structural explanation for the deficiency of S alpha 1-antitrypsin. *Nat Struct Biol* 1996;3:910–911.

73. Lewis JH, Iammarino RM, Spero JA, Hasiba U. Antithrombin Pittsburgh: an alpha1-antitrypsin variant causing hemorrhagic disease. *Blood* 1978;51:129–137.

74. Brennan SO, Sheat JM, Aiach M. Circulating proalbumin associated with a second case of antitrypsin Pittsburgh. *Clin Chim Acta* 1993;214:123–128.

75. Ghishan FK, Greene HL. Liver disease in children with PiZZ alpha 1-antitrypsin deficiency. *Hepatology* 1988;8:307–310.

76. Iezzoni JC, Gaffey MJ, Stacy EK, Normansell DE. Hepatocytic globules in end-stage hepatic disease: relationship to alpha1-antitrypsin phenotype. *Am J Clin Pathol* 1997;107:692–697.

77. Lima LC, Matte U, Leistner S, et al. Molecular analysis of the Pi*Z allele in patients with liver disease. *Am J Med Genet* 2001;**104**:287–290.

78. Lomas DA, Evans DL, Finch JT, Carrell RW. The mechanism of Z alpha 1-antitrypsin accumulation in the liver. *Nature* 1992;**357**:605–607.

79. Sveger T. Liver disease in alpha1-antitrypsin deficiency detected by screening of 200,000 infants. *N Engl J Med* 1976;**294**:1316–1321.

80. Lilley GL, Keren DF. A case of alpha1-antitrypsin deficiency. *Tech Sample Clin Immunol* 1986:C1–3.

81. Jack C, Evens CC. Three cases of alpha 1-antitrypsin deficiency in the elderly. *Postgrad Med J* 1991;**67**:840–842.

82. Seersholm N. Epidemiology of emphysema in subjects with severe alpha 1-antitrypsin deficiency. *Dan Med Bull* 2002;**49**:145–158.

83. Lohr HF, Schlaak JF, Dienes HP, Lorenz J, Meyer zum Buschenfelde KH, Gerken G. Liver cirrhosis associated with heterozygous alpha-1-antitrypsin deficiency type Pi MS and autoimmune features. *Digestion* 1995;**56**:41–45.

84. Graziadei IW, Joseph JJ, Wiesner RH, Therneau TM, Batts KP, Porayko MK. Increased risk of chronic liver failure in adults with heterozygous alpha1-antitrypsin deficiency. *Hepatology.* 1998;**28**:1058–1063.

85. Hutchison DC, Tobin MJ, Cook PJ. Alpha 1-antitrypsin deficiency: clinical and physiological features in heterozygotes of Pi type SZ. A survey by the British Thoracic Association. *Br J Dis Chest* 1983;**77**:28–34.

86. Tobin MJ, Cook PJ, Hutchison DC. Alpha 1-antitrypsin deficiency: the clinical and physiological features of pulmonary emphysema in subjects homozygous for Pi type Z. A survey by the British Thoracic Association. *Br J Dis Chest* 1983;**77**:14–27.

87. Larsson C, Dirksen H, Sundstrom G, Eriksson S. Lung function studies in asymptomatic individuals with moderately (Pi SZ) and severely (Pi Z) reduced levels of alpha1-antitrypsin. *Scand J Respir Dis* 1976;**57**:267–280.

88. Piitulainen E, Tornling G, Eriksson S. Effect of age and occupational exposure to airway irritants on lung function in non-smoking individuals with alpha 1-antitrypsin deficiency (PiZZ). *Thorax.* 1997;**52**:244–248.

89. Dahl M, Nordestgaard BG, Lange P, Vestbo J, Tybjaerg-Hansen A. Molecular diagnosis of intermediate and severe alpha(1)-antitrypsin deficiency: MZ individuals with chronic obstructive pulmonary disease may have lower lung function than MM individuals. *Clin Chem* 2001;**47**:56–62.

90. Klaassen CH, de Metz M, van Aarssen Y, Janssen J. Alpha(1)-antitrypsin deficiency as a result of compound heterozygosity for the Z and M(Heerlen) alleles. *Clin Chem* 2001;**47**:978–979.

91. Canva V, Piotte S, Aubert JP, et al. Heterozygous M3M$_{malton}$ alpha1-antitrypsin deficiency associated with end-stage liver disease: case report and review. *Clin Chem* 2001;**47**:1490–1496.

92. Lee JH, Brantly M. Molecular mechanisms of alpha1-antitrypsin null alleles. *Respir Med* 2000;**94**(Suppl C):S7–11.

93. Cook L, Janus ED, Brenton S, Tai E, Burdon J. Absence of alpha-1-antitrypsin (Pi Null Bellingham) and the early onset of emphysema. *Aust N Z J Med* 1994;**24**:263–269.

94. Sandhaus RA. Alpha(1)-antitrypsin deficiency therapy : pieces of the puzzle. *Chest* 2001;**119**:676–678.

95. Wencker M, Fuhrmann B, Banik N, Konietzko N. Longitudinal follow-up of patients with alpha(1)-protease inhibitor deficiency before and during therapy with IV alpha(1)-protease inhibitor. *Chest* 2001;**119**:737–744.

96. Crystal RG, Brantly ML, Hubbard RC, Curiel DT, States DJ, Holmes MD. The alpha 1-antitrypsin gene and its mutations. Clinical consequences and strategies for therapy. *Chest* 1989;**95**:196–208.

97. Sveger T, Thelin T. A future for neonatal alpha1-antitrypsin screening? *Acta Paediatr* 2000;**89**:628–631.

98. Abboud RT, Ford GT, Chapman KR. Alpha1-

antitrypsin deficiency: a position statement of the Canadian Thoracic Society. *Can Respir J* 2001;8:81–88.

99. Mullins CD, Huang X, Merchant S, Stoller JK. The direct medical costs of alpha(1)-antitrypsin deficiency. *Chest* 2001;**119**:745–752.

100. Alkins SA, O'Malley P. Should health-care systems pay for replacement therapy in patients with alpha(1)-antitrypsin deficiency? A critical review and cost-effectiveness analysis. *Chest* 2000;**117**:875–880.

101. Ritchie RF, Palomaki GE, Neveux LM, Navolotskaia O, Ledue TB, Craig WY. Reference distributions for the positive acute phase serum proteins, alpha1-acid glycoprotein (orosomucoid), alpha1-antitrypsin, and haptoglobin: a practical, simple, and clinically relevant approach in a large cohort. *J Clin Lab Anal* 2000;**14**:284–292.

102. Ritchie RF, Palomaki GE, Neveux LM, Navolotskaia O. Reference distributions for the positive acute phase proteins, alpha1-acid glycoprotein (orosomucoid), alpha1-antitrypsin, and haptoglobin: a comparison of a large cohort to the world's literature. *J Clin Lab Anal* 2000;**14**:265–270.

103. Lundholt BK, Madsen MW, Lykkesfeldt AE, Petersen OW, Briand P. Characterization of a nontumorigenic human breast epithelial cell line stably transfected with the human estrogen receptor (ER) cDNA. *Mol Cell Endocrinol* 1996;**119**:47–59.

104. Jeppson JO, Laurell CB, Franzen B. Agarose gel electrophoresis. *Clin Chem* 1979;**25**:629–638.

105. Leppert PC, Yu SY. Effect of parturition on serum alpha 1-antiprotease (alpha 1-antitrypsin). *Clin Chem* 1993;**39**:905–906.

106. Johnson AM, Rohlfs EM, Silverman LM. Chapter 20. Proteins. In: Burtis CA, Ashwood, ER, eds. *Tietz textbook of clinical chemistry*. Philadelphia: WB Saunders, 1999.

107. Chen RJ, Chen CK, Chang DY, et al. Immunoelectrophoretic differentiation of alpha-fetoprotein in disorders with elevated serum alpha-fetoprotein levels or during pregnancy. *Acta Oncol* 1995;**34**:931–935.

108. Henriksen HJ, Petersen MU, Pedersen FB. Serum alpha-1-acid glycoprotein (orosomucoid) in uremic patients on hemodialysis. *Nephron* 1982;**31**:24–26.

109. Philip AG, Hewitt JR. Alpha 1-acid glycoprotein in the neonate with and without infection. *Biol Neonate* 1983;**43**:118–124.

110. Mariotti R, Musumeci G, De Carlo M, et al. Acute-phase reactants in acute myocardial infarction: impact on 5-year prognosis. *Ital Heart J* 2001;**2**:294–300.

111. Billingsley GD, Walter MA, Hammond GL, Cox DW. Physical mapping of four serpin genes: alpha 1-antitrypsin, alpha 1-antichymotrypsin, corticosteroid-binding globulin, and protein C inhibitor, within a 280-kb region on chromosome I4q32.1. *Am J Hum Genet* 1993;**52**:343–353.

112. Lilja H, Cockett AT, Abrahamsson PA. Prostate specific antigen predominantly forms a complex with alpha 1-antichymotrypsin in blood. Implications for procedures to measure prostate specific antigen in serum. *Cancer* 1992;**70**:230–234.

113. Eriksson S, Lindmark B, Lilja H. Familial alpha 1-antichymotrypsin deficiency. *Acta Med Scand* 1986;**220**:447–453.

114. Miyake H, Hara S, Nomi M, Arakawa S, Kamidono S, Hara I. Value of prostate specific antigen alpha1-antichymotrypsin complex for the detection of prostate cancer in patients with a PSA level of 4.1–10.0 ng/mL: comparison with PSA-related parameters. *Int J Urol* 2001;**8**:589–593.

115. Burtin P, Grabar P. Nomenclature and identification of the normal human serum proteins. In: Bier M, ed. *Electrophoresis theory, methods and applications*. New York and London: Academic Press, 1989.

116. Thomas RM, Schiano TD, Kueppers F, Black M. Alpha1-antichymotrypsin globules within hepatocytes in patients with chronic hepatitis C and cirrhosis. *Hum Pathol* 2000;**31**:575–577.

117. Abraham CR. Reactive astrocytes and alpha1-antichymotrypsin in Alzheimer's disease. *Neurobiol Aging* 2001;**22**:931–936.

118. Petersen CM. Alpha 2-macroglobulin and pregnancy zone protein. Serum levels, alpha 2-macroglobulin receptors, cellular synthesis and aspects of function in relation to immunology. *Dan Med Bull* 1993;**40**:409–446.

119. Sottrup-Jensen L. Role of internal thiol esters in the alpha-macroglobulin-proteinase binding mechanism. *Ann N Y Acad Sci* 1994;**737**:172–187.

120. Gouin-Charnet A, Laune D, Granier C, et al. alpha2-Macroglobulin, the main serum antiprotease, binds beta2-microglobulin, the light chain of the class I major histocompatibility complex, which is involved in human disease. *Clin Sci (Lond)* 2000;**98**:427–433.

121. Grinnell F, Zhu M, Parks WC. Collagenase-1 complexes with alpha2-macroglobulin in the acute and chronic wound environments. *J Invest Dermatol* 1998;**110**:771–776.

122. Kurdowska AK, Geiser TK, Alden SM, et al. Activity of pulmonary edema fluid interleukin-8 bound to alpha(2)-macroglobulin in patients with acute lung injury. *Am J Physiol Lung Cell Mol Physiol* 2002;**282**:L1092–1098.

123. Legres LG, Pochon F, Barray M, Gay F, Chouaib S, Delain E. Evidence for the binding of a biologically active interleukin-2 to human alpha 2-macroglobulin. *J Biol Chem* 1995;**270**:8381–8384.

124. Binder RJ, Karimeddini D, Srivastava PK. Adjuvanticity of alpha 2-macroglobulin, an independent ligand for the heat shock protein receptor CD91. *J Immunol* 2001;**166**:4968–4972.

125. DeStefano A, Hoffman M. The effect of alpha 2 macroglobulin–proteinase complexes on macrophage Ia expression *in vivo*. *Immunol Invest* 1991;**20**:33–43.

126. Osada T, Ookata K, Athauda SB, Takahashi K, Ikai A. The active site titration of proteinases by using alpha 2-macroglobulin and high-performance liquid chromatography. *Anal Biochem* 1992;**207**:76–79.

127. Strickland DK, Ashcom JD, Williams S, Burgess WH, Migliorini M, Argraves WS. Sequence identity between the alpha 2-macroglobulin receptor and low density lipoprotein receptor-related protein suggests that this molecule is a multifunctional receptor. *J Biol Chem* 1990;**265**:17401–17404.

128. Borth W. Alpha 2-macroglobulin, a multifunctional binding protein with targeting characteristics. *FASEB J* 1992;**6**:3345–3353.

129. Koster MN, Dermaut B, Cruts M, et al. The alpha2-macroglobulin gene in AD: a population-based study and meta-analysis. *Neurology* 2000;**55**:678–684.

130. Zill P, Burger K, Behrens S, et al. Polymorphisms in the alpha-2 macroglobulin gene in psychogeriatric patients. *Neurosci Lett.* 2000;**294**:69–72.

131. Wang X, Luedecking EK, Minster RL, Ganguli M, DeKosky ST, Kamboh MI. Lack of association between alpha2-macroglobulin polymorphisms and Alzheimer's disease. *Hum Genet* 2001;**108**:105–108.

132. Cliver SP, Goldenberg RL, Neel NR, Tamura T, Johnston KE, Hoffman HJ. Neonatal cord serum alpha 2-macroglobulin and fetal size at birth. *Early Hum Dev* 1993;**33**:201–206.

133. James K, Merriman J, Gray RS, Duncan LJ, Herd R. Serum alpha 2-macroglobulin levels in diabetes. *J Clin Pathol* 1980;**33**:163–166.

134. Brissenden JE, Cox DW. alpha 2-Macroglobulin in patients with obstructive lung disease, with and without alpha 1-antitrypsin deficiency. *Clin Chim Acta* 1983;**128**:241–248.

135. Langlois MR, Delanghe JR. Biological and clinical significance of haptoglobin polymorphism in humans. *Clin Chem* 1996;**42**:1589–1600.

136. Giblett ER. Recent advances in heptoglobin and transferrin genetics. *Bibl Haematol* 1968;**29**:10–20.

137. Janik B. *High resolution electrophoresis and immunofixation of serum proteins on cellulosic media.* Ann Arbor: Gelman Sciences, 1985.

138. van Lente F. The diagnostic utility of haptoglobin. *Clin Immunol Newsl* 1991;**11**:135–138.

139. Daniels JC, Larson DL, Abston S, Ritzmann SE. Serum protein profiles in thermal burns. II. Protease inhibitors, complement factors, and c-reactive protein. *J Trauma* 1974;**14**:153–162.

140. Daniels JC, Larson DL, Abston S, Ritzmann SE. Serum protein profiles in thermal burns. I. Serum electrophoretic patterns, immunoglobulins, and transport proteins. *J Trauma* 1974;14:137–152.

141. Bowman BH. Haptoglobin. In: Bowman BH, ed. *Hepatic plasma proteins.* San Diego: Academic Press, 1993.

142. Smithies O. Zone electrophoresis in starch gels: group variations in the serum proteins of normal human adults. *Biochem J* 1955;61:629–641.

143. Nakhoul FM, Marsh S, Hochberg I, Leibu R, Miller BP, Levy AP. Haptoglobin genotype as a risk factor for diabetic retinopathy. *JAMA* 2000;284:1244–1245.

144. Koch W, Latz W, Eichinger M, et al. Genotyping of the common haptoglobin Hp 1/2 polymorphism based on PCR. *Clin Chem* 2002;48:1377–1382.

145. Kasvosve I, Gomo ZA, Gangaidzo IT, et al. Reference range of serum haptoglobin is haptoglobin phenotype-dependent in blacks. *Clin Chim Acta* 2000;296:163–170.

146. Fitzmaurice M, Valenzuela R, Winkelman EI. Serum protein electrophoresis pattern associated with decreased haptoglobin as a poor prognostic indicator in severe liver disease. *Am J Clin Pathol* 1989;91:365(Abstr).

147. Baumann H, Onorato V, Gauldie J, Jahreis GP. Distinct sets of acute phase plasma proteins are stimulated by separate human hepatocyte-stimulating factors and monokines in rat hepatoma cells. *J Biol Chem* 1987;262:9756–9768.

148. Baumann H, Richards C, Gauldie J. Interaction among hepatocyte-stimulating factors, interleukin 1, and glucocorticoids for regulation of acute phase plasma proteins in human hepatoma (HepG2) cells. *J Immunol* 1987;139:4122–4128.

149. Ritzmann SE, Daniels JC. Serum electrophoresis and total serum proteins. In: Ritzmann SE, Daniels JC, eds. *Serum protein abnormalities: diagnostic and clinical aspects.* New York: Alan R Liss, 1982.

150. Wallach J. *Interpretation of diagnostic tests.* Philadelphia: Lippincott Williams & Wilkins, 2000.

151. Blanco A, Solis P, Guisasola JA, Arranz E, Telleria JJ, Blanco C. Absence of fibronectin in a 40-day-old child who died as a result of septicemia with disseminated intravascular coagulation. *Sangre (Barc)* 1989;34:59–62.

152. Bevilacqua MP, Amrani D, Mosesson MW, Bianco C. Receptors for cold-insoluble globulin (plasma fibronectin) on human monocytes. *J Exp Med* 1981;153:42–60.

153. Makogonenko E, Tsurupa G, Ingham K, Medved L. Interaction of fibrin(ogen) with fibronectin: further characterization and localization of the fibronectin-binding site. *Biochemistry* 2002;41:7907–7913.

154. Chavarria ME, Lara-Gonzalez L, Gonzalez-Gleason A, Sojo I, Reyes A. Maternal plasma cellular fibronectin concentrations in normal and preeclamptic pregnancies: a longitudinal study for early prediction of preeclampsia. *Am J Obstet Gynecol* 2002;187:595–601.

155. Sobel RA, Mitchell ME. Fibronectin in multiple sclerosis lesions. *Am J Pathol* 1989;135:161–168.

156. Kamboh MI, Ferrell RE. Human transferrin polymorphism. *Hum Hered* 1987;37:65–81.

157. Kasvosve I, Delanghe JR, Gomo ZA, et al. Transferrin polymorphism influences iron status in blacks. *Clin Chem* 2000;46:1535–1539.

158. Wallach J. *Interpretation of diagnostic tests.* Philadelphia: Lippincott Williams & Wilkins, 2000.

159. Matsubara M, Odagaki E, Morioka T, Nakagawa K. The clinical significance of the measurement of plasma transferrin as a growth factor. II. The changes in various endocrine status. *Nippon Naibunpi Gakkai Zasshi* 1987;63:669–674.

160. Smithies O. Variations in serum beta-globulins. *Nature* 1957;180:1482–1483.

161. Beckman G, Beckman L, Sikstrom C. Transferrin C subtypes in different ethnic groups. *Hereditas* 1980;92:189–192.

162. Giblett E, Hickman C, Smithies O. Serum transferrins. *Nature* 1959;183:1589–1590.

163. Stibler H. The normal cerebrospinal fluid proteins identified by means of thin-layer

isoelectric focusing and crossed immunoelectrofocusing. *J Neurol Sci* 1978;36:273–288.

164. Stibler H, Allgulander C, Borg S, Kjellin KG. Abnormal microheterogeneity of transferrin in serum and cerebrospinal fluid in alcoholism. *Acta Med Scand* 1978;204:49–56.

165. van Eijk HG, van Noort WL, Kroos MJ, van der Heul C. Analysis of the iron-binding sites of transferrin by isoelectric focussing. *J Clin Chem Clin Biochem* 1978;16:557–560.

166. Arndt T. Carbohydrate-deficient transferrin as a marker of chronic alcohol abuse: a critical review of preanalysis, analysis, and interpretation. *Clin Chem* 2001;47:13–27.

167. Stibler H. Carbohydrate-deficient transferrin in serum: a new marker of potentially harmful alcohol consumption reviewed. *Clin Chem* 1991;37:2029–2037.

168. Royse VL, Greenhill E, Morley CG, Jensen DM. Microheterogeneity of human transferrin as revealed by agarose gel electrophoresis with an iron-specific stain. *Clin Chem* 1986;32:1983.

169. Wuyts B, Delanghe JR, Kasvosve I, Wauters A, Neels H, Janssens J. Determination of carbohydrate-deficient transferrin using capillary zone electrophoresis. *Clin Chem* 2001;47: 247–255.

170. Nandapalan V, Watson ID, Swift AC. Beta-2-transferrin and cerebrospinal fluid rhinorrhoea. *Clin Otolaryngol* 1996;21:259–264.

171. Ryall RG, Peacock MK, Simpson DA. Usefulness of beta 2-transferrin assay in the detection of cerebrospinal fluid leaks following head injury. *J Neurosurg* 1992;77:737–739.

172. Zaret DL, Morrison N, Gulbranson R, Keren DF. Immunofixation to quantify beta 2-transferrin in cerebrospinal fluid to detect leakage of cerebrospinal fluid from skull injury. *Clin Chem* 1992;38:1908–1912.

173. Keir G, Zeman A, Brookes G, Porter M, Thompson EJ. Immunoblotting of transferrin in the identification of cerebrospinal fluid otorrhoea and rhinorrhoea. *Ann Clin BioChem* 1992;29(Pt 2):210–213.

174. Tripathi RC, Millard CB, Tripathi BJ, Noronha A. Tau fraction of transferrin is present in human aqueous humor and is not unique to cerebrospinal fluid. *Exp Eye Res* 1990;50: 541–547.

175. Sloman AJ, Kelly RH. Transferrin allelic variants may cause false positives in the detection of cerebrospinal fluid fistulae. *Clin Chem* 1993;39: 1444–1445.

176. Hershko C, Bar-Or D, Gaziel Y, et al. Diagnosis of iron deficiency anemia in a rural population of children. Relative usefulness of serum ferritin, red cell protoporphyrin, red cell indices, and transferrin saturation determinations. *Am J Clin Nutr* 1981;34:1600–1610.

177. Arranz-Pena ML, Gonzalez-Sagrado M, Olmos-Linares AM, Fernandez-Garcia N, Martin-Gil FJ. Interference of iodinated contrast media in serum capillary zone electrophoresis. *Clin Chem* 2000;46:736–737.

178. Bossuyt X, Peetermans WE. Effect of piperacillin-tazobactam on clinical capillary zone electrophoresis of serum proteins. *Clin Chem* 2002;48:204–205.

179. Jurczyk K, Wawrzynowicz-Syczewska M, Boron-Kaczmarska A, Sych Z. Serum iron parameters in patients with alcoholic and chronic cirrhosis and hepatitis. *Med Sci Monit* 2001;7:962–965.

180. Kalender B, Mutlu B, Ersoz M, Kalkan A, Yilmaz A. The effects of acute phase proteins on serum albumin, transferrin and haemoglobin in haemodialysis patients. *Int J Clin Pract* 2002;56:505–508.

181. Hromec A, Payer J Jr, Killinger Z, Rybar I, Rovensky J. Congenital atransferrinemia. *Dtsch Med Wochenschr* 1994;119:663–666.

182. Hamill RL, Woods JC, Cook BA. Congenital atransferrinemia. A case report and review of the literature. *Am J Clin Pathol* 1991;96: 215–218.

183. Rifai N, Bachorik P, Albers J. Chapter 25. Lipids, lipoproteins, and apolipoproteins. In: Burtis CA, Ashwood, E.R., ed. *Tietz textbook of clinical chemistry.* Philadelphia: WB Saunders, 1999.

184. Wieme R. *Agar gel electrophoresis.* Amsterdam: Elsevier, 1965.

185. Barnum SR, Fey G, Tack BF. Biosynthesis and genetics of C3. *Curr Top Microbiol Immunol* 1990;**153**:23–43.

186. Tack BF, Morris SC, Prahl JW. Third component of human complement: structural analysis of the polypeptide chains of C3 and C3b. *Biochemistry* 1979;**18**:1497–1503.

187. McLean RH, Bryan RK, Winkelstein J. Hypomorphic variant of the slow allele of C3 associated with hypocomplementemia and hematuria. *Am J Med* 1985;**78**:865–868.

188. Taschini P, Addison N. 'Pseudomonoclonal' band caused by genetic polymorphism of the third component of complement. *ASCP Check Sample Immunopathol* 1988:1.

189. Platel D, Bernard A, Mack G, Guiguet M. Interleukin 6 upregulates TNF-alpha-dependent C3-stimulating activity through enhancement of TNF-alpha specific binding on rat liver epithelial cells. *Cytokine* 1996;**8**:895–899.

190. Tateda K, Matsumoto T, Yamaguchi K. Acute induction of interleukin-6 and biphasic changes of serum complement C3 by carrageenan in mice. *Mediators Inflamm* 1998;**7**:221–223.

191. Platel D, Guiguet M, Briere F, Bernard A, Mack G. Human interleukin-6 acts as a co-factor for the up-regulation of C3 production by rat liver epithelial cells. *Eur Cytokine Netw* 1994;**5**:405–410.

192. Katz Y, Revel M, Strunk RC. Interleukin 6 stimulates synthesis of complement proteins factor B and C3 in human skin fibroblasts. *Eur J Immunol* 1989;**19**:983–988.

193. Hirumi E, Safarian Z, Blessum C. Effect of sample age on electrophoretic resolution of C3 complement. *Clin Chem* 1995;**41**:S150(Abstr).

194. Smalley DL, Mayer R, Gardner C. Evaluation of capillary zone electrophoresis assessment of beta proteins. *Clin Lab Sci* 1999;**12**:262–265.

195. Blanchong CA, Zhou B, Rupert KL, et al. Deficiencies of human complement component C4A and C4B and heterozygosity in length variants of RP-C4-CYP21-TNX (RCCX) modules in caucasians. The load of RCCX genetic diversity on major histocompatibility complex-associated disease. *J Exp Med* 2000;**191**:2183–2196.

196. Fan Q, Uring-Lambert B, Weill B, Gautreau C, Menkes CJ, Delpech M. Complement component C4 deficiencies and gene alterations in patients with systemic lupus erythematosus. *Eur J Immunogenet* 1993;**20**:11–21.

197. Painter PC, Cope JY, Smith JL. Chapter 50. Reference information for the clinical laboratory. In: Burtis CA, Ashwood ER, ed. *Tietz textbook of clinical chemistry*. Philadelphia: WB Saunders, 1999.

198. Handin RI, Laposata M. Case 2 – 1993 – A 72-year-old woman with a coagulopathy and bilateral thigh masses. *N Engl J Med* 1993;**328**: 121–128.

199. Register LJ, Keren DF. Hazard of commercial antiserum cross-reactivity in monoclonal gammopathy evaluation. *Clin Chem* 1989;**35**: 2016–2017.

200. Pepys MB. C-reactive protein fifty years on. *Lancet* 1981;**1**:653–657.

201. Miettinen AK, Heinonen PK, Laippala P, Paavonen J. Test performance of erythrocyte sedimentation rate and C-reactive protein in assessing the severity of acute pelvic inflammatory disease. *Am J Obstet Gynecol* 1993;**169**:1143–1149.

202. Teisala K, Heinonen PK. C-reactive protein in assessing antimicrobial treatment of acute pelvic inflammatory disease. *J Reprod Med* 1990;**35**: 955–958.

203. Abdelmouttaleb I, Danchin N, Ilardo C, et al. C-Reactive protein and coronary artery disease: additional evidence of the implication of an inflammatory process in acute coronary syndromes. *Am Heart J* 1999;**137**:346–351.

204. Lehtinen M, Laine S, Heinonen PK, et al. Serum C-reactive protein determination in acute pelvic inflammatory disease. *Am J Obstet Gynecol* 1986;**154**:158–159.

205. Lacour AG, Gervaix A, Zamora SA, et al. Procalcitonin, IL-6, IL-8, IL-1 receptor antagonist and C-reactive protein as identificators of serious bacterial infections in children with fever without localising signs. *Eur J Pediatr* 2001;**160**:95–100.

206. Gendrel D, Raymond J, Coste J, et al.

Comparison of procalcitonin with C-reactive protein, interleukin 6 and interferon-alpha for differentiation of bacterial vs. viral infections. *Pediatr Infect Dis J* 1999;18:875–881.

207. Grutzmeier S, Sandstrom E. C-reactive protein levels in HIV complicated by opportunistic infections and infections with common bacterial pathogens. *Scand J Infect Dis* 1999;31: 229–234.

208. Aufweber E, Jorup-Ronstrom C, Edner A, Hansson LO. C-reactive protein sufficient as screening test in bacterial vs. viral infections. *J Infect* 1991;23:216–220.

209. Sheldon J, Riches P, Gooding R, Soni N, Hobbs JR. C-reactive protein and its cytokine mediators in intensive-care patients. *Clin Chem* 1993;39: 147–150.

210. Magadle R, Weiner P, Sotzkover A, Berar-Yanay N. C-reactive protein and vascular disease. *Isr Med Assoc J* 2001;3:50–52.

211. Canova CR, Courtin C, Reinhart WH. C-reactive protein (CRP) in cerebro-vascular events. *Atherosclerosis.* 1999;147:49–53.

212. Ridker PM, Cushman M, Stampfer MJ, Tracy RP, Hennekens CH. Plasma concentration of C-reactive protein and risk of developing peripheral vascular disease. *Circulation* 1998;97:425–428.

213. Jacquiaud C, Bastard JP, Delattre J, Jardel C. C reactive protein: a band mistaken as monoclonal gammapathy in acute inflammatory disease. *Ann Med Interne (Paris)* 1999;150:357–358.

214. Yu A, Pira U. False increase in serum C-reactive protein caused by monoclonal IgM-lambda: a case report. *Clin Chem Lab Med* 2001;39: 983–987.

215. Frazer JK, Capra JD. Chapter 3. Immunoglobulins: structure and function. In: Paul WE, ed. *Fundamental immunology.* Philadelphia: Lippincott-Raven, 1999.

216. Bradwell AR, Carr-Smith HD, Mead GP, et al. Highly sensitive, automated immunoassay for immunoglobulin free light chains in serum and urine. *Clin Chem* 2001;47:673–680.

217. Katzmann JA, Clark RJ, Abraham RS, et al. Serum reference intervals and diagnostic ranges for free kappa and free lambda immunoglobulin light chains: relative sensitivity for detection of monoclonal light chains. *Clin Chem* 2002;48: 1437–1444.

218. Taschini PA, Addison NM. An atypical electrophoretic pattern due to a selective polyclonal increase of IgG4 subclass. *ASCP Check Sample Immunopathol* 1991:IP91–95.

219. Batuman V, Sastrasinh M, Sastrasinh S. Light chain effects on alanine and glucose uptake by renal brush border membranes. *Kidney Int* 1986;30:662–665.

220. Schiff R, Schiff SE. Chapter 11. Flow cytometry for primary immunodeficiency diseases. In: Keren DF, McCoy JJP, Carey JL, eds. *Flow cytometry in clinical diagnosis.* Chicago: ASCP Press, 2001.

221. Kyle RA. Classification and diagnosis of monoclonal gammopathies. In: Rose NR, ed. *Manual of clinical laboratory immunology.* Washington, DC: American Society for Microbiology, 1986.

222. Levinson SS, Keren DF. Immunoglobulins from the sera of immunologically activated persons with pairs of electrophoretic restricted bands show a greater tendency to aggregate than normal. *Clin Chim Acta* 1989;182:21–30.

223. Keren DF, Morrison N, Gulbranson R. Evolution of a monoclonal gammopathy (MG) documented by high-resolution electrophoresis (HRE) and immunofixation (IFE). *Lab Med* 1994; 25:313–317.

224. Tagliabue A, Boraschi D, Villa L, et al. IgA-dependent cell-mediated activity against enteropathogenic bacteria: distribution, specificity, and characterization of the effector cells. *J Immunol* 1984;133:988–992.

225. Tagliabue A, Nencioni L, Villa L, Keren DF, Lowell GH, Boraschi D. Antibody-dependent cell-mediated antibacterial activity of intestinal lymphocytes with secretory IgA. *Nature* 1983;306:184–186.

226. Cunningham-Rundles C, Zhou Z, Mankarious S, Courter S. Long-term use of IgA-depleted intravenous immunoglobulin in immunodeficient subjects with anti-IgA antibodies. *J Clin Immunol* 1993;13:272–278.

227. Yamaguchi M, Sayama K, Yano K, et al. IgE enhances Fc epsilon receptor I expression and IgE-dependent release of histamine and lipid mediators from human umbilical cord blood-derived mast cells: synergistic effect of IL-4 and IgE on human mast cell Fc epsilon receptor I expression and mediator release. *J Immunol* 1999;**162**:5455–5465.

228. Pawankar R, Ra C. IgE-Fc epsilonRI-mast cell axis in the allergic cycle. *Clin Exp Allergy* 1998;**28**(Suppl 3):6–14.

229. Sandor M, Lynch RG. The biology and pathology of Fc receptors. *J Clin Immunol* 1993;**13**:237–246.

230. MacGlashan D Jr. Anti-IgE antibody therapy. *Clin Allergy Immunol* 2002;**16**:519–532.

231. Imai S, Tezuka H, Fujita K. A factor of inducing IgE from a filarial parasite prevents insulin-dependent diabetes mellitus in nonobese diabetic mice. *Biochem Biophys Res Commun* 2001;**286**:1051–1058.

232. Selassie FG, Stevens RH, Cullinan P, et al. Total and specific IgE (house dust mite and intestinal helminths) in asthmatics and controls from Gondar, Ethiopia. *Clin Exp Allergy* 2000;**30**:356–358.

233. Gounni AS, Lamkhioued B, Ochiai K, et al. High-affinity IgE receptor on eosinophils is involved in defence against parasites. *Nature* 1994;**367**:183–186.

234. Corte G, Tonda P, Cosulich E, et al. Characterization of IgD. I. Isolation of two molecular forms from human serum. *Scand J Immunol* 1979;**9**:141–149.

235. Sinclair D, Cranfield T. IgD myeloma: a potential missed diagnosis. *Ann Clin Biochem* 2001;**38**:564–565.

236. de Dios Garcia-Diaz J, Alvarez-Blanco MJ. High IgD could be a nonpathogenetic diagnostic marker of the hyper-IgD and periodic fever syndrome. *Ann Allergy Asthma Immunol* 2001;**86**:587.

237. Summaries for patients. Can genetics help diagnose the hyper-IgD and periodic fever syndrome. *Ann Intern Med* 2001;**135**:S–36.

238. Picco P, Gattorno M, Di Rocco M, Buoncompagni A. Non-steroidal anti-inflammatory drugs in the treatment of hyper-IgD syndrome. *Ann Rheum Dis* 2001;**60**:904.

239. de Hullu JA, Drenth JP, Struyk AP, van der Meer JW. Hyper-IgD syndrome and pregnancy. *Eur J Obstet Gynecol Reprod Biol* 1996;**68**:223–225.

240. Haraldsson A, Weemaes CM, de Boer AW, Bakkeren JA. Clinical and immunological follow-up in children with hyper-IgD syndrome. *Immunodeficiency* 1993;**4**:63–65.

5

Approach to pattern interpretation in serum

As with any clinical laboratory method, control serum samples must be run daily. Commercially available lyophilized controls are convenient and ensure adequate migration of the sample. Note that C3 may not provide a discrete band in some lyophilized preparations, altering the appearance of the β-region. Densitometric scanning or electropherograms (for capillary zone electrophoresis) of control serum provide objective criteria to determine whether the percentage of the major bands is within an acceptable range. Although some laboratories interpret gel-based serum protein electrophoresis without densitometric scanning, the information available from densitometry is useful in quality control, confirming impressions from direct visual examination of the gel and following patients with monoclonal gammopathies (M-proteins).[1] Below is a brief list of the approaches I recommend when interpreting protein electrophoretic patterns.

1. Use all the information available, both clinical and laboratory.
2. Use a consistent logical approach.
3. Know the basics (Chapter 4).
4. Know the artifacts: application problems, hemolysis, incomplete clotting, drug-binding effects, improper storage, radiocontrast dye effect on capillary zone electrophoresis (CZE), improperly labeled specimens, and contaminated specimens.
5. With gel-based methods use a densitometric scan for adjunctive information, but not as the sole means to evaluate a specimen.
6. When encountering a pattern you do not understand, call the clinician for more clinical information.

INITIAL PROCESSING OF THE SAMPLE

When the patient's blood sample is received in the laboratory with a request for serum protein electrophoresis, the serum should be allowed to clot, separated from the clot and then refrigerated. The longer the sample sits at room temperature (with the many neutrophil proteases being released), the more some key analytes (such as complement) deteriorate. The sample should be checked to be sure the correct procedures have been followed in special circumstances. For example, if the clinician has requested that the sample be examined for a cryoglobulin, be sure the sample has been drawn into a prewarmed syringe, transported to the laboratory at 37°C and then centrifuged at 37°C. If the sample was not handled properly, redraw the patient.[2]

Previous electrophoretic results are of great importance in following patients with M-proteins. Therefore, as part of our initial specimen processing, we check the laboratory file for any previous electrophoretic studies performed on the patient. These files, together with the present material, are reviewed by the pathologist at the time of sign-out. It is recommended that clinicians provide pertinent information on the requisition slip. Even in a hospital laboratory situation, however, it is often difficult to obtain history; in a reference laboratory (off-site) the lack of history can be frustrating. This is why the person reviewing the electrophoretic material should be aggressive about calling clinicians when an unusual pattern occurs, or when the present material is inconsistent with previous reports. For example, if a patient has had an M-protein documented by serum protein electrophoresis and immunological studies (characterizing the M-protein) and the present sample shows no evidence of a monoclonal process, one must be very skeptical as to the correct identification of one of the samples (how sure are you that the original was the correct sample?). This requires contacting the clinician and obtaining another sample, as well as a little detective work to find out where the sample mix-up occurred.

OVERVIEW OF THE ELECTROPHORETIC STRIP

On gel-based techniques, the basic information is in the gel itself. The densitometric scan provides useful objective measurable data. Therefore, when examining agarose or acetate gels, I prefer to examine the gel itself first to review the overall migration and staining of the bands. A study of the densitometric values for the control serum ensures reliability for these values for the patient samples of that run.

A consistent approach in examining serum

protein electrophoretic gels helps to avoid missing subtle alterations. Obvious findings should be noted and discussed (Fig. 5.1). Migration should be consistent from one sample to the next. By reading down the strip, comparing the same protein band from sample-to-sample, the examiner, in effect, uses adjacent samples as visual controls. This allows one to more easily note changes in concentration or subtle migration differences, such as anodal slurring of albumin caused by drug or bilirubin binding. After reviewing albumin, I proceed cathodally, observing the α_1-region for increase or decrease in concentration and alteration of electrophoretic mobility of the α_1-antitrypsin band (Fig. 5.2). By reading down the gel, one is more likely to detect a double α_1-band or a shift in the migration of that band as will be seen with α_1-antitrypsin Pittsburgh (see Chapter 4) than if one reads across a specimen on the gel.

In the α_2- and β-regions, monoclonal gammo-

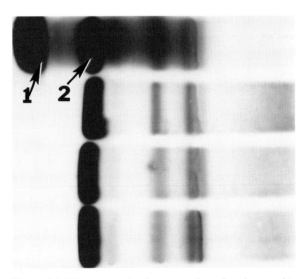

Figure 5.1 When comparing the top sample to the others on this gel, there is an obvious problem. The albumin band (1) migrates much too far toward the anode. There is a hazy zone between it and a second large band (2). Further, there is a densely staining haze between the other bands and the γ-region stains very lightly. This is due to a double application to this gel. One was at the normal location, and the other was anodal to it. The first albumin band (1) results from the anodal application and the second (2) results from the correct application. Bisalbuminemia would not have this great a separation and would not have the haze between the bands. (Paragon SPE2 system stained with Paragon Violet.)

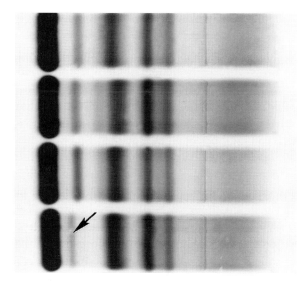

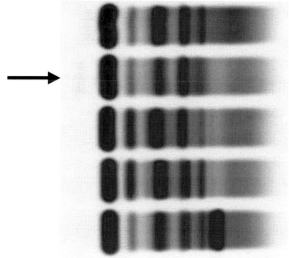

(a)

Figure 5.2 By reading down the bands of this gel, one is struck by the α_1-antitrypsin band in the bottom lane (arrow). It stains more weakly than those above it. Further, it has a much greater anodal migration than the others. This is due to a homozygous PiFF (fast) (typed by isoelectric focusing; clinically, they do not develop liver or respiratory disease). Note that there is a slight difference in the migrations of the α_1-antitrypsin bands in the three top samples because of the different M isoforms which have no clinical significance. (Paragon SPE2 system stained with Paragon Violet.)

pathies may be cloaked by the presence of other bands. One advantage of procedures that provide crisp discrimination of the β_1- and β_2-regions is that one can recognize the usual appearance of transferrin and C3 in the many normal cases examined. However, to take proper advantage of this improved resolution, I believe the interpreter must read down the gel, comparing the same band in many samples. When there is a change in the size or shape of either band, performing an immunofixation is recommended to be certain that a monoclonal gammopathy is not present. Figure 5.3 shows the serum protein electrophoresis from the 2001-SPE02 survey sample from the College of American Pathologists. Because of the crisp resolution of the technique used, and by comparing the C3 bands in the five cases shown, there is a subtle distortion of the band in the second sample (survey sample). It is broader and more diffuse than the sharp C3 bands above and below it. This automatically results in an immunofixation in our

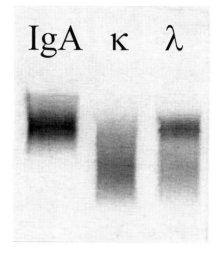

(b)

Figure 5.3 (a) Five serum protein electrophoresis patters are shown. The second sample is from the College of American Pathologists (CAP) Survey 2001-SPE02. Reading across this pattern, it may look normal. However, if you look at the C3 band (β_2) in all five samples I think you will note a difference in the indicated sample. I have not put an arrow on this band because it would draw too much attention to it. The other four C3 bands are sharp distinct bands. This one is not. It is a faintly staining broad band. My laboratory always checks this band by performing an immunofixation. (Sebia $\beta_{1,2}$ gel.) (b) The immunofixation in this case for IgA, κ and λ demonstrates an IgA λ monoclonal gammopathy. (Sebia immunofixation.)

laboratory that demonstrated the IgA λ monoclonal gammopathy. Both high-resolution gel-based methods and CZE methods lend themselves to quantify the β₁- and β₂-regions thereby providing objective evidence that an abnormality is present.

After I have compared each band on the strip, I briefly turn the gel 90° and look at them again. This only takes a minute (since I have already formed an impression about any alterations seen). In addition to picking up several β-region monoclonal gammopathies, this examination technique improves detection of α₁-antitrypsin variants, transferrin variants, or oligoclonal banding. I know of no objective justification for this; perhaps an analogy is when one turns one's head to the side when looking at some paintings to try to achieve a different perspective. Try it.

With CZE, there are no readily available methods that compare with reading down the electrophoretic gel. Since each sample is processed individually through a small capillary, there will be subtle variations in migration from one sample to the next. However, CZE has the advantage of not having to rely upon a protein stain like Amido Black or Ponceau S. These dyes are taken up to varying extents depending on the degree of glycosylation of the protein. Therefore, the specific α₁-antitrypsin band can be distinguished from the α₁-acid glycoprotein (orosomucoid) band with this technique. Alterations in migration, or duplicate

α₁-antitrypsin bands can be detected by paying close attention to this region of the electropherogram.

With regard to detecting subtle β-region monoclonal gammopathies, CZE offers high-resolution and crisp transferrin and C3 bands. By paying close attention to the sharpness of these two analytes, one can pick up small monoclonal gammopathies by detecting changes in and between these two peaks. Figure 5.4 demonstrates the same subtle monoclonal gammopathy by CZE that was shown in Fig. 5.3. Although one cannot read down the gel, both currently available CZE systems provide a control overlay that allows one to compare the crispness of the bands. Even immunosubtraction on such a subtle case is able to identify the monoclonal gammopathy as an IgA λ (Fig. 5.5).

INTERPRETATION OF THE INDIVIDUAL PATIENT'S SAMPLE

After the entire strip has been examined and abnormalities noted, the pattern for each patient is reviewed and compared with the densitometric scan. When an abnormality such as an increase in the α₁-region is suggested by the initial review, it is useful to have objective, quantitative corroboration by the densitometric scan. However, with gel-based methods I rely more on the visual inspection

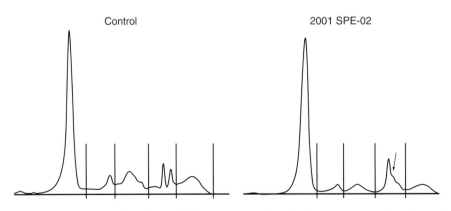

Figure 5.4 The same College of American Pathologists (CAP) Survey sample from Fig. 5.3 is demonstrated on this capillary zone electropherogram (right) with a control serum on the left. Here, I indicate the location of the monoclonal gammopathy (arrow). (Paragon CZE 2000.)

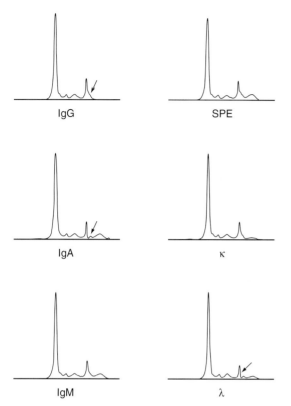

IgG   SPE

IgA   κ

IgM   λ

Figure 5.5 This is the immunosubtraction on the same survey sample as in Fig. 5.3 (College of American Pathologists (CAP) Survey 2001-SPE02). In this case the monoclonal gammopathy is seen as a bridge (note the arrow in the IgG lane) between the normal sized transferrin band and the small C3 band. The bridge is only absent in the IgA and λ lanes, indicating that this is an IgA λ monoclonal gammopathy. (Paragon CZE 2000.)

of the gels than the densitometric measurements, because the eye is more sensitive to subtle alterations in migration and staining intensity than is the densitometric instrumentation presently available for gels. In contrast, the electropherograms of CZE are a direct representation of the information from the procedure. With both CZE and some densitometers, one can alter the magnification and pick up tiny band shifts barely visualized on inspection of the strip (Fig. 5.6).

When the electrophoretic strip, densitometer information, appended clerical information, and our laboratory file for that patient have been reviewed, an interpretation is made for each sample. We have developed coded phrases for many commonly observed patterns and have entered these into the laboratory computer (Table 5.1). Please feel free to use or modify any of these interpretations for your laboratory. I do not force unusual diagnoses into these pigeon holes and occasionally write a report about any unique finding.

In some cases, interpretation cannot be completed until the clinician has been contacted. I am quick to contact clinicians for a variety of reasons: when I have a pattern that is unusual and does not fit into a typical disease pattern, or when there has been a clinically significant change in the pattern from a previous sample, such as the loss of an M-protein. I also have my client services department contact the clinician to cancel needless repetitions of serum protein electrophoresis (often, several individuals order tests on a patient, or there is a misunderstanding of what can be learned from serial protein electrophoresis studies). Canceling such needless tests has always been good laboratory practice, but because the laboratory is now viewed as a cost center, this is an absolute necessity for cost-efficient management.

SERUM PATTERN DIAGNOSIS

Pattern recognition is an excellent screening technique for a wide variety of abnormalities, and for suggesting or confirming a clinical diagnosis. There are analogies to tissue pathology pattern recognition; for example, a collection of chronic inflammatory cells may suggest the diagnosis of tuberculosis, fungal infection, autoimmune disease, or lymphoma. To complete the diagnosis, it is necessary to correlate the pattern and clinical information with additional testing such as acid-fast stains, fungal stains, autoantibody testing, or lymphocyte surface marker assays. Similarly, the finding of an unusual electrophoretic pattern suggesting elevated hormone levels should be followed up to establish the cause, which may range from birth control pills or pregnancy to a steroid hormone-producing neoplasm. As always, understanding the clinical situation is critical to providing a useful interpretation. For example, including 'birth control pill effect' or 'pregnancy pattern' as

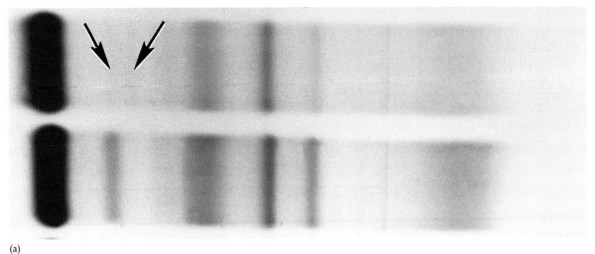

(a)

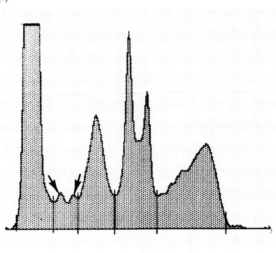

(b)

Figure 5.6 (a) The top sample shows two faintly stained bands in the α_1-region (arrows) compared to the normal α_1-region in the other serum. (Panagel stained with Amido Black.) (b) A densitometric scan indicates that the α_1-region abnormality observed in (a) can be seen (arrows) by enhancing the sensitivity of the densitometer.

part of the interpretation of the electrophoretic pattern for a middle-aged man (whose age and sex were not provided, but which one did not bother to look up) may be a source of humor on the clinical ward, but will do little to enhance the laboratory's credibility.

LIVER DISEASE PATTERNS

Cirrhosis

The electrophoretic changes that occur with advanced cirrhosis were among the earliest of the clinically relevant patterns described.[3–6] It is no surprise that the many proteins which the liver normally synthesizes will be decreased when there has been extensive loss or damage to hepatocytes. However, this is complicated by the fact that during early injury (for example, in hepatitis), some acute-phase reactants such as α_1-antitrypsin will be increased. Whether a particular analyte is increased or decreased will depend on the balance of inflammation (increased) versus the ever decreasing number of hepatocytes synthesizing the protein. The pattern consists of hypoalbuminemia (typically with anodal slurring due to the binding of bilirubin), increased (occasionally normal, or

Table 5.1 Serum protein electrophoresis interpretation templates

Normal pattern

There is a decrease in all parameters suggesting hemodilution.

One measured value is slightly outside the normal range. The pattern is otherwise within normal limits.

Decreased albumin.

There is blurring of the anodal margin of albumin. This change may be caused by binding of a drug or bilirubin to albumin. This may also be seen during heparin therapy where α_1-lipoprotein may migrate anodally to albumin.

Markedly reduced α_1-antitrypsin band. Recommend: α_1-antitrypsin measurement and phenotype studies to exclude congenital deficiency.

Abnormal α_1-antitrypsin band. Recommend: α_1-antitrypsin measurement and phenotype studies to exclude congenital deficiency.

Relatively low α_1-antitrypsin band. Recommend: α_1-antitrypsin measurement to exclude congenital deficiency.

Increased α_1- and α_2-globulins, decreased albumin and transferrin, consistent with acute inflammation.

Relative elevation of acute phase proteins suggesting a reactive condition.

Increased α_2-globulin and C3 band suggest subacute inflammation.

Increased acute-phase reactants and polyclonal increase in immunoglobulins suggest active, chronic inflammation.

Decreased haptoglobin with a slow α_2 band is likely due to hemoglobin–haptoglobin complex. This pattern suggests hemolysis. Hemolysis may occur as a result of venepuncture technique.

Haptoglobin is decreased. This is consistent with a hemolytic process or hereditary deficiency.

Increase in α_2-globulin consistent with estrogen effect.[a]

Increase in α_2- and β-globulins consistent with estrogen effect.[a]

The transferrin band is increased. If this is not due to estrogen effect, an evaluation for iron deficiency is warranted.[a]

Increased β_1-band may be due to increased transferrin in iron deficiency, or to the presence of an exogenous agent such as a radiocontrast dye. Recommend: correlate with history and iron studies if warranted.[b]

The pattern reveals β–γ bridging. This suggests an increase in IgA. This may occur in liver disease, intestinal or respiratory infections, and rheumatoid arthritis.

Decreased albumin and haptoglobin with a polyclonal increase in γ-globulins and β–γ bridging. This is consistent with liver disease.

Decreased total protein with low albumin and γ-globulin. This is a mild protein loss pattern.

Very low albumin, transferrin and γ-globulin with increased α_2-globulin. This protein loss pattern suggests renal disease (nephrotic syndrome) or gastrointestinal protein loss.

A small restriction is present at the application point on the gel. This may indicate the presence of a cryoglobulin. Recommend: evaluate serum for cryoglobulin on a new sample drawn in a 37°C tube and transported to the laboratory at 37°C.

Polyclonal increase in immunoglobulins is consistent with chronic inflammation.

Borderline polyclonal increase in γ-globulins suggesting a chronic reactive condition.

Table 5.1 – continued

There is a polyclonal increase in γ-globulins with oligoclonal banding. This pattern may be seen in a variety of conditions, most often chronic infections, occasionally autoimmune diseases and less commonly lymphoproliferative processes.

A few tiny oligoclonal bands are seen in the γ-region. These bands can be seen with infections and autoimmune diseases. Occasionally, they have been noted as part of lymphoproliferative processes.

One of the oligoclonal bands is more prominent than the rest. Recommend: urine protein electrophoresis now and repeat serum studies in 3–6 months.

Borderline hypogammaglobulinemia. This may be seen in some chronic lymphoproliferative conditions.

Hypogammaglobulinemia is present. In adults, this may reflect the presence of immune deficiency, chemotherapy or B-cell neoplasm. Recommendation: immunofixation of urine or quantification of serum free κ and λ chains to detect monoclonal free light chains (or Bence Jones protein). In addition, immunofixation of the serum and measurement of IgG, IgA, and IgM will help to rule out a monoclonal process.

A tiny restriction is seen in the γ-region. This may indicate the presence of a lymphoproliferative process. Recommend: immunofixation of urine or quantification of serum free κ and λ chains to detect monoclonal free light chains (or Bence Jones protein) and repeat serum electrophoresis in 6–12 months to see if the process regresses.

A monoclonal gammopathy is present. This indicates the presence of a lymphoproliferative process. Recommend: serum and urine immunofixation.

Recommend: immunofixation of urine or quantification of serum free κ and λ chains to detect monoclonal free light chains (or Bence Jones protein) if light chain disease is part of the differential diagnosis.

The monoclonal gammopathy is identified as _____ (specify type) and measures _____ (specify quantity measured by densitometry or electropherograms measurement).

The monoclonal gammopathy persists with no significant change from the previous study.

The monoclonal gammopathy persists significantly greater than on the previous study.

The monoclonal gammopathy persists significantly less than on the previous sample.

The normal γ-globulins appear suppressed.

To identify the monoclonal protein, please call (telephone number) to add an immunofixation to this sample. No redraw is necessary.

Unusual electrophoretic pattern indicates that the sample is denatured. Please resubmit.

Repeat immunofixation is not recommended to follow patients with previously characterized M-proteins.

Serum and 24-h urine electrophoresis with measurement of the M-protein will provide quantitative information to follow this patient. Alternatively, a serum measurement of serum free κ and λ chains will provide a useful assessment of free light chains associated with the M-protein.

---

[a]Use for women suspected of having estrogen effect.
[b]Use only for capillary zone electrophoresis.
These suggested sign-outs should be adjusted for the needs and communication style of each institution. They are not all inclusive. Often, I will combine different sign-outs as needed. Not infrequently, a narrative report is required when a sample does not fit into a particular category.

even low) α_1-antitrypsin, polyclonal hypergamma-globulinemia, beta-gamma bridging, and often, decreased haptoglobin (Figs 5.7 and 5.8).[7]

This relatively complex pattern is the result of several pathophysiological alterations. Synthesis of albumin and other proteins is affected by the number of hepatocytes remaining, their current state of health (damage by ethanol, toxins, or biological agents), and the nutritional and metabolic status of the individual resulting from diet and hormonal changes.[8-11] Clearly, when sufficient hepatocytes are damaged, synthesis of albumin will be decreased, but this is not necessarily the cause of hypoalbuminemia in most cases of cirrhosis.

Although considerable loss of hepatocytes has occurred in patients with cirrhosis and ascites, the large reserve capacity of the liver can result in a normal or even elevated synthesis of serum albumin in some of these individuals.[12] Therefore, hypoalbuminemia in many individuals results from the altered distribution of albumin due to the presence of ascites. Attempts to remove the ascites result in an absolute loss of albumin. It is not useful

to perform serum protein electrophoresis on ascites fluid.[13]

The complex pathophysiology of hypoalbuminemia prevents the interpreter from equating the level of albumin with prognosis.[14] When hyperbilirubinemia accompanies the liver disease, the bilirubin binds to albumin, causing an anodal slurring of this band (Figs 5.7 and 5.8). In addition to albumin, other hepatocyte-derived proteins including α_1-acid glycoprotein (orosomucoid) and haptoglobin are often decreased.[7] Whereas transthyretin (prealbumin) is decreased consistently in cirrhosis and serves as a sensitive indicator of the level of hepatocyte function,[15] serum protein electrophoresis is too insensitive for meaningful quantification of this band, even with the excellent resolution found with newer CZE techniques.

Although transferrin is often decreased in cirrhosis, this is difficult to detect on serum protein electrophoresis because of increased IgA. The β–γ bridging in cirrhosis results from a marked polyclonal increase in the level of IgA that migrates in

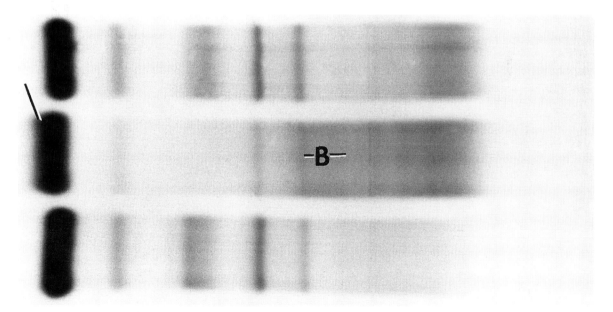

Figure 5.7 Center serum has a classic cirrhosis pattern. Albumin is slightly decreased and moved anodally (indicated) due to bilirubin binding. The α_2-region is quite low. Prominent β–γ bridging (–B–) is seen together with a polyclonal increase in γ-globulin. Note that the top sample has a diffuse polyclonal increase in the slow γ-region. The sample in the bottom lane is normal. (Panagel stained with Amido Black.)

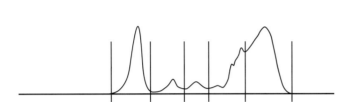

Fraction	Rel %		g/dl	
ALBUMIN	23.8	---	2.07	---
ALPHA 1	5.9		0.52	
ALPHA 2	5.1	–	0.45	
BETA	10.5	+++	1.01	+++
GAMMA	46.6	+++	4.06	+++

Reference ranges

	Rel %	g/dl
ALBUMIN	52.6 – 68.9	3.80 – 5.20
ALPHA 1	3.6 – 8.1	0.30 – 0.60
ALPHA 2	5.3 – 12.2	0.40 – 0.90
BETA	8.3 – 14.3	0.60 – 1.10
GAMMA	8.2 – 18.6	0.60 – 1.40

TP: 6.40 – 8.20 A/G: 1.20 – 2.20

Figure 5.8 This capillary zone electropherogram demonstrates the features of cirrhosis. Albumin is decreased and the band itself is broadened toward the anode, likely because of bilirubin binding. Although the α_2-region is in the normal range, it is at the lower end. There is a prominent β–γ bridging and a polyclonal increase in γ-globulin. (Paragon CZE 2000.)

this region and tends to obscure other β-region bands. There is also a considerable polyclonal increase in IgM and IgG levels that completes the 'bridge' and is responsible for the broad increase in γ-globulin. Overall, the increase in polyclonal immunoglobulins likely relates to an immune response against enteric antigens.[16–18] In addition, patients with accompanying cholestasis have elevated secretory IgA levels in serum.[19]

In addition to α_1-antitrypsin, α_2-macroglobulin may also increase in cirrhosis. Both may function to inhibit proteases released during the ongoing tissue damage in many of these patients. α_2-Macroglobulin helps restore serum oncotic pressure that declines due to hypoalbuminemia.[20] The α_2-macroglobulins may be increased as a result of the hyperestrogenic effect seen in patients with cirrhosis. There is also usually an effect on the lipoproteins in cirrhosis. α_1-Lipoprotein (HDL) levels are usually decreased while β_1-lipoprotein (LDL) levels are increased during biliary obstruction with jaundice.[7,21] However, these are not readily measured by serum protein electrophoresis. Biochemical assays for cholesterol and low-density lipoproteins will provide more precise information about serum lipids in those patients.

Hepatitis

During active hepatic injury, relevant changes of the serum protein electrophoresis pattern occur prior to the development of cirrhosis. However, these are too non-specific to be interpreted as a hepatitis pattern without supportive clinical and/or laboratory information.

Although considerable ongoing injury results from inflammation, the concentrations of the proteins associated with an acute inflammatory reaction pattern (discussed later) may be elevated or may be in the normal range because they are produced by the liver. Specifically, α_1-acid glycoprotein (orosomucoid), α_1-antichymotrypsin, and C-reactive protein are usually in the normal range while haptoglobin is either normal or reduced. Transferrin, which is usually decreased in the acute inflammatory reaction pattern, is in the normal or elevated range in most of these patients. Only α_1-antitrypsin follows the usual acute reaction pattern by being elevated in patients with active hepatic injury.[22] Obviously, some exceptions to these findings occur, such as occasional elevations of C-reactive protein (especially early in the disease).

Most patients with hepatitis have normal serum

albumin levels. However, an anodal slurring of albumin results from attachment of bilirubin (Figs 5.9 and 5.10). As in cirrhosis, the α_1-lipoproteins are usually decreased, but unlike cirrhosis the β_1-lipoproteins show no characteristic changes during hepatitis.

Examination of the γ-globulin region can supply useful information about hepatitis patients. A polyclonal gammopathy, occasionally with oligoclonal bands suggesting the presence of particular expansion of a few of the B-cell clones is often seen during clinically active hepatitis (Figs 5.9 and 5.10). The polyclonal increase in patients with chronic active hepatitis tends to be especially prominent in the slow γ-region. Patients with acute hepatitis usually have a polyclonal gammopathy initially that may become more pronounced if the disease progresses but usually regresses with clinical improvement.[22–24]

The polyclonal expansion may have two, three, or more small discrete bands that reflect an oligo-clonal gammopathy. These bands reflect the greater expansion of those particular B-cell clones and are further evidence of the polyclonal nature of the process. When immunofixation is performed on such a sample, both κ- and λ-bands will usually be identified. The polyclonal gammopathy is partly in response to the inciting agent (typically hepatitis B or C). The better prognosis, found in individuals whose polyclonal and oligoclonal gammopathy regresses, parallels recovery of normal hepatocyte structure with restoration of more normal intra-hepatic blood flow.[25,26] Some clonal expansions in the γ-region may be especially prominent, producing areas of restricted mobility in the slow γ-region that may be mistaken for monoclonal gammo-pathies.[27] Indeed, monoclonal gammopathies do occur with increased frequency in patients with hepatitis (see below). However, a small monoclonal gammopathy superimposed on a polyclonal pattern may reflect a transient process (Fig. 5.11).

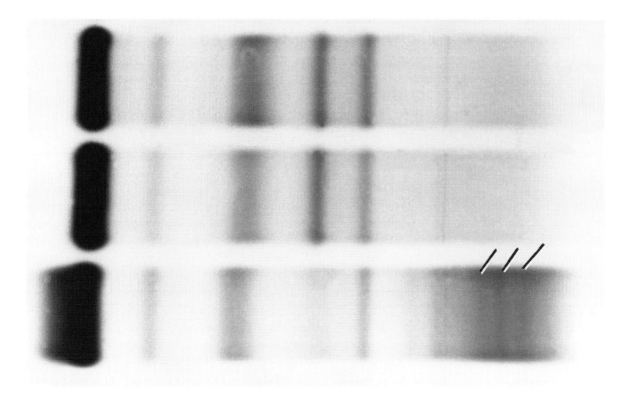

Figure 5.9 Bottom serum is from a patient with hepatitis. The anodal slurring is caused by bilirubin binding to albumin. In the slow γ-region, several prominent bands (indicated) are superimposed on a polyclonal (diffuse) increase in γ-globulin. There is no β–γ bridging. The top two lanes contain normal serum. (Panagel stained with Amido Black.)

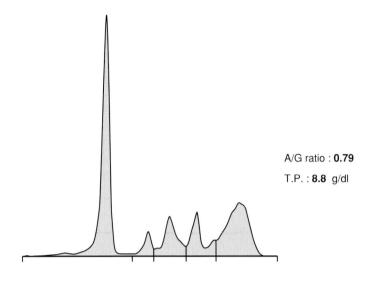

A/G ratio : **0.79**

T.P. : **8.8** g/dl

Serum protein electrophoresis

Fractions	%		Ref. %	Conc.	Ref. conc.
Albumin	44.0	<	45.3 – 67.7	3.9	3.4 – 5.2
Alpha 1	4.5		2.9 – 6.8	0.4	0.2 – 0.4
Alpha 2	12.2		6.2 – 14.9	1.1	0.5 – 1.0
Beta	11.4		8.1 – 18.0	1.0	0.6 – 1.1
Gamma	27.9	>	8.8 – 24.5	2.5	0.6 – 1.6

Figure 5.10 Capillary zone electropherogram from a patient with active hepatitis. The α_1-region is at the upper limit of normal and the concentration of the α_2-region is slightly increased. There is a prominent β–γ bridging with polyclonal and oligoclonal increase in the γ-region. The irregularity and slightly angled peaks indicate the oligoclonal banding. (Sebia Capillarys.)

No particular pattern will allow the interpreter to identify the causative agent of the hepatitis. This should be accomplished by appropriate serological and molecular investigations. The currently available cadre of assays for specific antibodies, antigens and virus-specific nucleic acid allows determination of the probable etiological agent, as well as often assisting the clinician in determining the infectivity of the patient. The most serious error I have seen in interpretation of these patterns has been to 'overcall' the polyclonal slow γ increase as a monoclonal gammopathy. Sometimes, aberrant bound (i.e. intact immunoglobulin) κ/λ ratios can develop in these patients, although the vast majority have a ratio close to 2:0.

Patients with hepatitis C virus are particularly prone to develop Type III cryoglobulinemia (see Chapter 6).[28,29] Some of these patients may go on to have one clone predominate and switch to a Type II cryoglobulin.[30] In one study, 11 per cent of patients with hepatitis C virus infection developed an M-protein.[31] The likelihood of this occurring increased when the disease was caused by hepatitis C genotype 2a/c.[31] If a large γ band develops, I recommend characterizing it immediately by immunofixation, testing the serum for the presence of a cryoglobulin and evaluating the urine for the presence of monoclonal free light chains (MFLC; see Chapter 7).

BILIARY OBSTRUCTION

Biliary obstruction is an end result of a wide variety of disease processes, which may involve the hepatic parenchyma proper, such as in primary biliary

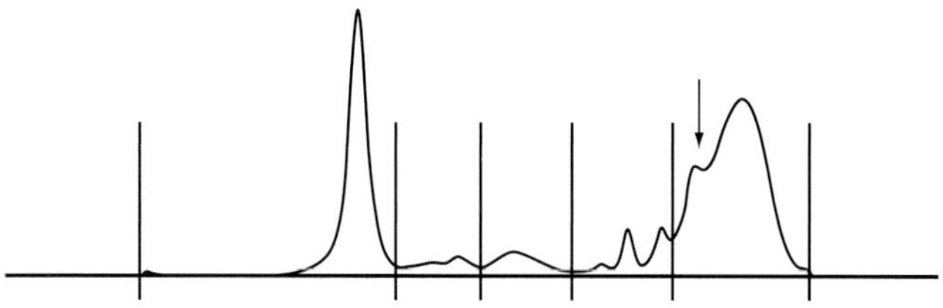

Fraction	Rel %	g/dl
ALBUMIN	29.1 ---	2.74 ---
ALPHA 1	4.6	0.43
ALPHA 2	5.5	0.52
BETA	7.7 –	0.72
GAMMA	53.0 +++	4.99 +++

Reference Ranges

	Rel %	g/dl
ALBUMIN	52.6 – 68.9	3.80 – 5.20
ALPHA 1	3.6 – 8.1	0.30 – 0.60
ALPHA 2	5.3 – 12.2	0.40 – 0.90
BETA	8.3 – 14.3	0.60 – 1.10
GAMMA	8.2 – 18.6	0.60 – 1.40

TP: 6.40 – 8.20 A/G: 1.20 – 2.20

Figure 5.11 Capillary zone electropherogram from a patient with hepatitis C infection. In addition to decreased albumin and a broad polyclonal increase in γ-globulin, a modest-sized monoclonal band is present (arrow). On cases with these features I note the presence of the monoclonal gammopathy superimposed on a polyclonal pattern. I recommend performing a urine test to rule out the presence of monoclonal free light chains, but caution the clinician that monoclonal gammopathies in the presence of a polyclonal pattern could indicate the presence of a transient, reactive process. Because of this, I recommend a repeat electrophoresis of serum in 3–6 months to see if the process resolves. (Paragon CZE 2000.)

cirrhosis, or which may result from external obstruction of the biliary tree by stones, inflammation, or neoplasm. The main effect on the serum protein electrophoresis pattern is anodal slurring of albumin because of its association with bilirubin. Additional irregularities may include acute-phase reaction, elevated C3, increased β_1-lipoprotein, decreased α-lipoprotein, and occasionally elevated γ-globulin, especially in patients with primary biliary cirrhosis (who usually have an elevated IgM level).[32–35] However these findings do not allow a specific diagnosis. Although the albumin band is broadened (slurred) toward the anode because of the binding of bilirubin, it is normal or only slightly decreased in concentration and IgA is only slightly elevated (therefore, no β–γ bridge is seen). These features are helpful in the differential diagnosis from the cirrhosis and hepatitis patterns.

RENAL DISEASE PATTERN

When renal disease is severe enough to produce the nephrotic syndrome, serum protein electrophoresis demonstrates a characteristic pattern consisting of a low serum albumin, a diffuse slurring of α_1-lipoprotein anodal to albumin (possibly due to heparin's activation of lipoprotein lipase; see Chapter 4), decreased or low normal α_1 globulin, and decreased γ-globulin. The α_2- and β-regions are often elevated (Figs 5.12 and 5.13).[36–38] Examining the selectivity index, however, provides better clinical information about ultimate outcome for the patients than the electrophoretic pattern of their serum proteins (see Chapter 7).[39]

The nephrotic pattern results from loss of serum proteins through the damaged nephron and the body's attempts to restore the oncotic pressure by overproduction of large proteins that do not pass through the damaged glomeruli. Nephrotic syndrome is defined as proteinuria (> 3 g/24 h.1.73 m^2 of body surface area), hypoproteinemia, edema, and hyperlipidemia. Considerably less proteinuria can produce a clinical picture of nephrosis, especially in children having a relapse of renal disease.[40] Electrophoresis of serum will only detect the severe cases of renal damage and, as mentioned above, is of little help in defining the specific site of damage in the

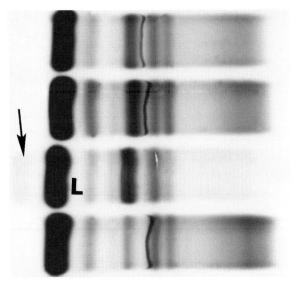

Figure 5.12 The third sample from the top is from a patient with the nephrotic syndrome. Albumin is decreased in concentration, although this is difficult to appreciate on this Paragon Violet-stained sample. There is anodal slurring of albumin with a faint haze (arrow) anodal to albumin due to heparin's activation of lipoprotein lipase. This liberates free fatty acids that bind to α_1-lipoprotein causing it to migrate anodal to albumin. Note the especially clear α_1-lipoprotein region (L) in this sample. The α_2-globulin is increased because of the large molecules (α_2-macroglobulin and haptoglobin) in this region. Once again, this increase is difficult to appreciate in this gel. In contrast, the γ-globulin region is decreased. (Paragon SPE2 gel stained with Paragon Violet.)

nephron. However, examination of the urine by routine urinalysis, study of specific proteins and/or electrophoresis of urine (see Chapter 7) can help to determine the site and extent of the renal damage.

Transthyretin, albumin, α_1-acid glycoprotein, α_1-antitrypsin, transferrin and smaller proteins are lost into the urine. Depending on the degree of damage, somewhat larger proteins such as IgG may also be lost. The loss of albumin and other proteins into the urine decreases the oncotic pressure, resulting in edema. As a compensatory mechanism, the synthesis of serum proteins is increased. Because they are relatively large proteins, α_2-macroglobulin and β_1-lipoprotein (see Chapter 4) do not readily pass through even moderately damaged glomeruli and are retained in the serum. Many such patients will have α_2-macroglobulin as the major serum protein. For this reason, dye-binding techniques give an inaccurate assessment of serum albumin levels in nephrotic patients. Dyes such as bromcresol green bind to α_2-macroglobulin as well as to albumin. Whereas in normal serum the binding of bromcresol green to α_2-macroglobulin is a relatively trivial false addition (about 5 per cent) to the quantification of albumin and a relatively constant factor, it results in a gross overestimation of the serum albumin in nephrotic patients.

Fraction	Rel %		g/dl	
ALBUMIN	39.2	---	2.00	---
ALPHA 1	8.8	+	0.45	
ALPHA 2	28.5	+++	1.45	+++
BETA	14.3		0.73	
GAMMA	9.2		0.47	---

Reference ranges

	Rel %	g/dl
ALBUMIN	52.6 – 68.9	3.80 – 5.20
ALPHA 1	3.6 – 8.1	0.30 – 0.60
ALPHA 2	5.3 – 12.2	0.40 – 0.90
BETA	8.3 – 14.3	0.60 – 1.10
GAMMA	8.2 – 18.6	0.60 – 1.40

TP: 6.40 – 8.20 A/G: 1.20 – 2.20

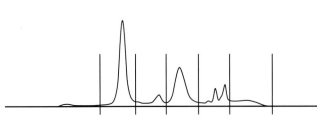

Figure 5.13 Capillary zone electropherogram demonstrates more clearly some of the quantitative features of a nephrotic pattern. Albumin and γ-globulin are decreased with a prominent increase in α_2-globulin. (Paragon CZE 2000.)

Elevated cholesterol in patients with nephrotic syndrome is directly related to the increased β_1-lipoprotein.[41,42] Because of the self-aggregation that occurs with β_1-lipoprotein at elevated concentrations (Chapter 4), its electrophoretic mobility is slowed and it appears as an irregular band cathodal to its usual location using gel-based methods.

Samples from patients receiving hemodialysis may be incompletely clotted by the heparin that they are given during their treatment. In addition to a band in the fibrinogen region, patients receiving heparin have the α_1-lipoprotein band migrating anodal to albumin and the β_1-lipoprotein band migrates in the α_2 region owing to activation of lipoprotein lipase and the binding of heparin to β_1-lipoprotein respectively (see Chapter 4).[43,44]

GASTROINTESTINAL PROTEIN LOSS

Damage to the gastrointestinal tract will affect the serum protein electrophoresis pattern in a variety of ways. Acute damage from invasive bacteria or acute exacerbation of inflammatory bowel disease produces an acute-phase reaction pattern. Similarly, dehydration due to agents such as *Vibrio cholera* or staphylococcal enterotoxin B will result in an increase in the concentration of all serum proteins. Patients with inflammatory bowel disease usually have a polyclonal increase in the γ-globulin region reflecting the chronic inflammatory nature of that condition. Also, depending on the extent and severity of the process, absorption of nutrients, including amino acids, may be impaired and serum protein may be lost into the lumen of the bowel. One cannot distinguish between ulcerative colitis and Crohn's disease by this technique.

Protein-losing enteropathy may occur at any age as the result of a wide variety of pathological processes including: gluten-sensitive enteropathy, acute enteritis, Whipple's disease, and Henoch–Schönlein purpura. The serum protein electrophoresis pattern usually displays hypoalbuminemia, occasionally with decreased γ-globulins.[45] α_2-Macroglobulin may be increased as in the

nephrotic pattern but, in contrast, it is unusual to have an elevated β_1-lipoprotein in patients with protein-losing enteropathies. Whicher[42] suggested that the amount of protein loss into the gastrointestinal tract can be estimated by measuring the α_1-antitrypsin excretion in stool. Obviously, the features of protein-losing enteropathy are far too non-specific for serum protein electrophoresis patterns to suggest a specific diagnosis. An absolute distinction between a nephrotic and a protein-losing enteropathy pattern cannot be made reliably just from looking at the electrophoresis pattern. Clinical history and ancillary laboratory information are needed for the diagnostic process.

PROTEIN LOSS THROUGH THERMAL INJURY

Thermal injury to the skin produces a large surface area through which serum protein is lost. In the first few days following thermal injury, there is an increase in acute-phase reactants, including C-reactive protein and α_1-antitrypsin along with α_2-macroglobulin (the last is not typically an acute-phase reactant – its rise may reflect renal loss of other proteins – see below), while albumin is decreased.[46] At the same time, patients with thermal injury experience a rapid fall in the levels of transthyretin (prealbumin) that reach their nadir at 6 days post-burn.[47] The administration of low dose insulin-like growth factor-1/binding protein-3 intravenously to children with burns covering > 40 per cent of their skin has been shown to increase the levels of constitutive serum proteins such as transthyretin and transferrin, while attenuating the increases in α_1-antitrypsin, α_1-acid glycoprotein, and haptoglobin.[48]

The early thermal injury pattern differs from the nephrotic and protein-losing enteropathy patterns in that the last two usually show decreased α_1-globulin.[46] After burn injury, protein is also lost into the urine. Initially the pattern of loss resembles a mild glomerular proteinuria that may transform to a tubular proteinuria (see Chapter 7).[49]

Studies of serum from burn patients with sodium

dodecyl sulfate–polyacrylamide gel electrophoresis (SDS-PAGE) have demonstrated abnormal bands that react with antibodies against haptoglobin. These haptoglobin-like fractions correlate with the presence of immunosuppressive factors in the serum of these patients.[50] This finding may help to account for the increased susceptibility to infections that patients with thermal injury suffer.

ACUTE-PHASE REACTION PATTERN

During acute episodes of tissue damage (infection, tissue injury, tumor necrosis) with or without inflammation, elevation typically occurs in a group of hepatocyte-derived proteins called the acute-phase reactants. Reference distributions for the major acute phase proteins have been reported by Ritchie et al.[51–53] The presence of these acute-phase reactants often parallels clinical features including: fever, leukocytosis, muscle proteolysis, and a negative nitrogen balance.[54] In addition, a corresponding decrease occurs in other proteins. The typical serum protein electrophoresis pattern of acute-phase reaction contains a slightly low albumin, elevated α_1-globulin with slight anodal slurring, elevated α_2-globulin, decreased transferrin and a small mid-γ band (due to C-reactive protein) (Figs 5.14 and 5.15). Late in the course of the acute-

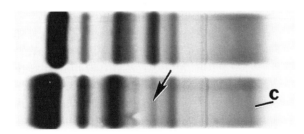

Figure 5.14 The bottom serum is from a patient with a classic acute-phase reaction. Albumin is slurred toward the anode, likely because of the antibiotic this patient was receiving (the patient did not have an elevated bilirubin). The α_1- and α_2-regions are increased and the transferrin band is decreased (arrow). In the γ-region, a small band indicates the presence of C-reactive protein (–C). Some fine black speckling between the α_1- and α_2 interregion is an artifact caused by precipitation of Paragon Violet stain. (Paragon SPE2 system stained with Paragon Violet.)

phase reaction (sometimes termed subacute), C3 is elevated.

Some of the components of the acute-phase reactants such as C-reactive protein, ceruloplasmin and serum amyloid A, which undergo major increases in concentration, do not produce dramatic effects on the serum protein electrophoresis pattern because their concentrations are too low. Indeed, increased serum amyloid A has no effect at all, and C-reactive protein produces only a minor (albeit occasionally troublesome) band in the γ-region of gel-based techniques (see below). Serum amyloid protein A has been suggested as a good marker for acute viral infections, whereas C-reactive protein shows little change (measured immunochemically) in most viral infections.[55] Using highly sensitive C-reactive protein assays, a strong correlation has been found between elevated levels of C-reactive protein and the risk of myocardial infarction.[56] However, these levels are too low to be measured by serum protein electrophoresis. A major protein of the acute-phase reaction is fibrinogen but this is not seen on serum protein electrophoresis.

The acute-phase response results from the effect of cytokines, released from inflammatory cells (mainly macrophages), on protein synthesis by hepatocytes. Several cytokines play major roles. Interleukin-1 (IL-1), interleukin-6 (IL-6), tumor necrosis factor-α (TNF-α) and tumor necrosis factor-β (TNF-β) have all been identified as prime inducers of the acute-phase proteins produced in the liver.[54,57] However, they play different roles. Interleukin-6 is a major stimulator of the acute-phase proteins whereas TNF-α and TNF-β increase production of α_1-antichymotrypsin while lowering production of haptoglobin.[58,59] *In vivo*, the overall acute phase response will vary from one situation to another, depending on the cytokines released, the individual's response to them, the duration of the stimulus, and other factors such as coexistent hemolysis (hemoglobin will bind to haptoglobin).

The typical time-course of an acute-phase response following a single episode of tissue injury is outlined in Table 5.2. Using gel-based methods, the anodal end of the α_1-antitrypsin band initially

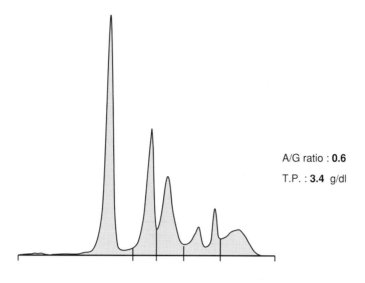

A/G ratio : **0.6**

T.P. : **3.4** g/dl

Serum protein electrophoresis

Fractions	%		Ref. %	Conc.	Ref. conc.
Albumin	37.5	<	45.3 – 67.7	1.3	3.4 – 5.2
Alpha 1	20.5	>	2.9 – 6.8	0.7	0.2 – 0.4
Alpha 2	18.9	>	6.2 – 14.9	0.6	0.5 – 1.0
Beta	11.9		8.1 – 18.0	0.4	0.6 – 1.1
Gamma	11.2		8.8 – 24.5	0.4	0.6 – 1.6

Figure 5.15 Capillary zone electropherogram from a patient with acute-phase reaction. Albumin is decreased, although there is no noticeable slurring toward the anode. The percentage of α_1- and α_2-globulins is increased, the transferrin band is decreased, and the entire β-region concentration is decreased. The γ-globulin is also low, a feature sometimes seen in acute-phase reaction. (Sebia Capillarys.)

has a diffuse increase in density caused by the rapid increase of α_1-acid glycoprotein. However, when using CZE, the α_1-acid glycoprotein band can be distinguished as a shoulder to the α_1-antitrypsin band (Fig. 5.16). C-Reactive protein is not detectable by serum protein electrophoresis in control specimens (normal concentration < 2 mg/dl), but with a vigorous acute-phase reaction, it may be seen as a small band in the mid-γ-region on gel-based systems (Fig. 5.14). As such, it may be confused with a small monoclonal gammopathy. By observing the presence of the other elements of the acute-phase pattern, noting other laboratory values and appropriate clinical history, one will not be led astray. If there is any doubt, a serum immunofixation reaction will quickly demonstrate that the band is not an immunoglobulin (see Chapter 3). The increase in serum

concentration of C-reactive protein can be quite dramatic after some inflammatory stimuli (as much as 1000-fold).[60] Quantification of the C-reactive protein (by immunochemical means) has been recommended to detect early infection in patients with leukemia who often do not demonstrate typical granulocyte responses to the infection.[61] I have not observed a C-reactive protein band in CZE samples of acute-phase reaction.

By 12–24 h after the onset of inflammation, α_1-antitrypsin and haptoglobin levels have usually increased (Table 5.2). Because α_1-antitrypsin levels increase during acute tissue damage even in patients with α_1-antitrypsin deficiency, the presence of acute inflammation may obscure the presence of this deficiency (see Chapter 4). In such individuals it is important to perform a concomitant measurement of C-reactive protein to be

Table 5.2 Time course[a] of positive acute-phase reactants which affect the high-resolution electrophoresis pattern

Protein	Earliest elevation	Peak elevation (h)
C-reactive protein	6–12 h	48–72
α_1-acid glycoprotein	12–24 h	48–72
α_1-antitrypsin	24 h	72–96
α_1-antichymotrypsin	24 h	72–96
Haptoglobin	24 h	72–96
Fibrinogen	24 h	72–96
C3	4–7 days	–

[a]Following a single acute episode of tissue damage. Note that many diseases have ongoing tissue injury, which results in more complicated, overlapping elevations of several components. Data modified from Ritchie and Whicher,[62] and Fischer et al.[63]

certain that the patient does not have evidence of an acute-phase pattern. Pregnancy or the use of birth control pills is usually accompanied by an elevation of α_1-antitrypsin, but these also produce an elevation in transferrin, which does not usually occur in an acute-phase reaction (iron deficiency being an important exception). Haptoglobin is variable in the acute-phase reaction. Although it is usually increased 24 h after an acute episode of tissue injury, if hemolysis accompanies the acute tissue injury then haptoglobin will bind hemoglobin and this product will be removed by the reticuloendothelial system, resulting in a decreased haptoglobin level. Fibrinogen is increased within 24 h of the injury, but one needs to quantify fibrinogen in *plasma* to measure this. Less consistently, elevations in C3, hemopexin, ceruloplasmin, and Gc globulin are seen within a week after acute tissue damage.[52,53,62–64]

As noted in Table 5.2, the elevation in C3 occurs later during the inflammatory response and I refer to it as an indicator of a later or subacute stage of the inflammation. C3, however, is an inconsistent marker because both synthesis and catabolism may be increased. During the inflammatory process, complement is activated by the alternative and/or classical pathway (depending on the initial mode of stimulation – see Chapter 4), therefore, the increased C3 produced may be used up in the process.[65]

Fraction	Rel %		g/dl	
ALBUMIN	39.9	---	2.51	---
ALPHA 1	14.3	+++	0.90	+++
ALPHA 2	21.2	+++	1.34	+++
BETA	12.4		0.78	
GAMMA	12.2		0.77	

Reference ranges

	Rel %	g/dl
ALBUMIN	52.6 – 68.9	3.80 – 5.20
ALPHA 1	3.6 – 8.1	0.30 – 0.60
ALPHA 2	5.3 – 12.2	0.40 – 0.90
BETA	8.3 – 14.3	0.60 – 1.10
GAMMA	8.2 – 18.6	0.60 – 1.40

TP: 6.40 – 8.20 A/G: 1.20 – 2.20

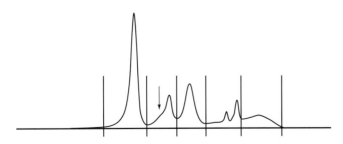

Figure 5.16 Capillary zone electropherogram from a patient with an acute-phase reaction, In addition to the features mentioned in Figure 5.15, this serum has a prominent anodal shoulder to α_1-antitrypsin due to an increase in α_1-acid glycoprotein (orosomucoid) (arrow). (Paragon CZE 2000.)

Several proteins consistently decrease in concentration following an episode of acute tissue injury. These 'negative' acute-phase reactants are useful guides in distinguishing an acute-phase pattern from estrogen effect. Typically, albumin and transferrin decrease within a few days following injury (Table 5.3). However, the concurrent presence of an iron deficiency anemia can produce an elevation of transferrin which would produce an atypical pattern.[65] Although transthyretin (prealbumin) and α_1-lipoprotein also decline during an acute-phase reaction, such a decrease is not reliably detected by serum protein electrophoresis because of the small quantity of transthyretin and the diffuse staining and wide variation of α_1-lipoprotein in the general population.

The acute-phase pattern is non-specific in that it can be seen following a wide variety of tissue injuries; however, when some clinical history is provided, its detection can be useful in confirming clinical impressions, in timing an internal injury event, and in predicting infections in patients with impaired leukocyte responses such as individuals with leukemia or others receiving chemotherapy.[66] In patients with autoimmune diseases, C-reactive protein levels have been useful as a more objective indication of disease activity than other tests such as erythrocyte sedimentation rate.[54] Interestingly, this is not true in systemic lupus erythematosus

where C-reactive protein usually does not correlate with inflammation. An increasing C-reactive protein in patients with systemic lupus erythematosus may indicate a coexistent bacterial infection.[67]

PROTEIN ABNORMALITIES IN AUTOIMMUNE DISEASE

While early studies held some promise for the use of electrophoresis in the diagnosis or prognosis of autoimmune diseases,[68,69] it has become clear that the electrophoretic findings in these conditions are too non-specific and varied to provide much useful information. There is often a polyclonal increase in γ-globulin. Further, during active autoimmune disease, circulating immune complexes are often accompanied by an oligoclonal pattern. An acute-phase reaction is seen during acute exacerbations; the α_2-globulin fraction may be altered with haptoglobin binding to hemoglobin during episodes of intravascular hemolysis.[70] Overall, screening for autoimmune diseases and for following particular patients, specific autoantibody titers and evidence of *in vivo* complement activation such as CH_{50}, C3 and C4 levels are more useful than serum protein electrophoresis.

Table 5.3 Time course[a] of negative acute-phase reactants that affect the high-resolution electrophoresis pattern

Protein	Earliest decline (days)	Lowest levels (days)
Transthyretin (prealbumin)	1	1–3
Transferrin	1–2	3–4
Albumin	2–3	–

[a]Following a single acute episode of tissue damage. Note that many diseases have ongoing tissue injury, which results in more complicated, overlapping elevations of several components. Data modified from Ritchie and Whicher,[62] and Fischer et al.[63]

PROTEIN PATTERNS IN HYPERESTROGENISM

The alterations in hormone balance that occur during pregnancy, in patients taking birth control pills or estrogen supplements, and in patients with estrogen-producing neoplasms cause changes in many of the serum proteins seen by serum protein electrophoresis. During pregnancy, albumin and IgG levels are decreased by 20–25 per cent while α_1-antitrypsin and transferrin levels are increased by about 66 per cent (Fig. 5.17).[70] The same protein alterations are seen in patients taking birth control pills. Whereas patients with preeclampsia may have a decreased transferrin level.[71,72]

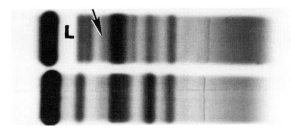

Figure 5.17 The top serum is an acute-phase reaction from a patient with a PiMS α_1-antitrypsin variant. The albumin in the top sample has moved more toward the anode (compare it with the albumin below) and the α_1-lipoprotein (L) region is relatively clear. The inter-α_1–α_2 region, which contains α_1-antichymotrypsin (arrow), is increased and the transferrin band is decreased. The moderate polyclonal increase in the γ-region of the top lane suggests that there is a chronic inflammatory process in this patient as well. The bottom sample has a densely staining α_1-lipoprotein region, a heavy staining α_1-antitrypsin band, and α_2-globulin and transferrin stain densely. This combination is consistent with the estrogen effect. In an acute-phase reaction, the transferrin would be decreased along with the α_1-lipoprotein. However, the densely staining transferrin band should be viewed with suspicion. Even in patients with iron deficiency or on estrogen therapy, the transferrin band should not stain this darkly. Immunofixation studies of this case demonstrated that the dark transferrin region was due to an IgA monoclonal gammopathy. (Paragon SPE2 system stained with Paragon Violet).

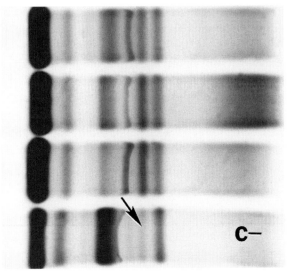

Figure 5.18 The second sample from the top shows a moderate diffuse polyclonal increase in γ-globulins that is typical of chronic inflammatory reactions. The bottom sample shows an acute-phase reaction with a decreased albumin (no anodal slurring is seen in contrast to Fig. 5.14), increased α_1-globulin with a more anodal migration than usual (possibly reflecting a fast variant of α_1-antitrypsin), increased α_2-globulin, marked decrease in transferrin (arrow), a faint C-reactive protein band (C–), and a very tiny γ-restriction directly beneath the 'C'. (Paragon SPE2 system stained with Paragon Violet.)

γ-GLOBULIN PATTERNS

Increased γ-globulin

DIFFUSE POLYCLONAL

A broad, diffuse increase in the γ-globulin region is the result of a polyclonal plasma cell response to chronic antigenic stimulation. These may be seen in patients with a wide variety of chronic infections, autoimmune diseases, chronic inflammatory bowel diseases, inflammatory lung diseases, and in cirrhosis (see above) (Figs 5.18 and 5.19). Occasionally polyclonal gammopathies can be large, producing bands that give an impression of monoclonality (Fig. 5.20). Usually, with experience using a particular system, the broad base of the polyclonal pattern will be distinctive enough for a correct interpretation. When in doubt, however, performing an immunofixation is recommended because some monoclonal gammopathies (especially IgA

and IgM) have a broad migration. When polyclonal gammopathies are extremely large, they may produce an increase in serum viscosity (increased viscosity is not limited to Waldenström's macroglobulinemia or myeloma).[73]

RESTRICTED POLYCLONAL

Polyclonal increases in one isotype of immunoglobulin or one subclass can give a broad area of restriction (Fig. 5.21). Polyclonal increases in IgG4 subclass have been reported to mimic a monoclonal gammopathy. IgG4 typically is increased under conditions of chronic antigen exposure and can be enhanced by the presence of glucocorticoids.[74,75] Polyclonal increases in IgG4 have been reported in patients with filarial infections, allergies, individuals receiving allergy immunotherapy, patients with Wegener's granulomatosis or Sjögren's syndrome.[76–84] By performing an immunofixation or immunosubtraction, the

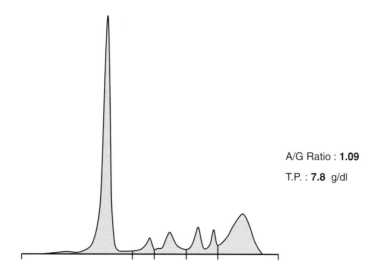

A/G Ratio : **1.09**

T.P. : **7.8** g/dl

Serum protein electrophoresis

Fractions	%		Ref. %	Conc.	Ref. Conc.
Albumin	52.2		45.3 – 67.7	4.1	3.4 – 5.2
Alpha 1	4.4		2.9 – 6.8	0.3	0.2 – 0.4
Alpha 2	8.2		6.2 – 14.9	0.6	0.5 – 1.0
Beta	10.4		8.1 – 18.0	0.8	0.6 – 1.1
Gamma	24.8	>	8.8 – 24.5	1.9	0.6 – 1.6

Figure 5.19 Capillary zone electropherogram (Sebia Capillarys) with a straightforward polyclonal increase in γ-globulin.

polyclonal nature of this increase can be demonstrated.

Polyclonal increases in IgM can present a difficult diagnostic problem (Fig. 5.22). They must be distinguished from monoclonal lesions by either immunofixation or immunosubtraction. The case in Fig. 5.22 was polyclonal. Recently, a benign lymphoproliferation has been demonstrated with a polyclonal increase in IgM. Persistent polyclonal B-cell lymphocytosis (PPBL) presents with a polyclonal increase in IgM, low to normal levels of IgA and IgG in association with a polyclonal B-cell lymphocytosis.[85] The condition occurs more often in women, and especially in those that are cigarette smokers.[85] PPBL has been reported in identical twins, is associated with HLA-DR7 phenotype and in three-quarters of the patients, a +i(3Q) chromosomal abnormality.[85,86] The peripheral blood of these patients contains bilobed lymphocytes where the two lobes are joined by a delicate chromatin bridge.[87] About half of the patients have splenomegaly and lymphoid aggregates in bone marrow express FMC7.[86] Immunophenotyping in the case reported by Woessner et al.[87] demonstrated these cells to be CD19+, HLA-DR+, IgM+, CD3–, CD5–, CD23–, κ+, and λ+ (the last two markers prove the polyclonal nature of the proliferation).[87] The lymphocytosis persists for years, and although it is still unclear whether this is a premalignant or benign condition, aggressive treatment has been discouraged.[85,88]

Polyclonal increase in IgM has also been reported in the X-linked hyper-IgM immunodeficiency syndrome.[89] In addition to polyclonal increases in IgM, these patients have hypogammaglobulinemia. As its name implies, these cases present during infancy with recurrent infections, diarrhea and often require parenteral nutrition to combat failure to thrive.[90–97] The defect is caused by a mutation in the CD40 ligand (CD40L), which results in

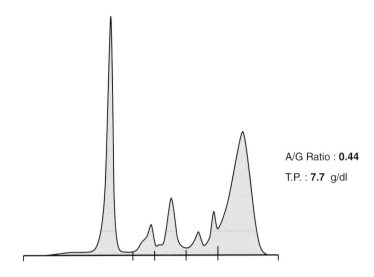

A/G Ratio : **0.44**

T.P. : **7.7** g/dl

Serum protein electrophoresis

Fractions	%		Ref. %	Conc.	Ref. Conc.
Albumin	30.5	<	45.3 – 67.7	2.3	3.4 – 5.2
Alpha 1	5.5		2.9 – 6.8	0.4	0.2 – 0.4
Alpha 2	10.5		6.2 – 14.9	0.8	0.5 – 1.0
Beta	9.0		8.1 – 18.0	0.7	0.6 – 1.1
Gamma	44.5	>	8.8 – 24.5	3.4	0.6 – 1.6

Figure 5.20 Capillary zone electropherogram (Sebia Capillarys) of a patient with decreased albumin and a marked polyclonal increase in γ-globulin. Because the capillary zone electropherogram patterns have relatively small γ-globulin regions compared with gel-based systems, large polyclonal increases can have the appearance of monoclonal gammopathies. When there is any uncertainty, I recommend performing an immunofixation.

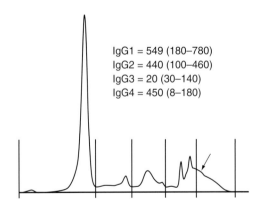

IgG1 = 549 (180–780)
IgG2 = 440 (100–460)
IgG3 = 20 (30–140)
IgG4 = 450 (8–180)

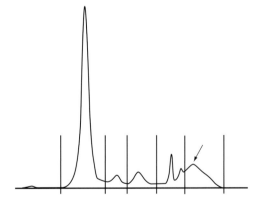

Figure 5.21 This capillary zone electropherogram (Paragon CZE 2000) demonstrates a rounded area of restriction in the β-γ region (arrow). Whereas most monoclonal gammopathies have a more narrow restriction, I have seen cases of IgA and IgD monoclonal gammopathies with similar appearances. In this case, the restriction was caused by a polyclonal increase in IgG4 subclass.

Figure 5.22 This capillary zone electropherogram (Paragon CZE 2000) demonstrates a rounded area of restriction in the mid-γ-region (arrow). I recommend performing an immunofixation or immunosubtraction on such restrictions. They are unusual, could be monoclonal gammopathies, and should have the polyclonal nature documented. This is a polyclonal increase of IgM.

disturbed antibody production and poor T-cell function.[98]

DIFFUSE OLIGOCLONAL

The prefix 'oligo' is Greek and means 'having few'. In a variety of benign and malignant conditions, a few clones of B cells proliferate in excess. Sometimes these are in response to an obvious stimulus such as in viral hepatitis or in bacterial endocarditis. In those cases, the responses are transient, although they may persist for months. Other oligoclonal expansions are the result of significantly altered immunoregulatory mechanisms, such as angioimmunoblastic lymphadenopathy where aberrant T cell clones (usually CD4+) result in massive polyclonal B-cell proliferation.[99] These oligoclonal responses were once thought to represent immune complexes, but it is now clear that they reflect the immunoglobulin products of a few clones of B cells/plasma cells. Patients with chronic lymphocytic leukemia and chronic B-cell lymphoma may have hypogammaglobulinemia and oligoclonal bands (Fig. 5.23)

Occasionally, a combination of several factors is responsible for an oligoclonal pattern. For example, transplant patients have profound immunosuppression from the required therapy and they suffer from a variety of infections because of this suppression. Also profoundly immunosuppressed by their disease, acquired immune-deficiency syndrome (AIDS) patients have a complex pattern with occasionally massive polyclonal increases in γ-globulin, often accompanied by oligoclonal, or occasionally monoclonal gammopathies (Figs 5.24 and 5.25).[100–106] Some of these bands have been shown to have specificity for human immunodeficiency virus antigens.[107] Taichman et al.[108] reported a positive correlation between the presence of oligoclonal bands and the degree of anti-human immunodeficiency virus (HIV) antibody positivity. In their study, they found that 43 per cent of patients that were HIV antibody positive had oligoclonal bands by serum protein electrophoresis. These patterns are the result of immune suppression due to decreased CD4 lymphocytes and extraordinary antigenic stimulation by the many infections they contract.

In other instances, patients with autoimmune diseases can have complex serum protein electrophoresis patterns because they may have aberrant immunoregulation with T-cell dysfunction, be receiving a variety of immunosuppressive regimens, suffer from infections related to the immunosuppression, and may be receiving plasmapheresis or γ-globulin replacement therapy.

Oligoclonal gammopathies have been described in the serum of patients with circulating immune complexes.[109–111] Kelly et al.[109,110] pointed out that during an oligoclonal B-cell expansion due to an infection, individual clones do not always achieve equivalent serum concentrations. Furthermore, the peak response of one clone may not occur synchronously with those of others. Thus, a particular sample from such a patient may show only the larger peak that could be mistaken for a monoclonal gammopathy that was part of a lymphoproliferative process.[109,112]

Oligoclonal bands may be particularly prominent in chronic hepatitis B and C.[25,113] These are other instances of infectious diseases where M-proteins have been reported.[114–118] One must be cautious, however, when describing an M-protein in a polyclonal background, or a background with oligoclonal expansion. In some infectious diseases, the monoclonal gammopathy is transient. The M-protein may persist for as long as 6 months.[119] Monoclonal gammopathies may be especially prominent in bacterial endocarditis caused by

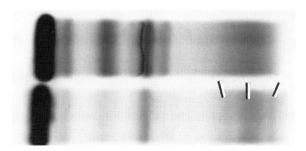

Figure 5.23 The bottom sample has a hypogammaglobulinemia with three or four oligoclonal bands (three are indicated). The sample above has a normal γ-globulin region for comparison. (Paragon SPE2 system stained with Paragon Violet.)

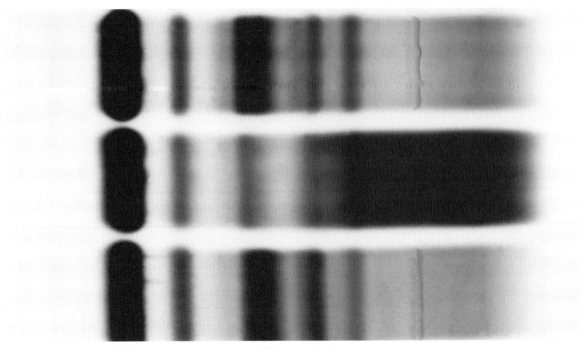

Figure 5.24 The middle sample shows a massive polyclonal expansion of γ-globulins in a patient with acquired immunodeficiency syndrome (AIDS). This massive expansion is also seen in patients with hepatitis and occasionally in angioimmunoblastic lymphadenopathy. This pattern is not diagnostic of AIDS, but this, or a similar polyclonal γ increase with or without oligoclonal bands should cause one to include these conditions in the differential diagnosis. (Paragon SPE2 system stained with Paragon Violet.)

staphylococcal species.[112,120,121] When one of the oligoclonal bands is appreciably larger than the rest, I note its presence and recommend a urine protein electrophoresis to rule out monoclonal free light chains (Bence Jones proteins) and a repeat serum electrophoresis in 3–6 months to see if the process resolves.[112,122,123]

It is important not to dismiss the potential importance of a monoclonal gammopathy in a patient with an infectious disease. Infections are a frequent feature of patients with multiple myeloma. Monoclonal gammopathies in AIDS patients have been seen together with myeloma and lymphomas (the latter often in association with Epstein–Barr virus).[124] When monoclonal gammopathies are identified in serum, performing urine immunofixation electrophoresis (IFE) is recommended to determine if a Bence Jones protein is present and following the serum immunoglobulin concentrations to determine if the process regresses or evolves.

POLYCLONAL GAMMOPATHY IN NEOPLASTIC DISEASES

Less well appreciated, however, is that polyclonal increases can also be seen in leukemias and lymphomas. In some cases, specific mechanisms can be identified that explain the response. In others, the presence of the polyclonal gammopathy has prognostic significance. For example, Surico et al.[125] reported that the presence of a polyclonal hypergammaglobulinemia at the onset of acute myeloid leukemia was associated with an increased number of cases that achieved complete remission. In one case of chronic lymphocytic leukemia that produced interleukin-2 (IL-2), a polyclonal increase in γ-globulins was present, in contrast to the usual moderately hypogammaglobulinemia in that condition.[126] In one case of mantle cell lymphoma, the polyclonal hypergammaglobulinemia was associated with the presence of autoreactive T-cells that were thought to provide a polyclonal stimulus for B cells.[127] Polyclonal B-cell lymphocytosis

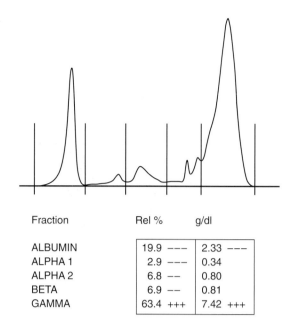

Fraction	Rel %	g/dl
ALBUMIN	19.9 ―――	2.33 ―――
ALPHA 1	2.9 ―――	0.34
ALPHA 2	6.8 ――	0.80
BETA	6.9 ――	0.81
GAMMA	63.4 +++	7.42 +++

Figure 5.25 Capillary zone electropherogram (Paragon CZE 2000) of serum from a patient with a massive polyclonal increase in γ-globulins. This patient had a co-infection with human immunodeficiency virus and hepatitis C virus.

accompanied by a polyclonal increase in IgG was reported in four patients with hematological and clinical features of hairy cell leukemia.[128] The authors viewed this to be a nonmalignant counterpart to hairy cell leukemia.[128]

Angiofollicular lymph node hyperplasia is characterized histologically by lymph node hyperplasia with hypocellular germinal centers, prominent polyclonal plasma cells, and prominent immunoblasts. This histopathological complex is not specific and has been reported in patients with autoimmune syndromes, acquired immunodeficiency syndrome, Wiskott–Aldrich syndrome, and Takatsuki's syndrome (an unusual dysproteinemic condition).[129,130]

Angioimmunoblastic lymphadenopathy (AILD) is characterized by lymphadenopathy, hepatosplenomegaly, high fever, and histologically by obliteration of normal lymph node architecture, prominent immunoblasts, plasma cells, and vascular proliferation with finely arborized vessels.[131–133] The serum often demonstrates marked polyclonal increases in γ-globulin (especially IgG and IgA), often with an oligoclonal pattern. Occasional cases with cryoglobulinemia and, uncommonly, monoclonal proteins have been demonstrated.[134,135] The polyclonal increase in IgA will often result in a β–γ bridging. This, together with the polyclonal increase in IgG, often with a decrease in albumin and increase in α_1-antitrypsin will produce a pattern that looks like that seen in patients with cirrhosis. The polyclonal increase in γ-globulins correlates with the extent of IL-6 production by the lymphoma cells in these patients. Hsu et al.[136] found that patients with AILD who did not produce IL-6 in their lymphoma cells did not have the polyclonal gammopathy. Late in the course of AILD, however, immunological suppression may result from increased suppressor T-cell function.[137]

In AILD, abnormal T-lymphocytes are thought to promote B-cell hyperactivity. Clinically, the disease progresses with a median survival time less than 2 years after diagnosis, often evolving into immunoblastic lymphoma (Fig. 5.26).[133] As mentioned above, one must be wary of assuming that a polyclonal or oligoclonal process is always part of an infectious process. This assumption has occasionally delayed the institution of chemotherapy.[138]

CHRONIC LYMPHOCYTIC LEUKEMIA

In chronic lymphocytic leukemia and well-differentiated lymphocytic lymphoma, the γ-globulin region is usually decreased in concentration. Some cases, however, display small monoclonal or occasionally small oligoclonal bands, often on the background of hypogammaglobulinemia. With the increased sensitivity afforded by serum immunofixation, cases of lymphocytic leukemia with two, three or more monoclonal proteins (oligoclonal expansions) have been described.[139]

POST TRANSPLANT LYMPHOPROLIFERATIVE DISORDER

Patients that have received allogeneic organ transplants are maintained on various regimens of immunosuppression to inhibit rejection of the graft. About 2 per cent of these patients develop a lymphoproliferative disorder associated with Epstein–Barr virus.[140–142] Many of these lesions will

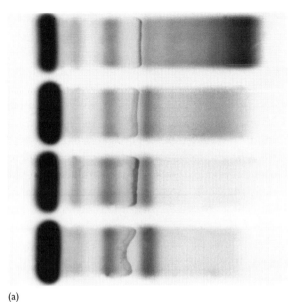

(a)

Figure 5.26 (a) The top serum is from a patient who had angioimmunoblastic lymphadenopathy that evolved to immunoblastic lymphoma. The prominent β–γ bridging together with the marked diffuse increase in γ-globulins, and the slightly decreased albumin, give this serum the appearance of a cirrhotic pattern. There are also several small areas of restriction (oligoclonal bands) in this γ-region (indicated on the immunofixation in (b). SPE, serum protein electrophoresis. (Paragon SPE1 system stained with Paragon Violet.) (b) Immunofixation of the top serum from (a) identifies several of the oligoclonal bands (indicated) as IgG, IgM, κ, and λ. This emphasizes the altered polyclonal B-cell proliferation that occurs in some cases of immunoblastic lymphoma. SPE, serum protein electrophoresis. (Paragon immunofixation stained with Paragon Violet; anode at the top.) Photographs contributed by Dorothy Wilkins.

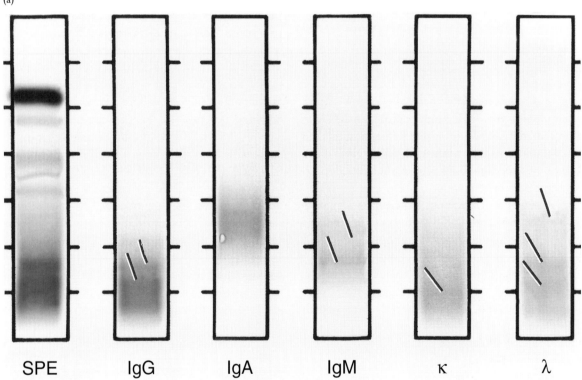

| SPE | IgG | IgA | IgM | κ | λ |

(b)

regress when the clinician decreases the immunosuppressive therapy. Recently, the use of anti-CD20 (rituximab) has been shown to be effective in treating post-transplantation lymphoproliferative disorder (PTLD).[143]

The type of immunosuppression has a major effect on when the disorder begins. For example, patients receiving either cyclosporine A or monoclonal anti-T3 (OKT3) can develop PTLD relatively early (1 month after transplantation), whereas individuals receiving neither of these drugs may take years before PTLD develops.[144] The disease is often regional, but may be widespread. The latter has the worse prognosis.

Both monoclonal and oligoclonal gammopathies have been found in these individuals. Myara et al.[105] found that 35 per cent of their 76 heart transplant patients had oligoclonal bands; 15 per cent had monoclonal components. The most common protein band found was IgG and it was seven times more common than IgM; IgA was not found in their series.[105] Oligoclonal bands in these patients are usually present in low concentration, and are usually transient. Monoclonal gammopathies in these patients may precede the demonstration of PTLD.[145,146]

Monoclonal gammopathies are discussed in Chapter 6.

Decreased γ-globulin

γ-Globulin is decreased in several clinically important circumstances (Table 5.4). Decreased γ-globulin may be seen as part of a protein loss

Table 5.4 Conditions to consider with isolated hypogammaglobulinemia

Children	X-linked agammaglobulinemia
	Transient hypogammaglobulinemia of the newborn
	Pre-B-cell deficiency
	Autosomal recessive immunodeficiency
Adults	Common variable immunodeficiency[a]
	Light chain disease
	Chronic lymphocytic leukemia
	Well-differentiated lymphocytic lymphoma
	Chemotherapy
	Plasma exchange

[a]Common variable immunodeficiency is usually detected in young adults.

pattern, either renal or gastrointestinal loss (see above). Isolated hypogammaglobulinemia, however, with other serum protein fractions in the normal range, implies an immunological abnormality. A low γ-globulin in an individual older than 2 years of age may indicate an abnormal immune system, and should be pursued. The immunodeficiency may be congenital or acquired owing to suppression by a neoplasm or suppression by chemotherapy. Even individuals in their eighth and ninth decades should have normal γ-regions (assessed by densitometric scan) (Fig. 5.27). Knowledge of the age and sex of the patient are of great importance in evaluating a patient with hypogammaglobulinemia.

BRUTON'S X-LINKED AGAMMAGLOBULINEMIA

Bruton's X-linked agammaglobulinemia (XLA) is a congenital deficiency of B cells in males which occurs once in each 100 000 live male births.[147,148] It is caused by mutations in the gene coding for

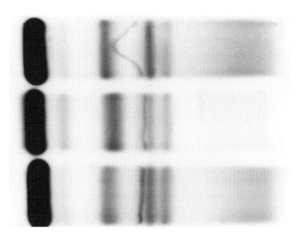

Figure 5.27 The middle sample has a very significant finding which is frequently overlooked: isolated hypogammaglobulinemia. The albumin and other major protein bands are normal (perhaps transferrin is slightly decreased). Therefore, the decrease in γ-globulin likely relates to a dysfunction of the immune system and not merely to protein loss (renal or gastrointestinal). The densitometric scan of this region was 0.40 (lower normal is 0.62 on our system). Evaluation of the urine from this patient demonstrated κ light chain disease. I recommend that any case where the γ-region is below our cut-off, has urine evaluated for the presence of monoclonal free light chains. (Paragon SPE2 system stained with Paragon Violet.)

Bruton's agammaglobulinemia tyrosine kinase.[149] Their bone marrow and peripheral blood B-lymphocytes are markedly decreased.[150] The few circulating B lymphocytes they do possess may express the CD20 surface marker, but not CD21.[147] A maturational arrest occurs between the cytoplasmic μ negative and positive stages.[150] Because maternal IgG crosses the placenta, the newborn with XLA has a normal γ-globulin until about 6 months of age. IgG has a half-life of about 21 days, therefore, after about 6 months the maternal IgG has been catabolized. By that age, most normal children will begin to synthesize their own IgG. Patients with XLA will not be able to synthesize IgG and begin to suffer from recurrent pyogenic, bacterial infections because of the lack of opsonins and complement-fixing antibody.[89]

TRANSIENT HYPOGAMMAGLOBULINEMIA OF INFANCY (THI)

THI must be considered in the differential diagnosis of hypogammaglobulinemia pattern in an infant. This condition has been associated with a delayed maturation of helper CD4 lymphocytes.[151–153] They possess normal numbers of peripheral blood B lymphocytes. Some patients with transient hypogammaglobulinemia of infancy develop mucosal infections (usually respiratory), however, they do not suffer from the pattern of recurrent pyogenic infections seen in patients with XLA.

COMMON VARIABLE IMMUNODEFICIENCY DISEASE

Common variable immunodeficiency disease (CVID) was the most common form of acquired immunodeficiency disease in adults prior to the AIDS epidemic. It usually manifests with diarrhea or respiratory tract infections in young adults. The disease is uncommon in childhood or in the elderly. Despite the similarity in name to acquired immune deficiency, the pathogenesis is not related to HIV. The electrophoretic pattern shows extreme hypogammaglobulinemia (usually less than 250 mg/dl of IgG). These patients respond well to γ-globulin replacement and judicious antibiotic therapy.[154] The cell surface marker patterns in these

patients vary from one to another. This indicates that CVID is really a cluster of diseases which share the end-stage characteristic of extreme hypogammaglobulinemia.[155] Many cases of CVID demonstrate benign lymphoid lesions, including atypical lymphoid hyperplasia, reactive lymphoid hyperplasia and chronic granulomatous inflammation.[156]

LYMPHOPROLIFERATIVE DISEASE

As discussed in Chapter 6, light chain disease, chronic lymphocytic lymphoma and well-differentiated lymphocytic lymphoma are commonly associated with isolated hypogammaglobulinemia (Table 5.4). This is usually a much more subtle decrease than the hypogammaglobulinemia associated with the immunodeficiency diseases mentioned above. When the deficiency in γ-globulin is profound, intravenous γ-globulin therapy has been reported to decrease bacterial infections.[157] Therefore, in adults, any decrease in the γ-globulin region below our two standard deviation limits causes us to note its existence in a narrative signout. We also recommend performing immunofixation on a urine sample to rule out the possibility of a monoclonal free light chain (Bence Jones protein).

IATROGENIC HYPOGAMMAGLOBULINEMIA

Some forms of therapy suppress the immune system. Chemotherapy may be given for a variety of autoimmune and neoplastic diseases. Depending on dose and duration, this may decrease the γ-globulin levels. Plasma exchange is now commonly used for autoimmune diseases and for hyperviscosity syndromes. Immediately after plasma exchange, most plasma components are decreased, but they usually recover within a day or two. It takes about 2 weeks for IgG to return to its former value, whereas IgA and IgM often return to their reference values within a week of plasma exchange.[158] Therefore, when hypogammaglobulinemia is detected on protein electrophoresis, clinical correlation is important to help avoid confusion as to the significance of this finding for the patient.

REFERENCES

1. Keren DF, Alexanian R, Goeken JA, Gorevic PD, Kyle RA, Tomar RH. Guidelines for clinical and laboratory evaluation of patients with monoclonal gammopathies. *Arch Pathol Lab Med* 1999;**123**:106–107.

2. Kallemuchikkal U, Gorevic PD. Evaluation of cryoglobulins. *Arch Pathol Lab Med* 1999;**123**:119–125.

3. Kawai T. Clinical application of serum protein fractions using paper electrophoresis. 3. Liver cirrhosis. *Rinsho Byori* 1965;**13**:502–505.

4. Ravel R. Serum protein electrophoresis in cirrhosis. *Am J Gastroenterol* 1969;**52**:509–514.

5. Mincis M, Guimaraes RX, Chaud Sobrinho J, Wachslicht H. Serum protein electrophoresis in Laennec's and postnecrotic cirrhosis. *AMB Rev Assoc Med Bras* 1972;**18**:447–452.

6. Zuberi SJ, Lodi TZ. Serum protein electrophoresis in healthy subjects and patients with liver disease. *J Pak Med Assoc* 1978;**28**:140–141.

7. Hallen J, Laurell CB. Plasma protein pattern in cirrhosis of the liver. *Scand J Clin Lab Invest Suppl* 1972;**124**:97–103.

8. Tessari P, Barazzoni R, Kiwanuka E, Davanzo G, De Pergola G, Orlando R, Vettore M, Zanetti M. Impairment of albumin and whole body postprandial protein synthesis in compensated liver cirrhosis. *Am J Physiol Endocrinol Metab* 2002;**282**:E304–311.

9. Rothschild MA, Oratz M, Schreiber SS. Albumin synthesis. *Int Rev Physiol* 1980;**21**:249–274.

10. Rothschild MA, Oratz M, Schreiber SS. Effects of nutrition and alcohol on albumin synthesis. *Alcohol Clin Exp Res* 1983;**7**:28–30.

11. Rothschild MA, Oratz M, Schreiber SS, Mongelli J. The effects of ethanol and hyperosmotic perfusates on albumin synthesis and release. *Hepatology* 1986;**6**:1382–1385.

12. Ballmer PE, Walshe D, McNurlan MA, Watson H, Brunt PW, Garlick PJ. Albumin synthesis rates in cirrhosis: correlation with Child–Turcotte classification. *Hepatology* 1993;**18**:292–297.

13. Haglund U, Hulten L. Protein patterns in serum and peritoneal fluid in Crohn's disease and ulcerative colitis. *Acta Chir Scand* 1976;**142**:160–164.

14. Cozzolino G, Francica G, Lonardo A, Cigolari S, Cacciatore L. Lack of correlation between the laboratory findings and a series of steps in the clinical severity of chronic liver disease. *Ric Clin Lab* 1984;**14**:641–648.

15. Agostoni A, Marasini B, Stabilini R, Del Ninno E, Pontello M. Multivariate analysis of serum protein assays in chronic hepatitis and postnecrotic cirrhosis. *Clin Chem* 1974;**20**:428–429.

16. Mutchnick MG, Keren DF. *In vitro* synthesis of antibody to specific bacterial lipopolysaccharide by peripheral blood mononuclear cells from patients with alcoholic cirrhosis. *Immunology* 1981;**43**:177–182.

17. Genesca J, Gonzalez A, Torregrosa M, Mujal A, Segura R. High levels of endotoxin antibodies contribute to hyperglobulinemia of cirrhotic patients. *Am J Gastroenterol* 1998;**93**: 664–665.

18. Berger SR, Helms RA, Bull DM. Cirrhotic hyperglobulinemia: increased rates of immunoglobulin synthesis by circulating lymphoid cells. *Dig Dis Sci* 1979;**24**:741–745.

19. Fukuda Y, Imoto M, Hayakawa T. Serum levels of secretory immunoglobulin A in liver disease. *Am J Gastroenterol* 1985;**80**:237–241.

20. Housley J. Alpha-2-macroglobulin levels in disease in man. *J Clin Pathol* 1968;**21**:27–31.

21. Gjone E, Norum KR. Plasma lecithin-cholesterol acyltransferase and erythrocyte lipids in liver disease. *Acta Med Scand* 1970;**187**:153–161.

22. Kindmark CO, Laurell CB. Sequential changes of the plasma protein pattern in inoculation hepatitis. *Scand J Clin Lab Invest Suppl* 1972;**124**:105–115.

23. Ciobanu V, Zalaru MC, Steinbruh L, Pambuccian G, Popescu C. Immunoelectrophoretic analysis of serum gamma globulins in chronic hepatitis and hepatic cirrhosis. *Med Interna (Bucur)* 1973;**25**: 285–305.

24. Hirayama C, Tominaga K, Irisa T, Nakamura M. Serum gamma-globulins and hepatitis-associated antigen in blood donors, chronic liver disease and primary hepatoma. *Digestion* 1972;7:257–265.

25. Tsianos EV, Di Bisceglie AM, Papadopoulos NM, Costello R, Hoofnagle JH. Oligoclonal immunoglobulin bands in serum in association with chronic viral hepatitis. *Am J Gastroenterol* 1990;85:1005–1008.

26. Perez Losada J, Losa Garcia JE, Alonso Ralero L, Lerma Marquez J, Sanchez Sanchez R. Oligoclonal gammapathy and infection by hepatitis B virus. *An Med Interna* 1998;15:397–398.

27. Demenlenaere L, Wieme RJ. Special electrophoretic anomalies in the serum of liver patients: a report of 1145 cases. *Am J Dig Dis* 1961;6:661–675.

28. Staak JO, Glossmann JP, Diehl V, Josting A. Hepatitis-C-virus-associated cryoglobulinemia. Pathogenesis, diagnosis and treatment. *Med Klin* 2002;97:601–608.

29. Kayali Z, Buckwold VE, Zimmerman B, Schmidt WN. Hepatitis C, cryoglobulinemia, and cirrhosis: a meta-analysis. *Hepatology* 2002;36:978–985.

30. Agnello V. The etiology and pathophysiology of mixed cryoglobulinemia secondary to hepatitis C virus infection. *Springer Semin Immunopathol* 1997;19:111–129.

31. Andreone P, Zignego AL, Cursaro C, et al. Prevalence of monoclonal gammopathies in patients with hepatitis C virus infection. *Ann Intern Med* 1998;129:294–298.

32. Hickman PE, Dwyer KP, Masarei JR. Pseudohyponatraemia, hypercholesterolaemia, and primary biliary cirrhosis. *J Clin Pathol* 1989;42:167–171.

33. Lindgren S. Accelerated nonproteolytic cleavage of C3 in plasma from patients with primary biliary cirrhosis. *J Lab Clin Med* 1989;114:655–661.

34. Loginov AS, Reshetniak VI, Chebanov SM, Matiushin BN. Clinical significance of gel electrophoresis of blood lipoproteins in primary biliary cirrhosis. *Klin Lab Diagn* 1996:33–35.

35. Teramoto T, Kato H, Hashimoto Y, Kinoshita M, Toda G, Oka H. Abnormal high density lipoprotein of primary biliary cirrhosis analyzed by high performance liquid chromatography. *Clin Chim Acta* 1985;149:135–148.

36. Jensen H. Plasma protein and lipid pattern in the nephrotic syndrome. *Acta Med Scand* 1967;182:465–473.

37. Markiewicz K. Serum lipoprotein pattern in the nephrotic syndrome. *Pol Med J* 1966;5:737–743.

38. Ghai CL, Chugh KS, Kumar S, Chhuttani PN. Electrophoretic pattern of serum proteins in secondary renal amyloidosis with nephrotic syndrome. *J Assoc Physicians India* 1966;14:141–144.

39. Bazzi C, Petrini C, Rizza V, Arrigo G, D'Amico G. A modern approach to selectivity of proteinuria and tubulointerstitial damage in nephrotic syndrome. *Kidney Int* 2000;58:1732–1741.

40. Robson AM, Loney LC. Nephrotic syndrome. In: Conn RB, ed. *Current diagnosis*. Philadelphia: WB Saunders, 1985.

41. Appel GB, Blum CB, Chien S, Kunis CL, Appel AS. The hyperlipidemia of the nephrotic syndrome. Relation to plasma albumin concentration, oncotic pressure, and viscosity. *N Engl J Med* 1985;312:1544–1548.

42. Whicher JT. The interpretation of electrophoresis. *Br J Hosp Med* 1980;24:348–356.

43. Su LD, Keren DF. The effects of exogenous free fatty acids on lipoprotein migration in serum high-resolution electrophoresis: addition of free fatty acids improves visualization of normal and abnormal alpha1-antitrypsin. *Am J Clin Pathol* 1998;109:262–267.

44. Pearson JP, Keren DF. The effects of heparin on lipoproteins in high-resolution electrophoresis of serum. *Am J Clin Pathol* 1995;104:468–471.

45. Reif S, Jain A, Santiago J, Rossi T. Protein losing enteropathy as a manifestation of Henoch–Schonlein purpura. *Acta Paediatr Scand* 1991;80:482–485.

46. Latha B, Ramakrishnan KM, Jayaraman V, Babu M. Action of trypsin:chymotrypsin (Chymoral forte DS) preparation on acute-phase proteins following burn injury in humans. *Burns* 1997;**23**(Suppl 1):S3–7.

47. Moody BJ. Changes in the serum concentrations of thyroxine-binding prealbumin and retinol-binding protein following burn injury. *Clin Chim Acta* 1982;**118**:87–92.

48. Spies M, Wolf SE, Barrow RE, Jeschke MG, Herndon DN. Modulation of types I and II acute phase reactants with insulin-like growth factor-1/binding protein-3 complex in severely burned children. *Crit Care Med* 2002;**30**:83–88.

49. Shakespeare PG, Coombes EJ, Hambleton J, Furness D. Proteinuria after burn injury. *Ann Clin BioChem* 1981;**18**:353–360.

50. Huang WH. Study of the abnormal plasma proteins after burn injury. *Zhonghua Zheng Xing Shao Shang Wai Ke Za Zhi* 1992;**8**:89–92, 163.

51. Ritchie RF, Palomaki GE, Neveux LM, Navolotskaia O, Ledue TB, Craig WY. Reference distributions for the positive acute phase serum proteins, alpha1-acid glycoprotein (orosomucoid), alpha1-antitrypsin, and haptoglobin: a practical, simple, and clinically relevant approach in a large cohort. *J Clin Lab Anal* 2000;**14**:284–292.

52. Ritchie RF, Palomaki GE, Neveux LM, Navolotskaia O. Reference distributions for the positive acute phase proteins, alpha1-acid glycoprotein (orosomucoid), alpha1-antitrypsin, and haptoglobin: a comparison of a large cohort to the world's literature. *J Clin Lab Anal* 2000;**14**:265–270.

53. Ritchie RF, Palomaki GE, Neveux LM, Navolotskaia O, Ledue TB, Craig WY. Reference distributions for the negative acute-phase serum proteins, albumin, transferrin and transthyretin: a practical, simple and clinically relevant approach in a large cohort. *J Clin Lab Anal* 1999;**13**:273–279.

54. Thompson D, Milford-Ward A, Whicher JT. The value of acute phase protein measurements in clinical practice. *Ann Clin BioChem* 1992;**29**(Pt 2):123–131.

55. Nakayama T, Sonoda S, Urano T, Yamada T, Okada M. Monitoring both serum amyloid protein A and C-reactive protein as inflammatory markers in infectious diseases. *Clin Chem* 1993;**39**:293–297.

56. Mariotti R, Musumeci G, De Carlo M, et al. Acute-phase reactants in acute myocardial infarction: impact on 5-year prognosis. *Ital Heart J* 2001;**2**:294–300.

57. Ikeda T, Kawakami K, Fujita J, Bandoh S, Yamadori I, Takahara J. Thymic carcinoma associated with a high serum level of interleukin 6 diagnosed through the evaluation for asymptomatic elevation of acute-phase reactants. *Intern Med* 1998;**37**:414–416.

58. Kushner I. Regulation of the acute phase response by cytokines. *Perspect Biol Med* 1993;**36**:611–622.

59. Jiang SL, Samols D, Sipe J, Kushner I. The acute phase response: overview and evidence of roles for both transcriptional and post-transcriptional mechanisms. *Folia Histochem Cytobiol* 1992;**30**:133–135.

60. Kushner I. C-reactive protein in rheumatology. *Arthritis Rheum* 1991;**34**:1065–1068.

61. Schofield KP, Voulgari F, Gozzard DI, Leyland MJ, Beeching NJ, Stuart J. C-reactive protein concentration as a guide to antibiotic therapy in acute leukaemia. *J Clin Pathol* 1982;**35**:866–869.

62. Ritchie RF, Whicher JT. Acute phase proteins: markers of inflammation. *Clin Chem News* 1993;**19**(Suppl):7.

63. Fischer CL, Gill C, Forrester MG, Nakamura R. Quantitation of 'acute-phase proteins' postoperatively. Value in detection and monitoring of complications. *Am J Clin Pathol* 1976;**66**:840–846.

64. Killingsworth LM. Plasma protein patterns in health and disease. *CRC Crit Rev Clin Lab Sci* 1979;**3**:1–30.

65. Jolliff C. Agarose gel electrophoresis in the acute and chronic inflammatory states. *Clin Immunol Newsl* 1991;**11**:132–135.

66. Smith SJ, Bos G, Esseveld MR, van Eijk HG, Gerbrandy J. Acute-phase proteins from the liver and enzymes from myocardial infarction; a quantitative relationship. *Clin Chim Acta* 1977;**81**:75–85.

67. Becker GJ, Waldburger M, Hughes GR, Pepys MB. Value of serum C-reactive protein measurement in the investigation of fever in systemic lupus erythematosus. *Ann Rheum Dis* 1980;**39**:50–52.

68. Labac E, Tirzonalis A, Gossart J. Electrophoresis of blood proteins in collagen diseases. *Acta Gastroenterol Belg* 1959;**22**:544–549.

69. Stava Z. Serum proteins in scleroderma. *Dermatologica* 1958;**117**:147–153.

70. Ganrot PO. Variation of the concentrations of some plasma proteins in normal adults, in pregnant women and in newborns. *Scand J Clin Lab Invest Suppl* 1972;**124**:83–88.

71. Horne CH, Howie PW, Goudie RB. Serum-alpha2-macroglobulin, transferrin, albumin, and IgG levels in preeclampsia. *J Clin Pathol* 1970;**23**:514–516.

72. Hubel CA, Kozlov AV, Kagan VE, et al. Decreased transferrin and increased transferrin saturation in sera of women with preeclampsia: implications for oxidative stress. *Am J Obstet Gynecol* 1996;**175**:692–700.

73. Vladutiu AO, Roach BM, Farahmand SM. Polyclonal gammopathy with marked increase in serum viscosity. *Clin Chem* 1991;**37**:1788–1793.

74. Aalberse RC, Schuurman J. IgG4 breaking the rules. *Immunology* 2002;**105**:9–19.

75. Akdis CA, Blesken T, Akdis M, Alkan SS, Heusser CH, Blaser K. Glucocorticoids inhibit human antigen-specific and enhance total IgE and IgG4 production due to differential effects on T and B cells in vitro. *Eur J Immunol* 1997;**27**:2351–2357.

76. Brouwer E, Tervaert JW, Horst G, et al. Predominance of IgG1 and IgG4 subclasses of anti-neutrophil cytoplasmic autoantibodies (ANCA) in patients with Wegener's granulomatosis and clinically related disorders. *Clin Exp Immunol* 1991;**83**:379–386.

77. Segelmark M, Wieslander J. ANCA and IgG subclasses. *Adv Exp Med Biol* 1993;**336**:71–75.

78. Mellbye OJ, Mollnes TE, Steen LS. IgG subclass distribution and complement activation ability of autoantibodies to neutrophil cytoplasmic antigens (ANCA). *Clin Immunol Immunopathol* 1994;**70**:32–39.

79. Lindstrom FD, Eriksson P, Tejle K, Skogh T. IgG subclasses of anti-SS-A/Ro in patients with primary Sjogren's syndrome. *Clin Immunol Immunopathol* 1994;**73**:358–361.

80. Shakib F, Pritchard DI, Walsh EA, et al. The detection of autoantibodies to IgE in plasma of individuals infected with hookworm (*Necator americanus*) and the demonstration of a predominant IgG1 anti-IgE autoantibody response. *Parasite Immunol* 1993;**15**:47–53.

81. Shiddo SA, Huldt G, Nilsson LA, Ouchterlony O, Thorstensson R. Visceral leishmaniasis in Somalia. Significance of IgG subclasses and of IgE response. *Immunol Lett* 1996;**50**:87–93.

82. Suzuki S, Kida S, Ohira Y, et al. A case of Sjogren's syndrome accompanied by lymphadenopathy and IgG4 hypergammaglobulinemia. *Ryumachi* 1993;**33**:249–254.

83. van Nieuwkoop JA, Brand A, Radl J, Skvaril F. Increased levels of IgG4 subclass in 5 patients with acquired respiratory disease. *Int Arch Allergy Appl Immunol* 1982;**67**:61–65.

84. Taschini PA, Addison NM. An atypical electrophoretic pattern due to a selective polyclonal increase of IgG4 subclass. *ASCP Check Sample Immunopathol* 1991:IP91–95.

85. Troussard X, Mossafa H, Flandrin G. Persistent polyclonal B-cell lymphocytosis. *Ann Biol Clin (Paris)* 2002;**60**:273–280.

86. Carr R, Fishlock K, Matutes E. Persistent polyclonal B-cell lymphocytosis in identical twins. *Br J Haematol* 1997;**96**:272–274.

87. Woessner S, Florensa L, Espinet B. Bilobulated circulating lymphocytes in persistent polyclonal B-cell lymphocytosis. *Haematologica* 1999;**84**:749.

88. Mossafa H, Malaure H, Maynadie M, et al. Persistent polyclonal B lymphocytosis with binucleated lymphocytes: a study of 25 cases. Groupe Français d'Hematologie Cellulaire. *Br J Haematol* 1999;**104**:486–493.

89. Schiff R, Schiff SE. Chapter 11. Flow cytometry for primary immunodeficiency diseases. In: Keren DF, McCoy JJP, Carey JL, eds. *Flow cytometry in clinical diagnosis*. Chicago: ASCP Press, 2001.

90. Schneider LC. X-linked hyper IgM syndrome. *Clin Rev Allergy Immunol* 2000;**19**:205–215.

91. Tsuge I, Matsuoka H, Nakagawa A, et al. Necrotizing toxoplasmic encephalitis in a child with the X-linked hyper-IgM syndrome. *Eur J Pediatr* 1998;**157**:735–737.

92. Leiva LE, Junprasert J, Hollenbaugh D, Sorensen RU. Central nervous system toxoplasmosis with an increased proportion of circulating gamma delta T cells in a patient with hyper-IgM syndrome. *J Clin Immunol* 1998;**18**:283–290.

93. Miller ML, Algayed IA, Yogev R, Chou PM, Scholl PR, Pachman LM. Atypical *Pneumocystis carinii* pneumonia in a child with hyper-IgM syndrome. *Pediatr Pathol Lab Med* 1998;**18**:71–78.

94. Chang MW, Romero R, Scholl PR, Paller AS. Mucocutaneous manifestations of the hyper-IgM immunodeficiency syndrome. *J Am Acad Dermatol* 1998;**38**:191–196.

95. Hostoffer RW, Berger M, Clark HT, Schreiber JR. Disseminated *Histoplasma capsulatum* in a patient with hyper IgM immunodeficiency. *Pediatrics* 1994;**94**:234–236.

96. Iseki M, Anzo M, Yamashita N, Matsuo N. Hyper-IgM immunodeficiency with disseminated cryptococcosis. *Acta Paediatr* 1994;**83**:780–782.

97. Di Santo JP, de Saint Basile G, Durandy A, Fischer A. Hyper-IgM syndrome. *Res Immunol* 1994;**145**:205–209.

98. Lobo FM, Scholl PR, Fuleihan RL. CD40 ligand-deficient T cells from X-linked hyper-IgM syndrome carriers have intrinsic priming capability. *J Immunol* 2002;**168**:1473–1478.

99. Willenbrock K, Roers A, Seidl C, Wacker HH, Kuppers R, Hansmann ML. Analysis of T-cell subpopulations in T-cell non-Hodgkin's lymphoma of angioimmunoblastic lymphadenopathy with dysproteinemia type by single target gene amplification of T cell receptor-beta gene rearrangements. *Am J Pathol* 2001;**158**:1851–1857.

100. Franciotta D, Zardini E, Bono G, Brustia R, Minoli L, Cosi V. Antigen-specific oligoclonal IgG in AIDS-related cytomegalovirus and toxoplasma encephalitis. *Acta Neurol Scand* 1996;**94**:215–218.

101. Zeman AZ, Keir G, Luxton R, Thompson EJ. Serum oligoclonal IgG is a common and persistent finding in multiple sclerosis, and has a systemic source. *QJM* 1996;**89**:187–193.

102. Frankel EB, Greenberg ML, Makuku S, Kochwa S. Oligoclonal banding in AIDS and hemophilia. *Mt Sinai J Med* 1993;**60**:232–237.

103. Zeman A, McLean B, Keir G, Luxton R, Sharief M, Thompson E. The significance of serum oligoclonal bands in neurological diseases. *J Neurol Neurosurg Psychiatry* 1993;**56**:32–35.

104. Pisa P, Cannon MJ, Pisa EK, Cooper NR, Fox RI. Epstein–Barr virus induced lymphoproliferative tumors in severe combined immunodeficient mice are oligoclonal. *Blood* 1992;**79**:173–179.

105. Myara I, Quenum G, Storogenko M, Tenenhaus D, Guillemain R, Moatti N. Monoclonal and oligoclonal gammopathies in heart-transplant recipients. *Clin Chem* 1991;**37**:1334–1337.

106. Amadori A, Gallo P, Zamarchi R, et al. IgG oligoclonal bands in sera of HIV-1 infected patients are mainly directed against HIV-1 determinants. *AIDS Res Hum Retrovirus* 1990;**6**:581–586.

107. Papadopoulos NM, Costello R, Ceroni M, Moutsopoulos HM. Identification of HIV-specific oligoclonal immunoglobulins in serum of carriers of HIV antibody. *Clin Chem* 1988;**34**:973–975.

108. Taichman DB, Bayer K, Senior M, Goodman DB, Kricka LJ. Oligoclonal immunoglobulins in HIV-antibody-positive serum. *Clin Chem* 1988;**34**:2377.

109. Kelly RH, Hardy TJ, Shah PM. Benign monoclonal gammopathy: a reassessment of the problem. *Immunol Invest* 1985;**14**:183–197.

110. Kelly RH, Scholl MA, Harvey VS, Devenyi AG. Qualitative testing for circulating immune complexes by use of zone electrophoresis on agarose. *Clin Chem* 1980;**26**:396–402.

111. Levinson SS, Keren DF. Immunoglobulins from the sera of immunologically activated persons with pairs of electrophoretic restricted bands show a greater tendency to aggregate than normal. *Clin Chim Acta* 1989;**182**:21–30.

112. Keren DF, Morrison N, Gulbranson R. Evolution of a monoclonal gammopathy (MG) documented by high-resolution electrophoresis (HRE) and immunofixation (IFE). *Lab Med* 1994; 25:313–317.

113. Papadopoulos NM, Tsianos EV, Costello R. Oligoclonal immunoglobulins in serum of patients with chronic viral hepatitis. *J Clin Lab Anal* 1990;4:180–182.

114. Perrone A, Deramo MT, Spaccavento F, Santarcangelo P, Favoino B, Antonaci S. Hepatitis C virus (HCV) genotypes, human leucocyte antigen expression and monoclonal gammopathy prevalence during chronic HCV infection. *Cytobios* 2001;106(Suppl 1):125–134.

115. Schott P, Pott C, Ramadori G, Hartmann H. Hepatitis C virus infection-associated non-cryoglobulinaemic monoclonal IgMkappa gammopathy responsive to interferon-alpha treatment. *J Hepatol* 1998;29:310–315.

116. Andreone P, Gramenzi A, Cursaro C, Bernardi M, Zignego AL. Monoclonal gammopathy in patients with chronic hepatitis C virus infection. *Blood* 1996;88:1122.

117. Gattoni A, Cecere A, Romano C, Di Martino P, Caiazzo R, Rippa A. Case report of a monoclonal gammopathy in a patient with chronic hepatitis: effects of beta-IFN treatment. *Panminerva Med* 1996;38:175–178.

118. Heer M, Joller-Jemelka H, Fontana A, Seefeld U, Schmid M, Ammann R. Monoclonal gammopathy in chronic active hepatitis. *Liver* 1984;4:255–263.

119. Kanoh T. Fluctuating M-component level in relation to infection. *Eur J Haematol* 1989;42:503–504.

120. Struve J, Weiland O, Nord CE. *Lactobacillus plantarum* endocarditis in a patient with benign monoclonal gammopathy. *J Infect* 1988;17:127–130.

121. Larrain C. Transient monoclonal gammopathies associated with infectious endocarditis. *Rev Med Chil* 1986;114:771–776.

122. Keshgegian AA. Prevalence of small monoclonal proteins in the serum of hospitalized patients. *Am J Clin Pathol* 1982;77:436–442.

123. Keshgegian AA. Oligoclonal banding in sera of hospitalized patients. *Clin Chem* 1992;38:169.

124. Chang KL, Flaris N, Hickey WF, Johnson RM, Meyer JS, Weiss LM. Brain lymphomas of immunocompetent and immunocompromised patients: study of the association with Epstein–Barr virus. *Mod Pathol* 1993;6:427–432.

125. Surico G, Muggeo P, Muggeo V, Lucarelli A, Novielli C, Conti V, Rigillo N. Polyclonal hypergammaglobulinemia at the onset of acute myeloid leukemia in children. *Ann Hematol* 1999;78:445–448.

126. Mouzaki A, Matthes T, Miescher PA, Beris P. Polyclonal hypergammaglobulinaemia in a case of B-cell chronic lymphocytic leukaemia: the result of IL-2 production by the proliferating monoclonal B cells? *Br J Haematol* 1995;91:345–349.

127. Hirokawa M, Lee M, Kitabayashi A, et al. Autoreactive T cell-dependent polyclonal hypergammaglobulinemia in mantle cell lymphoma. *Leuk Lymphoma* 1994;14:509–513.

128. Machii T, Yamaguchi M, Inoue R, et al. Polyclonal B-cell lymphocytosis with features resembling hairy cell leukemia-Japanese variant. *Blood* 1997;89:2008–2014.

129. McCarty MJ, Vukelja SJ, Banks PM, Weiss RB. Angiofollicular lymph node hyperplasia (Castleman's disease). *Cancer Treat Rev* 1995;21:291–310.

130. Menke DM, Camoriano JK, Banks PM. Angiofollicular lymph node hyperplasia: a comparison of unicentric, multicentric, hyaline vascular, and plasma cell types of disease by morphometric and clinical analysis. *Mod Pathol* 1992;5:525–530.

131. Ohsaka A, Saito K, Sakai T, Mori S, Kobayashi Y, Amemiya Y, Sakamoto S, Miura Y. Clinicopathologic and therapeutic aspects of angioimmunoblastic lymphadenopathy-related lesions. *Cancer* 1992;69:1259–1267.

132. Attygalle A, Al-Jehani R, Diss TC, et al. Neoplastic T cells in angioimmunoblastic T-cell lymphoma express CD10. *Blood* 2002;99:627–633.

133. Strupp C, Aivado M, Germing U, Gattermann N, Haas R. Angioimmunoblastic lymphadenopathy (AILD) may respond to thalidomide treatment: two case reports. *Leuk Lymphoma* 2002;**43**: 133–137.

134. Schultz DR, Yunis AA. Immunoblastic lymphadenopathy with mixed cryoglobulinemia. A detailed case study. *N Engl J Med* 1975;**292**: 8–12.

135. Chodirker WB, Komar RR. Angioimmunoblastic lymphadenopathy in a child with unusual clinical and immunologic features. *J Allergy Clin Immunol* 1985;**76**:745–752.

136. Hsu SM, Waldron JA Jr, Fink L, et al. Pathogenic significance of interleukin-6 in angioimmunoblastic lymphadenopathy-type T-cell lymphoma. *Hum Pathol* 1993;**24**: 126–131.

137. Steinberg AD, Seldin MF, Jaffe ES, et al. NIH conference. Angioimmunoblastic lymphadenopathy with dysproteinemia. *Ann Intern Med* 1988;**108**:575–584.

138. Whitten RO, Zutter M, Iaci-Hall J, Odell M, Kidd P. Oligoclonal immunoglobulin heavy chain gene rearrangement in a childhood immunoblastic lymphoma. Presentation as a polyphenotypic atypical lymphoproliferative reaction. *Am J Clin Pathol* 1990;**93**:286–293.

139. Tienhaara A, Irjala K, Rajamaki A, Pulkki K. Four monoclonal immunoglobulins in a patient with chronic lymphocytic leukemia. *Clin Chem* 1986;**32**:703–705.

140. Jain A, Nalesnik M, Reyes J, et al. Posttransplant lymphoproliferative disorders in liver transplantation: a 20-year experience. *Ann Surg.* 2002;**236**:429–436.

141. Nepomuceno RR, Snow AL, Robert Beatty P, Krams SM, Martinez OM. Constitutive activation of Jak/STAT proteins in Epstein–Barr virus-infected B-cell lines from patients with posttransplant lymphoproliferative disorder. *Transplantation* 2002;**74**:396–402.

142. Dunphy CH, Gardner LJ, Grosso LE, Evans HL. Flow cytometric immunophenotyping in posttransplant lymphoproliferative disorders. *Am J Clin Pathol* 2002;**117**:24–28.

143. Berney T, Delis S, Kato T, et al. Successful treatment of posttransplant lymphoproliferative disease with prolonged rituximab treatment in intestinal transplant recipients. *Transplantation* 2002;**74**:1000–1006.

144. Craig FE, Gulley ML, Banks PM. Posttransplantation lymphoproliferative disorders. *Am J Clin Pathol* 1993;**99**:265–276.

145. Badley AD, Portela DF, Patel R, et al. Development of monoclonal gammopathy precedes the development of Epstein–Barr virus-induced posttransplant lymphoproliferative disorder. *Liver Transpl Surg* 1996;**2**:375–382.

146. Stevens SJ, Verschuuren EA, Pronk I, et al. Frequent monitoring of Epstein–Barr virus DNA load in unfractionated whole blood is essential for early detection of posttransplant lymphoproliferative disease in high-risk patients. *Blood* 2001;**97**:1165–1171.

147. Conley ME. B cells in patients with X-linked agammaglobulinemia. *J Immunol* 1985;**134**: 3070–3074.

148. Conley ME, Rohrer J, Minegishi Y. X-linked agammaglobulinemia. *Clin Rev Allergy Immunol* 2000;**19**:183–204.

149. Vihinen M, Kwan SP, Lester T, et al. Mutations of the human BTK gene coding for bruton tyrosine kinase in X-linked agammaglobulinemia. *Hum Mutat* 1999;**13**: 280–285.

150. Noordzij JG, de Bruin-Versteeg S, Comans-Bitter WM, et al. Composition of precursor B-cell compartment in bone marrow from patients with X-linked agammaglobulinemia compared with healthy children. *Pediatr Res* 2002;**51**:159–168.

151. Siegel RL, Issekutz T, Schwaber J, Rosen FS, Geha RS. Deficiency of T helper cells in transient hypogammaglobulinemia of infancy. *N Engl J Med* 1981;**305**:1307–1313.

152. Kilic SS, Tezcan I, Sanal O, Metin A, Ersoy F. Transient hypogammaglobulinemia of infancy: clinical and immunologic features of 40 new cases. *Pediatr Int* 2000;**42**:647–650.

153. Rosefsky JB. Transient hypogammaglobulinemia of infancy. *Acta Paediatr Scand* 1990;**79**: 962–963.

154. Busse PJ, Razvi S, Cunningham-Rundles C. Efficacy of intravenous immunoglobulin in the prevention of pneumonia in patients with common variable immunodeficiency. *J Allergy Clin Immunol* 2002;**109**:1001–1004.

155. Cunningham-Rundles C. Common variable immunodeficiency. *Curr Allergy Asthma Rep* 2001;**1**:421–429.

156. Sander CA, Medeiros LJ, Weiss LM, Yano T, Sneller MC, Jaffe ES. Lymphoproliferative lesions in patients with common variable immunodeficiency syndrome. *Am J Surg Pathol* 1992;**16**:1170–1182.

157. Besa EC. Recent advances in the treatment of chronic lymphocytic leukemia: defining the role of intravenous immunoglobulin. *Semin Hematol* 1992;**29**:14–23.

158. Filomena CA, Filomena AP, Hudock J, Ballas SK. Evaluation of serum immunoglobulins by protein electrophoresis and rate nephelometry before and after therapeutic plasma exchange. *Am J Clin Pathol* 1992;**98**:243–248.

6

Conditions associated with monoclonal gammopathies

A monoclonal gammopathy is defined as the electrophoretically and antigenically homogeneous protein product of a single clone of B lymphocytes and/or plasma cells that has proliferated beyond the constraints of normal control mechanisms. Monoclonal gammopathies are detected in serum and/or urine from individuals with a wide variety of neoplastic, potentially neoplastic, neurological and infectious conditions. The monoclonal gammopathy is produced by a single clone of plasma cells or B lymphocytes. Several terms have been used to refer to monoclonal gammopathies: paraprotein, dysproteinemia, monoclonal gammo-pathy, monoclonal component, and M-protein.[1,2] The guidelines for clinical and laboratory evaluation of patients with monoclonal gammopathies

has recommended the use of the term 'M-protein' and I will use it in this chapter interchangeably with monoclonal gammopathy.[3]

DIFFERENTIATION OF B LYMPHOCYTES

The discovery that there are major subpopulations of lymphocytes resulted from careful observations of immune deficiency in humans and a serendipi-tous discovery in bursectomized chickens. In 1952, Bruton[4] described a child that suffered from recurrent infections with pyogenic bacteria. Using serum protein electrophoresis, then a relatively new clinical laboratory test, he demonstrated that this child's serum lacked a γ-globulin region. The patient was treated successfully by administering γ-globulin parenterally. Such patients are now known to have Bruton's X-linked agammaglobulinemia (XAG), which is caused by deficient maturation of the B lymphocytes as a result of the absence of Bruton's tyrosine kinase (BTK).[5] A few years after Bruton's report, Glick et al.[6] discovered that removal of the cloacal Bursa of Fabricius early in the life of chickens produced agammaglobulinemia similar to that found in Bruton's patients. Both bursectomized chickens and individuals with XAG lack plasma cells in their tissues, and neither has germinal centers in their lymphoid tissues. Nonetheless, normal numbers of peripheral blood lymphocytes are present, and the patients do not have problems with viral, fungal, or intracellular bacterial infections. We now know that the remaining lymphocytes are the normal peripheral blood T-lymphocytes, which play the main role in host defense against viral, fungal, and intracellular bacterial infections. The B-lymphocytes are so named because in the chicken they are derived from the *Bursa of Fabricius*. In humans, B- and T-cells originate in the bone marrow. However, T-cells must be processed subsequently in the thymus gland, while B-lymphocytes are subsequently processed in the fetal liver and bone marrow.

Differentiation of B-lymphocytes from hematopoietic stem cells to plasma cells is a continuous process, however, it is useful to divide the process into maturation before and after activation by antigen. The earliest stage of B-lymphocyte development occurs in the fetal liver and continues in adult life in the bone marrow (Table 6.1).[7] The availability of cluster designation (CD) monoclonal antibody markers to lymphocyte surface antigens has dramatically improved the delineation of the stages of B-lymphocyte maturation and fine-tuned the characterization of lymphoproliferative disorders. At the stem cell stage CD34 is expressed.[8] Following this, the earliest B-lymphocytes (pro-B cells) express CD19 (a marker present at all stages of B-cell maturation), and HLA-DR (an HLA Class II antigen prominently expressed on B-cells) in addition to CD34. These early B- lymphocytes also contain the enzyme terminal deoxyribonucleotidyl transferase (TdT) in the nucleus. At this time, the B cells have already begun the process of immunoglobulin gene rearrangement that will result in the production of antibody directed against a single epitope. Even before the pre-B cell stage, the cell has rearranged its μ chain gene, although the μ heavy chain cannot yet be detected in the cytoplasm.[9,10] These 'pre-pre-B' cells usually express the major histocompatability antigen HLA-DR, and have surface markers that are recognized by CD10 (common acute lymphoblastic leukemia antigen, CALLA), CD19 and CD20 (dim).[11]

Table 6.1 Maturation of B lymphocytes before activation by antigen

Stage of maturation	Markers	Immunoglobulin gene
Stem cell	CD34	Germline
Pro-B-cell	CD34, CD19, TdT, HLA-DR	Germline
Pre-pre-B-cell	CD34, CD19, CD20 (dim), CD10, TdT, HLA-DR	Rearranged
Pre-B-cell	CD10, CD19, CD20, HLA-DR, cμ[a]	Rearranged
Mature B-cell	CD19, CD20, CD21, HLA-DR, sIgD and/or sIgM[b]	Rearranged

[a]cμ, cytoplasmic μ-heavy chain.
[b]sIgD and/or sIgM, surface IgD and/or surface IgM.

Around the eighth week of human gestation, large lymphoid cells with a small amount of detectable cytoplasmic μ heavy chain but no detectable light chains are present in the fetal liver. These are termed 'pre-B' cells, they do not show surface expression of the μ chain at this stage.[12,13] Pre-B cells already have selected the variable region that will be part of the immunoglobulin heavy chain that their plasma cell progeny will eventually produce. Pre-B cells divide at a rapid rate (generation time is about 12 h), leading to the production of small pre-B lymphocytes that still contain cytoplasmic μ. At this stage, allelic exclusion occurs wherein either the κ or λ gene is selected for production. Although there will be subsequent switches in heavy chain expression, the light chain remains constant for this clone. On the surface of the cell, one finds CD10, HLA-DR, CD19, and CD20. A small subset of pre-B cells also express CD5 (usually a T-cell marker).[7]

The next stage of development can be recognized in the fetus by the tenth to twelfth week of gestation.[12] These mature B cells contain surface whole IgM and/or IgD molecules with the selected light chain, but they no longer contain the cytoplasmic μ chain. The variable regions of both the light and heavy chain are the same as those in the immunoglobulins ultimately produced by the plasma cell progeny of these B cells. Although IgD and/or IgM are the major surface isotypes at this stage, they will not usually be the final isotype produced by this clone. At this point they lose their CD10 marker while they continue to express CD19, CD20 and, when they pass into the circulation, CD21.[7]

The genes that direct the expression of the heavy chain undergo rearrangement as the B-lymphocytes mature to plasma cells. Whereas the light chain is always the same, these mature B cells may express more than one isotype on their surface; understanding this helps to explain the occurrence of some biclonal gammopathies. In most biclonal gammopathies, both monoclonal proteins have the same light chain type. When the light chains are the same, it is likely that we are seeing expression of two heavy chain genes by the same clone. This recapitulates the events seen during development. When the light chain types differ, the monoclonal components of a biclonal gammopathy truly arise from different clones.[14]

Following contact between B-lymphocytes and antigen with the appropriate costimulation by macrophages and helper T cells, the B-lymphocytes become activated. At this level of maturity, B-lymphocytes migrate to secondary lymphoid tissues where further differentiation occurs in the lymphoid follicles.[7,15] Here, the surface antigen expression varies depending upon the location of the B-lymphocytes within the lymph node (Table 6.2).

The final differentiation of activated B cells to plasma cells results in loss of expression of surface immunoglobulin, HLA-DR, CD20, and CD40. Normal bone marrow plasma cells are strongly positive for CD38 and CD79a (an immunoglobulin-associated antigen), express CD19 and CD45 weakly, but lack expression of CD56.[16–19]

Table 6.2 CD markers in lymph nodes[a]

Location within lymph node	Markers
Mantle zone and marginal zone	CD21, CD22, CD23, CD24, CD39, CDw76,
Follicular center	CD10, CD38(dim), CD77, CD22(dim), CD23(dim), CD24(dim), CD79a(dim)
Pan lymph node	CD19, CD20, CD37, CD40, CD45, CD72, CD74, CD275, CDw78

[a]Data from Pirruccello and Aoun.[7]

CONDITIONS ASSOCIATED WITH MONOCLONAL GAMMOPATHIES

Monoclonal gammopathies are found in a wide variety of conditions, benign lymphoproliferative disorders, malignant lymphoproliferative disorders, infectious conditions, neuropathies, and poorly understood conditions. At one end of the spectrum of conditions associated with monoclonal gammopathies is multiple myeloma; at the other end is monoclonal gammopathy of undetermined significance (MGUS) (Table 6.3).

Most monoclonal gammopathies found in serum are classified as MGUS. They usually have no specific symptoms (although they may eventually progress to a malignant lymphoproliferative condi-

Table 6.3 Electrophoretic patterns associated with monoclonal gammopathies

Condition	Typical serum electrophoretic pattern
I. Multiple myeloma	
IgG	Large γ spike
IgM	Broad spike at origin
IgA	Broad β spike
IgD	Small γ or β spike
IgE	Small β spike
Light chain disease (κ or λ)	Hypogammaglobulinemia, occasional β spike
Heavy chain disease (α, μ, γ)	Broad β band
Biclonal (double) gammopathy	Two γ spikes
Nonsecretory	Low γ region, normal
II. Waldenström's macroglobulinemia	
IgM	Broad band near origin
IgA	Broad band near origin or slightly β
IgG	Broad band near origin or slightly γ
III. B-lymphoproliferative disorders	
Chronic lymphocytic leukemia	Low γ, small band
Well-differentiated lymphocytic lymphoma	Low γ, small band
IV. Amyloidosis (AL)	
With multiple myeloma	γ Spike
Not associated with myeloma	Normal, low γ
V. Monoclonal gammopathy with other clinical correlation	
Autoimmune diseases	Small γ spike
Neuropathy	Small usually IgM β–γ restriction
Infectious disease	Small γ spike usually with a diffuse or oligoclonal increase in γ-globulin
VI. Monoclonal gammopathy of undetermined significance	Small γ spike

tion). Malignant lymphoproliferative conditions associated with monoclonal gammopathies involve bone marrow, soft tissues and produce immunoglobulin products that may have peculiar characteristics, such as self-aggregation (resulting in hyperviscosity), cryoprecipitation (occasionally producing life-threatening vascular problems), or specific reactivity (producing neuropathies, coagulation defects or problems with various laboratory tests). Consequently, the clinical signs and symptoms that precipitate the performance of screening tests for malignant lymphoproliferative disorders associated with monoclonal gammopathies are varied (Table 6.4).

There is not a precise cut-off in terms of gravimetric amount of the monoclonal gammopathy for the clinical laboratory to separate malignant from 'benign' monoclonal gammopathies. The sensitivity of electrophoretic techniques to detect M-proteins varies from one method to another. Methods with higher resolution can detect subtle M-proteins. M-proteins located in the β-region may

be obscured by transferrin, C3 or β-lipoprotein. Techniques that provide a crisp transferrin and C3 band are more likely to disclose the presence of subtle M-proteins. Finding small monoclonal proteins in the serum can be critically important in cases of light chain myeloma (Fig. 6.1).

The vast majority of the M-proteins found are MGUS (see below). They are present in asymptomatic individuals and have no apparent clinical significance at the time of their detection and hence fit into the MGUS category established by Kyle.[20–22] Some may, however, progress to clinically significant lymphoproliferative disorders and are therefore associated with an increased risk in mortality.[23] Others are associated with infectious diseases and usually are evanescent.[24–26]

Because of the heterogeneity of the conditions associated with monoclonal gammopathies, differences in their occurrence by race, as well as the differences in sensitivity of the electrophoretic techniques used, estimates of their prevalence vary in the medical literature. Kyle[27] reported the prevalence of

Table 6.4 Clinical features associated with monoclonal gammopathies

Clinical feature	Possible monoclonal-associated disorder
None	Monoclonal gammopathy of undetermined significance
Back pain	Myeloma
Osteolytic lesions	Myeloma
Unexplained fatigue	Myeloma, Waldenström's
Elevated sedimentation rate	Myeloma, Waldenström's
Nephrotic syndrome	Myeloma (monoclonal free light chain – Bence Jones protein), Amyloidosis, MIDD[a]
Infections associated with immunoglobulin deficiency	Myeloma
Congestive heart failure	Amyloidosis, MIDD[a]
Carpal tunnel syndrome	Amyloidosis
Dizziness	Waldenström's
Anemia	Myeloma, Waldenström's
Peripheral neuropathy	Monoclonal anti-myelin-associated glycoprotein

[a]MIDD, monoclonal immunoglobulin deposition disease. This includes light chain deposition disease (LCDD), heavy chain deposition disease (HCDD), and a combination of the two.

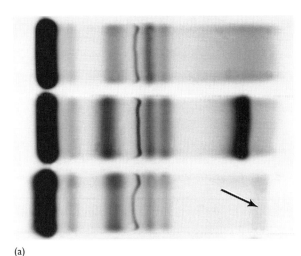

(a)

Figure 6.1 (a) Serum protein electrophoresis of three serum samples. The top sample is normal. The middle sample contains an obvious mid-γ-monoclonal gammopathy. The bottom sample has a normal albumin, but a decreased γ-globulin and a very small slow γ-restriction (arrow). Although the slow γ-restriction is very small in the bottom lane, the presence of decreased γ-globulin makes this band highly suspicious. (Paragon SPE2 system stained with Paragon Violet.) (b) Immunofixation of urine from the patient shown in (a). A large slow γ-band is present (arrow), which at the 100-fold concentrated urine reacts only with κ. A second much fainter band is seen toward the anode (monomer and dimer monoclonal free light chains are not uncommon). In the center of the κ (K) band, one sees an antigen excess effect (X). (Paragon system stained with Paragon Violet; anode at the top.)

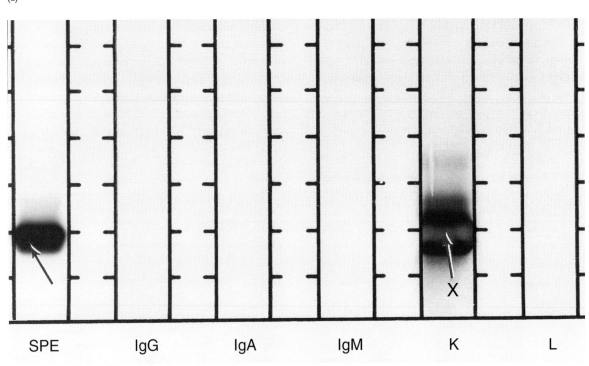

SPE   IgG   IgA   IgM   K   L

X

(b)

monoclonal gammopathies to be about 1.5 per cent of the population over 50 years of age and about 3 per cent of the population over 70 years of age. Looking at different age groups, Axelsson et al.[25] found monoclonal gammopathies in 2 per cent of Swedish subjects between the ages of 70 years and 79 years and 5.7 per cent in the subjects older than 80 years. Cohen et al.[28] evaluated samples from individuals older than 70 years of age using the Paragon SPE2 system (Beckman Coulter, Fullerton, CA, USA) and reported that of 1732 individuals, 8.4 per cent of blacks and 3.8 per cent of whites had monoclonal gammopathies. This finding is notable because multiple myeloma also occurs approximately twice as often in blacks than in whites.[29,30] Lastly, using a high-resolution acetate method, Aguzzi detected monoclonal gammopathies in 7–8 per cent of patients over 55 years of age.[31]

Nonetheless, these studies all point out the common occurrence of M-proteins and agree that the occurrence of monoclonal gammopathies progressively increases with age.

MULTIPLE MYELOMA

Clinical picture in myeloma

Multiple myeloma is a malignant neoplasm, expressed mainly as a proliferation of plasma cells, that usually presents with bone marrow involvement.[32-34] While the plasma cells form the vast majority of the malignant cells, remnants of the parent B-cell clone and other B-clonal precursors persist and play a role in the evolution of the neoplastic process.[35-37] Typically, these malignant plasma cells synthesize and secrete considerable amounts of monoclonal whole immunoglobulin or fragments of immunoglobulin and/or monoclonal free light chains (MFLC; Bence Jones proteins). Depending on their structure and concentration, these M-proteins may increase the viscosity of blood or precipitate in various tissues or form a cryoglobulins.[38-41] Rare cases of non-secretory myeloma produce no detectable immunoglobulin in serum or urine when studied by conventional electrophoretic techniques. About 15 per cent of

cases of multiple myeloma make only MFLC and may be mistaken for non-secretory myeloma if the urine is not examined because monomeric MFLC are more readily detected in urine than in serum by electrophoretic techniques. However, a new quantitative immunoassay can measure free light chains (unbound to heavy chains) in both serum and urine. As discussed below, this technique will improve detection of cases of light chain disease and should provide an improvement over 24-h urine samples to follow the tumor burden of these patients.

Patients typically present with fatigue, bone pain and pathological fractures. The prominent bone marrow involvement in this disease is associated with lytic lesions in the ribs, vertebrae, skull, and long bones. The bone involvement results in hypercalcemia and a normochromic, normocytic anemia. In addition to its effects on the bone marrow, monoclonal free light chains damage the kidneys and other organs occasionally with deposition of MFLC or amyloid AL (amyloid associated with immunoglobulin light chain; see chapter 7).[42-47] Suppression of normal γ-globulins in advanced cases is associated with recurrent bacterial infections.[48] Ong et al.[49] looked at the medical histories of 127 patients diagnosed with multiple myeloma to see which symptoms were associated with patients who were diagnosed quickly and which with those in whom the diagnosis was delayed. As shown in Table 6.5,

Table 6.5 Presenting signs and symptoms of multiple myeloma[a]

Symptom or sign	Delayed diagnosis	Diagnosis at first presentation
Bone pain/fractures	11 (23%)	59 (74%)[b]
Malaise	9 (19%)	37 (46%)
Infection/fever	4 (9%)	8 (10%)
Anemia	12 (26%)	47 (59%)[b]
Hypercalcemia	1 (2%)	18 (23%)[b]
Hyperviscosity/bleeding	1 (2%)	7 (9%)
Elevated ESR	11 (23%)	16 (20%)

[a]Data from Ong et al.[49] ESR, erythrocyte sedimentation rate.
[b]Chi-square significance $P \leq 0.001$.

those with delayed diagnosis were significantly less likely to have bone pain and/or fractures, anemia or hypercalcemia.[49]

Multiple myeloma accounts for about 1 per cent of all malignancies and 10 per cent of hematopoietic malignancies.[50] Although the incidence increases dramatically with age, multiple myeloma has been reported (rarely) in a few children.[51-53] However, the clinical picture is often unusual, lacking key features such as anemia, elevated calcium (in about one-third of the patients), and lytic lesions. It has been better documented in a few young adults, but even this is exceptionally rare.[54] Individuals under the age of 40 years account for only 2 per cent of myeloma cases and those under 30 years comprise just 0.3 per cent.[55]

Nonetheless, the possibility of this disease should not be ignored when appropriate symptoms are present.[56-58] A comparison of the clinical presentations and laboratory features of the disease revealed no significant differences between those that present with multiple myeloma prior to 50 years of age and those who present after age 50 years.[59] Although prognosis for younger patients does not generally differ from that of older individuals, young patients with good renal function and low β_2-microglobulin levels have a longer survival than the older patients and may benefit from aggressive therapy.[55,59] The overwhelming majority of children with monoclonal gammopathies do not have multiple myeloma. They usually have B-cell lymphoproliferative lesions that are sometimes related to the presence of primary and secondary immunodeficiency diseases, occasionally to autoimmune diseases and also to hematological malignancies other than myeloma.[60-63]

There have been dramatic improvements in the therapy offered for myeloma patients in the past few years. The therapy offered depends on the stage of the disease (see below). Depending on their performance status, newly diagnosed cases of multiple myeloma may now be treated with stem cell transplantation rather than the previous chemotherapeutic approach with melphalan and prednisone.[50] Kyle[64] recommends that patients under the age of 70 years be considered for autologous

peripheral blood stem cell transplantation. He notes that the main issues at the present time for this therapy are the ability of chemotherapeutic agents to destroy the tumor cells in the patient and the need to remove myeloma cells and their precursors from the harvested stem cells that are given back to the patient to reconstitute the bone marrow.[64] This treatment has been highly successful in some cases (Fig. 6.2). In addition to more conventional chemotherapeutic agents, thalidomide and new immunomodulatory agents show promise.[65] Nonetheless, despite these new techniques and aggressive chemotherapy, the prognosis is poor for most cases.[50]

Staging

Prognosis in myeloma is highly dependent on clinical and laboratory features. The system devised by Durie and Salmon[66] is still largely used to predict prognosis (Table 6.6). Based on their criteria, individuals with stage I, II and III disease had median survivals of 38, 35 and 13 months, respectively.[67] Using a multivariate analysis of 265 patients with multiple myeloma, Cherng and coworkers[68] found that the three most important factors were plasmacytosis (> 30 per cent), hypercalcemia (> 11.5 mg/dl), and hypoalbuminemia (< 3.5 g/dl).[69] Other factors predicting an unfavorable course include elevated alkaline phosphatase, hyperuricemia, renal insufficiency, and male gender.[70]

Flow cytometry studies are also improving our ability to stage patients with multiple myeloma. Patients with plasmablasts positive for CD56 had a 63-month overall survival compared with 22 months for those that had CD56-negative plasmablasts.[71] Others have shown that hyperdiploidy on initial bone marrow examination has a significantly better overall survival than diploid or hypodiploid tumor cells.[72] This area is still somewhat controversial, however, because although an Eastern Cooperative Oncology Group (ECOG) study was able to confirm the survival advantage of hyperdiploid versus diploid, it was unable to confirm the worse prognosis of the hypodiploid cases.[73] Clearly, the issue of staging will be undergoing changes.

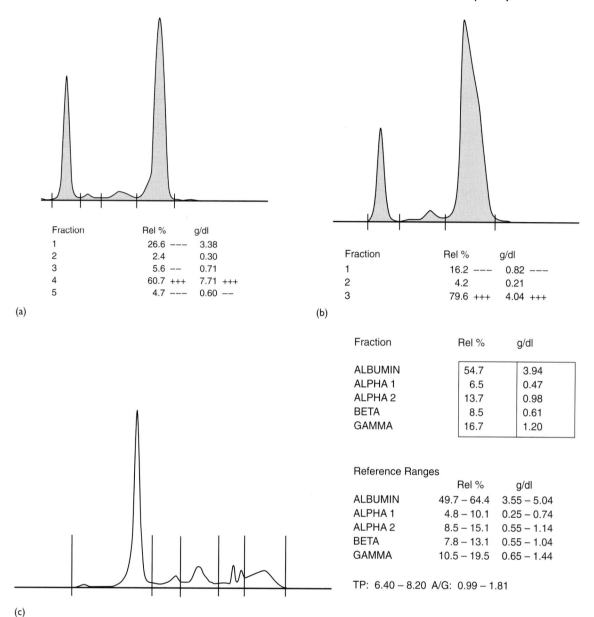

Fraction	Rel %	g/dl
1	26.6 ---	3.38
2	2.4	0.30
3	5.6 --	0.71
4	60.7 +++	7.71 +++
5	4.7 ---	0.60 --

(a)

Fraction	Rel %	g/dl
1	16.2 ---	0.82 ---
2	4.2	0.21
3	79.6 +++	4.04 +++

(b)

Fraction	Rel %	g/dl
ALBUMIN	54.7	3.94
ALPHA 1	6.5	0.47
ALPHA 2	13.7	0.98
BETA	8.5	0.61
GAMMA	16.7	1.20

Reference Ranges

	Rel %	g/dl
ALBUMIN	49.7 – 64.4	3.55 – 5.04
ALPHA 1	4.8 – 10.1	0.25 – 0.74
ALPHA 2	8.5 – 15.1	0.55 – 1.14
BETA	7.8 – 13.1	0.55 – 1.04
GAMMA	10.5 – 19.5	0.65 – 1.44

TP: 6.40 – 8.20 A/G: 0.99 – 1.81

(c)

Figure 6.2 (a) Densitometric scan of serum protein electrophoresis from a patient with multiple myeloma prior to receiving a bone marrow transplantation. The massive monoclonal gammopathy in the β-region measures 7.71 g/dl. (Scan from Paragon SPE2 gel stained with Amido Black.) (b) Urine from pre-transplant analysis demonstrated a massive monoclonal free light chain (MFLC) (> 5 g/24 h). (c) Capillary zone electropherogram of serum protein electrophoresis from same patient as in (a) 9 years after bone marrow transplant. The patient is clinically well with no sign of multiple myeloma. The urine is negative for MFLC.

Epidemiology

There has not been a change in the rates of development of multiple myeloma for the past half century.[74] The annual incidence of multiple myeloma is 5–10 per 100 000.[75] In 2002, estimates from the American Cancer Society predicted 14 600 new cases of multiple myeloma in the USA.[76] Multiple myeloma remains twice as common among blacks (mean age of onset

Table 6.6 Myeloma staging criteria[a]

Stage	Criteria
I	Must have all of the following:
	Hemoglobin > 10g/dl
	Serum calcium < 12 mg/dl
	No lytic lesions or solitary plasmacytoma on radiographs
	Relatively low quantity of M-protein:
	IgG < 5g/dl
	IgA < 3g/dl
	Urine monoclonal free light chains < 4 g/24 h
II	Lacks criteria for stage III, but exceeds criteria for stage I
III	Has criteria for stage I, and has any one or more of the following:
	Hemoglobin < 8.5g/dl
	Serum calcium > 12 mg/dl
	Advanced lytic bone lesions
	Relatively large quantity of M-protein:
	IgG > 7 g/dl
	IgA > 5 g/dl
	Urine monoclonal free light chains = 12 g/24h
	Subclassification:
	A. serum creatinine < 2.0 mg/dl
	B. serum creatinine = 2.0 mg/dl

[a]Table from Durie and Salmon.[66]

67 years) than among whites (mean age of onset 71 years) and is somewhat more common in men than women.[29,30,75] These differences may relate to socioeconomic factors including diet and dietary supplement use (such as vitamin C).[77,78] There is a clear association of multiple myeloma with radiation exposure in Japanese survivors of the atomic bomb, radium dial workers, and even among radiologists.[79–83] Although there has been a report that individuals who work at nuclear power plants may have an increased risk for myeloma, there does not appear to be a risk for those living near the plants.[84]

Neither does there appear to be a risk of developing multiple myeloma associated with the small doses of radiation one receives through diagnostic radiology procedures.[85]

Genetics and cytogenetics

There are suggestions of a genetic predisposition to the development of monoclonal gammopathies. Many studies have recorded multiple myeloma in first degree relatives.[86–99] Individuals that report having first-degree relatives with multiple myeloma have a 3.7 times greater chance of developing multiple myeloma than those that do not share this family history.[86] Broader epidemiological evidence that genetics plays a role derives from Bowden et al. who documented that monoclonal gammopathies are more often detected in Americans than in a Japanese cohort.[100] However, while genetic differences may play a role, the American population had a larger proportion of older individuals, which may have accounted for some of their results.

Cytogenetic abnormalities have been found in multiple myeloma.[101] The most typical translocations occur at the site of the immunoglobulin heavy chain locus in switch regions.[102,103] Fluorescence *in situ* hybridization (FISH) probes are available to detect the presence of translocations on chromosome 11 at position q13.[104] Cytogenetics may be especially useful in cases of non-secretory myeloma where M-proteins are not demonstrable in serum or urine by traditional electrophoretic methods.[105] Karyotypic instability has been detected at the earliest stages of multiple myeloma and increases along with progression of the disease.[106] These cytogenetic abnormalities may eventually explain why myeloma plasma cells do not undergo their usual programmed death by apoptosis.[48]

Cellular features of myeloma

Most neoplastic plasma cells in multiple myeloma lack surface immunoglobulins despite the secretions they produce. Bone marrow plasma cells from patients with multiple myeloma express CD45, CD38, and CD79a but, in contrast to normal plasma cells, they have an aberrant expression of CD56 (the natural killer cell antigen) and CD19 (CD19–/CD56+ or CD19–/CD56–).[16,17] This surface marker information may be useful in the differential diagnosis between multiple myeloma and MGUS. Sezer et al.[17] noted that all MGUS patients in their study contained both phenotypically normal (CD19+/CD56–) plasma cells and aberrant monoclonal populations (CD19–/CD56+ and/or CD19–/CD56–). The ratio of the normal to all bone marrow plasma cells was always 20 per cent. In contrast, individuals with multiple myeloma had either no phenotypically normal plasma cells (CD19+/CD56–), or when they were present always totaled less than 20 per cent of the bone marrow plasma cells.[17]

A histological correlate of this immaturity and malignant behavior is the presence of large atypical cells that may be seen, occasionally with binucleate cells, in the involved marrow. However, atypical nuclei and multinucleate plasma cells are insufficient criteria to absolutely distinguish between proliferating polyclonal plasma cells in chronic osteomyelitis (for example) and monoclonal plasma cells of multiple myeloma.

Although multiple myeloma is mainly a disease of malignant plasma cells, both circulating B-lymphocytes and even pre-B cells in the bone marrow can be found that express the specific idiotype of the monoclonal protein being produced by the myeloma cells from that patient.[107–110] The finding of such precursor cells may seem surprising considering that the neoplasm is characterized by mature-appearing plasma cells. However, this is entirely consistent with our understanding of the variable maturation by neoplastic cells. For example, a squamous cell carcinoma may consist mainly of mature keratinized cells, yet one is not at all surprised to find some immature non-keratinizing elements in the same neoplasm. Indeed, it would be difficult to explain why a neoplasm of mature plasma cells would produce such a rapidly fatal disease if there were not progenitor cells that formed the 'silent' proliferative compartment of the neoplasm.

There are some markers that can help to detect the B-cell precursors of the neoplastic plasma cells in multiple myeloma. Unlike normal activated B cells, the circulating B cells in patients with multiple myeloma express adhesion molecules including β_1-integrins that may facilitate transportation of these cells as they metastasize (localize) in the bone marrow and other sites.[111] Despite the fact that the neoplastic B cells circulate, a case of myeloma in a 27-year-old pregnant woman whose child did not show evidence of myeloma after a 2-year follow-up suggests that these B cells do not readily pass through the placenta.[112] Although circulating CD19 lymphocytes are increased in myeloma patients, circulating mature plasma cells in these patients may be difficult to detect except in the extreme case of plasma cell leukemia (see below).

Although T-lymphocytes in myeloma are not thought to be involved directly in the neoplastic process, there are notable alterations in their peripheral blood subpopulations because they are affected by the myeloma process. The production of transforming growth factor β (TGF-β) by the myeloma cells suppresses the normal proliferation and blastogenic response of T-lymphocytes to interleukin-2 (IL-2).[113] This observation explains the decreased responsiveness of peripheral blood T-lymphocytes from patients with multiple myeloma to lectin stimulation and also the decrease in the number of CD4-positive cells.[114,115]

Individuals with multiple myeloma who have normal levels of CD4+ and CD19+ cells at the time of diagnosis have a better prognosis than those with decreased levels.[116,117] This observation has been extended by a recent ECOG study where the peripheral blood lymphocyte levels were followed from baseline to the end of a chemotherapy regimen. They found that when patients undergoing chemotherapy maintained the baseline numbers of CD3+, CD4+, CD8+ and CD19+ cells in the peripheral blood, they had a better long-term survival than those that did not.[118]

Cytokines play a major role in the growth and drug responsiveness of myeloma cells.[119,120] Both interleukin-6 (IL-6) and insulin-like growth factor

1 (IGF-1) are important in the proliferation and drug resistance of myeloma and may operate by upregulating telomerase activity of the malignant cells.[121] This interferes with dexamethasone-induced apoptosis in the myeloma cells.[122] Interestingly the plasma cells themselves can produce IL-6 along with the bone marrow cells. While not thought of as a currently used marker of the disease, the level of IL-6 in serum has been shown to correlate inversely with survival.[123] Patients with multiple myeloma also have higher serum levels of TGF-β than do patients with MGUS.[124] Interleukin 1 (IL-1) and IL-6 have both been implicated either as a marker or as part of the process resulting in conversion of MGUS to myeloma.[119] In addition to correlating with conversion of MGUS to myeloma, IL-6, tumor necrosis factor-α (TNF-α) and soluble interleukin-2 receptor (sIL-2r) have been used in experimental studies as markers of cases of known myeloma that have an aggressive advanced disease.[125] Again, these markers are not currently recommended to distinguish between these conditions.

Circulating plasma cells and plasma cell leukemia

Under normal circumstances, plasma cells are not detected in the peripheral blood. However, patients with multiple myeloma and patients with amyloid AL may have circulating monoclonal plasma cells.[126,127] Their presence in concentrations ≥ 4 per cent of immunoglobulin positive cells or ≥ 4 × 10⁶/l is a negative prognostic indicator and suggests the presence of active multiple myeloma.[126,127] The mere presence of plasma cells in the circulation of patients with multiple myeloma, however, does not equate with the diagnosis of plasma cell leukemia.

Plasma cell leukemia is an uncommon occurrence in myeloma (about 2 per cent of cases), making it a rare disease indeed.[128–130] To qualify, a patient needs to have an absolute plasma cell count of ≥ 2 × 10⁹/l and ≥ 20 per cent plasma cells in the peripheral

blood.[129–134] Two-thirds of cases of plasma cell leukemia do not have evidence of multiple myeloma at the time of presentation and are termed primary plasma cell leukemia.[129,131] The remainder of cases are usually the end-stage of a case of multiple myeloma.[129,131] Perhaps because of the late stage of presentation in the secondary form, plasma cell leukemia has a reputation for an aggressive course.[21]

Monoclonal gammopathies in multiple myeloma

IMMUNOGLOBULIN ISOTYPES IN MULTIPLE MYELOMA

The prevalence of the major immunoglobulin isotypes among large studies of monoclonal gammopathies roughly parallels the concentration of that immunoglobulin isotype in serum. For example, in his large series of patients with myeloma, Kyle found an IgG monoclonal protein in about 55–60 per cent of cases, IgA in about 25 per cent, and κ light chain was twice as frequent as λ (consistent with the normal 2:1 ratio of κ to λ in the serum).[135] About 15 per cent of cases only secrete MFLC, IgD accounts for about 1–2 per cent of cases, IgM in 0.5 per cent, and IgE is exceedingly rare.[136]

The vast majority of IgM monoclonal gammopathies associated with lymphoproliferative disorders are considered separately as Waldenström's in the Kyle series, but they typically comprise about 25–30 per cent of cases of monoclonal gammopathies.[135] Similar findings for the occurrence of major isotypes in multiple myeloma have been reported in smaller studies.[137,138] There are, however, some notable exceptions to the generalization that the occurrence of M-proteins in myeloma follows their concentration in the serum. Among the subclasses of IgG, Schur et al.[139] found significantly fewer cases of IgG2 than would be predicted by its concentration in the serum. Similar observations have been made about the infrequency of IgA2 monoclonal gammopathies.[140]

In most cases of myeloma, electrophoretic findings are straightforward. The characteristic densely staining spike typically occurs in the γ-region, near the origin, and in the β-region, for IgG, IgM, and IgA monoclonal proteins, respectively (Fig. 6.3). Note, however, that monoclonal gammopathies may migrate in the α-region (uncommonly), and they may bind to other serum proteins thereby altering their migration on electrophoretic gels.[141]

IGG AND IGA MYELOMA

IgG myeloma proteins are almost always 160 kDa monomers that only rarely produce clinical symptoms of hyperviscosity (although the measured viscosity of serum may be elevated in most patients with multiple myeloma).[142-144] IgA myeloma proteins can occur as monomers or as polymers with variable molecular weight (monomers, dimers, trimers and tetramers in the same myeloma serum).[145,146] Because these molecules can self-aggregate, they have been known to cause problems with hyperviscosity.[145,147-149] This polymerization may result in the presence of two or three M-protein peaks on the serum protein electrophoresis (Fig. 6.4). Since most of the IgA monoclonal gammopathies migrate in the β-region, they may be masked by the C3, β1 lipoprotein and transferrin bands on serum protein electrophoresis. Because of these concerns, the *Guidelines for Clinical and Laboratory Evaluation of Patients*

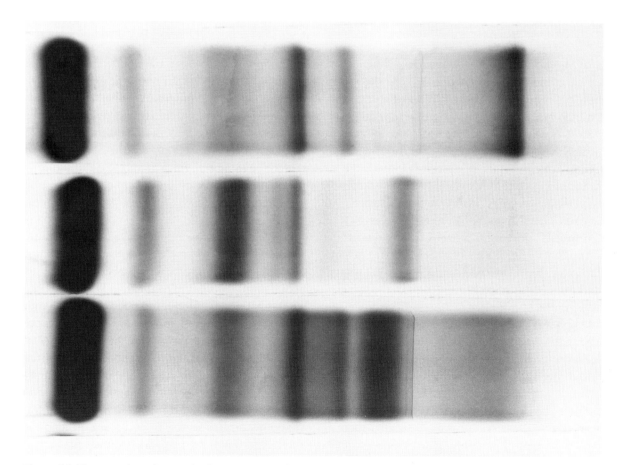

Figure 6.3 Three samples with monoclonal proteins in typical positions for their heavy chain class. The top sample has an IgG monoclonal protein migrating in the mid-γ-region. Middle sample has an IgM monoclonal protein near the origin. Bottom sample has a broad IgA monoclonal protein just cathodal to the C3 band. Although these are typical locations for monoclonal proteins of these isotypes, they may migrate at a variety of locations from α₂-region to the slow γ-region. (Panagel system stained with Coomassie Blue.)

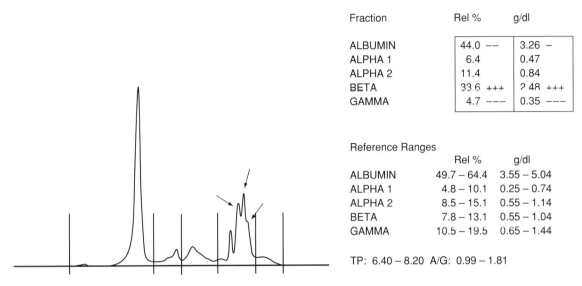

Fraction	Rel %	g/dl
ALBUMIN	44.0 −−	3.26 −
ALPHA 1	6.4	0.47
ALPHA 2	11.4	0.84
BETA	33.6 +++	2.48 +++
GAMMA	4.7 −−−	0.35 −−−

Reference Ranges

	Rel %	g/dl
ALBUMIN	49.7 – 64.4	3.55 – 5.04
ALPHA 1	4.8 – 10.1	0.25 – 0.74
ALPHA 2	8.5 – 15.1	0.55 – 1.14
BETA	7.8 – 13.1	0.55 – 1.04
GAMMA	10.5 – 19.5	0.65 – 1.44

TP: 6.40 – 8.20 A/G: 0.99 – 1.81

Figure 6.4 Capillary zone electrophoresis demonstrates three peaks (arrows) of an IgA κ monoclonal gammopathy. (Paragon CZE 2000.)

with Monoclonal Gammopathies recommends the use of immunofixation in cases where the serum protein electrophoresis is negative, but where there is a high index of suspicion.[3] In practice, however, in the laboratory we often do not know the clinical history. Therefore, anytime a distortion in the β-region protein bands is observed, I perform an immunofixation to rule out a subtle M-protein. By using the pentavalent (Penta) antisera available on semiautomated immunofixation gels, this can be done quickly and, using otherwise unused portions of the Penta gels, at very little cost. By performing this extra step I have detected cases of M-proteins where the major portion was found in the urine.

IGM MYELOMA

Although unusual, IgM may be the main immunoglobulin in cases of multiple myeloma, accounting for about 0.5 per cent of cases.[150,151] Clinically, these cases resemble the typical case of multiple myeloma but may have an increased propensity to hyperviscosity.[152] Typically, there is suppression of the uninvolved immunoglobulin classes, IgG and IgA.[153] IgM myeloma is associated with plasma cell proliferations as opposed to the more typical IgM monoclonal gammopathies (part of Waldenström's macroglobulinemia) in which

lymphoplasmacytoid cells are seen.[152] The use of immunophenotyping to confirm the presence of the plasma cell surface antigen CD38 is helpful to distinguish this condition from Waldenström's macroglobulinemia.[153,154] Another useful marker for making this distinction is the translocation t(11;14)(q13;q32), which is present in the vast majority of cases of IgM myeloma.[155]

IGD MYELOMA

IgD myelomas are uncommon, accounting for only 1–2 per cent of cases, but they have some characteristics of which one must be aware to avoid misdiagnosis.[138,156,157] Most IgD monoclonal gammopathies migrate in the γ-region, 25 per cent of cases have β-region spikes, and in one case the monoclonal component was in the α₂-region (Fig. 6.5).[157] IgD myeloma may be missed because the monoclonal gammopathy can be relatively small and it may be hidden among the normal α₂- or β-region proteins. A high index of suspicion about subtle distortions in the β- and γ-region helps in detecting them. Individuals with IgD myeloma tend to have a worse prognosis than most other isotypes because of the increased chance of renal involvement due to the high proportion of cases with monoclonal free light chains, amyloidosis and

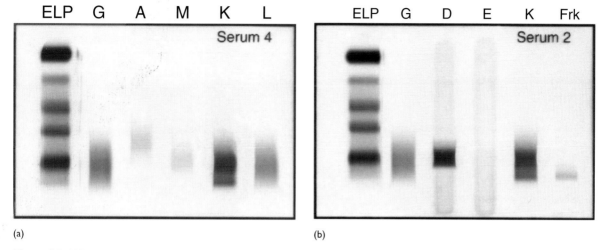

Figure 6.5 (a) Immunofixation of serum with mid-γ M-protein in the ELP lane. The M-protein is matched by a broad restriction in the κ (K) lane, but in none of the other lanes. (Sebia Immunofixation). (b) Immunofixation of serum from case shown in (a). Here, the IgD and κ lanes both demonstrate this IgD κ monoclonal gammopathy. The anti-free κ demonstrates that the extra band in the κ lane is due to κ monoclonal free light chain. (Sebia Immunofixation). This case was contributed by Chrissie Dyson.

extramedullary plasmacytomas, although, rarely, they may have good responses to chemotherapy.[138,156–159] Cytoplasmic crystalline inclusions and amyloidosis have been described in patients with IgD myeloma.[160,161] The latter may require immunohistochemical identification. Although an IgD monoclonal gammopathy is usually associated with a lymphoproliferative process, IgD MGUS does occur.[159] Cutaneous plasmacytomas have been reported rarely in patients with IgD myeloma.[162] Older literature reported the κ/λ ratio of cases with IgD myeloma to be 1:9 (as mentioned above, it is 2:1 for most other monoclonal gammopathies, paralleling the usual κ/λ in the serum); however, a more recent study of 53 cases at the Mayo Clinic revealed a ratio of κ/λ of 1:2 in their cases.[156] In any event, λ clearly predominates.

To follow the tumor burden in IgD myelomas, measurement of IgD may be performed by a variety of techniques, including nephelometry, radial immunodiffusion and enzyme immunoassay.[163,164] The IgD measurements may be more useful than the typical densitometric or electropherogram measurements of the M-protein because accuracy decreases with smaller quantities of the protein measured. IgD myeloma should always be suspected when a light chain is identified in the serum by immunofixation or immunoelectrophoresis but there is no corresponding heavy chain (γ, μ, or α). One should also have a higher index of suspicion for an IgD monoclonal gammopathy in cases containing λ than for κ spikes (reflecting the disproportionate κ/λ ratio in the occurrence of IgD myelomas).

IGE MYELOMA

IgE myeloma is rare (38 cases were reported in the most recent tally I found).[136] The reported patients have been younger than is typical for myeloma, and the disease pursues a relatively rapid course. Patients with IgE myeloma are more likely to have MFLC, anemia and plasma cell leukemia than patients with other isotypes (Table 6.7).[135] The average survival time from time of diagnosis is 16 months compared with 30 months for non-IgE multiple myeloma.[135,165] Osteolytic lesions have been found in 14 of 28 patients where skeletal information was available.[165] Other boney lesions include osteoporosis, vertebral collapse and one case with osteoblastic lesions.[165,166] In addition, two

Table 6.7 Characteristic features of IgE myeloma at clinical presentation[a]

Feature	IgE myeloma	Non-IgE myeloma
Monoclonal free light chain	59%	49%
Anemia	86%	62%
Plasma cell leukemia	18%	1–2%
Mean survival	16 months	30 months
κ/λ Ratio	1.8	2.0
α_2-Region migration by electrophoresis	2/29 cases	n/a
β-Region migration by electrophoresis	5/29 cases	n/a
γ-Region migration by electrophoresis	22/29 cases	n/a

[a]Data from Macro et al.[165] and Kyle.[135]

cases with hyperviscosity have been reported.[167,168] Small, occasionally diffuse peaks have been described in the few cases where the IgE monoclonal protein migrates in the β-region.[169,170] Macro et al.'s summary of 29 published cases where electrophoretic data was available is shown in Table 6.7.[165] As with IgD myelomas, when I detect a monoclonal light chain in the serum without a corresponding heavy chain (using the more usual IgG, IgA and IgM antisera) I perform an immunofixation to rule out an IgE M-protein.

MONOCLONAL FREE LIGHT CHAINS AND LIGHT CHAIN MYELOMA

As discussed in Chapter 7, plasma cells produce heavy and light chains separately. Under normal circumstances, these chains combine and are secreted as intact immunoglobulin molecules bearing two identical heavy chains and two identical light chains. Free light chains are produced in excess by normal plasma cells. These light chains are secreted as monomers (mainly κ) with a molecular mass of about 25 kDa, or as dimers (mainly λ) with a molecular mass of 50 kDa.[171–173] These polyclonal free light chains (PFLC) are usually reabsorbed by the proximal convoluted tubules where they are catabolysed. As a result, little PFLC finds its way into urine under normal circumstances. PFLC proteinuria results when there is a polyclonal increase in immunoglobulin production, glomerular damage or tubular damage. In those circumstances, by electrophoresis on concentrated urine, PFLC migrate broadly in the β- and γ-regions. Yet, by immunofixation, they may produce a ladder pattern that could be confused with a monoclonal gammopathy (see Chapter 7 for discussion of ladder pattern) (Fig. 6.6). In contrast, when MFLC are present, they usually migrate as one, two or more discrete bands in the β- and γ-regions (Fig. 6.7). The specific isotype may be determined by immunofixation. The Guidelines recommended that both urine protein electrophoresis and immunofixation be performed on concentrated urine as the screening test for MFLC. Urine protein electrophoresis alone is too insensitive to reliably detect some of the small MFLC that may be associated with amyloid AL (Fig. 6.8).

For the past several years, the standard has been to perform electrophoresis and immunofixation on concentrated urine to detect and identify the MFLC. After identification, it is important to measure the amount of MFLC to estimate the tumor burden. For this, a 24-h urine specimen is submitted for electrophoresis and protein determination. The MFLC peak is integrated by densitometry and the final result is expressed as mg/24 h (see Chapter 7). Using these electrophoretic techniques, Kyle[135] demonstrated that

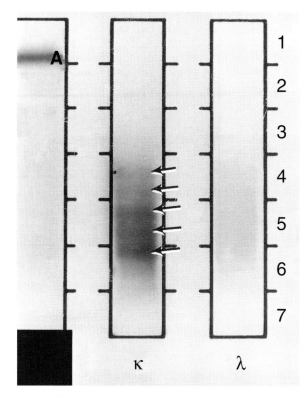

Figure 6.6 Immunofixation of a urine concentrated 100-fold with a mild protein loss pattern indicated by the presence of a small amount of albumin (A) in the far left lane (protein electrophoresis fixed in acid). Although no other protein bands are visible in the left lane, there are multiple bands visible in the κ immunofixation electrophoresis lane (arrows), and a faintly staining diffuse haze in the λ lane. This is a classic ladder pattern that occurs in any urine containing *polyclonal* free light chains. The bands are more often seen in κ, although they are also seen, on occasion, with λ. The bands with this polyclonal pattern are evenly spaced (like the rungs on a ladder). I interpret this pattern as 'Negative for Bence Jones protein'. Although the bands vary somewhat in their staining intensity from one to another, none shows the type of dense staining or antigen excess effect of the monoclonal free light chain (MFLC) protein bands in Fig. 6.7. (Paragon system stained with Paragon Violet; anode at the top.)

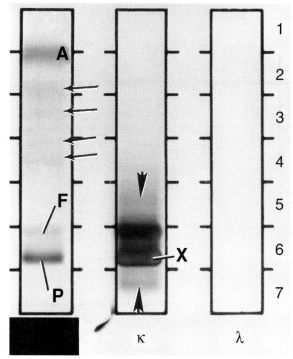

Figure 6.7 Immunofixation of urine concentrated 100-fold. The far left lane (protein electrophoresis fixed in acid) shows a weakly staining albumin band (A) along with a few weakly staining α_2- and β-region bands (small arrows) most consistent with a tubular proteinuria. The γ-region shows one prominent band (P) and one faint band (F). Immunofixation for κ shows two strong bands in with the same migration as the prominent and faint bands in the adjacent lane. Note that the center of the prominent band shows a slight antigen excess effect (X). Slightly above the faint band and slightly below the prominent band are lightly staining areas (large arrowheads), which may represent other forms of the monoclonal proteins (breakdown products, post-translational modifications, or forms associated with intact heavy chains). Note that this immunofixation only shows κ and λ reactions, therefore, we needed to perform immunofixation for heavy chains to be certain that these were not intact monoclonal proteins (in this case they were not). Contrast this pattern of true κ monoclonal free light chain (Bence Jones protein) with a classic 'ladder pattern' in Fig. 6.6. (Paragon system stained with Paragon Violet; anode at the top.)

about 60 per cent of cases of multiple myeloma produce MFLC (formerly termed Bence Jones proteins) in addition to the intact monoclonal immunoglobulin, or exclusively when present as part of light chain myeloma. The presence of MFLC produces a more dire prognosis than their absence.[174–180] Occasionally, patients with small or modest-sized monoclonal gammopathies in the serum will have massive amounts of MFLC in the urine. This is why urine must always be studied (Figure 6.9). Weber et al.[176] studied the course of 101 asymptomatic patients with multiple myeloma who were detected by chance while they lacked symptoms and found that the presence of MFLC excretion of > 50 mg/24 h in the urine was a significant independent variable that indicated treatment

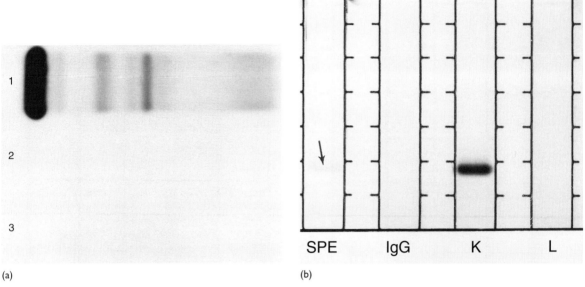

(a)   (b)

Figure 6.8 (a) The top sample (1) is a lyophilized serum for reference. The bottom two samples of urine concentrated 100-fold (2, 3) have no protein bands visible. (Paragon SPE2 system stained with Paragon Violet.) (b) Immunofixation of the urine sample labeled 2 from (a). In the serum protein electrophoresis (SPE) lane a very faint band is barely discernable (arrow), yet the immunofixation using anti-κ shows an obvious κ monoclonal free light chain. (Paragon system stained with Paragon Violet; anode at the top.)

even when the patients lacked lytic skeletal lesions. Further, Knudsen et al.[181] reported that myeloma patients with renal failure who have relatively low amounts of MFLC in their urine have a better chance of achieving an improvement in their renal function with chemotherapy than patients with high quantities of MFLC.

In most cases of light chain multiple myeloma, the serum protein electrophoresis shows hypogammaglobulinemia and only a small M-protein spike because most of the MFLC passes through the glomerular basement membrane into the urine. However, in cases of tetrameric light chain disease, or aggregates of light chains such as the monoclonal λ hexameric aggregates (trimolecular complex of λ dimers) reported by Abraham et al.,[182] a large M-protein is seen in the serum that contains only light chain and no heavy chain.[183–186] Nonetheless, in the vast majority of cases of MFLC, urine protein electrophoresis and immunofixation lead to the correct diagnosis.

In some patients, the presence of a polyclonal increase in serum γ-globulins does not rule out multiple myeloma in the presence of an appropri-

ate clinical picture: lytic lesions, anemia, hypercalcemia, and elevated creatinine. Urine must always be studied to rule out light chain multiple myeloma. Patients with multiple myeloma are more predisposed to develop infections and the antigenic stimulation from such infections may result in a polyclonal increase in γ-globulins.[187]

While detection of MFLC by electrophoresis and immunofixation has been useful, it will likely be supplemented by assays that allow quantification of total (polyclonal and/or monoclonal) free light chains (FLC) in serum or urine. Using sensitive nephelometric techniques to measure FLC, Bradwell et al.[173] found that the vast majority of patients with multiple myeloma produce excessive amounts of the FLC type that corresponds to the intact M-protein (i.e. free κ chains in an IgG κ M-protein) that can be detected in serum (Freelite; The Binding Site Ltd, Birmingham, UK; see Chapter 7). Katzmann et al.[188] established 0.26–1.65 as the diagnostic interval of FLC κ/λ in serum. Their 95 per cent reference interval for κ FLC in serum was 3.3–19.4 mg/l and for λ was 5.7–26.3 mg/l. This method was more sensitive

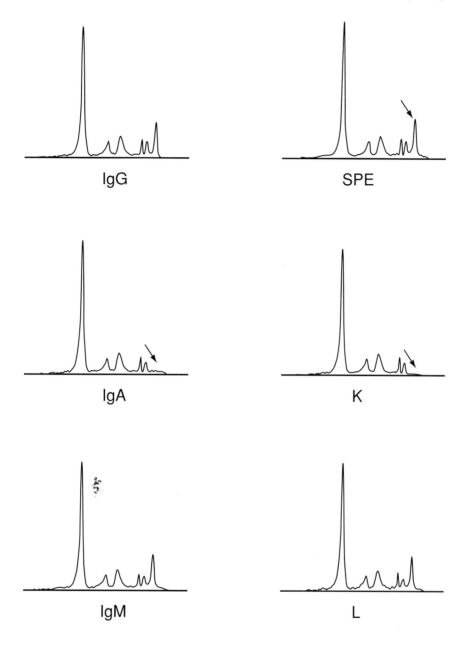

(a)

Figure 6.9 (a) Immunosubtraction of serum from a patient with a modest-sized monoclonal gammopathy in the γ-region (arrow in serum protein electrophoresis (SPE) box, top right). When the serum was preincubated with antisera against the immunoglobulin indicated in each box, only the antisera against IgA (arrow) and κ (arrow) removed the monoclonal protein. (Paragon CZE 2000 immunosubtraction.)

(Continues over page)

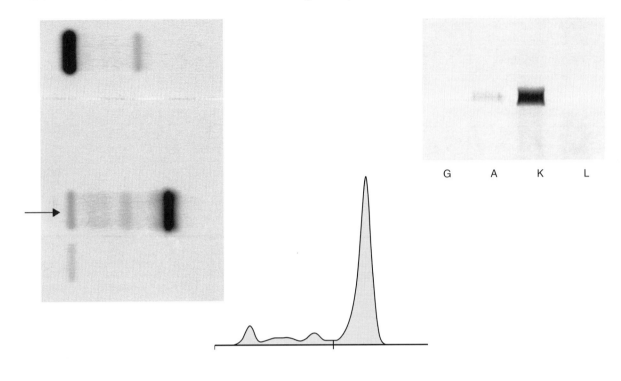

(b)

Figure 6.9 (*contd*) (b) Composite figure showing the protein electrophoresis on urine concentrated 50-fold (arrow indicates sample) from the patient identified in (a) as having an IgA κ monoclonal gammopathy. The pattern shows a small amount of albumin and several small proteins in the β- and γ-regions consistent with tubular damage. The densitometric scan demonstrates the massive amount of monoclonal protein present. The immunofixation to the right of the densitometric scan has dense staining only in the κ (K) lane with a tiny amount of IgA staining. Although both the small amount of intact IgA κ and the large amount of κ monoclonal free light chain (MFLC) migrate in the same location, the entire peak is integrated to measure the MFLC. (Sebia $\beta_{1,2}$ gel and Sebia Immunofixation.)

than current immunofixation methods of serum and urine to detect free light chains in patients with amyloid AL.[188] Mariën et al.[189] compared capillary zone electrophoresis (CZE) with determination of FLC in frozen sera from 54 patients confirmed to contain M-proteins by immunofixation. They found that sera from all 16 patients studied had abnormal κ/λ and increased absolute values for the relevant FLC.[189]

The method to measure FLC in serum is a powerful new tool that will improve our ability to detect and follow many of these patients. It will be much easier to obtain the serum sample than the 24-h urine to follow the MFLC level in these patients. At the time of writing, however, 24-h urine is still the standard, although our laboratory is beginning to use the free light chain assay to follow myeloma patients. The main limitation of the FLC assay is its inability to distinguish between MFLC and polyclonal FLC of the same isotype. Electrophoresis and immunofixation do this by the presence of restricted mobility. Therefore the demonstration of probable monoclonality is accomplished by statistical comparison of reference ranges. In cases where there is a polyclonal increase along with MFLC, a combination of electrophoresis, immunofixation and FLC measurement may prove to be ideal.

NON-SECRETORY MYELOMA

Non-secretory myeloma has been reported in from 1 to 4 per cent of patients with multiple myeloma.[190-194] These are cases of monoclonal plasma cell proliferations in which no monoclonal protein can be detected in the serum or urine by conventional electrophoretic techniques. The plasma cells in this condition may contain a monoclonal immunoglobulin, which may be detected by immunohistochemical analysis and occasionally ultrastructural studies.[195,196] A model of non-secretion by plasma cells is offered by Mott cells (plasma cells with large intracytoplasmic inclusions of immunoglobulin). They have a partial or complete block of the secretion of their immunoglobulins, resulting in distended endoplasmic reticulum.[197] Indeed, this phenomenon seems to be responsible for at least one report of non-secretory IgA κ myeloma where immunohistochemistry was used to demonstrate the trapped intracellular M-protein.[198]

Some cases of non-secretory myeloma result from synthesis of structurally abnormal molecules that may be degraded intracellularly.[196] Within the plasma cells from these patients, one may detect truncated or even intact M-proteins, but they are not secreted in sufficient quantity to be found in the serum or urine by conventional electrophoresis and immunofixation.[199] Molecular evidence of abnormal immunoglobulin formation in non-secretory myeloma was presented by Cogné and Guglielmi[200] who demonstrated a truncated 42 kDa γ_1 heavy chain that lacked a variable region because of a 2-base pair (bp) deletion. Cytogenetically, the vast majority of non-secretory myeloma cases have the same t(11;14)(q13;q32) translocation seen with other types of multiple myeloma.[155]

Non-secretory multiple myeloma presents clinically with similar symptoms as cases of secretory myeloma, yet they lack the renal involvement (due to the absence of MFLC production) and lack the convenient M-protein in serum and/or urine to confirm the diagnosis.[131] To avoid delay in diagnosis, special diligence is needed when confronted with symptoms of hypercalcemia in patients with radiographic and other features consistent with multiple myeloma but no M-protein in serum or urine.[105,201] Typically, the diagnosis is made when a bone marrow biopsy and aspirate are obtained to explain the hypercalcemia and lytic skeletal lesions. Similarly, because of the lack of a measurable M-protein, the response to therapy has been also followed by bone marrow studies.[191]

Occasional cases of acquired non-secretory myeloma secondary to chemotherapy have been reported. These can be a particular problem because the M-protein is an important measure of tumor burden.[202] The decreased, or absent synthesis in the 'natural' or chemotherapy-acquired non-secretory myeloma may relate to a defective B-cell response to differentiation signals.[203]

Another unusual situation is that of pseudo-nonsecretory myeloma described in two patients where no M-protein was found in serum or urine, but where it was found by *in vitro* studies of the neoplastic plasma cells.[204] In those cases, the authors concluded that the secreted M-protein was rapidly catabolized and deposited in the kidneys and other organs.[204]

The FLC nephelometric assay improves our ability to detect most cases of non-secretory myeloma. Drayson et al.[205] studied serum from patients with non-secretory myeloma using the FLC assay and detected abnormal ratios of free κ/λ in 19 of 28 patients. This indicates that non-secretory myeloma may really be a pauci-secretory disease in most of the cases described where small amounts of MFLC are secreted.

BICLONAL GAMMOPATHIES

Biclonal gammopathies (or double gammopathies) are uncommon, but not rare. They occur in about 3–5 per cent of patients with conditions associated with monoclonal gammopathies.[27,135,206] Because of their unusual appearance, biclonal gammopathies can be somewhat confusing. Clinically, there is no difference between biclonal gammopathies and monoclonal gammopathies in terms of outcome for the patient.[206] There have been two cases reported in the past few years where biclonal gammopathies (an uncommon lesion) were associated with the occurrence of angioimmunoblastic T-lymphomas (another unusual lesion).[207-210] This is probably a coincidence, but it deserves notice. However, in patients with multiple myeloma and a biclonal gammopathy, it is important to measure both clones during follow-up because the clinical response to the two clones may be asynchronous.[211]

The term 'biclonal' implies that the plasma cell neoplasm arose from two separate clones of B lymphocytes. As such, they should always have a different variable region and may differ in light chain class. There is no reason to look at variable regions in clinical laboratory testing. Biclonal gammopathies may originate from a single clone or separate clones.[212,213]

The presence of different light chains is a priori evidence that the neoplasm represents the product of two separate clones, which would be true 'biclonal' gammopathies (Fig. 6.10). However, most reported cases and most that we have seen in our laboratory have had the same light chain type with two different heavy or heavy chain subclass types. Some heavy chain classes occur more frequently in double gammopathies with IgG–IgM occurring most frequently followed by IgG–IgA, IgG–IgG (detected by their different electrophoretic mobilities), and IgA–IgM.[214] These may represent one clone of B-lymphocytes that is expressing two different heavy chain constant regions (similar to the stage in B-cell ontogeny where both IgM and another heavy chain class are present on the surface of the B cell). Therefore, for cases with the same light chain type one may use the term 'double gammopathy,' which recognizes the fact that two distinct proteins are seen but does not imply that they resulted from two separate clones.

The reasons for the occurrence of double gammopathies are better understood by examining the structure of the immunoglobulin heavy chain constant-region gene sequence (Fig. 6.11). When a cell switches from expressing one heavy chain gene to another, such as from μ to α_1, the intervening genes (in this case $c\mu$–$c\alpha_1$) are deleted and cannot be expressed later.[215] The remaining segments, however, could be selected and subsequently expressed. This occurs normally during B-lymphocyte maturation where B cells often bear both surface IgM and another heavy chain isotype. Hammarstrom et al.[140] noted that in gammopathies with two heavy chain

classes but only one light chain type, there may be preferential switches to explain the frequency of the two heavy chain types observed. Fortunately, other than being confused by the initial electrophoretic pattern itself; there is no known clinical significance to the demonstration of double (or even true biclonal) gammopathies. Most of these patients have 'monoclonal gammopathy of undetermined significance' (see below), while others have B-cell lymphoproliferative disorders or multiple myeloma.

Oligoclonal expansions due to infections, often seen as part of hepatitis C virus infection or acquired immunodeficiency syndrome may produce two, three or more M-protein peaks.[216] It is not clear where one of these peaks ceases to be a prominent oligoclonal band and becomes an M-protein. When one peak of an oligoclonal process appears appreciably larger than the rest, I measure it and recommend that the clinician perform a urine study to rule out MFLC, and follow the serum sample in 3–6 months to see if the process regresses (Fig. 6.12).

Immunosuppression in multiple myeloma and B-cell lymphoproliferative disorders

In multiple myeloma and chronic lymphocytic leukemia, concomitant suppression of the normal immunoglobulin secretion is a key feature recognized by examining the electrophoretic pattern (Fig. 6.13).[217,218] The decrease in production of polyclonal immunoglobulins predisposes these patients to recurrent bacterial infections and may be a cause of death.[218]

The production of normal immunoglobulins is the result of a balanced interaction between B cells, helper T-cells, cytokines, and antigen presenting cells. Peripheral blood lymphocytes from patients with multiple myeloma exhibit a mechanism extrinsic to the B cells that mediates an arrest in terminal B-lymphocyte maturation.[114] These data are also consistent with a profound decrease (as much as 20- to 600-fold) of the normal polyclonal B-lymphocytes in the circulation of patients with

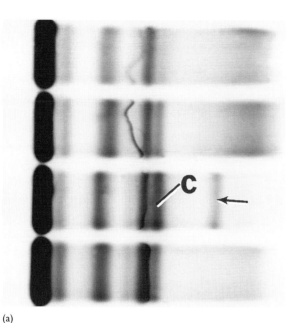

(a)

Figure 6.10 (a) The sample in the third lane has one relatively small band in the γ-region (arrow). However, by comparing the C3 regions of all of the samples, one may note that the C3 region of this sample is considerably darker (C−). Therefore, despite the less than optimal separation of the different protein bands on this gel, one should be suspicious that a second monoclonal band may also be lurking in this area. (Paragon SPE2 system stained with Paragon Violet.) (b) Immunofixation of the serum from (a) demonstrates a true biclonal gammopathy (IgA κ and IgM λ). These small bands were not associated with clinical symptoms and in this case are considered biclonal gammopathies of undetermined significance (BiGUS?). (Paragon system stained with Paragon Violet; anode at the top.)

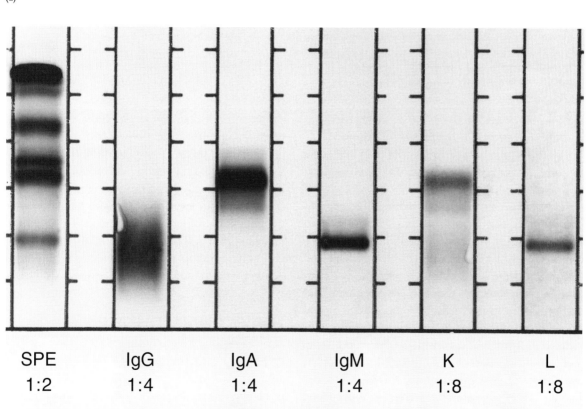

SPE	IgG	IgA	IgM	K	L
1:2	1:4	1:4	1:4	1:8	1:8

(b)

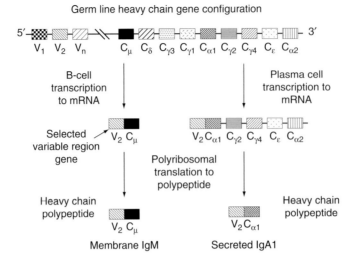

Figure 6.11 Schematic representation of heavy chain gene rearrangement during B-cell maturation. In B lymphocytes, the C μ chain is usually selected to be the heavy chain isotype expressed on the lymphocyte surface membrane. During further maturation to a plasma cell, another heavy chain gene is selected (in this case C α₁. While intervening genes are deleted (C μ, C d, C γ₃, and C γ₁ in this case) during maturation of a particular clone, the remaining heavy chain genes are still available and may be selected for expression at a later time in maturation. This could result in the double gammopathies; that is, two heavy chains that originated from a single clone. Note that the variable region gene (actually a combination of genes) selected is the same in the B-cell surface membrane and in the eventual immunoglobulin product secreted by the plasma cell.

multiple myeloma, implying the existence of a suppressive influence on B-lymphocyte proliferation.[219] Interestingly, the number of B cells does not seem to correlate with disease status or the concentration of the monoclonal protein.[219]

The decrease in normal immunoglobulins that occurs in myeloma patients has variously been hypothesized as reflecting excessive suppressor-T cell activities, deficient helper T-cell numbers and function, decreased numbers of pre-B cells, unusual macrophage products or dysfunctional natural killer (NK) cells.[220–223] In studies using mitogen stimulation of peripheral blood mononuclear cells from patients with multiple myeloma and reduced serum immunoglobulin levels, Walchner and Wick[218] identified CD8+, CD11b+, Leu-8– T cells as playing a role in suppressing polyclonal immunoglobulin secretion.

In evaluating serum protein electrophoresis, after I measure the M-protein in the γ-region, I note the amount of non-M-protein γ-globulin. If the number is less than 0.25 g/dl I note that there is

suppression of the normal γ-globulin content. This number has not been derived through rigorous investigation. It reflects the cut off of 0.25 g/dl that has been used as a marker for common variable immune deficiency (CVID). In patients with CVID, those with IgG levels over 0.3 g/dl have adequate lymphocyte proliferation responses to *S. aureus* and *E. coli*, whereas those with IgG levels < 0.125 g/dl have markedly decreased responses.[224]

HEAVY CHAIN DISEASE

α Heavy chain disease

α Heavy chain disease is extremely uncommon, especially in the Western world. It occurs most frequently in the Middle East and Mediterranean region.[225,226] The disease has an onset in much younger patients than myeloma or most B-cell lymphoproliferative diseases. It usually develops

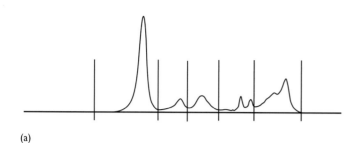

Fraction	Rel %	g/dl
Albumin	42.7	2.60
Alpha 1	8.1	0.50
Alpha 2	11.9	0.73
Beta	9.2	0.56
Gamma	28.0	1.71

TP: 6.40 – 8.20 A/G: 1.20 – 2.20

(a)

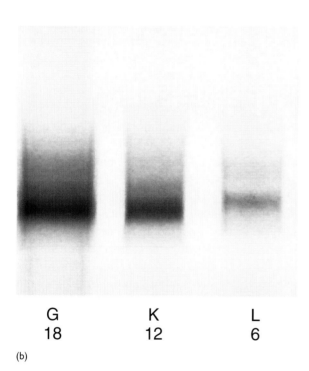

G	K	L
18	12	6

(b)

Figure 6.12 (a) Electropherogram illustrating a monoclonal gammopathy superimposed on a modest polyclonal and oligoclonal increase in γ-globulins. The M-protein was an IgG κ, but there were several smaller bands including an IgG λ on immunofixation. (b) Immunofixation on serum from the same case demonstrating a definite IgG κ band, a smaller IgG λ band and at least a second λ band anodal to the obvious one. The presence of these multiple small bands indicates the process is more likely either reactive to an infection or part of a systemic dysregulation of the immune system than a precursor to multiple myeloma.

in the third decade, but has been reported as early as 9 years of age.[227] Using immunoelectrophoresis or immunoselection techniques (see Chapter 3), free α heavy chains have been demonstrated in the serum, urine and intestinal fluids.[227–229] Clinically, the patients present with diarrhea (often with steatorrhea), malabsorption and weight loss. Usually, there is slow progression of their disease during the early phase and, importantly, cures have been reported when treated at this stage.[230]

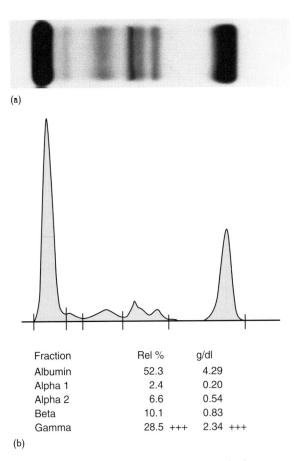

(a)

(b)

Fraction	Rel %	g/dl
Albumin	52.3	4.29
Alpha 1	2.4	0.20
Alpha 2	6.6	0.54
Beta	10.1	0.83
Gamma	28.5 +++	2.34 +++

Figure 6.13 (a) Serum protein electrophoresis with a large monoclonal gammopathy with suppression of the normal γ-globulins. (b) Densitometric scan of sample from (a) provides a clear view of the virtual absence of the normal γ-globulins.

Unfortunately it may progress to high-grade lymphoma.[16]

The tissue distribution of α heavy chain disease roughly parallels that of the normal distribution of IgA along the gut and bronchial mucosa, but it has been reported in the respiratory tract and other locations.[231–233] Histologically, the gastrointestinal tract (occasionally the respiratory tract or other location) is involved with a lymphoplasmacytic proliferation that results in the chronic diarrhea and malabsorption.[234] The T-cells present within these lesions may occasionally cause them to mimic a T-cell lymphoma.[235] Immunohistology demonstrates the presence of α heavy chains in the cytoplasm of these cells without corresponding light chains.[236]

Unlike multiple myeloma, α heavy chain disease is usually difficult to diagnose in the clinical laboratory because serum protein electrophoresis may fail to disclose the presence of the abnormal protein. Immunofixation will show heavy chain restriction without corresponding light chain restriction. The same laboratory picture can be found in some cases of IgA myeloma, where their light chains are not detected by some antisera.[237] An antigen excess situation may also cause one to falsely identify a patient with multiple myeloma as having α heavy chain disease. This error can be avoided by recalling the vastly different clinical pictures of these two diseases. α Heavy chain occurs in adolescents and young adults, whereas IgA multiple myeloma is found in middle-aged and older adults. Also, α heavy chain presents with complaints related to the gastrointestinal tract and lacks bone lesions, whereas multiple myeloma rarely presents with gastrointestinal complaints and usually has bone lesions. Therefore, when the clinical picture does not match the interpretation of α heavy chain, take a step back and redo the immunofixation with another manufacturer's anti-light chain antisera and/or adjust the dilution of the patient's serum used in the immunofixation to avoid antigen excess situations.

In as many as half of the patients with true α heavy chain, the α chains are either not secreted, or secreted in tiny amounts by the neoplastic cells.[238] Further, gene deletions in α heavy chain disease may influence either the amino acid sequence of the secreted chain or the ability of the neoplastic cells to secrete it at all.[239] Many patients with α heavy chain disease have unremarkable serum protein electrophoresis patterns.[27] For those cases with visible alterations, a broad β-band is the most typically seen on serum protein electrophoresis, however, discrete bands and γ-migrating bands have been reported.[228] Since there is no monoclonal light chain by definition, the disease ideally is established by demonstrating the lack of light chain in the presence of excessive heavy α chains. An immunoselection technique has been used to assist in this diagnosis (see Chapter 3 for details). Briefly, anti-κ and anti-λ antisera are mixed into the agarose (as is done for radial immunodiffusion) and standard immunoelectrophoresis is carried out.[240] Since intact immunoglobulin molecules contain either κ or λ chains, they will precipitate around the sample well, the α chains will form a β-migrating precipitin arc with the anti-α antisera.[241]

γ Heavy chain disease

Whereas α heavy chain disease usually has symptoms relating to the mucosal surfaces, γ heavy chain disease (Franklin's disease) has systemic symptoms more reminiscent of lymphoma: weakness, fatigue, anemia (sometimes autoimmune hemolytic), generalized lymphadenopathy, hepatosplenomegaly, pleural effusions, and ascites. In addition, there is often edema of the uvula and soft palate.[16,242,243] Further, γ heavy chain disease is found in the Western world and that occurs in middle-aged or older individuals (more often in men than women).[243-245] In the peripheral blood lymphocytes, lymphocytoid plasma cells or plasma cells may be found that contain a truncated γ-molecule in the cytoplasm, but no identifiable light chains.[246-249] Histological examination will demonstrate infiltrates with atypical lymphocytes and plasma cells in the lymph nodes involved and in bone marrow.[16,243,250]

Similar to α heavy chain disease, in γ heavy chain disease the electrophoretic pattern often is non-specific, giving a relatively broad band anywhere from the α₂- to the γ-region (Fig. 6.14). γ Heavy chain disease has been reported to occur as a biclonal lesion together with other B-cell neoplasms.[251,252] A low γ-globulin region on serum protein electrophoresis in a patient with a normal or elevated total IgG level could suggest this diagnosis.[253] Because of the deceptively benign appearances of the serum protein electrophoretic patterns, clinicians must be alerted that immunofixation should be requested as the screening test for patients suspected of having heavy chain disease.[251] Sun *et al.* reported a modified immunoselection technique that takes advantage of the simplicity of the immunofixation procedure to make the identification technique more readily available in clinical laboratories (see Chapter 3 for details).[254]

μ Heavy chain disease

This is an extremely rare B-cell lymphoproliferative disorder that has shown a similar clinical picture to γ chain disease, with the addition of hypogammaglobulinemia in μ heavy chain disease.[16,27,255,256] Unlike γ heavy chain disease, peripheral lymphadenopathy is uncommon, whereas hepatosplenomegaly is usually present. Serum protein electrophoresis may be negative and immunofixation or immunoselection are usually needed to identify the μ heavy chains in serum. The μ heavy chains are fragments that lack a variable region or light chain.[16,257] Interestingly, the urine of these cases demonstrates a MFLC, most often a κ light chain in the urine but no μ heavy chain (the latter was found in the serum).[16,255] It is not clear why μ heavy chain is absent in the urine in these cases. The clinical course of the disease is slowly progressive, as illustrated by one of the cases initially termed 'benign' monoclonal gammopathy, but which evolved within 3 years to become an aggressive lymphoproliferative malignancy.[257,258] The misnomer 'benign' is a good reminder of the value of Kyle's term MGUS.

WALDENSTRÖM'S MACROGLOBULINEMIA

Waldenström's macroglobulinemia is a low-grade B-cell lymphoplasmacytic disorder in which a relatively large IgM monoclonal gammopathy is present in serum.[259] It occurs at a rate of 6.1 cases per million in white men and 2.5 per million in white women; it occurs much less commonly in black individuals.[260] As with multiple myeloma, the incidence increases sharply with age such that the incidence rate per million over the age of 70 years is 25.7 versus only 0.4 at the age of 40 years.[261,262] The disease typically involves the bone marrow, lymph nodes, occasionally with hepatomegaly and/or splenomegaly (Table 6.8).[263] Despite the bone marrow involvement, however, these patients do not have lytic skeletal lesions. Uncommonly,

Table 6.8 Presenting clinical features of Waldenström's macroglobulinemia[a]

Clinical feature	Percentage present at presentation
Fatigue	70
Lymphadenopathy	22
Splenomegaly	18
Hepatomegaly	13
Extranodal involvement	6
Bleeding tendency	17
Infection	17
Hyperviscosity syndrome	12
Cardiac failure	25

[a]Data from Kyrtsonis et al.[263]

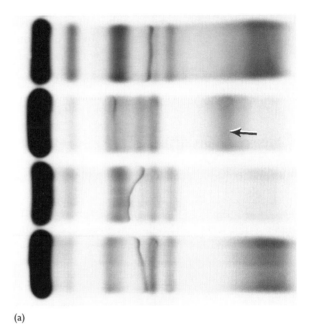

(a)

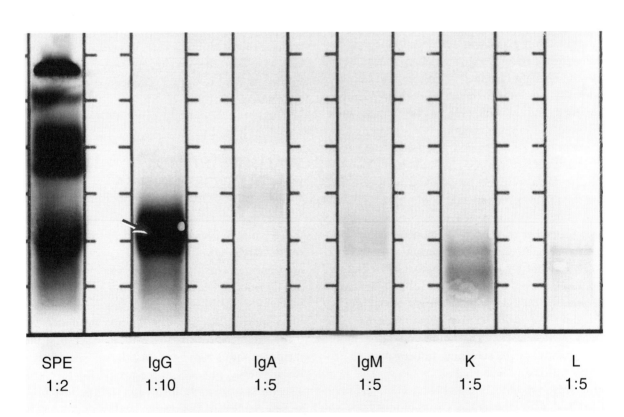

SPE	IgG	IgA	IgM	K	L
1:2	1:10	1:5	1:5	1:5	1:5

(b)

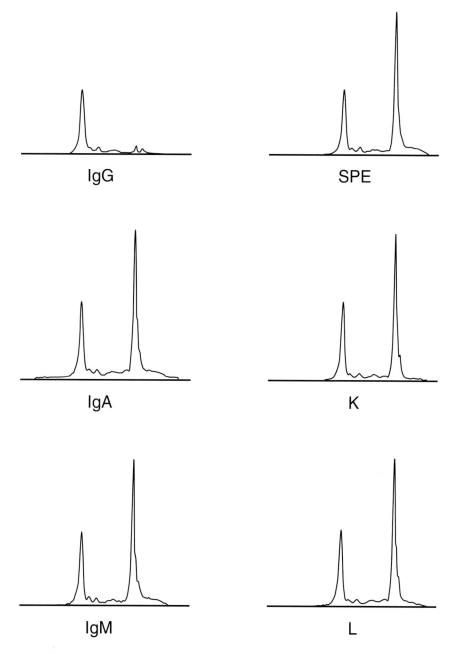

IgG

SPE

IgA

K

IgM

L

(c)

Figure 6.14 (a) The top lane shows a serum with an acute-phase reaction and a polyclonal increase in γ-globulins. The serum in the second lane has a somewhat diffuse restriction (arrow) in the fast γ-region. It stands out especially well, despite being rather small because the remainder of the γ-region is decreased. The third sample has a borderline low/normal γ-region. The bottom lane shows a polyclonal increase in γ-globulin. SPE, serum protein electrophoresis. (SPE2 system stained with Paragon Violet.) (b) Immunofixation demonstrates the diffuse band in the fast γ-region is IgG (arrow), but at 1:5 dilution, no such band is seen in the K or L lanes. This is presumptive evidence for γ heavy chain disease, however, an immunoselection procedure which precipitates the intact IgG molecules and demonstrates the free IgG would be needed to absolutely confirm this impression. (Paragon system stained with Paragon Violet.) (c) Immunosubtraction of the same serum from (a) and (b) demonstrates that the large monoclonal gammopathy can only be removed by the antisera against IgG, not by either the anti-κ or anti-λ antisera. Further supportive evidence of this case being γ heavy chain disease. (Paragon CZE 2000.) Case contributed by Dr Gary Assarian.

other major organs are involved.[263–267] Worse prognosis is associated with older age of onset (≥ 60 years), males, anemia, neutropenia, albumin levels lower than 4 g/dl (40 g/l), and thrombocytopenia.[268,269]

Clinically patients usually present with fatigue and weakness caused by anemia (normochromic and normocytic), elevated erythrocyte sedimentation rate and hyperviscosity.[270–272] Hyperviscosity, a key feature of this disease, produces significant neurological complaints, cardiac insufficiency, and resultant vascular insufficiency throughout the body. They have an increased bleeding tendency, often resulting in epistaxis and cutaneous purpuric lesions.[273] Uncommonly, amyloidosis has been reported.[274–281]

Whereas the hyperviscosity caused by IgM is usually attributed to the large molecular weight of this molecule, other features such as axial asymmetry of the molecule, self-association, and extremely high concentrations likely explain the hyperviscosity associated with the other isotypes.[282] As mentioned above, rare patients with IgM monoclonal proteins have the clinical picture of multiple myeloma, including lytic skeletal lesions, and should be treated accordingly.[152]

The clinical course of Waldenström's macroglobulinemia is more indolent than that in multiple myeloma, although exceptional cases have been noted.[283] Hyperviscosity usually causes significant clinical problems when the IgM concentration is greater than 2.0 g/dl (20 g/l).[135] Histologically, the bone marrow is infiltrated with lymphoplasmacytoid cells, the monoclonality of which can be demonstrated by the immunohistochemistry. Immunophenotyping has been useful to confirm the diagnosis of Waldenström's macroglobulinemia. Harris et al.[284] reported that 90 per cent of cases had the phenotype of post-germinal center cells: CD19+, CD20+, CD5–, CD10–, and CD23–.[285]

On serum protein electrophoresis, a typical case of Waldenström's macroglobulinemia demonstrates a monoclonal gammopathy either at or near the origin of gel-based techniques and in the β-γ region of CZE (Fig. 6.15). This electrophoretic behavior reflects both the isoelectric point of IgM

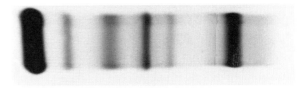

Figure 6.15 Serum protein electrophoresis from a patient with Waldenström's macroglobulinemia. The prominent M-protein migrates just cathodal to the origin. (Panagel System stained with Coomassie Blue.)

and a molecular mass of around 1000 kDa and its predilection to self-aggregate. It should be noted that, occasionally, cases of IgM with a half molecule 7S associated with a single light chain weighing only 64 kDa may occur.[286] Immunoglobulin quantification usually demonstrates a several-fold increase of IgM. While IgG and IgA typically are in the normal range, hypogammaglobulinemia has been reported in as many as one-quarter of the cases.[263] From 2.0 to 3.0 g/dl (20–30 g/l) have been suggested as cut-off levels for the diagnosis of Waldenström's macroglobulinemia.[285,287,288] Monoclonal free light chains (Bence Jones proteins) may be found in about half of the patients and cryoglobulinemia is present in about 5 per cent.[263] As discussed below, under cryoglobulinemia, the presence of an IgM κ M-protein in cryoglobulin is strongly associated with both hepatitis C and type II cryoglobulinemia. Although IgM is overwhelmingly the major immunoglobulin class associated with this condition, rare cases have been reported with IgG and IgA monoclonal proteins.[289–293]

Laboratories that use immunoelectrophoresis (IEP) may experience problems typing the light chain because of the umbrella effect (see Chapter 3). Characterization of the M-protein by immunofixation or immunosubtraction is usually straightforward. About once a year, I need to use 2-mercaptoethanol (2-ME) to break up a molecule into a form that is easier to deal with. Even with CZE, 2-ME is occasionally needed (Fig. 6.16).[294]

Finally, some of the IgM monoclonal proteins have been found to have specific reactivity, most often against self antigens (autoimmune). In a

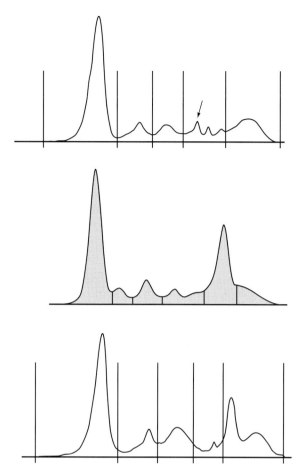

Figure 6.16 Composite figure showing original electropherogram on a patient with Waldenström's macroglobulinemia but an unimpressive M-spike (arrow in top figure) (Paragon CZE 2000). The middle figure is the densitometric scan from the Sebia $\beta_{1,2}$ gel that demonstrates the prominent M-protein peak. The bottom figure is the same serum re-assayed by capillary zone electrophoresis following treatment of the serum with 2-mercaptoethanol.

study of 57 patients with Waldenström's macroglobulinemia, Jonsson et al.[295] reported that 51 per cent of them had clinical and/or serological autoimmune findings: autoimmune hemolytic anemia, rheumatoid arthritis, parietal cell antibodies and IgM-anticardiolipin. Occasionally, the reactivity of these M-proteins explains specific symptoms, such as IgM antibodies against myelin basic protein in patients with neurological symptoms.[296] Another was associated with lupus

anticoagulant activity and gastrointestinal malabsorption possibly caused by ischemia.[297]

Monoclonal gammopathies in lymphoma and leukemia

Protein electrophoresis of serum and urine may provide useful information about B-cell lymphoproliferative processes. As discussed above, patients with multiple myeloma have clonal B-cell precursors with surface immunoglobulin and even pre-B cells with cytoplasmic μ of the same idiotype (roughly equivalent with reactivity), implying that although the neoplasm manifests mainly as monoclonal plasma cells, this is a disease of the entire lineage of that particular B-cell clone.[35–37,109,298] B-cell chronic lymphocytic leukemia (CLL) and myeloma represent cells from the same lineage at various stages of maturation; although myeloma seems to involve the transformation of a pluripotent stem cell, whereas, CLL seems to involve a more mature, terminally committed cell.[299,300] The stage of a given B-cell neoplasm is not irreversibly fixed, and may change during the course of an illness. Not surprisingly then, B cell neoplasms such as chronic lymphocytic leukemia may transform to a predominantly plasma cell neoplasm producing the same heavy and light chain types found on the B cells, and some cases of myeloma evolve into an aggressive lymphoproliferative phase, characterized by rapidly enlarging soft-tissue masses.[301,302]

When less sensitive techniques were available to evaluate serum and urine from patients with B-cell lymphoproliferative disorders, monoclonal gammopathies were detected only in a small number of cases.[303] However, the current widespread use of immunofixation demonstrates that a large number of patients with chronic lymphocytic leukemia will have at least one and some will have a few (oligoclonal) clonal immunoglobulin products demonstrable in serum and/or urine.[304] Using immunofixation and high-resolution agarose gel electrophoresis Berstein et al.[303] detected monoclonal gammopathies in the serum from 36 of 111 patients with CLL. This is not a new finding. In

1909, Decastello detected monoclonal free light chains in the urine from a patient with CLL.[305] Not surprisingly, the monoclonal gammopathy in the serum and/or urine of patients with CLL corresponds to the molecules expressed on the cell surfaces.[306–308] B-lymphocytes from patients with CLL can be induced, by Epstein–Barr virus or mitogens such as phorbol ester, to differentiate into immunoglobulin-secreting cells *in vitro*.[309] Other CLL cells have been shown to spontaneously secrete monoclonal light chain or monoclonal whole immunoglobulins.[310] Further, the demonstration of a monoclonal gammopathy in those individuals was associated with a decrease in median survival from 103 months for those without the monoclonal gammopathy to 63 months for those with one.[303] Interestingly, that decrease was independent of the clinical stage of the patients.

Other B-cell lymphoproliferative processes have also been reported to show monoclonal proteins in serum or MFLC in urine. Nodular lymphoma, Burkitt's lymphoma, lymphoplasmacytoid leukemia with hairy-cell morphology, and even angioimmunoblastic T-cell lymphoma (AITL) evolving into an immunoblastic lymphoma have had monoclonal proteins demonstrated by a combination immunofixation.[311–322] It should be noted, however, that in most cases of AITL the serum contains a polyclonal increase in γ-globulins.[323] One extraordinary case of T-cell acute lymphoblastic leukemia has been reported in a child with a monoclonal gammopathy.[324] Molecular studies on that case demonstrated both rearrangement of the T cell receptor β gene and the immunoglobulin heavy chain genes.[324]

As mentioned above, similar to patients with multiple myeloma, those with B-cell lymphoproliferative disorders may also suffer hypogammaglobulinemia and have difficulty synthesizing immunoglobulins in response to infectious diseases.[11,325,326] Intravenous γ-globulin has been recommended to ameliorate this problem.[327] As in multiple myeloma, mechanisms have been hypothesized for the decreased immunoglobulin production against normal stimuli. One mechanism involves impairment by the neoplastic B cells of the function of T lymphocytes to facilitate proper maturation of uninvolved B lymphocytes.[328] Another hypothesis notes that CD95 is upregulated on the uninvolved plasma cells of patients with CLL. Interaction of the CD95 ligand (CD95L) produced by the neoplastic B cells with this receptor leads to death of normal plasma cells via apoptosis.[328] Paradoxically, even with their hypogammaglobulinemia, patients with CLL are more likely than the general population to develop autoantibodies, and especially those against hematologic cells (i.e., autoimmune hemolytic anemias and immune thrombocytopenias).[11,329,330]

MONOCLONAL GAMMOPATHY ASSOCIATED WITH TISSUE DEPOSITION: AMYLOIDOSIS AND NON-AMYLOID MONOCLONAL IMMUNOGLOBULIN DEPOSITION DISEASE

Amyloidosis

Amyloidosis is another major clinical condition associated with monoclonal gammopathies. Amyloid is a general term that literally means 'starchlike'. It is defined by the tinctorial quality of the staining reaction in paraffin-embedded tissue. Amyloid stains with Congo Red and has a characteristic yellow-green birefringence under polarized light. Several different proteins may result in this deposition; therefore, amyloidosis may be caused by a wide variety of conditions. On an ultrastructural level, the amyloid fibrils of serum amyloid A (see below) are 7.5–8.0 nm wide whereas those for transferrin-associated familial amyloidic polyneuropathy are 13.0 nm wide.[331] All forms of amyloid have the β-pleated sheet structure that accounts for the Congo Red birefringence.[331]

Amyloid may be broadly classified as acquired or inherited. The most common form of acquired amyloidosis is amyloid AL that has an incidence of 0.9 per 100 000 person years.[332] Amyloid AL results in the deposition of MFLC systemically as amyloid fibrils. Tissue deposition of AL preferentially involves the tongue, heart, gastrointestinal tract, blood vessels (including glomerular capillaries), tendons, skin, and peripheral nerves. The clinical picture in these patients parallels the sites of involvement, with macroglossia, congestive heart failure, carpal tunnel syndrome, purpura, renal failure, and peripheral neuropathy as prominent features. Further, the optimal sites (which should be judged on the symptoms for the individual case) for biopsies of suspected cases reflect distribution and availability of the site. These patients usually do not have bone pain or osteolytic lesions.[333] Kyle and Greipp[334] recorded 229 patients with AL of whom 47 (20.5 per cent) had multiple myeloma; they found that the presence of myeloma did not contribute to prediction of survival at 1 year. Using a five-band electrophoretic technique, Kyle and Greipp found a discrete band in only 40 per cent of their patients with amyloidosis while demonstrating monoclonal protein in 68 per cent of the sera and MFLC in 70 per cent of the urine by immunoelectrophoresis.[334] Immunofixation is more sensitive and may detect a monoclonal component in as many as 95 per cent of amyloid AL cases.[335,336] Prognosis varies with the type of light chain involved. The presence of monoclonal λ light chains in the urine of patients with primary systemic amyloidosis was found to have an average 12-month survival compared with the 30-month survival of patients with κ chain excretion and 35 months for those with no monoclonal protein in the urine.[337]

Kyle[338] cautions that amyloid AL needs to be ruled out when a patient has unexplained renal failure, congestive heart disease, peripheral neuropathies, hepatomegaly, or malabsorption. Amyloid AL is always associated with clonal plasma cell proliferation, however, in 15 per cent of them a monoclonal gammopathy is not detected in either the urine or serum even when immunofixation is used to enhance sensitivity.[20,135,339]

When the serum and urine are negative in an individual suspected of having amyloid AL, one further study should be performed. Katzmann et al.[188] have demonstrated that the presence of abnormal free κ/λ ratio in serum using the nephelometric FLC assay is more sensitive than immunofixation in detecting patients with amyloid AL as well as in patients with light chain deposition disease.

The other form of acquired amyloidosis is reactive systemic or secondary amyloid A.[340] This form of amyloid is found in patients with a wide variety of chronic inflammatory processes and is composed of the an acute-phase reactant protein, serum amyloid A (SAA).[341] Its synthesis in both hepatic and extrahepatic locations is stimulated by IL-1, IL-6 and TNF.[341] The deposition of SAA is a dynamic process during the inflammation such that the use of anti-inflammatory agents to keep SAA concentration below 10 mg/l may allow the amyloid deposition to regress and the organ involved to recover its function.[340]

The hereditary form of amyloidosis may be caused by deposition of a wide variety of structurally abnormal forms of plasma proteins: transthyretin (prealbumin), apolipoprotein A-I, apolipoprotein A-II, fibrinogen, lysozyme, gelsolin, and cystatin C.[342] Variants of transthyretin are the most common forms of autosomal dominant systemic amyloidosis typically resulting in neuropathies (although some do involve the heart or kidneys).[343] It is important to distinguish between hereditary forms of amyloid, amyloid AL, and amyloid AA because the treatment, prognosis and genetic counseling differ. Although the presence of an M-protein in serum or urine is helpful, it does not confirm that the patient has amyloid AL. As discussed below, monoclonal gammopathy of undetermined significance is common and may be an innocent bystander in the serum of a patient with a hereditary form of amyloid. Therefore, if immunohistochemistry of the tissue involved does not confirm the presence of immunoglobulin light chain, genetic studies should be performed.[343]

Non-amyloid monoclonal immunoglobulin deposition disease

Light chains are made in excess by plasma cells, pass readily through the glomerulus and are reabsorbed mainly by the proximal convoluted tubules. About 1–2 g of protein can be reabsorbed by normal tubules in a 24-h period. Since the amount of polyclonal free light chains rarely exceeds this, only trivial quantities of polyclonal free light chains find their way into urine daily. These can be visualized by performing immunofixation only after concentrating the urine much greater than the typical 50- to 100-fold used in the clinical laboratory.

Large amounts of MFLC overwhelm the low-affinity receptors present on the brush border of the proximal convoluted tubules that normally bind the polyclonal free light chains and begin them on their journey to lysosomal acid hydrolysis.[344] In patients with large amounts of MFLC, these proteins may deposit in glomeruli. Absorption of the excess amount of MFLC occasionally damages the proximal convoluted tubule cells. Sanders and Booker[344] note that the proximal convoluted tubule damage and the myeloma cast nephropathy are different events.

Although amyloidosis has been classically associated with systemic dysfunction due to deposition of amyloid in the involved tissues, non-amyloid light chains (and occasionally truncated heavy chains) can also deposit in glomeruli and other organs, resulting in disturbed function.[345,346] In this situation, where amyloid cannot be demonstrated (negative for Congo Red staining), but light chains can, the term light chain deposition disease (LCDD) has been used.[345] Because we now recognize that occasionally truncated heavy chains may also be involved either alone or with light chains the terms HCDD and LHCDD have been added.[46,345,347–349] Since these terms become clumsy, I prefer the 'lumpers' designation of monoclonal immunoglobulin deposition disease (MIDD).[335] Serum and urine protein electrophoresis and urine immunofixation demonstrate the presence of a monoclonal gammopathy in 70–85 per cent of patients with MIDD.[335] Often, hypogammaglobulinemia is notable in the serum.

Although many cases of MIDD occur in patients with multiple myeloma,[345,347] it can be a part of MGUS or may present with clinical manifestations,

such as restrictive cardiomyopathy, or more commonly, renal disease.[345,347,350] Reports of tumoral non-amyloidotic monoclonal immunoglobulin light chain deposits in lymphoid and pulmonary tissue may be an early presentation of LCDD.[351] Amyloidosis AL is more likely to occur in the absence of multiple myeloma than is MIDD; however, the clinical overlap of the two conditions is considerable. In Buxbaum's series, 13 per cent of patients with amyloidosis AL and 20 per cent of patients with LCDD had neither serum nor urine monoclonal component (Fig. 6.17).[352] Light chain deposition disease is present much less often than either amyloidosis or BJP (Bence Jones protein) cast nephropathy. In an autopsy series of 57 patients with myeloma, Iványi[353] found that 32 per cent had MFLC cast nephropathy, 11 per cent had renal amyloidosis and only 5 per cent (three patients) had light chain deposition nephropathy. The severity of the renal involvement varies widely with the survival related to the diagnosis of the underlying plasma cell process.[345]

In those cases where there is no monoclonal component in serum or urine the diagnosis of MIDD can be a challenge. Light chain deposition disease has been described in cases of non-secretory myeloma, later called 'pseudo-nonsecretory myeloma' because of the demonstration *in vitro*

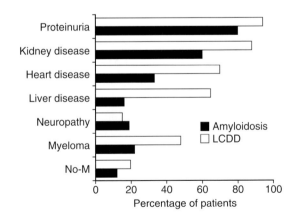

Figure 6.17 Percentages of patients with the indicated clinical features are shown for those with amyloid AL (black bars) and those with light chain deposition disease (LCDD; white bars). Data from Table 2 of Buxbaum.[352]

that the bone marrow plasma cells produce a defective monoclonal protein.[204] As mentioned above, in those cases with a high index of suspicion but a negative immunofixation of serum and urine, measurement of the free κ/λ ratio in serum using the nephelometric assay specific for polyclonal free light chains may demonstrate the presence of an abnormal ratio.[188,205]

SOLITARY PLASMACYTOMA

Solitary plasmacytomas may be located in bone in extramedullary locations or, rarely, at both sites.[354] There is a major difference in outcome depending on whether the bone is involved. Whereas all of these patients are probably part of the same spectrum of disease, patients with solitary plasmacytomas of bone have a worse prognosis than individuals with extramedullary plasmacytomas (typically the head and neck, especially in the upper respiratory tract).[355] This may partly reflect the difficulty in excluding involvement of other bones at the time of diagnosis; since in multiple myeloma, the bone marrow involvement is often patchy.[356] The best outcome seems to be among individuals with primary lymph node plasmacytomas, where none of the 25 cases progressed to multiple myeloma.[357]

Although solitary plasmacytoma involving bone may be an early presentation for multiple myeloma, Frassica et al.[358] reported that they had a much better 5-year survival (74 per cent) than did patients with multiple myeloma (18 per cent). They recommend the use of aggressive radiotherapy for these lesions. Monoclonal proteins are found in 56 per cent of serum screens (using a low-resolution electrophoresis method) of these patients.[358] The presence of a monoclonal gammopathy, however, is not required for a diagnosis of either solitary plasmacytoma of bone or extramedullary plasmacytoma. Typically, monoclonal gammopathies in both groups of patients are relatively small. If a serum or urine monoclonal protein is present, follow-up electrophoresis is useful to gauge response to the

radiotherapy. When the monoclonal protein disappears after radiotherapy, a long-term disease-free survival can be anticipated.[359,360] Liebross et al.[360] found that plasmacytoma patients with non-secretory lesions and those with a monoclonal gammopathy that persisted both had about a 60 per cent chance of developing multiple myeloma as opposed to only about 20 per cent of patients with a monoclonal gammopathy that disappeared.

MONOCLONAL GAMMOPATHIES NOT ASSOCIATED WITH B-LYMPHOPROLIFERATIVE DISORDERS

Autoimmune disease

Monoclonal gammopathies occasionally have been detected in patients with autoimmune diseases. In some cases, such as monoclonal anti-rheumatoid factor, anti-nuclear antibody, lupus anticoagulant, anti-platelet, anti-neutrophil, and anti-insulin, the specificity of the autoantibody is known.[361-367] In other cases, the relationship of the monoclonal antibody to the autoimmune disease is not known. However, removal of the monoclonal antibody by plasmapheresis has been reported to result in clinical improvement in patients with monoclonal gammopathies associated with polymyositis (where monoclonal antibodies have been detected in the sarcolemmal basement membrane).[368] Of course, other antibodies and non-immunoglobulin molecules with significant biological activity are also removed by this process. In some patients with Sjögren's syndrome, IgM monoclonal proteins have been associated with plasma cell infiltrates in the salivary glands.[369]

In addition to autoantibodies, monoclonal proteins have reactivity to other common antigens. A wide variety of reactivities of monoclonal proteins that have been determined include bacterial proteins, cardiolipin, polysaccharides, viral antigens, and other major serum proteins including isoenzymes, albumin, and α_1-antitrypsin.[370-373]

Monoclonal gammopathies and neuropathies

About 10 per cent of patients with idiopathic peripheral neuropathies have monoclonal gammopathies in their serum.[374,375] Looking at it from another perspective, almost half of the individuals with macroglobulinemia have clinical evidence of neuropathy.[376] The underlying process may be benign or malignant.[375] Yet, because monoclonal gammopathies are relatively common, coincidental occurrence of this finding with a patient that has peripheral neuropathy from another cause needs to be ruled out.[375] This can be accomplished by looking at the specific reactivity of the monoclonal protein. The neuropathies are usually sensorimotor, but may be limited to motor disturbances. IgM κ is the most common M-protein detected, although other isotypes have been described. Whereas the relationship between the monoclonal gammopathy and the peripheral neuropathy is unclear in most cases, autoreactivity with myelin has been shown in some.[377–384]

Specific reactivity against myelin-associated glycoprotein (MAG; 100 kDa) has been characterized in many cases.[385–387] Demyelinating peripheral neuropathy is associated with IgM monoclonal anti-MAG, whereas antibody reactivity against ganglioside antigens GM1 is most closely associated with motor neuropathies, and anti-GD1b, anti-sulfatide and anti-chondroitin sulfate reactivity has been associated with sensory neuropathies.[375,388,389] Further, the monoclonal antibody HNK-1 has been shown to react with an epitope similar to that recognized by IgM anti-MAG.[390] *In vitro* studies suggest that the anti-GM1 antibodies are a key participant in facilitating blood–nerve barrier dysfunction.[391]

While most cases of monoclonal gammopathy associated with peripheral neuropathies are sporadic, familial occurrence of polyneuropathy has been reported.[392,393] It is important to distinguish between hereditary neuropathies and those associated with monoclonal gammopathies directed against MAG because the latter may respond to interventional drug therapy while hereditary neuropathies lacking these antibodies do not.[394]

Even though most of the recorded cases have an IgM monoclonal antibody, IgA and IgG monoclonal isotypes have also been described in patients with polyneuropathy, especially those associated with osteosclerotic myeloma (POEMS syndrome; see below).[375,395] Histologically, nerve biopsy may reveal loss of both myelinated and unmyelinated nerve fibers, and in cases associated with macroglobulinemia, diffuse infiltration by lymphoplasmacytic B-cells.[396] The monoclonal antibodies have been found in widened lamellae of myelinated fibers by immunoelectron microscopy.[397,398]

The quantity of these monoclonal proteins in serum may be quite small, requiring electrophoretic techniques of high-resolution and immunofixation for adequate demonstration; routine five-band electrophoresis may miss the monoclonal band.[399] Vrethem et al.[400] have documented that immunofixation, overall, is superior to routine agarose electrophoresis in detecting small monoclonal gammopathies in these patients. The correct diagnosis is important because plasma exchange has been shown to be helpful in treating some neuropathies associated with MGUS.[401] Surprisingly, considering that IgM is located mainly in the intravascular compartment and would be expected to decline more rapidly with plasma exchange than would IgG or IgA, plasma exchange has been more effective with IgA and IgG monoclonal proteins than with IgM monoclonals.[401] This may reflect a difference in the pathogenesis of neuropathy associated with IgM than that associated with IgG or IgA. IgM monoclonal proteins associated with polyneuropathy can result in a complement mediated demyelination.[402] In contrast, a controlled study of intravenous immunoglobulin in demyelinating neuropathy associated with IgM anti-MAG concluded that this form of therapy had only a modest benefit to less than 20 per cent of their patients.[403]

POEMS SYNDROME

Another association of monoclonal gammopathies and neuropathies is the POEMS syndrome, also called Crow–Fukase syndrome.[404–406] The acronym

POEMS stands for *p*eripheral neuropathy, *o*rgano-megaly (usually hepatosplenomegaly although lymphadenopathy has been included as an alternative), *e*ndocrine dysfunction (including diabetes mellitus, thyroid dysfunctions, impotence, virilization, gynecomastia and infertility), *m*onoclonal gammopathy and *s*kin changes (hyperpigmentation).[407] The disease tends to occur in a younger age group than multiple myeloma, with a mean of 51 years reported in a study of 99 patients.[407] Median survival was 165 months in that study.

The presenting signs and/or symptoms in most of these cases relate to neuropathy or weakness.[408] The POEMS syndrome is associated with the rare osteosclerotic variant of multiple myeloma. Most patients with myeloma develop lytic skeletal lesions that result in bone pain and pathologic fractures. However, about 3 per cent of myeloma patients have osteosclerotic lesions and do not manifest bone pain. These individuals usually have a single or multiple sites of osteosclerotic bone lesions, but do not have diffuse osteosclerotic bone involvement. This distinction is important because Lacy et al.[409] report that patients with widespread bone lesions have a more aggressive clinical course typical of multiple myeloma rather than the more indolent course of POEMS syndrome. The monoclonal proteins that occur with the POEMS syndrome are usually small IgA or IgG λ monoclonal gammopathies that do not have specificity for myelin.[409] However, there is no clinical value in distinguishing the more common osteosclerotic myeloma (with one or a few sites involved) from POEMS syndrome because they have a similar clinical course.[408]

MONOCLONAL GAMMOPATHY OF UNDETERMINED SIGNIFICANCE

The incidence of both polyclonal and monoclonal gammopathies increases with advancing age.[410] A monoclonal gammopathy is demonstrable in the serum of about 1 per cent of individuals over the age of 25 years.[25,411] The incidence of MGUS increases with time such that 2 per cent of individuals over the age of 50 years and 3 per cent over 70 will have one. Another study found 10 per cent of the ambulatory elderly population have a demonstrable monoclonal gammopathy.[20,31,412] Although the cause of this increased incidence of monoclonal gammopathies with age is unknown, it is clear that immunoregulatory capability also declines with age.[413,414] Deficiencies of regulatory T-suppressor activity could allow emergence of clonal proliferations, resulting in monoclonal gammopathies.[415] The risk of an MGUS to progress to either multiple myeloma or a related B-cell lymphoproliferative process was reported by Kyle et al. to be about 1 per cent per year.[20] However, that may be a relatively high estimate because the MGUS cases culled early on in that data used electrophoretic techniques of low resolution that may not have detected some of the small M-proteins detected by many laboratories today.

It is unclear how, or even if, a 'benign' monoclonal gammopathy evolves into a malignant process; hence Kyle coined the term monoclonal gammopathy of undetermined significance (MGUS) to account for this phenomenon in order to avoid the term 'benign' monoclonal gammopathy.[416] An MGUS is the presence of a monoclonal gammopathy in serum at a concentration of ≤ 3.0 g/dl (30 g/l) in patients with no or at most moderate amounts of MFLC in the urine and who do not have lytic lesions, anemia, hypercalcemia, or renal insufficiency (secondary to a monoclonal protein).[20] Bone marrow must contain less than 10 per cent plasma cells.[20]

In a review of 241 patients thought to have 'benign' monoclonal gammopathy, 24 per cent developed myeloma or related disorders after 20–35 years of follow-up (Table 6.9).[416] Because of this long-term potential problem, evaluation for a monoclonal gammopathy should be part of the workup of a potential bone marrow donor.[417] Recently, Kyle et al.[20] reported on the long-term follow-up of 1384 patients with MGUS diagnosed between 1960 and 1994. With 11 009 person-years of follow-up they found that 115 of the patients progressed to develop either multiple myeloma, or another B-cell lymphoproliferative disorder (Table

6.9). The greatest risk (compared with an age- and sex-matched population) of progression was for Waldenström's macroglobulinemia and multiple myeloma, and the overall risk of progression to some lymphoplasmacytic neoplasm was 7.3 times that of the controls.[20]

There have been many attempts to identify the individuals with MGUS who are most likely to progress. As mentioned above (see Multiple Myeloma section), the neoplastic cells from patients with multiple myeloma typically have translocations involving the immunoglobulin heavy chain locus in switch regions.[102,103] Some workers have suggested that 14q32 translocations and monosomy 13 suggest a transition phase from MGUS to multiple myeloma.[418,419] Unfortunately, translocations of the immunoglobulin heavy chain locus were found in 46 per cent, λ light chain translocations in 11 per cent, and t(11;14)(q13;q32) in 25 per cent of patients with MGUS and were not felt to be useful in early detection of progression.[420] At present, cytogenetics does not help to distinguish the subpopulation of MGUS patients who are at highest risk for progression of their disease. The clinical significance of these translocations for the prognosis of those patients is not known.

A variety of non-invasive markers including bone turnover markers, cytokines, and labeling studies have been reported with varying degrees of success.[124,421–423] None of these special markers has yet been embraced to evaluate MGUS patients. However, some more common clinical features of MGUS may be helpful, in identifying those individuals that might benefit from stricter monitoring. The presence of > 5 per cent bone marrow plasmacytosis, detectable MFLC, hypogammaglobulinemia, and increased sedimentation rate indicates an increased chance of MGUS progressing to a neoplastic lymphoplasmacytic process.[424] However, these factors are not agreed upon by all studies. Kyle et al.[20] did not find a clear association between suppression of the uninvolved immunoglobulin class and progression. They also did not find that the presence of an MFLC in the urine indicated progression. Yet, they did note that patients with IgM or IgA MGUS were more likely to develop a malignant lymphoplasmacytic process than patients that had an IgG MGUS.[20] They also found that the higher the concentration of the M-protein was, the more likely it was that the disease would progress.[20]

Smoldering multiple myeloma

To diagnose myeloma, one must document the presence of increased plasma cells, tissue involvement,

Table 6.9 Risk of monoclonal gammopathy of undetermined significance (MGUS) progressing[a]

Final diagnosis	Number of patients	Relative risk of progression[b]
Multiple myeloma	75	25
Lymphoma[c]	19	3.9
Amyloid AL	10	8.4
Macroglobulinemia	7	46
Chronic lymphocytic leukemia[d]	6	1.7
Plasmacytoma	1	8.5

[a]Data from Kyle et al.[20]
[b]Risk is compared with age- and sex-matched populations.
[c]Includes patients with IgM, IgA or IgG monoclonal gammopathy and lymphoma.
[d]Includes patients with IgM, IgA or IgG monoclonal gammopathy and chronic lymphocytic leukemia.

and monoclonality. Kyle and Greipp[334] noted that some patients with these features did not undergo progressive deterioration; they did not have anemia, lytic bone lesions, hypercalcemia, or renal failure. Even though the median initial monoclonal protein was 3.1 g/l, overt symptoms of myeloma did not develop for at least 5 years of follow-up. They termed the disease of these individuals 'smoldering multiple myeloma' and recommended following them closely without therapy. Other investigators also have reported a slow course. Kanoh[425] reported a case with 3–4 g/dl of IgG κ monoclonal protein and 10 per cent plasma cells in the bone marrow. Although the patient was mildly anemic (hemoglobin was 10.2 per cent), he was otherwise well and was followed with no disease progression for more than two decades. The type of monoclonal protein does not appear to be a distinguishing feature; even IgD MGUS cases have been described. The presence of > 10 per cent plasma cells in the bone marrow, detectable MFLC in urine and a monoclonal protein of the IgA isotype are useful hints that closer monitoring of patients with smoldering multiple myeloma may be indicated.[424]

It is not always possible to categorize patients as having myeloma or MGUS. There are many reported cases in the literature in which a patient with a small monoclonal protein was followed for several years, sometimes longer than two decades, before the condition 'evolved' into clear-cut myeloma. I have seen a case in which a solitary plasmacytoma was removed, and 17 years later a monoclonal protein of the same isotype was detected in the serum. Therefore, although the lymphoproliferative condition in most patients with MGUS will not evolve, it is important to follow these patients every 6–12 months with a serum and urine protein electrophoresis (depending on the location of their gammopathy) to determine if the disease is evolving. When a monoclonal gammopathy is detected for the first time, the patient needs to have a thorough physical examination, laboratory evaluation for hemoglobin, hematocrit, white blood cell count and differential, calcium, urine study for MFLC (note that the serum assay for free κ and λ may supersede this), bone marrow

examination, skeletal X-rays, and examination of tissue lesions for the conditions discussed above.[426–428]

MONOCLONAL GAMMOPATHIES IN INFECTIOUS DISEASES

Monoclonal gammopathies have been reported in association with infections.[425,429–432] Endocarditis is a particularly frequent clinical diagnosis in infections with monoclonal gammopathies.[433–435] Most of the monoclonal gammopathies associated with infectious diseases are transient, although some persist for more than 6 months.[24,436] More typically, patients with infections have oligoclonal gammopathies.[437] The reported monoclonal gammopathies may reflect the fact that during an oligoclonal expansion caused by an infection, individual clones may not always produce the same serum concentrations of antibody directed against the infectious agent. Therefore, the peak response of one clone may predominate to such an extent that it has the same electrophoretic appearance as that of a monoclonal gammopathy caused by a neoplastic lymphoproliferative process. I have seen many cases of oligoclonal gammopathies and occasionally monoclonal gammopathies in patients with acquired immune deficiency syndrome (AIDS). In patients with AIDS, relatively large M-proteins may be seen along with plasma cell hyperplasia as part of this infection.[438] When I see one or more M-proteins superimposed on a polyclonal background, in my interpretation, I caution that this may be reflecting a reactive process. I recommend urine immunofixation and a follow-up of the serum electrophoresis to determine if the process regresses or evolves into either a clearly infectious or definable neoplastic condition. Since infections are occasionally a presenting feature in patients with multiple myeloma, one cannot dismiss the potential importance of a monoclonal gammopathy in patients with infections. I look for indications that the monoclonal gammopathy is probably caused by an infection (Table 6.10). Despite the electrophoretic

Table 6.10 Transient prominent oligoclonal band versus monoclonal gammopathy

Factor	Transient process	Lymphoproliferative process
κ/λ Ratio	Usually normal range	Usually high or low
Uninvolved isotypes[a]	Usually normal or elevated	Often decreased
MFLC[b] in urine	No	Often present
Follow-up samples	M-protein resolves	M-protein persists/increases
Acute-phase reaction	Usually present	May be present with hypogammaglobulinemia

[a]If the isotype of the M-protein is IgG, the uninvolved isotypes would be IgA and IgM.
[b]MFLC, monoclonal free light chain.

appearance of an M-protein, infections usually have an oligoclonal expansion and therefore do not share the same features that an M-protein due to the presence of multiple myeloma or a lymphoproliferative process do.[24] Also, although patients with multiple myeloma and suppression of the normal IgG response may suffer from infections and develop an acute-phase pattern on serum protein electrophoresis, the M-proteins that arise from infectious processes do not have hypogammaglobulinemia. Thus, the presence of an acute-phase pattern in the presence of a normal or polyclonal increase in the uninvolved γ-globulins suggests that the process may be a transient reaction to infection.

This serves to reemphasize that monoclonal gammopathies do not equate with myeloma or uncontrolled B-cell proliferation. To help in understanding these processes, I recommend both clinical and laboratory follow-up with urine and serum specimens when monoclonal gammopathies are detected. Fortunately, one does not treat asymptomatic myeloma, whereas the infection will be treated with antibiotic therapy. The true nature of the monoclonal process will usually declare itself with careful follow-up. Finally, when the findings of any particular case are confusing, speak to the clinician to be certain that they understand the findings you have, while you learn more about their differential diagnosis to guide your interpretations.

MONOCLONAL GAMMOPATHIES IN IMMUNODEFICIENCY

A wide variety of congenital and acquired immunodeficiency conditions have been associated with the presence of oligoclonal and monoclonal gammopathies. Radl[439] pointed out that monoclonal gammopathies (homogeneous immunoglobulin components) in the elderly may represent a loss of immune regulatory function with T < B immune system imbalance. Monoclonal gammopathies of this sort also develop in experimental animals as they age or under the influence of antigenic stimulation.[440,441] The relationship of an immune system dysfunction to development of monoclonal gammopathies is supported by the findings in immunosuppressed allograft recipients who often develop post-transplant lymphoproliferative disorder (see Chapter 4).[442] Clearly, the immunosuppressive therapy permits infectious agents and/or spontaneous proliferation of B cell clones that will often regress when the immunosuppressive therapy is discontinued.[443–445] Monoclonal gammopathies in children are exceedingly rare but, when present, may indicate either a transient or congenital immune dysfunction.[60,61] Congenital immunodeficiency diseases such as Wiskott–Aldrich Syndrome, DiGeorge Syndrome, Nezelof Syndrome and severe combined immunodeficiency (SCID) are frequently associated with the presence of monoclonal immunoglobulins.[439] Therefore, in addition to the conditions described above, immune dysfunction may be the underlying cause of monoclonal and oligoclonal gammopathies.

CRYOGLOBULINS

Cryoglobulins are immunoglobulins that aggregate and precipitate or gel at temperatures lower than 37°C. Most are not monoclonal proteins. They have clinical importance because, in addition to problems caused by the underlying condition (such as multiple myeloma, hepatitis C, autoimmune disease, etc.), they can precipitate in blood vessels with life-threatening consequences. They were first recognized in association with multiple myeloma in 1933 by Wintrobe and Buel, but the term 'cryoglobulin' was introduced in 1947 by Lerner and Watson.[446–448] Depending on their primary disease association, they have been treated by various means, including ribaviron and α-interferon, cytotoxic drugs, steroids, plasmapheresis, and even colchicine.[449–454] Occasionally, the therapies seemed to trigger adverse events.[455,456] Cryoglobulins have been classified by Brouet into Types I, II, and III (Table 6.11).[457] When no underlying disease such as multiple myeloma is known to be present, the term 'essential' is used to describe the cryoglobulin. However, as we learn more about the etiology of cryoglobulins, the use of 'essential' has declined. For example, as discussed below, most cases of Type II cryoglobulins used to be thought of as essential cryoglobulins, but since the discovery of hepatitis C virus (HCV) it has become clear that HCV is a major cause of both Type II and Type III cryoglobulins.[458,459]

Type I cryoglobulins are most often seen in patients with lymphoproliferative diseases, especially multiple myeloma, Waldenström's macroglobulinemia, and monoclonal gammopathy of undetermined significance. IgM is the most frequent isotype encountered in this category followed by IgG, then IgA.[460] In Type I cryoglobulinemia, the monoclonal protein is usually present in large amounts (> 500 mg/dl) (Table 6.11). In the original data from Brouet et al.,[457] Type I cryoglobulinemia accounts for about 25 per cent of cases of cryoglobulinemia. However, the recent data from Trejo et al.[461] demonstrate that the extraordinary number of HCV infections have dramatically changed the percentage of cases that result from infectious diseases in recent years (Table 6.12) such that in their series only 7 per cent of cases were associated with hematologic conditions. Those numbers more likely reflect differences between the current case mix and that of the original Brouet data.

Type II cryoglobulins are also associated with monoclonal proteins, but are different from those of Type I. As shown in Table 6.11, these are usually present in smaller amounts than Type I cryoglobulin. Type II cryoglobulins are a mixture of an IgM (rarely IgA and IgG) usually κ monoclonal protein with rheumatoid factor activity (reacts with the Fc portion of IgG).[457,462,463] These proteins are present in lower concentration than Type I, and are most often found in patients with infectious conditions. They have occasionally been described in association with autoimmune or lymphoproliferative diseases. In the past decade, several studies have pointed to an extraordinarily high incidence of HCV infection in patients with Type II cryoglobulinemia.[459,464–474] Indeed, HCV was a major factor in the recent series by Trejo et al.[461] where they evaluated sera from 7043 samples sent to their immunology department between 1991 and 1999. Of the 443 patients with a cryocrit of ≥ 1 an extraordinary 321 (73 per cent) had HCV (Table 6.12). In contrast, hepatitis B infection accounted for only 15 (3 per cent) cases.

Type III cryoglobulins, the type most frequently encountered and unrelated to monoclonal proteins, consist of polyclonal rheumatoid factor, usually IgM, that reacts with polyclonal IgG.[460] Since both Types II and III cryoglobulinemias contain more than one type of immunoglobulin, they are termed mixed cryoglobulins. Currently, the vast majority of these patients are infected with HCV, although the exact percentage varies

Table 6.11 Classification of cryoglobulins[a]

Type	Cryoglobulin level (mg/dl)			Composition
	< 100	100–500	> 500	
I	10%	30%	60%	Monoclonal immunoglobulin
II	20%	40%	40%	Monoclonal (usually IgM, rarely IgG or IgA) and polyclonal IgG
III	80%	20%	0	Polyclonal IgM, IgG and (occasionally) IgA

[a]Data from Brouet et al.[457]

Table 6.12 Etiological factors in cryoglobulinemia[a]

Factor	Number of patients (%)
Infection	331 (75)
Hepatitis C virus (HCV)	321 (73)
Hepatitis B virus (HBV)	15 (3)
Human immunodeficiency virus (HIV)	29 (19)
Autoimmune disease	95 (24)
Primary Sjögren's syndrome	40 (9)
Systemic lupus erythematosus	30 (7)
Polyarteritis nodosa	7 (2)
Systemic sclerosis	6 (2)
Other (primary antiphospholipid syndrome, rheumatoid arthritis, autoimmune thyroitiditis, Horton arteritis, dermatomyositis, Henoch–Schonlein)	13 (2.9)
Hematologic disease	33 (7)
Non-Hodgkin's lymphoma	16 (4)
Chronic lymphocytic leukemia	3 (1)
Multiple myeloma	3 (1)
Hodgkin's lymphoma	2 (0.5)
Myelodysplasia	2 (0.5)
Waldenström macroglobulinemia	1 (0.2)
Castelman disease	1 (0.2)
Thrombocytopenic thrombotic purpura	1 (0.2)
Essential cryoglobulinemia	49 (11)

[a]Data from Trejo et al.[461]

widely from one study to the next. In addition to HCV, however, these cases may be associated with other infectious diseases or autoimmune diseases.[475-477] Interestingly, however, even the autoimmune diseases may have more than a passing relationship with the HCV infection.[478] Typically, Type III cryoglobulin is present in low concentration (< 100 mg/dl). Because of the relatively small amount of cryoglobulin in these cases, they prove the most difficult to characterize. Immunoblotting and two-dimensional electrophoresis studies have demonstrated that rather than a polyclonal IgM, the cryoglobulin is composed of a few clones of (oligoclonal) IgM with polyclonal IgG.[460,463,479]

It is not clear why cryoglobulins precipitate. Some have noted structural changes in the variable portions of the immunoglobulin heavy and light changes, an abundance of hydrophobic amino acids, unusual glycosylation, a decrease in galactose at the Fc portion of IgG, or a change in the CH3 domain glycosylation sites.[478,480-483] Unfortunately, there is no consensus on the mechanism, and it is likely there are several mechanisms depending on the individual proteins involved.

Cryoglobulins can have significant clinical consequences that occur secondary to the obstruction of blood vessels and/or to the vasculitis that results from the inflammatory effects of immune complex deposition.[460] Prominent signs and symptoms include: purpura (virtually always), arthralgias, renal disease (often membranoproliferative glomerulonephritis), peripheral neuropathy, hepatic involvement, abdominal pain (likely due to vessel involvement in the gastrointestinal tract), Raynaud's phenomenon, and leg ulcers.[446] The clinical consequences of cryoglobulins may depend as much on the temperature at which they precipitate as on their amount. Letendre and Kyle[484] described two patients with relatively small amounts of Type I cryoglobulin who had significant clinical consequences because they precipitated *in vitro* at temperatures higher than 25°C. The presence of cryoglobulinemia in patients with autoimmune disease, such as systemic lupus erythematosus,

significantly increases the likelihood that the individual will develop renal disease.[485]

In the clinical laboratory, a tiny amount of cryoglobulin may be detected in normal individuals. Levels up to 80 µg/ml may occur in controls.[446] When a cryoglobulin is detectable, but present in amounts < 2 per cent, we cannot characterize them. I report them as trace of cryoglobulin present, but too small to quantify. Types II and III (the mixed cryoglobulins) can present with a variety of laboratory findings, including rheumatoid factor activity and often a low level of C1q and C4 (although C3 is usually normal).[446,478] A definition of clinically symptomatic mixed cryoglobulinemia by Invernizzi et al.[486] provided a usable standard. Key laboratory features of their definition of mixed cryoglobulins are: the presence of a cryocrit > 1 per cent for at least 6 months, C4 less than 8 mg/dl, and a positive rheumatoid factor.[486] In addition, patients often have increased erythrocyte sedimentation rates and a normochromic normocytic anemia.[446]

Cryoglobulins may be missed by electrophoretic analysis, especially if they precipitate at relatively high temperatures – a feature with considerable clinical importance.[484] If proper precautions are not taken in handling the specimen, the cryoglobulin will precipitate during the clotting process and will be missed by electrophoresis or when the sample is placed in the cold. To detect cryoglobulins, the specimen should be drawn in a prewarmed (37°C) red top tube or syringe. To minimize transportation problems, ideally the patient would have their venepuncture in the laboratory. However, because of the condition of the patient or the location of the laboratory, this is often not practical. To maintain the temperature during transportation to the laboratory, some house officers place the sample close to the body such as under an armpit or in a pocket. As a more reasonable alternative when the patient cannot be readily transported to the laboratory, I recommend using a device to maintain that temperature. A commercial device is now available for this, facilitating temperature maintenance during transportation. One inexpensive solution is to use a thermos filled with a material to preserve the temperature at 37°C. For years, I recommended using sand in the thermos kept in a 37°C incubator. However, the sand can be messy, and a clever alternative is to put commercially available gel-filled plastic bags in the thermos (suggested by Linda Thomas at the University of Michigan Immunology Laboratory, Ann Arbor, MI, USA) instead of the sand. These gel packs are available as Gel-Ice (Pioneer Packaging Co., Kent, WA, USA) and as Polar Pack (Mid-Lands Chemical Co., DesMoines, IA, USA). This provides excellent thermal stability when compared with 37°C water in a styrofoam cup, and has resulted in improved yield of cryoglobulins (Fig. 6.18).[487]

After separating the clot, the specimen is split into two fractions. One is kept at 37°C while the other is placed at 4°C. Samples are examined daily for up to 7 days. When a precipitate is seen in the 4°C tube and not in the 37°C tube, it is reported as positive for cryoglobulin. The precipitate is most often a fluffy, white flocculant substance; however, some are crystalline or gelatinous (the latter are often Type III cryoglobulins and may take up to 7 days to form).[488] One may visualize cryoglobulins as small as 1–5 µg/ml.[488] The cryocrit is measured by placing the warm serum sample into a Wintrobe or other calibrated tube and incubate it at 4°C until the precipitate forms. Centrifugation at 2000 r.p.m. for 30 min in the cold provides the cryocrit.[488] Unfortunately, there is not a world-wide standard for the measurement of the cryoglobulin and even low cryocrits have been associated with severe disease whereas some high cryocrits are clinically asymptomatic.[478] An alternative to quantify the cryoglobulin involves washing the cryoprecipitate at least six times with ice-cold saline by repeated centrifugations and vortexing. Then one redissolves the cryoprecipitate in 37°C and performs a total protein determination on the redissolved precipitate. Unfortunately, considerable cryoglobulin may be lost in the washing procedure. Therefore, I prefer to use the simpler cryocrit.

After establishing the cryocrit, our laboratory characterizes cryoglobulins by immunofixation on the thoroughly washed cryoglobulin (as above). One needs to wash thoroughly in order to remove the polyclonal immunoglobulins that are in the serum but which are not part of the cryoglobulin. Redissolving cryoglobulins may be a problem. If the cryoglobulin does not readily dissolve upon reheating to 37°C, adjusting pH or salt concentrations can help. Grose[489] suggested adding 5 per cent acetic acid to a mixture of saline and the cryoprecipitate. When I examine the serum protein electrophoresis (SPE) lane (fixed in acid) on the immunofixation gel I can tell if the sample has been washed adequately by requiring that no albumin band be visible (Fig. 6.19). By immunofixation, a Type I cryoglobulin will have only the monoclonal protein present. A Type II cryoglobulin will have the monoclonal protein (usually IgM κ) and polyclonal IgG (Fig. 6.19). A Type III cryoglobulin will have polyclonal immunoglobulins present. Other laboratories use either gel diffusion or immunoblotting to characterize the cryoglobulins.[463,488,490]

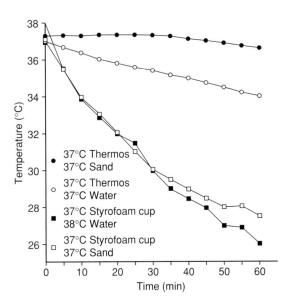

Figure 6.18 Thermal stability of water under the conditions indicated. When water is stored in a thermos containing sand, there is almost no change in temperature up to 1 h. Thus, unless the patient can have their venepuncture in the laboratory (the best solution, but one which is not practical in many circumstances), I recommend using a transportation method that reliably maintains the temperature at 37°C to transport blood samples from patients suspected of having a cryoglobulinemia.

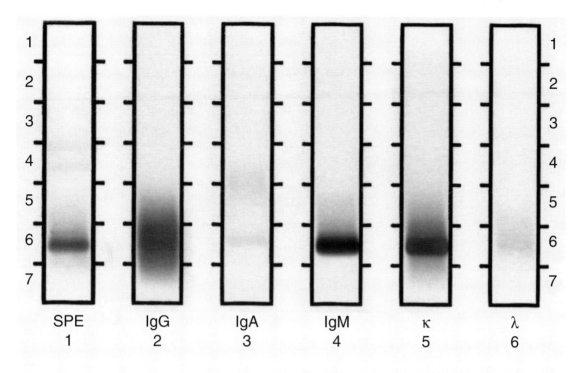

Figure 6.19 Immunofixation of thoroughly washed cryoprecipitate. Note that the serum protein electrophoresis (SPE) lane shows no band in the albumin position, indicating that the sample has been washed adequately. There is a diffuse broad staining in the IgG and faintly in the λ lanes. There is a dense band near the origin in the IgM and the κ lanes. In addition, there is some diffuse staining in the κ lane both anodal and cathodal to the band. This is a Type II cryoglobulin. It contains monoclonal IgM κ and polyclonal IgG (and of course the light chains bound to the polyclonal IgG). If this had been a Type I cryoglobulin, there should be no polyclonal antibody present. The wash step that removes unbound polyclonal antibodies is critical to making this distinction. (Paragon Immunofixation stained with Paragon Violet.)

When a Type II or III cryoglobulin is present, I recommend performing serology for HCV, rheumatoid factor and C4.

Serum protein electrophoresis in patients with Type I cryoglobulin usually shows the M-protein. With Type II cryoglobulins, an M-protein may or may not be seen by electrophoresis. It may be present in an amount so small that an immunofixation will be needed to identify it. Type III cryoglobulins will usually have a polyclonal increase, and sometimes a polyclonal and oligoclonal increase in the γ-region. However, serum protein electrophoresis is not an appropriate screening test for cryoglobulins. Normal patterns and even hypogammaglobulinemia have been reported in patients with cryoglobulinemia.[446,491–493]

In addition to the above patterns, on gel-based electrophoretic techniques, there may be a precipitate at the origin because of the cooler temperature of the gel allowing precipitation of some cryoglobulins (Fig. 6.20). Finding such a precipitate should prompt one to investigate the patient for cryoglobulin. Since the protein precipitates without any specific antisera being added, it may also produce confusing patterns on immunofixation of the serum. Therefore, when an immunofixation of a serum sample shows an origin precipitate in several lanes, I repeat the immunofixation, replacing one of the antisera with buffer or saline. When the precipitate recurs in this location, one is certain to be dealing with a spontaneously precipitating protein, usually a

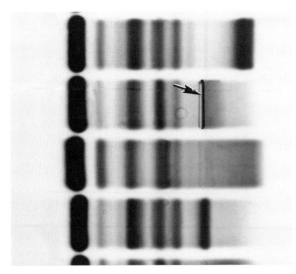

Figure 6.20 Serum protein electrophoresis of four samples. The top and bottom samples contain obvious M-proteins in the γ-region and near the origin, respectively. The third lane contains a large transferrin band that deserves an immunofixation to be certain it is not a monoclonal gammopathy. The second lane contains an origin artifact (arrow) that may indicate a cryoglobulin. Just cathodal to this origin artifact is a lightly staining, but distinct band likely representing a monoclonal cryoglobulin. (Paragon SPE2 system stained with Paragon Violet.)

cryoglobulin. At that point, a fresh specimen drawn and transported as recommended above will provide the best results to characterize the cryoglobulin.

Cryofibrinogens may be confused with cryoglobulins when the sample is drawn in a tube with an anticoagulant rather than the recommended red top (no anticoagulant present) tube. They result from a precipitation of fibrinogen with fibrin in the cold.[488] Detection of cryofibrinogen requires drawing a sample of blood into a prewarmed citrated tube and allowing it to sit overnight at 4°C. One should not use ethylenediaminetetraacetic acid (EDTA) because this may inhibit cryofibrinogen formation, neither should one use heparin because it may enhance cryoprecipitation of fibronectin.[488] A washed cryofibrinogen immunofixation may be used to characterize the nature of the precipitate with anti-fibrinogen antibody (Figure 6.21).

BANDS MISTAKEN FOR MONOCLONAL GAMMOPATHIES

One must maintain a high level of suspicion for monoclonal gammopathies when one finds any alteration in the electrophoretic pattern (even α-, or β-region changes). There are several alterations which can mimic monoclonal gammopathies that need to be considered.

Fibrinogen

One of our most common problems in dealing with serum protein electrophoresis is the presence of the fibrinogen band. This should not be present in a properly clotted specimen. However, it may be present owing to a variety of circumstances: a blood sample may not have been allowed to clot for a sufficient period of time; the patient may be on an anticoagulant which prevents complete clotting; the sample may have been collected in a tube containing an anticoagulant and then the fibrinogen band appears in the β–γ region of virtually all gel-based electrophoretic systems (Fig. 6.22). With the earlier Paragon CZE 2000 system 1.5 (Beckman Coulter), a fibrinogen band was not visible. In the more recent version of this system 1.6 and in the Sebia Capillarys system, fibrinogen also occurs at the β–γ region. There is no way to absolutely rule out a monoclonal gammopathy when detecting a band in that location by just examining the serum protein electrophoresis. If the tube containing the sample is examined, and a clot is present at the bottom, this suggests that some fibrinogen was present after the clot was removed. One may repeat the electrophoresis on the next run, if the band is gone (because the fibrinogen has now clotted), it is due to fibrinogen. Care is needed here because, if the band is due to a cryoprecipitating monoclonal gammopathy, this too would form a precipitate in the bottom of the tube (however, the precipitate from cryoglobulins *usually* looks different from the typical fibrin clot). When I suspect a fibrinogen band is present, I perform a Penta immunofixation (see Chapter 3). An alterna-

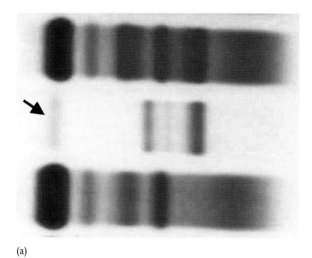

(a)

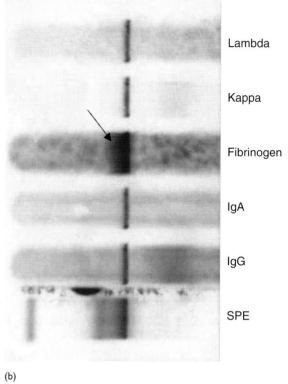

Lambda

Kappa

Fibrinogen

IgA

IgG

SPE

(b)

Figure 6.21 (a) Serum protein electrophoresis demonstrates three samples. The top sample has a large β–γ region band that had a cryoprecipitate form at 4°C. The second lane shows the results of the washed precipitate. Note that albumin is still present (arrow) indicating that it was not washed thoroughly. The bottom sample is a lyophilized control with no C3 band.
(b) Immunofixation of the washed precipitate from (a). A precipitate is at the origin of all the lanes. No specific precipitate is seen in any of the lanes with antisera against immunoglobulins. However, anti-fibrinogen gave a definite precipitate (arrow) indicating that this was a cryofibrinogen. The additional material seen in the SPE lane anodal to the origin may be another cryoprecipitating protein, fibronectin.

tive is to perform a three-lane immunofixation using antifibrinogen, anti-κ and anti-λ.

Genetic variants

As discussed in Chapter 4, transferrin, α_1-antitrypsin, and C3 have several possible alleles in the population and may give two bands or one band with an unusual migration. The typical situation is the heterozygote, with a band in the usual location for the particular protein and a second band of identical staining intensity immediately anodal or cathodal to it. These twin bands are usually obvious as to their nature. When they are seen in the α_1-antitrypsin area, I recommend that studies

be performed to determine the individual's phenotype for genetic counseling.

In the transferrin and C3 areas, however, things are often more confusing. First, this region is more likely to have monoclonal gammopathies than the α_1-region. Second, the band may be obscured by the presence of β_1-lipoprotein band. Consequently, it is more difficult to conclude with certainty that one is seeing a genetic variant as opposed to a small monoclonal gammopathy. I have occasionally seen small monoclonal gammopathies (some of which later become major monoclonal spikes and others already are responsible for large MFLC in the urine) that migrate precisely over, or next to, the transferrin and C3 bands. This causes these bands to be larger than usual or to have an altered migra-

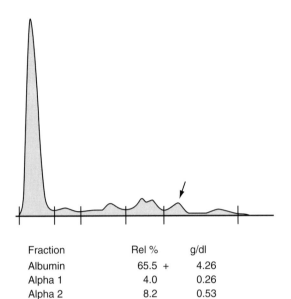

Fraction	Rel %		g/dl
Albumin	65.5	+	4.26
Alpha 1	4.0		0.26
Alpha 2	8.2		0.53
Beta	11.2	−	0.73
Gamma	11.1	−	0.72

Figure 6.22 Densitometric scan of serum with a fibrinogen band (indicated) at the β–γ region. In the mid-γ-region there is a small restriction that turned out to be κ monoclonal free light chain. (Densitometric scan of SPE2 gel.)

tion. Therefore, when I see any unusual band in the transferrin and C3 regions, although I suspect a genetic variant, I perform an immunofixation to rule out a possible monoclonal gammopathy.

C-reactive protein

As mentioned above, C-reactive protein normally is not seen on serum protein electrophoresis. However, in cases with strong acute-phase reaction patterns, a small band is often seen in the mid to slow γ-region on gel-based techniques. With CZE, I have not seen a C-reactive protein band even in cases with prominent acute-phase patterns. One may determine where it migrates on their gels by performing an immunofixation. With the availability of rapid immunofixation testing such as the Penta test, I always perform this procedure even when the band is in the typical C-reactive protein location. Patients with light chain multiple myeloma may have a small component of the M-protein in the γ-region of the serum. Further, the

immunosuppression these patients may exhibit will predispose them to infections that elicit an acute-phase pattern.

HEMOLYSIS

A hemoglobin–haptoglobin band usually migrates in the α_2- to β_1- region. It can resemble a monoclonal gammopathy. Usually, there is marked depletion of the normal haptoglobin band and the serum is red. If the serum is not red, or if the haptoglobin band is not depleted, perform an immunofixation to rule out a monoclonal gammopathy.

Radiocontrast dyes

On CZE, radiocontrast dyes create peaks anywhere from the α_2- to the γ-globulin region (see Chapter 2). Because of this, any restriction suggestive of an M-protein that has not previously been characterized must be proven to be an M-protein by immunofixation or immunosubtraction before reporting it as such. When a small restriction is seen, perform a Penta analysis to rule out an M-protein.

UNUSUAL EFFECTS OF MONOCLONAL GAMMOPATHIES ON LABORATORY TESTS

The monoclonal proteins themselves can play havoc with many clinical laboratory tests. For example, some M-proteins bind to enzymes such as lactate dehydrogenase (LDH) creating unusual migration; others have been associated with factitious hypercalcemia, unpredictable artifactual increases in serum iron levels (which can be especially problematic because these patients are often being studied for the presence of anemia), increases in organic phosphorus, and positive direct Coombs tests (due to passive adsorption of the monoclonal immunoglobulin onto the erythrocytes).[494–499] Several coagulopathies have been noted in patients with monoclonal gammopathy, including isolated factor X deficiency, acquired von Willebrand disease, a cryoglobulin that inhibited fibrin polymerization, and disseminated intravascular coagulation.[500–503] Unusual physiological manifestations can occur, for example the monoclonal protein that bound to insulin and produced recurrent severe hypoglycemia,[504,505] or the binding of an IgM monoclonal gammopathy to calcium resulting in a hypercalcemia while the ionized fraction of calcium was normal.[506] Interaction between an erythrocyte lysing reagent and an IgM monoclonal gammopathy resulted in inaccurate hemoglobin measurements in one patient with Waldenström's macroglobulinemia.[507] Because of their unique properties, cryoglobulins may be mistaken by electronic cell counters when the size of the precipitates fall within the ranges the instrument interprets as cellular components, resulting in elevated leukocyte or platelet counts.[508–512] Direct analysis of such blood smears will demonstrate suspicious aggregates of amorphous particles.[513]

REFERENCES

1. Pick AI, Shoenfeld Y, Frohlichmann R, Weiss H, Vana D, Schreibman S. Plasma cell dyscrasia. Analysis of 423 patients. *JAMA* 1979;**241**: 2275–2278.

2. Katzin WE. Cancer (multiple myeloma and related disorders). *Anal Chem* 1993;**65**: 382R–387R.

3. Keren DF, Alexanian R, Goeken JA, Gorevic PD, Kyle RA, Tomar RH. Guidelines for clinical and laboratory evaluation of patients with monoclonal gammopathies. *Arch Pathol Lab Med* 1999;**123**:106–107.

4. Bruton OC. Agammaglobulinemia. *Pediatrics* 1952;**9**:722–727.

5. Schiff R, Schiff SE. Chapter 11. Flow cytometry for primary immunodeficiency diseases. In: Keren DF, McCoy JJP, Carey JL, eds. *Flow cytometry in clinical diagnosis*. Chicago: ASCP Press, 2001.

6. Glick B, Chang TS, Jaap RG. The bursa of Fabricius and antibody production. *Poult Sci* 1956;**35**:224–232.

7. Pirruccello SJ, Aoun P. Chapter 3. Hematopoietic cell differentiation: monoclonal antibodies and cluster designation (CD)-defined hematopoietic cell antigens. In: Keren DF, McCoy JJP, Carey JL, eds. *Flow cytometry in clinical diagnosis*. Chicago: ASCP Press, 2001.

8. Ciudad J, Orfao A, Vidriales B, et al. Immunophenotypic analysis of CD19+ precursors in normal human adult bone marrow: implications for minimal residual disease detection. *Haematologica* 1998;**83**:1069–1075.

9. Korsmeyer SJ, Greene WC, Cossman J, et al. Rearrangement and expression of immunoglobulin genes and expression of Tac antigen in hairy cell leukemia. *Proc Natl Acad Sci USA* 1983;**80**:4522–4526.

10. Caligaris-Cappio F, Janossy G. Surface markers in chronic lymphoid leukemias of B cell type. *Semin Hematol* 1985;**22**:1–12.

11. Winkelstein A, Jordan, PS. Immune deficiencies in chronic lymphocytic leukemia and multiple myeloma. In: Ballow M, ed. *Intravenous immunoglobulin therapy today*. Totowa: The Humana Press Inc, 1992.

12. Hayward AR. Development of lymphocyte responses and interactions in the human fetus and newborn. *Immunol Rev* 1981;**57**:39–60.

13. Hofman FM, Danilovs J, Husmann L, Taylor CR. Ontogeny of B cell markers in the human fetal liver. *J Immunol* 1984;**133**:1197–1201.

14. Jensen GS, Mant MJ, Pilarski LM. Sequential maturation stages of monoclonal B lineage cells from blood, spleen, lymph node, and bone marrow from a terminal myeloma patient. *Am J Hematol* 1992;**41**:199–208.

15. Rudin CM, Thompson CB. B-cell development and maturation. *Semin Oncol* 1998;**25**:435–446.

16. Grogan TM, van Camp B, Kyle RA. Plasma cell neoplasms. In: Jaffe ES, Harris NL, Stein H, Vardiman JW, ed. *Pathology and genetics of tumours of haematopoietic and lymphoid tissues*. Lyon: IARC Press, 2001.

17. Sezer O, Heider U, Zavrski I, Possinger K. Differentiation of monoclonal gammopathy of undetermined significance and multiple myeloma using flow cytometric characteristics of plasma cells. *Haematologica* 2001;**86**:837–843.

18. van Riet I, De Waele M, Remels L, Lacor P, Schots R, van Camp B. Expression of cytoadhesion molecules (CD56, CD54, CD18 and CD29) by myeloma plasma cells. *Br J Haematol* 1991;79:421–427.

19. van Camp B, Durie BG, Spier C, et al. Plasma cells in multiple myeloma express a natural killer cell-associated antigen: CD56 (NKH-1; Leu-19). *Blood* 1990;76:377–382.

20. Kyle RA, Therneau TM, Rajkumar SV, et al. A long-term study of prognosis in monoclonal gammopathy of undetermined significance. *N Engl J Med* 2002;346:564–569.

21. Kyle RA. 'Benign' monoclonal gammopathy – after 20–35 years of follow-up. *Mayo Clin Proc* 1993;68:26–36.

22. Kyle RA, Rajkumar SV. Monoclonal gammopathies of undetermined significance. *Hematol Oncol Clin North Am* 1999;13:1181–1202.

23. Clark KT. Monoclonal gammopathy of undetermined significance. *J Insur Med* 1997;29:136–138.

24. Keren DF, Morrison N, Gulbranson R. Evolution of a monoclonal gammopathy (MG) Documented by high-resolution electrophoresis (HRE) and immunofixation (IFE). *Lab Med* 1994; 25:313–317.

25. Axelsson U, Bachmann R, Hallen J. Frequency of pathological proteins (M-components) in 6,995 sera from an adult population. *Acta Med Scand.* 1966;179:235–247.

26. Malik AA, Ganti AK, Potti A, Levitt R, Hanley JF. Role of *Helicobacter pylori* infection in the incidence and clinical course of monoclonal gammopathy of undetermined significance. *Am J Gastroenterol* 2002;97:1371–1374.

27. Kyle RA. Diagnostic criteria of multiple myeloma. *Hematol Oncol Clin N Am* 1992;6:347–358.

28. Cohen HJ, Crawford J, Rao MK, Pieper CF, Currie MS. Racial differences in the prevalence of monoclonal gammopathy in a community-based sample of the elderly. *Am J Med* 1998;104:439–444.

29. Ries LAG, Miller BA, Hankey BF. *SEER cancer statistics review, 1973–1991: tables and graphs, National Cancer Institute.* Bethesda: NIH, 1994.

30. Parker SL, Davis KJ, Wingo PA, Ries LA, Heath CW Jr. Cancer statistics by race and ethnicity. *CA Cancer J Clin* 1998;48:31–48.

31. Aguzzi F, Bergami MR, Gasparro C, Bellotti V, Merlini G. Occurrence of monoclonal components in general practice: clinical implications. *Eur J Haematol* 1992;48:192–195.

32. Alexanian R, Weber D, Liu F. Differential diagnosis of monoclonal gammopathies. *Arch Pathol Lab Med* 1999;123:108–113.

33. Kyle RA. Multiple myeloma. Diagnostic challenges and standard therapy. *Semin Hematol* 2001;38:11–14.

34. Kyle RA. Sequence of testing for monoclonal gammopathies. *Arch Pathol Lab Med* 1999;123:114–118.

35. Epstein J, Barlogie B, Katzmann J, Alexanian R. Phenotypic heterogeneity in aneuploid multiple myeloma indicates pre-B cell involvement. *Blood* 1988;71:861–865.

36. Barlogie B. Pathophysiology of human multiple myeloma – recent advances and future directions. *Curr Top Microbiol Immunol* 1992;182:245–250.

37. Taylor BJ, Pittman JA, Seeberger K, et al. Intraclonal homogeneity of clonotypic immunoglobulin M and diversity of nonclinical post-switch isotypes in multiple myeloma: insights into the evolution of the myeloma clone. *Clin Cancer Res* 2002;8:502–513.

38. Dimopoulos MA, Papadimitriou C, Sakarellou N, Athanassiades P. Complications and supportive therapy of multiple myeloma. *Baillieres Clin Haematol* 1995;8:845–852.

39. Vaswani SK, Sprague R. Pseudohyponatremia in multiple myeloma. *South Med J* 1993;86:251–252.

40. Avashia JH, Walsh TD, Valenzuela R, Fernando Quevedo J, Clough J. Pseudohypoproteinemia and multiple myeloma. *Cleve Clin J Med* 1990;57:298–300.

41. van Dijk JM, Sonnenblick M, Weissberg N,

Rosin A. Pseudohypercalcemia and hyperviscosity with neurological manifestations in multiple myeloma. *Isr J Med Sci* 1986;**22**: 143–144.

42. Michopoulos S, Petraki K, Petraki C, Dimopoulos MA. Light chain deposition disease of the liver without renal involvement in a patient with multiple myeloma related to liver failure and rapid fatal outcome. *Dig Dis Sci* 2002;**47**:730–734.

43. Grassi MP, Clerici F, Perin C, et al. Light chain deposition disease neuropathy resembling amyloid neuropathy in a multiple myeloma patient. *Ital J Neurol Sci* 1998;**19**:229–233.

44. Christou L, Hatzimichael EC, Sotsiou-Candila F, Siamopoulos K, Bourantas KL. A patient with multiple myeloma, amyloidosis and light-chain deposition disease in kidneys with a long survival. *Acta Haematol* 1999;**101**:202–205.

45. Baur A, Stabler A, Lamerz R, Bartl R, Reiser M. Light chain deposition disease in multiple myeloma: MR imaging features correlated with histopathological findings. *Skeletal Radiol* 1998;**27**:173–176.

46. Daliani D, Weber D, Alexanian R. Light–heavy chain deposition disease progressing to multiple myeloma. *Am J Hematol* 1995;**50**:296–298.

47. Comotti C, Mazzon M, Valli A, Rovati C, Vivaldi P. Light chain deposition nephropathy in multiple myeloma. *Contrib Nephrol* 1993;**105**: 133–138.

48. Bataille R, Harousseau JL. Multiple myeloma. *N Engl J Med* 1997;**336**:1657–1664.

49. Ong F, Hermans J, Noordijk EM, Wijermans PW, Kluin-Nelemans JC. Presenting signs and symptoms in multiple myeloma: high percentages of stage III among patients without apparent myeloma-associated symptoms. *Ann Hematol* 1995;**70**:149–152.

50. Rajkumar SV, Gertz MA, Kyle RA, Greipp PR. Current therapy for multiple myeloma. *Mayo Clin Proc* 2002;**77**:813–822.

51. Bernstein SC, Perez-Atayde AR, Weinstein HJ. Multiple myeloma in a child. *Cancer.* 1985;**56**:2143–2147.

52. Maeda K, Abesamis CM, Kuhn LM, Hyun BH. Multiple myeloma in childhood: report of a case with breast tumors as a presenting manifestation. *Am J Clin Pathol* 1973;**60**:552–558.

53. Prematilleke MN. Multiple myeloma in a child. *Ceylon Med J* 1987;**32**:143–145.

54. Hewell GM, Alexanian R. Multiple myeloma in young persons. *Ann Intern Med* 1976;**84**: 441–443.

55. Blade J, Kyle RA. Multiple myeloma in young patients: clinical presentation and treatment approach. *Leuk Lymphoma* 1998;**30**:493–501.

56. Cenac A, Pecarrere JL, Abarchi H, Devillechabrolle A, Moulias R. Juvenile multiple myeloma. A Nigerian case. *Presse Med* 1988;**17**:1849–1850.

57. Harster GA, Krause JR. Multiple myeloma in two young postpartum women. *Arch Pathol Lab Med* 1987;**111**:38–42.

58. Badwey TM, Murphy DA, Eyster RL, Cannon MW. Multiple myeloma in a 25-year-old woman. *Clin Orthop* 1993;**294**:290–293.

59. Corso A, Klersy C, Lazzarino M, Bernasconi C. Multiple myeloma in younger patients: the role of age as prognostic factor. *Ann Hematol* 1998;**76**:67–72.

60. Gerritsen EJ, van Tol MJ, Lankester AC, et al. Immunoglobulin levels and monoclonal gammopathies in children after bone marrow transplantation. *Blood.* 1993;**82**:3493–3502.

61. Gerritsen E, Vossen J, van Tol M, Jol-van der Zijde C, van der Weijden-Ragas R, Radl J. Monoclonal gammopathies in children. *J Clin Immunol* 1989;**9**:296–305.

62. Yoshida K, Minegishi Y, Okawa H, et al. Epstein–Barr virus-associated malignant lymphoma with macroamylasemia and monoclonal gammopathy in a patient with Wiskott–Aldrich syndrome. *Pediatr Hematol Oncol* 1997;**14**:85–89.

63. Bruce RM, Blaese RM. Monoclonal gammopathy in the Wiskott–Aldrich syndrome. *J Pediatr* 1974;**85**:204–207.

64. Kyle RA. Current therapy of multiple myeloma. *Intern Med* 2002;**41**:175–180.

65. Rajkumar SV, Hayman S, Gertz MA, et al. Combination therapy with thalidomide plus

dexamethasone for newly diagnosed myeloma. *J Clin Oncol* 2002;20:4319–4323.

66. Durie BG, Salmon SE. A clinical staging system for multiple myeloma. Correlation of measured myeloma cell mass with presenting clinical features, response to treatment, and survival. *Cancer* 1975;36:842–854.

67. Alexanian R, Balcerzak S, Bonnet JD, et al. Prognostic factors in multiple myeloma. *Cancer.* 1975;36:1192–1201.

68. Cherng NC, Asal NR, Kuebler JP, Lee ET, Solanki D. Prognostic factors in multiple myeloma. *Cancer* 1991;67:3150–3156.

69. Chen YH, Magalhaes MC. Hypoalbuminemia in patients with multiple myeloma. *Arch Intern Med* 1990;150:605–610.

70. Kaneko M, Kanda Y, Oshima K, et al. Simple prognostic model for patients with multiple myeloma: a single-center study in Japan. *Ann Hematol* 2002;81:33–36.

71. Sahara N, Takeshita A, Shigeno K, et al. Clinicopathological and prognostic characteristics of CD56-negative multiple myeloma. *Br J Haematol* 2002;117:882–885.

72. Smadja NV, Bastard C, Brigaudeau C, Leroux D, Fruchart C. Hypodiploidy is a major prognostic factor in multiple myeloma. *Blood* 2001;98: 2229–2238.

73. Greipp PR, Trendle MC, Leong T, et al. Is flow cytometric DNA content hypodiploidy prognostic in multiple myeloma? *Leuk Lymphoma* 1999;35:83–89.

74. Kyle RA, Beard CM, O'Fallon WM, Kurland LT. Incidence of multiple myeloma in Olmsted County, Minnesota: 1978 through 1990, with a review of the trend since 1945. *J Clin Oncol* 1994;12:1577–1583.

75. Mundy GR. Myeloma bone disease. *Eur J Cancer* 1998;34:246–251.

76. American Cancer Society. Multiple Myeloma Cancer Resource Center; 2002.

77. Baris D, Brown LM, Silverman DT, et al. Socioeconomic status and multiple myeloma among US blacks and whites. *Am J Public Health* 2000;90:1277–1281.

78. Brown LM, Gridley G, Pottern LM, et al. Diet and nutrition as risk factors for multiple myeloma among blacks and whites in the United States. *Cancer Causes Control* 2001;12:117–125.

79. Hayakawa N, Ohtaki M, Ueoka H, Matsuura M, Munaka M, Kurihara M. Mortality statistics of major causes of death among atomic bomb survivors in Hiroshima Prefecture from 1968 to 1982. *Hiroshima J Med Sci* 1989;38:53–67.

80. Shimizu Y, Kato H, Schull WJ. Studies of the mortality of A-bomb survivors. 9. Mortality, 1950–1985: Part 2. Cancer mortality based on the recently revised doses (DS86). *Radiat Res* 1990;121:120–141.

81. Matanoski GM, Seltser R, Sartwell PE, Diamond EL, Elliott EA. The current mortality rates of radiologists and other physician specialists: specific causes of death. *Am J Epidemiol* 1975;101:199–210.

82. Stebbings JH, Lucas HF, Stehney AF. Mortality from cancers of major sites in female radium dial workers. *Am J Ind Med* 1984;5:435–459.

83. Stebbings JH. Health risks from radium in workplaces: an unfinished story. *Occup Med* 2001;16:259–270.

84. Jablon S, Hrubec Z, Boice JD. *Cancer in populations living near nuclear facilities.* Washington, DC: DHHS–US Government Print Office, 1990.

85. Hatcher JL, Baris D, Olshan AF, et al. Diagnostic radiation and the risk of multiple myeloma (United States). *Cancer Causes Control* 2001;12:755–761.

86. Brown LM, Linet MS, Greenberg RS, et al. Multiple myeloma and family history of cancer among blacks and whites in the US. *Cancer* 1999;85:2385–2390.

87. Alexander LL, Benninghoff DL. Familial multiple myeloma. *J Natl Med Assoc* 1965;57: 471–475.

88. Alexander LL, Benninghoff DL. Familial multiple myeloma. II. Final pathological findings in two brothers and a sister. *J Natl Med Assoc* 1967;59: 278–281.

89. Robbins R. Familial multiple myeloma: the tenth reported occurrence. *Am J Med Sci* 1967;254: 848–850.

90. Barbieri D, Grampa A. Familial multiple myeloma: beta myeloma in two sisters. *Haematologica* 1972;58:565–566.

91. Wiedermann D, Urban P, Wiedermann B, Cidl K. Multiple myeloma in two brothers. Immunoglobulin levels among their relatives. *Neoplasma* 1976;23:197–207.

92. Law MI. Familial occurrence of multiple myeloma. *South Med J* 1976;69:46–48.

93. Shoenfeld Y, Berliner S, Shaklai M, Gallant LA, Pinkhas J. Familial multiple myeloma. A review of thirty-seven families. *Postgrad Med J* 1982;58:12–16.

94. Grosbois B, Gueguen M, Fauchet R, et al. Multiple myeloma in two brothers. An immunochemical and immunogenetic familial study. *Cancer* 1986;58:2417–2421.

95. Eriksson M, Hallberg B. Familial occurrence of hematologic malignancies and other diseases in multiple myeloma: a case-control study. *Cancer Causes Control* 1992;3:63–67.

96. Crozes-Bony P, Palazzo E, Meyer O, De Bandt M, Kahn MF. Familial multiple myeloma. Report of a case in a father and daughter. Review of the literature. *Rev Rheum Engl Ed* 1995;62:439–445.

97. Roddie PH, Dang R, Parker AC. Multiple myeloma in three siblings. *Clin Lab Haematol* 1998;20:191–193.

98. Grosbois B, Jego P, Attal M, et al. Familial multiple myeloma: report of fifteen families. *Br J Haematol* 1999;105:768–770.

99. Deshpande HA, Hu XP, Marino P, Jan NA, Wiernik PH. Anticipation in familial plasma cell dyscrasias. *Br J Haematol* 1998;103:696–703.

100. Bowden M, Crawford J, Cohen HJ, Noyama O. A comparative study of monoclonal gammopathies and immunoglobulin levels in Japanese and United States elderly. *J Am Geriatr Soc* 1993;41:11–14.

101. Zandecki M, Lai JL, Facon T. Multiple myeloma: almost all patients are cytogenetically abnormal. *Br J Haematol* 1996;94:217–227.

102. Lai JL, Michaux L, Dastugue N, Vasseur F, Daudignon A, Facon T, Bauters F, Zandecki M. Cytogenetics in multiple myeloma: a multicenter study of 24 patients with t(11;14)(q13;q32) or its variant. *Cancer Genet Cytogenet.* 1998;104:133–138.

103. Fenton JA, Pratt G, Rawstron AC, Morgan GJ. Isotype class switching and the pathogenesis of multiple myeloma. *Hematol Oncol* 2002;20:75–85.

104. Lynch HT, Sanger WG, Pirruccello S, Quinn-Laquer B, Weisenburger DD. Familial multiple myeloma: a family study and review of the literature. *J Natl Cancer Inst* 2001;93:1479–1483.

105. Abdalla IA, Tabbara IA. Nonsecretory multiple myeloma. *South Med J* 2002;95:761–764.

106. Pratt G. Molecular aspects of multiple myeloma. *Mol Pathol* 2002;55:273–283.

107. Bast EJ, van Camp B, Reynaert P, Wiringa G, Ballieux RE. Idiotypic peripheral blood lymphocytes in monoclonal gammopathy. *Clin Exp Immunol* 1982;47:677–682.

108. van Acker A, Conte F, Hulin N, Urbain J. Idiotypic studies on myeloma B cells. *Eur J Cancer* 1979;15:627–635.

109. Kubagawa H, Vogler LB, Capra JD, Conrad ME, Lawton AR, Cooper MD. Studies on the clonal origin of multiple myeloma. Use of individually specific (idiotype) antibodies to trace the oncogenic event to its earliest point of expression in B-cell differentiation. *J Exp Med* 1979;150:792–807.

110. Rasmussen T, Kastrup J, Knudsen LM, Johnsen HE. High numbers of clonal CD19+ cells in the peripheral blood of a patient with multiple myeloma. *Br J Haematol* 1999;105:265–267.

111. Helfrich MH, Livingston E, Franklin IM, Soutar RL. Expression of adhesion molecules in malignant plasma cells in multiple myeloma: comparison with normal plasma cells and functional significance. *Blood Rev* 1997;11:28–38.

112. Pajor A, Kelemen E, Mohos Z, Hambach J, Varadi G. Multiple myeloma in pregnancy. *Int J Gynaecol Obstet* 1991;35:341–342.

113. Cook G, Campbell JD, Carr CE, Boyd KS, Franklin IM. Transforming growth factor beta from multiple myeloma cells inhibits

proliferation and IL-2 responsiveness in T lymphocytes. *J Leukoc Biol* 1999;66:981–988.

114. Pilarski LM, Ruether BA, Mant MJ. Abnormal function of B lymphocytes from peripheral blood of multiple myeloma patients. Lack of correlation between the number of cells potentially able to secrete immunoglobulin M and serum immunoglobulin M levels. *J Clin Invest* 1985;75:2024–2029.

115. Lapena P, Prieto A, Garcia-Suarez J, et al. Increased production of interleukin-6 by T lymphocytes from patients with multiple myeloma. *Exp Hematol* 1996;24:26–30.

116. Kay NE, Leong T, Kyle RA, et al. Circulating blood B cells in multiple myeloma: analysis and relationship to circulating clonal cells and clinical parameters in a cohort of patients entered on the Eastern Cooperative Oncology Group phase III E9486 clinical trial. *Blood* 1997;90:340–345.

117. Kay NE, Leong T, Bone N, et al. T-helper phenotypes in the blood of myeloma patients on ECOG phase III trials E9486/E3A93. *Br J Haematol* 1998;100:459–463.

118. Kay NE, Leong TL, Bone N, et al. Blood levels of immune cells predict survival in myeloma patients: results of an Eastern Cooperative Oncology Group phase 3 trial for newly diagnosed multiple myeloma patients. *Blood* 2001;98:23–28.

119. Anderson KC, Lust JA. Role of cytokines in multiple myeloma. *Semin Hematol* 1999;36:14–20.

120. Merico F, Bergui L, Gregoretti MG, et al. Cytokines involved in the progression of multiple myeloma. *Clin Exp Immunol* 1993;92:27–31.

121. Akiyama M, Hideshima T, Hayashi T, et al. Cytokines modulate telomerase activity in a human multiple myeloma cell line. *Cancer Res* 2002;62:3876–3882.

122. Ogawa M, Nishiura T, Oritani K, et al. Cytokines prevent dexamethasone-induced apoptosis via the activation of mitogen-activated protein kinase and phosphatidylinositol 3-kinase pathways in a new multiple myeloma cell line. *Cancer Res* 2000;60:4262–4269.

123. Lauta VM. Interleukin-6 and the network of several cytokines in multiple myeloma: an overview of clinical and experimental data. *Cytokine* 2001;16:79–86.

124. Diamond T, Levy S, Smith A, Day P, Manoharan A. Non-invasive markers of bone turnover and plasma cytokines differ in osteoporotic patients with multiple myeloma and monoclonal gammopathies of undetermined significance. *Intern Med J* 2001;31:272–278.

125. Filella X, Blade J, Guillermo AL, Molina R, Rozman C, Ballesta AM. Cytokines (IL-6, TNF-alpha, IL-1alpha) and soluble interleukin-2 receptor as serum tumor markers in multiple myeloma. *Cancer Detect Prev* 1996;20:52–56.

126. Rajkumar SV, Greipp PR. Prognostic factors in multiple myeloma. *Hematol Oncol Clin N Am* 1999;13:1295–1314.

127. Pardanani A, Witzig TE, Schroeder G, et al. Circulating peripheral blood plasma cells as a prognostic indicator in patients with primary systemic amyloidosis. *Blood* 2003;101:827–830.

128. Theil KS, Thorne CM, Neff JC. Diagnosing plasma cell leukemia. *Lab Med* 1987;18:684–687.

129. Noel P, Kyle RA. Plasma cell leukemia: an evaluation of response to therapy. *Am J Med* 1987;83:1062–1068.

130. Garcia-Sanz R, Orfao A, Gonzalez M, et al. Primary plasma cell leukemia: clinical, immunophenotypic, DNA ploidy, and cytogenetic characteristics. *Blood* 1999;93:1032–1037.

131. Blade J, Kyle RA. Nonsecretory myeloma, immunoglobulin D myeloma, and plasma cell leukemia. *Hematol Oncol Clin N Am* 1999;13:1259–1272.

132. Bernasconi C, Castelli G, Pagnucco G, Brusamolino E. Plasma cell leukemia: a report on 15 patients. *Eur J Haematol Suppl* 1989;51:76–83.

133. Dimopoulos MA, Palumbo A, Delasalle KB, Alexanian R. Primary plasma cell leukaemia. *Br J Haematol* 1994;88:754–759.

134. Kyle RA, Maldonado JE, Bayrd ED. Plasma cell leukemia. Report on 17 cases. *Arch Intern Med* 1974;133:813–818.

135. Kyle RA. Multiple myeloma: review of 869 cases. *Mayo Clin Proc* 1975;**50**:29–40.

136. Jako JM, Gesztesi T, Kaszas I. IgE lambda monoclonal gammopathy and amyloidosis. *Int Arch Allergy Immunol* 1997;**112**:415–421.

137. Qiu WX, Huang NH, Wu GH. Immunological classification of 31 multiple myeloma patients. *Proc Chin Acad Med Sci Peking Union Med Coll* 1990;**5**:79–83.

138. Ameis A, Ko HS, Pruzanski W. M components – a review of 1242 cases. *Can Med Assoc J* 1976; **114**:889–892, 895.

139. Schur PH, Kyle RA, Bloch KJ, et al. IgG subclasses: relationship to clinical aspects of multiple myeloma and frequency distribution among M-components. *Scand J Haematol* 1974;**12**:60–68.

140. Hammarstrom L, Mellstedt H, Persson MA, Smith CI, Ahre A. IgA subclass distribution in paraproteinemias: suggestion of an IgG-IgA subclass switch pattern. *Acta Pathol Microbiol Immunol Scand (C)* 1984;**92**:207–211.

141. Vincent C, Bouic P, Revillard JP, Bataille R. Complexes of alpha 1-microglobulin and monomeric IgA in multiple myeloma and normal human sera. *Mol Immunol* 1985;**22**: 663–673.

142. Pruzanski W, Watt JG. Serum viscosity and hyperviscosity syndrome in IgG multiple myeloma. Report on 10 patients and a review of the literature. *Ann Intern Med* 1972;**77**: 853–860.

143. Pruzanski W, Russell ML. Serum viscosity and hyperviscosity syndrome in IGG multiple myeloma – the relationship of Sia test and to concentration of M component. *Am J Med Sci* 1976;**271**:145–150.

144. Seward CW, Osterland CK. Hyperviscosity in IgG myeloma: detection and treatment in community medical facilities. *Tex Med* 1974;**70**: 63–65.

145. Chandy KG, Stockley RA, Leonard RC, Crockson RA, Burnett D, MacLennan IC. Relationship between serum viscosity and intravascular IgA polymer concentration in IgA myeloma. *Clin Exp Immunol* 1981;**46**:653–661.

146. Vaerman JP, Langendries A, Vander Maelen C. Homogenous IgA monomers, dimers, trimers and tetramers from the same IgA myeloma serum. *Immunol Invest* 1995;**24**:631–641.

147. Kimizu K, Hamada A, Haba T, et al. A case of IgA-kappa multiple myeloma with hyperviscosity syndrome terminating in plasma cell leukemia. *Rinsho Ketsueki.* 1983;**24**:572–579.

148. Tuddenham EG, Whittaker JA, Bradley J, Lilleyman JS, James DR. Hyperviscosity syndrome in IgA multiple myeloma. *Br J Haematol* 1974;**27**:65–76.

149. Whittaker JA, Tuddenham EG, Bradley J. Hyperviscosity syndrome in IgA multiple myeloma. *Lancet* 1973;**2**:572.

150. De Gramont A, Grosbois B, Michaux JL, et al. IgM myeloma: 6 cases and a review of the literature. *Rev Med Interne* 1990;**11**:13–18.

151. Takahashi K, Yamamura F, Motoyama H. IgM myeloma – its distinction from Waldenstrom's macroglobulinemia. *Acta Pathol Jpn* 1986;**36**: 1553–1563.

152. Zarrabi MH, Stark RS, Kane P, Dannaher CL, Chandor S. IgM myeloma, a distinct entity in the spectrum of B-cell neoplasia. *Am J Clin Pathol* 1981;**75**:1–10.

153. Dierlamm T, Laack E, Dierlamm J, Fiedler W, Hossfeld DK. IgM myeloma: a report of four cases. *Ann Hematol* 2002;**81**:136–139.

154. Haghighi B, Yanagihara R, Cornbleet PJ. IgM myeloma: case report with immunophenotypic profile. *Am J Hematol* 1998;**59**:302–308.

155. Avet-Loiseau H, Garand R, Lode L, Harousseau JL, Bataille R. Translocation t(11;14)(q13;q32) is the hallmark of IgM, IgE, and nonsecretory multiple myeloma variants. *Blood* 2003;**101**: 1570–1571.

156. Blade J, Lust JA, Kyle RA. Immunoglobulin D multiple myeloma: presenting features, response to therapy, and survival in a series of 53 cases. *J Clin Oncol* 1994;**12**:2398–2404.

157. Jancelewicz Z, Takatsuki K, Sugai S, Pruzanski W. IgD multiple myeloma. Review of 133 cases. *Arch Intern Med* 1975;**135**:87–93.

158. Kyle RA. IgD multiple myeloma: a cure at 21 years. *Am J Hematol* 1988;**29**:41–43.

159. Blade J, Kyle RA. IgD monoclonal gammopathy with long-term follow-up. *Br J Haematol* 1994;88:395–396.

160. Jennette JC, Wilkman AS, Benson JD. IgD myeloma with intracytoplasmic crystalline inclusions. *Am J Clin Pathol* 1981;75:231–235.

161. Schaldenbrand JD, Keren DF. IgD amyloid in IgD-lambda monoclonal conjunctival amyloidosis. A case report. *Arch Pathol Lab Med* 1983;107:626–628.

162. Patel K, Carrington PA, Bhatnagar S, Houghton JB, Routledge RC. IgD myeloma with multiple cutaneous plasmacytomas. *Clin Lab Haematol* 1998;20:53–55.

163. Vladutiu AO. Immunoglobulin D: properties, measurement, and clinical relevance. *Clin Diagn Lab Immunol* 2000;7:131–140.

164. Levan-Petit I, Cardonna J, Garcia M, et al. Sensitive ELISA for human immunoglobulin D measurement in neonate, infant, and adult sera. *Clin Chem* 2000;46:876–878.

165. Macro M, Andre I, Comby E, et al. IgE multiple myeloma. *Leuk Lymphoma* 1999;32:597–603.

166. Rogers JS 2nd, Spahr J, Judge DM, Varano LA, Eyster ME. IgE myeloma with osteoblastic lesions. *Blood* 1977;49:295–299.

167. Proctor SJ, Chawla SL, Bird AG, Stephenson J. Hyperviscosity syndrome in IgE myeloma. *Br Med J (Clin Res Edn)* 1984;289:1112.

168. West N. Hyperviscosity syndrome in IgE myeloma. *Br Med J (Clin Res Edn)* 1984;289:1539.

169. Alexander RL Jr, Roodman ST, Petruska PJ, Tsai CC, Janney CG. A new case of IgE myeloma. *Clin Chem* 1992;38:2328–2332.

170. Rowan RM. Multiple myeloma: some recent developments. *Clin Lab Haematol* 1982;4:211–230.

171. Dul JL, Aviel S, Melnick J, Argon Y. Ig light chains are secreted predominantly as monomers. *J Immunol* 1996;157:2969–2975.

172. Shapiro AL, Scharff MD, Maizel JV, Uhr JW. Synthesis of excess light chains of gamma globulin by rabbit lymph node cells. *Nature* 1966;211:243–245.

173. Bradwell AR, Carr-Smith HD, Mead GP, et al. Highly sensitive, automated immunoassay for immunoglobulin free light chains in serum and urine. *Clin Chem* 2001;47:673–680.

174. Okada K, Oguchi N, Shinohara K, et al. BUN, Bence Jones protein, and chromosomal aberrations predict survival in multiple myeloma. *Rinsho Ketsueki* 1997;38:1254–1262.

175. Sirohi B, Powles R, Mehta J, et al. Complete remission rate and outcome after intensive treatment of 177 patients under 75 years of age with IgG myeloma defining a circumscribed disease entity with a new staging system. *Br J Haematol* 1999;107:656–666.

176. Weber DM, Dimopoulos MA, Moulopoulos LA, Delasalle KB, Smith T, Alexanian R. Prognostic features of asymptomatic multiple myeloma. *Br J Haematol* 1997;97:810–814.

177. Baldini L, Guffanti A, Cesana BM, et al. Role of different hematologic variables in defining the risk of malignant transformation in monoclonal gammopathy. *Blood* 1996;87:912–918.

178. Peest D, Coldewey R, Deicher H, et al. Prognostic value of clinical, laboratory, and histological characteristics in multiple myeloma: improved definition of risk groups. *Eur J Cancer* 1993;29A:978–983.

179. Dimopoulos MA, Moulopoulos A, Smith T, Delasalle KB, Alexanian R. Risk of disease progression in asymptomatic multiple myeloma. *Am J Med* 1993;94:57–61.

180. Nagai M, Kitahara T, Minato K, et al. Prognostic factors and therapeutic results in multiple myeloma. *Jpn J Clin Oncol* 1985;15:505–515.

181. Knudsen LM, Hjorth M, Hippe E. Renal failure in multiple myeloma: reversibility and impact on the prognosis Nordic Myeloma Study Group. *Eur J Haematol* 2000;65:175–181.

182. Abraham RS, Charlesworth MC, Owen BA, Benson LM, Katzmann JA, Reeder CB, Kyle RA. Trimolecular complexes of lambda light chain dimers in serum of a patient with multiple myeloma. *Clin Chem* 2002;48:1805–1811.

183. Inoue N, Togawa A, Yawata Y. Tetrameric Bence Jones protein – case report and review of

the literature. *Nippon Ketsueki Gakkai Zasshi* 1984;47:1456–1459.

184. Caggiano V, Dominguez C, Opfell RW, Kochwa S, Wasserman LR. IgG myeloma with closed tetrameric Bence Jones proteinemia. *Am J Med* 1969;47:978–985.

185. Hom BL. Polymeric (presumed tetrameric) lambda Bence Jones proteinemia without proteinuria in a patient with multiple myeloma. *Am J Clin Pathol* 1984;82:627–629.

186. Inoue S, Nagata H, Yozawa H, Terai T, Hasegawa H, Murao M. A case of IgG myeloma with tetrameric Bence Jones proteinemia and abnormal fibrin polymerization (author's transl). *Rinsho Ketsueki* 1980;21:200–207.

187. Duffy TP. The many pitfalls in the diagnosis of myeloma. *N Engl J Med* 1992;326:394–396.

188. Katzmann JA, Clark RJ, Abraham RS, et al. Serum reference intervals and diagnostic ranges for free kappa and free lambda immunoglobulin light chains: relative sensitivity for detection of monoclonal light chains. *Clin Chem* 2002;48:1437–1444.

189. Marien G, Oris E, Bradwell AR, Blanckaert N, Bossuyt X. Detection of monoclonal proteins in sera by capillary zone electrophoresis and free light chain measurements. *Clin Chem* 2002;48:1600–1601.

190. Kyle RA. The monoclonal gammopathies. *Clin Chem* 1994;40:2154–2161.

191. Bourantas K. Nonsecretory multiple myeloma. *Eur J Haematol* 1996;56:109–111.

192. Cavo M, Galieni P, Gobbi M, et al. Nonsecretory multiple myeloma. Presenting findings, clinical course and prognosis. *Acta Haematol* 1985;74:27–30.

193. Dreicer R, Alexanian R. Nonsecretory multiple myeloma. *Am J Hematol* 1982;13:313–318.

194. Rubio-Felix D, Giralt M, Giraldo MP, et al. Nonsecretory multiple myeloma. *Cancer.* 1987;59:1847–1852.

195. Kawada E, Shinonome S, Saitoh T, et al. Primary nonsecretory plasma cell leukemia: a rare variant of multiple myeloma. *Ann Hematol* 1999;78:25–27.

196. Preud'Homme JL, Hurez D, Danon F, Brouet JC, Seligmann M. Intracytoplasmic and surface-bound immunoglobulins in 'nonsecretory' and Bence-Jones myeloma. *Clin Exp Immunol* 1976;25:428–436.

197. Alanen A, Pira U, Lassila O, Roth J, Franklin RM. Mott cells are plasma cells defective in immunoglobulin secretion. *Eur J Immunol* 1985;15:235–242.

198. Raubenheimer EJ, Dauth J, Senekal JC. Non-secretory IgA kappa myeloma with distended endoplasmic reticulum: a case report. *Histopathology* 1991;19:380–382.

199. Doster DR, Folds J, Gabriel DA. Nonsecretory multiple myeloma. *Arch Pathol Lab Med* 1988;112:147–150.

200. Cogne M, Guglielmi P. Exon skipping without splice site mutation accounting for abnormal immunoglobulin chains in nonsecretory human myeloma. *Eur J Immunol* 1993;23:1289–1293.

201. Derk CT, Sandorfi N. Nonsecretory multiple myeloma masquerading as a new osteoporotic vertebral compression fracture in an older female. *J Am Geriatr Soc* 2002;50:1747–1748.

202. Kubota K, Kurabayashi H, Kawada E, Okamoto K, Tamura J, Shirakura T. Nonsecretion of myeloma protein in spite of an increase in tumor burden by chemotherapy. *Ann Hematol* 1991;63:232–233.

203. Shustik C, Michel R, Karsh J. Nonsecretory myeloma: a study on hypoimmunoglobulinemia. *Acta Haematol* 1988;80:153–158.

204. Matsuzaki H, Yoshida M, Akahoshi Y, Kuwahara K, Satou T, Takatsuki K. Pseudo-nonsecretory multiple myeloma with light chain deposition disease. *Acta Haematol* 1991;85:164–168.

205. Drayson M, Tang LX, Drew R, Mead GP, Carr-Smith H, Bradwell AR. Serum free light-chain measurements for identifying and monitoring patients with nonsecretory multiple myeloma. *Blood* 2001;97:2900–2902.

206. Kyle RA, Robinson RA, Katzmann JA. The clinical aspects of biclonal gammopathies. Review of 57 cases. *Am J Med* 1981;71:999–1008.

207. Tasaka T, Matsuhashi Y, Uehara E, Tamura T, Kuwajima M, Nagai M. Angioimmunoblastic

T-cell lymphoma presenting with rapidly increasing biclonal gammopathy. *Rinsho Ketsueki* 2000;**41**:1281–1284.

208. Ogasawara T, Yasuyama M, Kawauchi K. Biclonal light chain gammopathy in multiple myeloma – a case report. *Nihon Rinsho Meneki Gakkai Kaishi* 2002;**25**:170–176.

209. Miralles ES, Nunez M, Boixeda P, Moreno R, Bellas C, Ledo A. Transformed cutaneous T cell lymphoma and biclonal gammopathy. *Int J Dermatol* 1996;**35**:196–198.

210. Murakami M, Sugiura K. Biclonal gammopathy in a patient with immunoblastic lymphadenopathy like T cell lymphoma. *Rinsho Ketsueki* 1989;**30**:116–121.

211. Pizzolato M, Bragantini G, Bresciani P, et al. IgG1-kappa biclonal gammopathy associated with multiple myeloma suggests a regulatory mechanism. *Br J Haematol* 1998;**102**:503–508.

212. Goni F, Chuba J, Buxbaum J, Frangione B. A double monoclonal IgG1 kappa and IgG2 kappa in a single myeloma patient. Variation in clonal products and therapeutic responses. *J Immunol* 1988;**140**:551–557.

213. Finco B, Schiavon R. Multiple myeloma with serum IgG kappa and Bence Jones lambda biclonal gammopathy. *Clin Chem* 1987;**33**: 1305–1306.

214. Weinstein S, Jain A, Bhagavan NV, Scottolini AG. Biclonal IgA and IgM gammopathy in lymphocytic lymphoma. *Clin Chem* 1984; **30**:1710–1712.

215. Flanagan JG, Rabbitts TH. Arrangement of human immunoglobulin heavy chain constant region genes implies evolutionary duplication of a segment containing gamma, epsilon and alpha genes. *Nature* 1982;**300**:709–713.

216. Guastafierro S, Sessa F, Tirelli A. Biclonal gammopathy and platelet antibodies in a patient with chronic hepatitis C virus infection and mixed cryoglobulinemia. *Ann Hematol* 2000;**79**:463–464.

217. Schroeder HW Jr, Dighiero G. The pathogenesis of chronic lymphocytic leukemia: analysis of the antibody repertoire. *Immunol Today* 1994;**15**: 288–294.

218. Walchner M, Wick M. Elevation of CD8+CD11b+ Leu-8– T cells is associated with the humoral immunodeficiency in myeloma patients. *Clin Exp Immunol* 1997;**109**:310–316.

219. Pilarski LM, Mant MJ, Ruether BA, Belch A. Severe deficiency of B lymphocytes in peripheral blood from multiple myeloma patients. *J Clin Invest* 1984;**74**:1301–1306.

220. Broder S, Humphrey R, Durm M, et al. Impaired synthesis of polyclonal (non-paraprotein) immunoglobulins by circulating lymphocytes from patients with multiple myeloma. Role of suppressor cells. *N Engl J Med* 1975;**293**:887–892.

221. Bergmann L, Mitrou PS, Weber KC, Kelker W. Imbalances of T-cell subsets in monoclonal gammopathies. *Cancer Immunol Immunother* 1984;**17**:112–116.

222. Levinson AI, Hoxie JA, Matthews DM, Schreiber AD, Negendank WG. Analysis of the relationship between T cell subsets and *in vitro* B cell responses in multiple myeloma. *J Clin Lab Immunol* 1985;**16**:23–26.

223. Pilarski LM, Andrews EJ, Serra HM, Ruether BA, Mant MJ. Comparative analysis of immunodeficiency in patients with monoclonal gammopathy of undetermined significance and patients with untreated multiple myeloma. *Scand J Immunol* 1989;**29**:217–228.

224. Cunningham-Rundles S, Cunningham-Rundles C, Ma DI, et al. Impaired proliferative response to B-lymphocyte activators in common variable immunodeficiency. *Scand J Immunol* 1982;**15**: 279–286.

225. Seligmann M, Rambaud JC. Alpha-chain disease: an immunoproliferative disease of the secretory immune system. *Ann N Y Acad Sci* 1983;**409**: 478–485.

226. Seligmann M, Mihaesco E, Preud'homme JL, Danon F, Brouet JC. Heavy chain diseases: current findings and concepts. *Immunol Rev* 1979;**48**:145–167.

227. Joller PW, Joller-Jemelka HI, Shmerling DH, Skvaril F. Immunological and biochemical studies of an unusual alpha heavy chain protein in a 9-year-old boy. *J Clin Lab Immunol* 1984;**15**:167–172.

228. Guardia J, Rubies-Prat J, Gallart MT, et al. The evolution of alpha heavy chain disease. *Am J Med* 1976;60:596–602.

229. Martin IG, Aldoori MI. Immunoproliferative small intestinal disease. Mediterranean lymphoma and alpha heavy chain disease. *Br J Surg* 1994;81:20–24.

230. Novis BH, King HS, Gilinsky NH, Mee AS, Young G. Long survival in a patient with alpha-chain disease. *Cancer* 1984;53:970–973.

231. Stoop JW, Ballieux RE, Hijmans W, Zegers BJ. Alpha-chain disease with involvement of the respiratory tract in a Dutch child. *Clin Exp Immunol* 1971;9:625–635.

232. Tracy RP, Kyle RA, Leitch JM. Alpha heavy-chain disease presenting as goiter. *Am J Clin Pathol* 1984;82:336–339.

233. Florin-Christensen A, Doniach D, Newcomb PB. Alpha-chain disease with pulmonary manifestations. *Br Med J* 1974;2:413–415.

234. Fine KD, Stone MJ. Alpha-heavy chain disease, Mediterranean lymphoma, and immunoproliferative small intestinal disease: a review of clinicopathological features, pathogenesis, and differential diagnosis. *Am J Gastroenterol* 1999;94:1139–1152.

235. Lavergne A, Brocheriou I, Rambaud JC, et al. T-cell rich alpha-chain disease mimicking T-cell lymphoma. *Histopathology* 1997;30:394–396.

236. Isaacson PG, Dogan A, Price SK, Spencer J. Immunoproliferative small-intestinal disease. An immunohistochemical study. *Am J Surg Pathol* 1989;13:1023–1033.

237. Su L, Keren DF, Warren JS. Failure of anti-lambda immunofixation reagent mimics alpha heavy-chain disease. *Clin Chem* 1995;41:121–123.

238. Rambaud JC, Galian A, Danon FG, et al. Alpha-chain disease without qualitative serum IgA abnormality. Report of two cases, including a 'nonsecretory' form. *Cancer* 1983;51:686–693.

239. Cogne M, Preud'homme JL. Gene deletions force nonsecretory alpha-chain disease plasma cells to produce membrane-form alpha-chain only. *J Immunol* 1990;145:2455–2458.

240. Seligmann M, Mihaesco E, Hurez D, Mihaesco C, Preud'homme JL, Rambaud JC. Immunochemical studies in four cases of alpha chain disease. *J Clin Invest* 1969;48:2374–2389.

241. Al-Saleem TI, Qadiry WA, Issa FS, King J. The immunoselection technic in laboratory diagnosis of alpha heavy-chain disease. *Am J Clin Pathol* 1979;72:132–133.

242. Fermand JP, Brouet JC, Danon F, Seligmann M. Gamma heavy chain 'disease': heterogeneity of the clinicopathologic features. Report of 16 cases and review of the literature. *Medicine (Baltimore)* 1989;68:321–335.

243. Franklin EC, Lowenstein J, Bigelow B. Heavy chain disease – a new disorder of serum gamma-globulins: report of the first case. *Am J Med* 1964;37:332–350.

244. Bloch KJ, Lee L, Mills JA, Haber E. Gamma heavy chain disease – an expanding clinical and laboratory spectrum. *Am J Med* 1973;55:61–70.

245. Ellman LL, Bloch KJ. Heavy-chain disease. Report of a seventh case. *N Engl J Med* 1968;278:1195–1201.

246. Franklin EC, Kyle R, Seligmann M, Frangione B. Correlation of protein structure and immunoglobulin gene organization in the light of two new deleted heavy chain disease proteins. *Mol Immunol* 1979;16:919–921.

247. Alexander A, Steinmetz M, Barritault D, et al. gamma Heavy chain disease in man: cDNA sequence supports partial gene deletion model. *Proc Natl Acad Sci USA* 1982;79:3260–3264.

248. Alexander A, Barritault D, Buxbaum J. Gamma heavy chain disease in man: translation and partial purification of mRNA coding for the deleted protein. *Proc Natl Acad Sci USA* 1978;75:4774–4778.

249. Buxbaum JN, Alexander A, Olivier O. Gamma heavy chain disease in man: synthesis of a deleted gamma3 immunoglobulin by lymphoid cells in short and long term tissue culture. *Clin Exp Immunol* 1978;32:489–497.

250. Lennert K, Stein H, Kaiserling E. Cytological and functional criteria for the classification of malignant lymphomata. *Br J Cancer* 1975;31(Suppl 2):29–43.

251. Presti BC, Sciotto CG, Marsh SG. Lymphocytic

lymphoma with associated gamma heavy chain and IgM-lambda paraproteins. An unusual biclonal gammopathy. *Am J Clin Pathol* 1990;**93**:137–141.

252. Hudnall SD, Alperin JB, Petersen JR. Composite nodular lymphocyte-predominance Hodgkin disease and gamma-heavy-chain disease: a case report and review of the literature. *Arch Pathol Lab Med* 2001;**125**:803–807.

253. Takatani T, Morita K, Takaoka N, et al. Gamma heavy chain disease screening showing a discrepancy between electrophoretic and nephelometric determinations of serum gamma globulin concentration. *Ann Clin Biochem* 2002;**39**:531–533.

254. Sun T, Peng S, Narurkar L. Modified immunoselection technique for definitive diagnosis of heavy-chain disease. *Clin Chem* 1994;**40**:664.

255. Ballard HS, Hamilton LM, Marcus AJ, Illes CH. A new variant of heavy-chain disease (mu-chain disease). *N Engl J Med* 1970;**282**:1060–1062.

256. Fermand JP, Brouet JC. Heavy-chain diseases. *Hematol Oncol Clin N Am* 1999;**13**: 1281–1294.

257. Wahner-Roedler DL, Kyle RA. Mu-heavy chain disease: presentation as a benign monoclonal gammopathy. *Am J Hematol* 1992;**40**:56–60.

258. Campbell JK, Juneja SK. Test and teach. Number One hundred and four. Mu heavy chain disease (mu-HCD). *Pathology* 2000;**32**:202–203; 227.

259. Owen RG, Barrans SL, Richards SJ, et al. Waldenstrom macroglobulinemia. Development of diagnostic criteria and identification of prognostic factors. *Am J Clin Pathol* 2001;**116**:420–428.

260. Dimopoulos MA, Galani E, Matsouka C. Waldenstrom's macroglobulinemia. *Hematol Oncol Clin N Am* 1999;**13**:1351–1366.

261. Groves FD, Travis LB, Devesa SS, Ries LA, Fraumeni JF Jr. Waldenstrom's macroglobulinemia: incidence patterns in the United States, 1988–1994. *Cancer* 1998;**82**: 1078–1081.

262. Herrinton LJ, Weiss NS. Incidence of

Waldenstrom's macroglobulinemia. *Blood* 1993;**82**:3148–3150.

263. Kyrtsonis MC, Vassilakopoulos TP, Angelopoulou MK, et al. Waldenstrom's macroglobulinemia: clinical course and prognostic factors in 60 patients. Experience from a single hematology unit. *Ann Hematol* 2001;**80**:722–727.

264. Isaac J, Herrera GA. Cast nephropathy in a case of Waldenstrom's macroglobulinemia. *Nephron* 2002;**91**:512–515.

265. Kyrtsonis MC, Angelopoulou MK, Kontopidou FN, et al. Primary lung involvement in Waldenstrom's macroglobulinaemia: report of two cases and review of the literature. *Acta Haematol* 2001;**105**:92–96.

266. Yasui O, Tukamoto F, Sasaki N, Saito T, Yagisawa H, Uno A, Nanjo H. Malignant lymphoma of the transverse colon associated with macroglobulinemia. *Am J Gastroenterol* 1997;**92**:2299–2301.

267. Shimizu K, Fujisawa K, Yamamoto H, Mizoguchi Y, Hara K. Importance of central nervous system involvement by neoplastic cells in a patient with Waldenstrom's macroglobulinemia developing neurologic abnormalities. *Acta Haematol* 1993;**90**:206–208.

268. Facon T, Brouillard M, Duhamel A, et al. Prognostic factors in Waldenstrom's macroglobulinemia: a report of 167 cases. *J Clin Oncol* 1993;**11**:1553–1558.

269. Morel P, Monconduit M, Jacomy D, et al. Prognostic factors in Waldenstrom macroglobulinemia: a report on 232 patients with the description of a new scoring system and its validation on 253 other patients. *Blood* 2000;**96**:852–858.

270. Ciric B, VanKeulen V, Rodriguez M, Kyle RA, Gertz MA, Pease LR. Clonal evolution in Waldenstrom macroglobulinemia highlights functional role of B-cell receptor. *Blood* 2001;**97**:321–323.

271. Gertz MA, Fonseca R, Rajkumar SV. Waldenstrom's macroglobulinemia. *Oncologist* 2000;**5**:63–67.

272. Andriko JA, Aguilera NS, Chu WS, Nandedkar

MA, Cotelingam JD. Waldenstrom's macroglobulinemia: a clinicopathologic study of 22 cases. *Cancer* 1997;80:1926–1935.

273. Kyle RA, Gleich GJ, Bayrid ED, Vaughan JH. Benign hypergammaglobulinemic purpura of Waldenstrom. *Medicine (Baltimore)* 1971;50: 113–123.

274. Siami GA, Siami FS. Plasmapheresis and paraproteinemia: cryoprotein-induced diseases, monoclonal gammopathy, Waldenstrom's macroglobulinemia, hyperviscosity syndrome, multiple myeloma, light chain disease, and amyloidosis. *Ther Apher* 1999;3:8–19.

275. Zimmermann I, Gloor HJ, Ruttimann S. General AL-amyloidosis: a rare complication in Waldenstrom macroglobulinemia. *Schweiz Rundsch Med Prax* 2001;90:2050–2055.

276. Zatloukal P, Bezdicek P, Schimonova M, Havlicek F, Tesarova P, Slovakova A. Waldenstrom's macroglobulinemia with pulmonary amyloidosis. *Respiration* 1998;65:414–416.

277. Muso E, Tamura I, Yashiro M, Asaka Y, Kataoka Y, Nagai H, Takahashi T. Waldenstrom's macroglobulinemia associated with amyloidosis and membranous nephropathy. *Nippon Jinzo Gakkai Shi* 1993;35:1265–1269.

278. Gertz MA, Kyle RA, Noel P. Primary systemic amyloidosis: a rare complication of immunoglobulin M monoclonal gammopathies and Waldenstrom's macroglobulinemia. *J Clin Oncol* 1993;11:914–920.

279. Lindemalm C, Biberfeld P, Christensson B, et al. Bilateral pleural effusions due to amyloidosis in a case of Waldenstrom's macroglobulinemia. *Haematologica* 1988;73:407–409.

280. Ogami Y, Takasugi M, Soejima M, et al. Waldenstrom's macroglobulinemia associated with amyloidosis and crescentic glomerulonephritis. *Nephron* 1989;51:95–98.

281. Forget BG, Squires JW, Sheldon H. Waldenstrom's macroglobulinemia with generalized amyloidosis *Arch Intern Med* 1966;118:363–375.

282. Fudenberg HH, Virella G. Multiple myeloma and Waldenstrom macroglobulinemia: unusual presentations. *Semin Hematol* 1980;17:63–79.

283. Maly J, Tichy M, Blaha M, et al. A case of 'acute' Waldenstrom macroglobulinaemia. *Haematologia (Budapest)* 1984;17:125–130.

284. Harris NL, Jaffe ES, Diebold J, et al. The World Health Organization classification of neoplastic diseases of the hematopoietic and lymphoid tissues. Report of the Clinical Advisory Committee meeting, Airlie House, Virginia, November, 1997. *Ann Oncol* 1999;10: 1419–1432.

285. Owen RG, Johnson SA, Morgan GJ. Waldenstrom's macroglobulinaemia: laboratory diagnosis and treatment. *Hematol Oncol* 2000;18:41–49.

286. Imoto M, Ishikawa K, Yamamoto K, et al. Occurrence of heavy chain of 7S IgM half-molecule whose NH_2-terminal sequence is identical with that of kappa light chain sequence in patients with Waldenstrom macroglobulinemia. *Clin Chim Acta* 1999;282: 77–88.

287. Bennett JM, Catovsky D, Daniel MT, et al. Proposals for the classification of chronic (mature) B and T lymphoid leukaemias. French–American–British (FAB) Cooperative Group. *J Clin Pathol* 1989;42:567–584.

288. Kyle RA, Garton JP. The spectrum of IgM monoclonal gammopathy in 430 cases. *Mayo Clin Proc* 1987;62:719–731.

289. Tursz T, Brouet JC, Flandrin G, Danon F, Clauvel JP, Seligmann M. Clinical and pathologic features of Waldenstrom's macroglobulinemia in seven patients with serum monoclonal IgG or IgA. *Am J Med* 1977;63:499–502.

290. Morey M, Bargay J, Duran MA, Matamoros N. Waldenstrom's macroglobulinemia with IgG monoclonal component. *Med Clin (Barc)* 1992;98:436–437.

291. Gallango ML, Suinaga R, Ramirez M. An unusual case of Waldenstrom macroglobulinemia with half molecules of IgG in serum and urine. *Blut* 1984;48:91–97.

292. Fair DS, Schaffer S, Krueger RG. Development of monoclonal IgA and an apparent IgG in a patient with macroglobulinemia: sharing of individually

specific antigenic determinants among IgM, IgA, and IgG. *J Immunol* 1976;**117**:944–949.

293. McNutt DR, Fudenberg HH. IgG myeloma and Waldenstrom macroglobulinemia. Coexistence and clinical manifestations in one patient. *Arch Intern Med* 1973;**131**:731–736.

294. Keren DF, Gulbranson R, Carey JL, Krauss JC. 2-Mercaptoethanol treatment improves measurement of an IgMkappa M-protein by capillary electrophoresis. *Clin Chem* 2001;**47**: 1326–1327.

295. Jonsson V, Kierkegaard A, Salling S, et al. Autoimmunity in Waldenstrom's macroglobulinaemia. *Leuk Lymphoma* 1999;**34**:373–379.

296. Kira J, Inuzuka T, Hozumi I, et al. A novel monoclonal antibody which reacts with a high molecular weight neuronal cytoplasmic protein and myelin basic protein (MBP) in a patient with macroglobulinemia. *J Neurol Sci* 1997;**148**: 47–52.

297. Tait RC, Oogarah PK, Houghton JB, Farrand SE, Haeney MR. Waldenstrom's macroglobulinaemia secreting a paraprotein with lupus anticoagulant activity: possible association with gastrointestinal tract disease and malabsorption. *J Clin Pathol* 1993;**46**:678–680.

298. Boccadoro M, Van Acker A, Pileri A, Urbain J. Idiotypic lymphocytes in human monoclonal gammopathies. *Ann Immunol (Paris)* 1981; **132C**:9–19.

299. Labardini-Mendez J, Alexanian R, Barlogie B. Multiple myeloma and chronic lymphocytic leukemia. *Rev Invest Clin* 1997;**49**(Suppl 1): 28–33.

300. Barlogie B, Gale RP. Multiple myeloma and chronic lymphocytic leukemia: parallels and contrasts. *Am J Med* 1992;**93**:443–450.

301. Pines A, Ben-Bassat I, Selzer G, Ramot B. Transformation of chronic lymphocytic leukemia to plasmacytoma. *Cancer* 1984;**54**: 1904–1907.

302. Suchman AL, Coleman M, Mouradian JA, Wolf DJ, Saletan S. Aggressive plasma cell myeloma. A terminal phase. *Arch Intern Med* 1981;**141**: 1315–1320.

303. Bernstein ZP, Fitzpatrick JE, O'Donnell A, Han T, Foon KA, Bhargava A. Clinical significance of monoclonal proteins in chronic lymphocytic leukemia. *Leukemia* 1992;**6**:1243–1245.

304. Tienhaara A, Irjala K, Rajamaki A, Pulkki K. Four monoclonal immunoglobulins in a patient with chronic lymphocytic leukemia. *Clin Chem* 1986;**32**:703–705.

305. Decastello AV. Beitrage zur Kenntnis der Bence-Jonesschen Albuminurie. *Z Klin Med* 1909;**67**: 319–343.

306. Deegan MJ, Abraham JP, Sawdyk M, van Slyck FJ. High incidence of monoclonal proteins in the serum and urine of chronic lymphocytic leukemia patients. *Blood* 1984;**64**:1207–1211.

307. Fu SM, Winchester RJ, Feizi T, Walzer PD, Kunkel HG. Idiotypic specificity of surface immunoglobulin and the maturation of leukemic bone-marrow-derived lymphocytes. *Proc Natl Acad Sci USA* 1974;**71**:4487–4490.

308. Qian GX, Fu SM, Solanki DL, Rai KR. Circulating monoclonal IgM proteins in B cell chronic lymphocytic leukemia: their identification, characterization and relationship to membrane IgM. *J Immunol* 1984;**133**: 3396–3400.

309. Deegan MJ, Maeda K. Differentiation of chronic lymphocytic leukemia cells after *in vitro* treatment with Epstein–Barr virus or phorbol ester. I. Immunologic and morphologic studies. *Am J Hematol* 1984;**17**:335–347.

310. Hannam-Harris AC, Gordon J, Smith JL. Immunoglobulin synthesis by neoplastic B lymphocytes: free light chain synthesis as a marker of B cell differentiation. *J Immunol* 1980;**125**:2177–2181.

311. Murata T, Fujita H, Harano H, et al. Triclonal gammopathy (IgA kappa, IgG kappa, and IgM kappa) in a patient with plasmacytoid lymphoma derived from a monoclonal origin. *Am J Hematol* 1993;**42**:212–216.

312. Tursi A, Modeo ME. Monoclonal gammopathy of undetermined significance predisposing to *Helicobacter pylori*-related gastric mucosa-associated lymphoid tissue lymphoma. *J Clin Gastroenterol* 2002;**34**:147–149.

313. Pascali E. Monoclonal gammopathy and cold agglutinin disease in non-Hodgkin's lymphoma. *Eur J Haematol* 1996;56:114–115.

314. Bajetta E, Gasparini G, Facchetti G, Ferrari L, Giardini R, Delia D. Monoclonal gammopathy (IgM-κ) in a patient with Burkitt's type lymphoblastic lymphoma. *Tumori* 1984;70: 403–407.

315. Braunstein AH, Keren DF. Monoclonal gammopathy (IgM-kappa) in a patient with Burkitt's lymphoma. Case report and literature review. *Arch Pathol Lab Med* 1983;107: 235–238.

316. Heinz R, Stacher A, Pralle H, et al. Lymphoplasmacytic/lymphoplasmacytoid lymphoma: a clinical entity distinct from chronic lymphocytic leukaemia? *Blut* 1981;43:183–192.

317. Orth T, Treichel U, Mayet WJ, Storkel S, Meyer zum Buschenfelde KH. Reversible myelofibrosis in angioimmunoblastic lymphadenopathy. *Dtsch Med Wochenschr* 1994;119:694–698.

318. Offit K, Macris NT, Finkbeiner JA. Monoclonal hypergammaglobulinemia without malignant transformation in angioimmunoblastic lymphadenopathy with dysproteinemia. *Am J Med* 1986;80:292–294.

319. Toccanier MF, Kapanci Y. Lymphomatoid granulomatosis, polymorphic reticulosis and angioimmunoblastic lymphadenopathy with pulmonary involvement. Similar or different entities? *Ann Pathol* 1983;3:29–41.

320. Palutke M, McDonald JM. Monoclonal gammopathies associated with malignant lymphomas. *Am J Clin Pathol* 1973;60: 157–165.

321. Pascali E, Pezzoli A, Melato M, Falconieri G. Nodular lymphoma eventuating into lymphoplasmocytic lymphoma with monoclonal IgM/lambda cold agglutinin and Bence-Jones proteinuria. *Acta Haematol* 1980;64:94–102.

322. Yamasaki S, Matsushita H, Tanimura S, et al. B-cell lymphoma of mucosa-associated lymphoid tissue of the thymus: a report of two cases with a background of Sjögren's syndrome and monoclonal gammopathy. *Hum Pathol* 1998;29: 1021–1024.

323. Attygalle A, Al-Jehani R, Diss TC, et al. Neoplastic T cells in angioimmunoblastic T-cell lymphoma express CD10. *Blood* 2002;99: 627–633.

324. Yetgin S, Olcay L, Yel L, et al. T-ALL with monoclonal gammopathy and hairy cell features. *Am J Hematol* 2000;65:166–170.

325. Fernandez LA, MacSween JM, Langley GR. Immunoglobulin secretory function of B cells from untreated patients with chronic lymphocytic leukemia and hypogammaglobulinemia: role of T cells. *Blood* 1983;62:767–774.

326. Freeland HS, Scott PP. Recurrent pulmonary infections in patients with chronic lymphocytic leukemia and hypogammaglobulinemia. *South Med J* 1986;79:1366–1369.

327. Pangalis GA, Vassilakopoulos TP, Dimopoulou MN, Siakantaris MP, Kontopidou FN, Angelopoulou MK. B-chronic lymphocytic leukemia: practical aspects. *Hematol Oncol* 2002;20:103–146.

328. Sampalo A, Brieva JA. Humoral immunodeficiency in chronic lymphocytic leukemia: role of CD95/CD95L in tumoral damage and escape. *Leuk Lymphoma* 2002;43:881–884.

329. Caligaris-Cappio F. Biology of chronic lymphocytic leukemia. *Rev Clin Exp Hematol* 2000;4:5–21.

330. Pritsch O, Maloum K, Dighiero G. Basic biology of autoimmune phenomena in chronic lymphocytic leukemia. *Semin Oncol* 1998;25: 34–41.

331. Serpell LC, Sunde M, Blake CC. The molecular basis of amyloidosis *Cell Mol Life Sci* 1997;53: 871–887.

332. Kyle RA, Gertz MA. Primary systemic amyloidosis: clinical and laboratory features in 474 cases. *Semin Hematol* 1995;32:45–59.

333. Pick AI, Frohlichmann R, Lavie G, Duczyminer M, Skvaril F. Clinical and immunochemical studies of 20 patients with amyloidosis and plasma cell dyscrasia. *Acta Haematol* 1981;66: 154–167.

334. Kyle RA, Greipp PR. Amyloidosis (AL). Clinical

and laboratory features in 229 cases. *Mayo Clin Proc* 1983;58:665–683.

335. Dhodapkar MV, Merlini G, Solomon A. Biology and therapy of immunoglobulin deposition diseases. *Hematol Oncol Clin North Am* 1997;11:89–110.

336. Perfetti V, Garini P, Vignarelli MC, et al. Diagnostic approach to and follow-up of difficult cases of AL amyloidosis. *Haematologica* 1995;80:409–415.

337. Gertz MA, Kyle RA. Prognostic value of urinary protein in primary systemic amyloidosis (AL). *Am J Clin Pathol* 1990;94:313–317.

338. Kyle R. AL Amyloidosis: current diagnostic and therapeutic aspects. *Amyloid: J Protein Folding Disord* 2001;8:67–69.

339. Gertz MA, Lacy MQ, Dispenzieri A. Immunoglobulin light chain amyloidosis and the kidney. *Kidney Int* 2002;61:1–9.

340. Gillmore JD, Lovat LB, Persey MR, Pepys MB, Hawkins PN. Amyloid load and clinical outcome in AA amyloidosis in relation to circulating concentration of serum amyloid A protein. *Lancet* 2001;358:24–29.

341. Marhaug G, Dowton SB. Serum amyloid A: an acute phase apolipoprotein and precursor of AA amyloid. *Baillieres Clin Rheumatol* 1994;8:553–573.

342. Saraiva MJ. Sporadic cases of hereditary systemic amyloidosis. *N Engl J Med* 2002;346:1818–1819.

343. Lachmann HJ, Booth DR, Booth SE, et al. Misdiagnosis of hereditary amyloidosis as AL (primary) amyloidosis. *N Engl J Med* 2002;346:1786–1791.

344. Sanders PW, Booker BB. Pathobiology of cast nephropathy from human Bence Jones proteins. *J Clin Invest* 1992;89:630–639.

345. Buxbaum J, Gallo G. Nonamyloidotic monoclonal immunoglobulin deposition disease. Light-chain, heavy-chain, and light- and heavy-chain deposition diseases. *Hematol Oncol Clin North Am* 1999;13:1235–1248.

346. Rivest C, Turgeon PP, Senecal JL. Lambda light chain deposition disease presenting as an amyloid-like arthropathy. *J Rheumatol* 1993;20:880–884.

347. Buxbaum JN, Chuba JV, Hellman GC, Solomon A, Gallo GR. Monoclonal immunoglobulin deposition disease: light chain and light and heavy chain deposition diseases and their relation to light chain amyloidosis. Clinical features, immunopathology, and molecular analysis. *Ann Intern Med* 1990;112:455–464.

348. Gallo G, Picken M, Buxbaum J, Frangione B. The spectrum of monoclonal immunoglobulin deposition disease associated with immunocytic dyscrasias. *Semin Hematol* 1989;26:234–245.

349. Kambham N, Markowitz GS, Appel GB, Kleiner MJ, Aucouturier P, D'Agati VD. Heavy chain deposition disease: the disease spectrum. *Am J Kidney Dis.* 1999;33:954–962.

350. McAllister HA Jr, Seger J, Bossart M, Ferrans VJ. Restrictive cardiomyopathy with kappa light chain deposits in myocardium as a complication of multiple myeloma. Histochemical and electron microscopic observations. *Arch Pathol Lab Med* 1988;112:1151–1154.

351. Rostagno A, Frizzera G, Ylagan L, Kumar A, Ghiso J, Gallo G. Tumoral non-amyloidotic monoclonal immunoglobulin light chain deposits ('aggregoma'): presenting feature of B-cell dyscrasia in three cases with immunohistochemical and biochemical analyses. *Br J Haematol* 2002;119:62–69.

352. Buxbaum J. Mechanisms of disease: monoclonal immunoglobulin deposition. Amyloidosis, light chain deposition disease, and light and heavy chain deposition disease. *Hematol Oncol Clin North Am* 1992;6:323–346.

353. Ivanyi B. Frequency of light chain deposition nephropathy relative to renal amyloidosis and Bence Jones cast nephropathy in a necropsy study of patients with myeloma. *Arch Pathol Lab Med* 1990;114:986–987.

354. Miwa T, Kimura Y, Nonomura A, Kamide M, Furukawa M. Unusual case of plasma cell tumor with monoclonal gammopathy of the sino-nasal cavity and clavicular bone. *ORL J Otorhinolaryngol Relat Spec* 1993;55:45–48.

355. Lasker JC, Bishop JO, Wilbanks JH, Lane M. Solitary myeloma of the talus bone. *Cancer* 1991;68:202–205.

356. Woodruff RK, Malpas JS, White FE. Solitary plasmacytoma. II: Solitary plasmacytoma of bone. *Cancer* 1979;**43**:2344–2347.

357. Menke DM, Horny HP, Griesser H, et al. Primary lymph node plasmacytomas (plasmacytic lymphomas). *Am J Clin Pathol* 2001;**115**: 119–126.

358. Frassica DA, Frassica FJ, Schray MF, Sim FH, Kyle RA. Solitary plasmacytoma of bone: Mayo Clinic experience. *Int J Radiat Oncol Biol Phys.* 1989;**16**:43–48.

359. Dimopoulos MA, Goldstein J, Fuller L, Delasalle K, Alexanian R. Curability of solitary bone plasmacytoma. *J Clin Oncol* 1992;**10**: 587–590.

360. Liebross RH, Ha CS, Cox JD, Weber D, Delasalle K, Alexanian R. Solitary bone plasmacytoma: outcome and prognostic factors following radiotherapy. *Int J Radiat Oncol Biol Phys* 1998;**41**:1063–1067.

361. Johnson TL, Keren DF, Thomas LR. Indirect immunofluorescence technique to detect monoclonal antinuclear antibody. *Arch Pathol Lab Med* 1987;**111**:560–562.

362. Jonsson V, Svendsen B, Vorstrup S, et al. Multiple autoimmune manifestations in monoclonal gammopathy of undetermined significance and chronic lymphocytic leukemia. *Leukemia* 1996;**10**:327–332.

363. Nakase T, Matsuoka N, Iwasaki E, Ukyo S, Shirakawa S. CML with autoimmune thrombocytopenia observed in the course of IgG (kappa) type monoclonal gammopathy. *Rinsho Ketsueki* 1992;**33**:706–708.

364. Guardigni L, Rasi F, Nemni R, Lorenzetti I, Valzania F, Pretolani E. Autoimmune neuropathy in monoclonal gammopathy. *Recenti Prog Med* 1991;**82**:669–671.

365. Wasada T, Eguchi Y, Takayama S, Yao K, Hirata Y, Ishii S. Insulin autoimmune syndrome associated with benign monoclonal gammopathy. Evidence for monoclonal insulin autoantibodies. *Diabetes Care* 1989;**12**:147–150.

366. Kozuru M, Nakashima Y, Kurata T, Kaneko S, Ibayashi H. Autoimmune neutropenia associated with monoclonal gammopathy (author's transl). *Nippon Naika Gakkai Zasshi* 1979;**68**: 1306–1312.

367. Zlotnick A, Benbassat J. Monoclonal gammopathy associated with autoimmune disease. *Harefuah* 1976;**90**:424–427.

368. Kiprov DD, Miller RG. Polymyositis associated with monoclonal gammopathy. *Lancet* 1984;**2**:1183–1186.

369. Moutsopoulos HM, Tzioufas AG, Bai MK, Papadopoulos NM, Papadimitriou CS. Association of serum IgM kappa monoclonicity in patients with Sjogren's syndrome with an increased proportion of kappa positive plasma cells infiltrating the labial minor salivary glands. *Ann Rheum Dis* 1990;**49**:929–931.

370. Seligmann M, Brouet JC. Antibody activity of human monoclonal immunoglobulins. *Pathol Biol (Paris)* 1990;**38**:822–823.

371. Seligmann M, Brouet JC. Antibody activity of human myeloma globulins. *Semin Hematol* 1973;**10**:163–177.

372. Potter M. Myeloma proteins (M-components) with antibody-like activity. *N Engl J Med* 1971;**284**:831–838.

373. Laurell CB. Complexes formed *in vivo* between immunoglobulin light chain kappa, prealbumin and-or alpha-1-antitrypsin in myeloma sera. *Immunochemistry* 1970;**7**:461–465.

374. Vrethem M, Cruz M, Wen-Xin H, Malm C, Holmgren H, Ernerudh J. Clinical, neurophysiological and immunological evidence of polyneuropathy in patients with monoclonal gammopathies. *J Neurol Sci* 1993;**114**:193–199.

375. Latov N. Pathogenesis and therapy of neuropathies associated with monoclonal gammopathies. *Ann Neurol* 1995;**37**(Suppl 1): S32–42.

376. Nobile-Orazio E, Marmiroli P, Baldini L, et al. Peripheral neuropathy in macroglobulinemia: incidence and antigen-specificity of M proteins. *Neurology* 1987;**37**:1506–1514.

377. Freddo L, Ariga T, Saito M, et al. The neuropathy of plasma cell dyscrasia: binding of IgM M-proteins to peripheral nerve glycolipids. *Neurology* 1985;**35**:1420–1424.

378. Latov N, Sherman WH, Nemni R, et al. Plasma-

cell dyscrasia and peripheral neuropathy with a monoclonal antibody to peripheral-nerve myelin. *N Engl J Med* 1980;303:618–621.

379. Latov N, Braun PE, Gross RB, Sherman WH, Penn AS, Chess L. Plasma cell dyscrasia and peripheral neuropathy: identification of the myelin antigens that react with human paraproteins. *Proc Natl Acad Sci USA* 1981;78: 7139–7142.

380. Latov N. Plasma cell dyscrasia and motor neuron disease. *Adv Neurol* 1982;36:273–279.

381. Shy ME, Rowland LP, Smith T, et al. Motor neuron disease and plasma cell dyscrasia. *Neurology* 1986;36:1429–1436.

382. Driedger H, Pruzanski W. Plasma cell neoplasia with peripheral polyneuropathy. A study of five cases and a review of the literature. *Medicine (Baltimore)* 1980;59:301–310.

383. Dalakas MC, Engel WK. Polyneuropathy with monoclonal gammopathy: studies of 11 patients. *Ann Neurol* 1981;10:45–52.

384. Ilyas AA, Quarles RH, Dalakas MC, Fishman PH, Brady RO. Monoclonal IgM in a patient with paraproteinemic polyneuropathy binds to gangliosides containing disialosyl groups. *Ann Neurol* 1985;18:655–659.

385. Steck AJ, Murray N, Meier C, Page N, Perruisseau G. Demyelinating neuropathy and monoclonal IgM antibody to myelin-associated glycoprotein. *Neurology* 1983;33:19–23.

386. Meier C, Vandevelde M, Steck A, Zurbriggen A. Demyelinating polyneuropathy associated with monoclonal IgM-paraproteinaemia. Histological, ultrastructural and immunocytochemical studies. *J Neurol Sci* 1984;63:353–367.

387. Steck AJ, Murray N, Justafre JC, et al. Passive transfer studies in demyelinating neuropathy with IgM monoclonal antibodies to myelin-associated glycoprotein. *J Neurol Neurosurg Psychiatry* 1985;48:927–929.

388. Latov N, Hays AP, Donofrio PD, et al. Monoclonal IgM with unique specificity to gangliosides GM1 and GD1b and to lacto-*N*-tetraose associated with human motor neuron disease. *Neurology* 1988;38:763–768.

389. Daune GC, Farrer RG, Dalakas MC, Quarles RH. Sensory neuropathy associated with monoclonal immunoglobulin M to GD1b ganglioside. *Ann Neurol* 1992;31:683–685.

390. Burger D, Perruisseau G, Simon M, Steck AJ. Comparison of the *N*-linked oligosaccharide structures of the two major human myelin glycoproteins MAG and P0: assessment of the structures bearing the epitope for HNK-1 and human monoclonal immunoglobulin M found in demyelinating neuropathy. *J NeuroChem* 1992;58:854–861.

391. Kanda T, Iwasaki T, Yamawaki M, Tai T, Mizusawa H. Anti-GM1 antibody facilitates leakage in an *in vitro* blood-nerve barrier model. *Neurology* 2000;55:585–587.

392. Manschot SM, Notermans NC, van den Berg LH, Verschuuren JJ, Lokhorst HM. Three families with polyneuropathy associated with monoclonal gammopathy. *Arch Neurol* 2000;57:740–742.

393. Jensen TS, Schroder HD, Jonsson V, et al. IgM monoclonal gammopathy and neuropathy in two siblings. *J Neurol Neurosurg Psychiatry* 1988; 51:1308–1315.

394. Latov N. Neuropathy, heredity, and monoclonal gammopathy. *Arch Neurol* 2000;57:641–642.

395. Simmons Z, Bromberg MB, Feldman EL, Blaivas M. Polyneuropathy associated with IgA monoclonal gammopathy of undetermined significance. *Muscle Nerve* 1993;16:77–83.

396. Kanda T, Usui S, Beppu H, Miyamoto K, Yamawaki M, Oda M. Blood-nerve barrier in IgM paraproteinemic neuropathy: a clinicopathologic assessment. *Acta Neuropathol (Berl)* 1998;95:184–192.

397. Lach B, Rippstein P, Atack D, Afar DE, Gregor A. Immunoelectron microscopic localization of monoclonal IgM antibodies in gammopathy associated with peripheral demyelinative neuropathy. *Acta Neuropathol (Berl)* 1993;85: 298–307.

398. Vital C, Vital A, Deminiere C, Julien J, Lagueny A, Steck AJ. Myelin modifications in 8 cases of peripheral neuropathy with Waldenstrom's macroglobulinemia and anti-MAG activity. *Ultrastruct Pathol* 1997;21:509–516.

399. Zuckerman SJ, Pesce MA, Rowland LP, et al. An alert for motor neuron diseases and peripheral neuropathy: monoclonal paraproteinemia may be missed by routine electrophoresis. *Arch Neurol* 1987;**44**:250–251.

400. Vrethem M, Larsson B, von Schenck H, Ernerudh J. Immunofixation superior to plasma agarose electrophoresis in detecting small M-components in patients with polyneuropathy. *J Neurol Sci* 1993;**120**:93–98.

401. Dyck PJ, Low PA, Windebank AJ, et al. Plasma exchange in polyneuropathy associated with monoclonal gammopathy of undetermined significance. *N Engl J Med* 1991;**325**: 1482–1486.

402. Monaco S, Bonetti B, Ferrari S, et al. Complement-mediated demyelination in patients with IgM monoclonal gammopathy and polyneuropathy. *N Engl J Med* 1990;**322**: 649–652.

403. Dalakas MC, Quarles RH, Farrer RG, et al. A controlled study of intravenous immunoglobulin in demyelinating neuropathy with IgM gammopathy. *Ann Neurol* 1996;**40**:792–795.

404. Koike H, Sobue G. Crow–Fukase syndrome. *Neuropathology* 2000;**20**(Suppl):S69–72.

405. Araki T, Konno T, Soma R, et al. Crow–Fukase syndrome associated with high-output heart failure. *Intern Med* 2002;**41**:638–641.

406. Nakanishi T, Sobue I, Toyokura Y, et al. The Crow–Fukase syndrome: a study of 102 cases in Japan. *Neurology* 1984;**34**:712–720.

407. Dispenzieri A, Kyle RA, Lacy MQ, et al. POEMS syndrome: definitions and long-term outcome. *Blood* 2003;**101**:2496–2506.

408. Miralles GD, O'Fallon JR, Talley NJ. Plasma-cell dyscrasia with polyneuropathy. The spectrum of POEMS syndrome. *N Engl J Med* 1992;**327**: 1919–1923.

409. Lacy MQ, Gertz MA, Hanson CA, Inwards DJ, Kyle RA. Multiple myeloma associated with diffuse osteosclerotic bone lesions: a clinical entity distinct from osteosclerotic myeloma (POEMS syndrome). *Am J Hematol* 1997;**56**: 288–293.

410. Buckley CE 3rd, Dorsey FC. The effect of aging on human serum immunoglobulin concentrations. *J Immunol* 1970;**105**:964–972.

411. Fine JM, Lambin P, Leroux P. Frequency of monoclonal gammopathy ('M components') in 13,400 sera from blood donors. *Vox Sang* 1972;**23**:336–343.

412. Crawford J, Eye MK, Cohen HJ. Evaluation of monoclonal gammopathies in the 'well' elderly. *Am J Med* 1987;**82**:39–45.

413. Franceschi C, Cossarizza A. Introduction: the reshaping of the immune system with age. *Int Rev Immunol* 1995;**12**:1–4.

414. Makinodan T, Kay MM. Age influence on the immune system. *Adv Immunol* 1980;**29**: 287–330.

415. Radl J. Differences among the three major categories of paraproteinaemias in aging man and the mouse. A minireview. *Mech Ageing Dev* 1984;**28**:167–170.

416. Kyle RA. Monoclonal gammopathy of undetermined significance. Natural history in 241 cases. *Am J Med* 1978;**64**:814–826.

417. Peters SO, Stockschlader M, Zeller W, Mross K, Durken M, Kruger W, Zander AR. Monoclonal gammopathy of unknown significance in a bone marrow donor. *Ann Hematol* 1993;**66**:93–95.

418. Avet-Loiseau H, Li JY, Morineau N, Facon T, Brigaudeau C, Harousseau JL, Grosbois B, Bataille R. Monosomy 13 is associated with the transition of monoclonal gammopathy of undetermined significance to multiple myeloma. Intergroupe Francophone du Myelome. *Blood* 1999;**94**:2583–2589.

419. Avet-Loiseau H, Facon T, Daviet A, et al. 14q32 translocations and monosomy 13 observed in monoclonal gammopathy of undetermined significance delineate a multistep process for the oncogenesis of multiple myeloma. Intergroupe Francophone du Myelome. *Cancer Res* 1999;**59**: 4546–4550.

420. Fonseca R, Bailey RJ, Ahmann GJ, et al. Genomic abnormalities in monoclonal gammopathy of undetermined significance. *Blood* 2002;**100**:1417–1424.

421. Vejlgaard T, Abildgaard N, Jans H, Nielsen JL,

Heickendorff L. Abnormal bone turnover in monoclonal gammopathy of undetermined significance: analyses of type I collagen telopeptide, osteocalcin, bone-specific alkaline phosphatase and propeptides of type I and type III procollagens. *Eur J Haematol* 1997;58: 104–108.

422. Pecherstorfer M, Seibel MJ, Woitge HW, et al. Bone resorption in multiple myeloma and in monoclonal gammopathy of undetermined significance: quantification by urinary pyridinium cross-links of collagen. *Blood* 1997;90: 3743–3750.

423. Zheng C, Huang DR, Bergenbrant S, Sundblad A, et al. Interleukin 6, tumour necrosis factor alpha, interleukin 1beta and interleukin 1 receptor antagonist promoter or coding gene polymorphisms in multiple myeloma. *Br J Haematol* 2000;109:39–45.

424. Cesana C, Klersy C, Barbarano L, et al. Prognostic factors for malignant transformation in monoclonal gammopathy of undetermined significance and smoldering multiple myeloma. *J Clin Oncol* 2002;20:1625–1634.

425. Kanoh T. Fluctuating M-component level in relation to infection. *Eur J Haematol* 1989;42: 503–504.

426. Pasqualetti P, Casale R. Risk of malignant transformation in patients with monoclonal gammopathy of undetermined significance. *Biomed Pharmacother* 1997;51:74–78.

427. Pasqualetti P, Festuccia V, Collacciani A, Casale R. The natural history of monoclonal gammopathy of undetermined significance. A 5- to 20-year follow-up of 263 cases. *Acta Haematol* 1997;97:174–179.

428. Colls BM. Monoclonal gammopathy of undetermined significance (MGUS) – 31 year follow up of a community study. *Aust N Z J Med* 1999;29:500–504.

429. Larrain C. Transient monoclonal gammopathies associated with infectious endocarditis. *Rev Med Chil* 1986;114:771–776.

430. Arima T, Tsuboi S, Nagata K, Gyoten Y, Tanigawa T. An extremely basic monoclonal IgG in an aged apoplectic patient with prolonged

bacterial infection. *Acta Med Okayama* 1976;30: 209–214.

431. Crapper RM, Deam DR, Mackay IR. Paraproteinemias in homosexual men with HIV infection. Lack of association with abnormal clinical or immunologic findings. *Am J Clin Pathol* 1987;88:348–351.

432. Tsuji T, Tokuyama K, Naito K, Okazaki S, Shinohara T. A case report of primary hepatic carcinoma with prolonged HB virus infection and monoclonal gammopathy. *Gastroenterol Jpn* 1977;12:69–75.

433. Godeau P, Herson S, De Treglode D, Herreman G. Benign monoclonal immunoglobulins during subacute infectious endocarditis. *Coeur Med Interne.* 1979;18:3–12.

434. Herreman G, Godeau P, Cabane J, Digeon M, Laver M, Bach JF. Immunologic study of subacute infectious endocarditis through the search for circulating immune complexes. Preliminary results apropos of 13 cases. *Nouv Presse Med* 1975;4:2311–2314.

435. Struve J, Weiland O, Nord CE. *Lactobacillus plantarum* endocarditis in a patient with benign monoclonal gammopathy. *J Infect* 1988;17: 127–130.

436. Covinsky M, Laterza O, Pfeifer JD, Farkas-Szallasi T, Scott MG. An IgM lambda antibody to *Escherichia coli* produces false-positive results in multiple immunometric assays. *Clin Chem* 2000;46:1157–1161.

437. Kelly RH, Hardy TJ, Shah PM. Benign monoclonal gammopathy: a reassessment of the problem. *Immunol Invest* 1985;14:183–197.

438. Turbat-Herrera EA, Hancock C, Cabello-Inchausti B, Herrera GA. Plasma cell hyperplasia and monoclonal paraproteinemia in human immunodeficiency virus-infected patients. *Arch Pathol Lab Med* 1993;117:497–501.

439. Radl J. Monoclonal gammapathies. An attempt at a new classification. *Neth J Med* 1985;28: 134–137.

440. van den Akker TW, Brondijk R, Radl J. Influence of long-term antigenic stimulation started in young C57BL mice on the development of age-related monoclonal

gammapathies. *Int Arch Allergy Appl Immunol* 1988;87:165–170.

441. Radl J, Liu M, Hoogeveen CM, van den Berg P, Minkman-Brondijk RJ, Broerse JJ, Zurcher C, van Zwieten MJ. Monoclonal gammapathies in long-term surviving rhesus monkeys after lethal irradiation and bone marrow transplantation. *Clin Immunol Immunopathol* 1991;60:305–309.

442. Radl J, Valentijn RM, Haaijman JJ, Paul LC. Monoclonal gammapathies in patients undergoing immunosuppressive treatment after renal transplantation. *Clin Immunol Immunopathol* 1985;37:98–102.

443. Jain A, Nalesnik M, Reyes J, et al. Posttransplant lymphoproliferative disorders in liver transplantation: a 20-year experience. *Ann Surg* 2002;236:429–436.

444. Craig FE, Gulley ML, Banks PM. Posttransplantation lymphoproliferative disorders. *Am J Clin Pathol* 1993;99:265–276.

445. Stevens SJ, Verschuuren EA, Pronk I, et al. Frequent monitoring of Epstein–Barr virus DNA load in unfractionated whole blood is essential for early detection of posttransplant lymphoproliferative disease in high-risk patients. *Blood* 2001;97:1165–1171.

446. Dispenzieri A, Gorevic PD. Cryoglobulinemia. *Hematol Oncol Clin North Am* 1999;13: 1315–1349.

447. Wintrobe MM, Buell MV. Hyperproteinemia associated with multiple myeloma. *Bull Johns Hopkins Hosp* 1933;52:156–165.

448. Lerner AB, Watson CJ. Studies of cryoglobulins. Unusual purpura associated with the presence of a high concentration of cryoglobulin (cold precipitable serum globulins). *Am J Med Sci* 1947;2:410.

449. Durand JM, Cacoub P, Lunel-Fabiani F, et al. Ribavirin in hepatitis C related cryoglobulinemia. *J Rheumatol* 1998;25:1115–1117.

450. Akriviadis EA, Xanthakis I, Navrozidou C, Papadopoulos A. Prevalence of cryoglobulinemia in chronic hepatitis C virus infection and response to treatment with interferon-alpha. *J Clin Gastroenterol* 1997;25:612–618.

451. Mazzaro C, Colle R, Baracetti S, Nascimben F, Zorat F, Pozzato G. Effectiveness of leukocyte interferon in patients affected by HCV-positive mixed cryoglobulinemia resistant to recombinant alpha-interferon. *Clin Exp Rheumatol* 2002;20: 27–34.

452. Naarendorp M, Kallemuchikkal U, Nuovo GJ, Gorevic PD. Long-term efficacy of interferon-alpha for extrahepatic disease associated with hepatitis C virus infection. *J Rheumatol* 2001;28:2466–2473.

453. Monti G, Saccardo F, Rinaldi G, Petrozzino MR, Gomitoni A, Invernizzi F. Colchicine in the treatment of mixed cryoglobulinemia. *Clin Exp Rheumatol* 1995;13 Suppl 13:S197–199.

454. Dispenzieri A. Symptomatic cryoglobulinemia. *Curr Treat Options Oncol* 2000;1:105–118.

455. Friedman G, Mehta S, Sherker AH. Fatal exacerbation of hepatitis C-related cryoglobulinemia with interferon-alpha therapy. *Dig Dis Sci* 1999;44:1364–1365.

456. Yebra M, Barrios Y, Rincon J, Sanjuan I, Diaz-Espada F. Severe cutaneous vasculitis following intravenous infusion of gammaglobulin in a patient with type II mixed cryoglobulinemia. *Clin Exp Rheumatol* 2002;20:225–227.

457. Brouet JC, Clauvel JP, Danon F, Klein M, Seligmann M. Biologic and clinical significance of cryoglobulins. A report of 86 cases. *Am J Med* 1974;57:775–788.

458. Liu F, Knight GB, Agnello V. Hepatitis C virus but not GB virus C/hepatitis G virus has a role in type II cryoglobulinemia. *Arthritis Rheum* 1999;42:1898–1901.

459. Agnello V, Chung RT, Kaplan LM. A role for hepatitis C virus infection in type II cryoglobulinemia. *N Engl J Med* 1992;327: 1490–1495.

460. Trendelenburg M, Schifferli JA. Cryoglobulins are not essential. *Ann Rheum Dis.* 1998;57:3–5.

461. Trejo O, Ramos-Casals M, Garcia-Carrasco M, et al. Cryoglobulinemia: study of etiologic factors and clinical and immunologic features in 443 patients from a single center. *Medicine (Baltimore)* 2001;80:252–262.

462. Feiner HD. Relationship of tissue deposits of cryoglobulin to clinical features of mixed

cryoglobulinemia. *Hum Pathol* 1983;**14**: 710–715.

463. Musset L, Diemert MC, Taibi F, et al. Characterization of cryoglobulins by immunoblotting. *Clin Chem* 1992;**38**:798–802.

464. Ferri C, Greco F, Longombardo G, et al. Association between hepatitis C virus and mixed cryoglobulinemia. *Clin Exp Rheumatol* 1991;**9**: 621–624.

465. Agnello V. The etiology and pathophysiology of mixed cryoglobulinemia secondary to hepatitis C virus infection. *Springer Semin Immunopathol* 1997;**19**:111–129.

466. Dammacco F, Sansonno D. Antibodies to hepatitis C virus in essential mixed cryoglobulinaemia. *Clin Exp Immunol* 1992;**87**: 352–356.

467. Duvoux C, Tran Ngoc A, Intrator L, et al. Hepatitis C virus (HCV)-related cryoglobulinemia after liver transplantation for HCV cirrhosis. *Transpl Int* 2002;**15**:3–9.

468. Nagasaka A, Takahashi T, Sasaki T, et al. Cryoglobulinemia in Japanese patients with chronic hepatitis C virus infection: host genetic and virological study. *J Med Virol* 2001;**65**: 52–57.

469. Schott P, Hartmann H, Ramadori G. Hepatitis C virus-associated mixed cryoglobulinemia. Clinical manifestations, histopathological changes, mechanisms of cryoprecipitation and options of treatment. *Histol Histopathol* 2001;**16**:1275–1285.

470. Della Rossa A, Tavoni A, Baldini C, Bombardieri S. Mixed cryoglobulinemia and hepatitis C virus association: ten years later. *Isr Med Assoc J* 2001;**3**:430–434.

471. Lunel F, Musset L. Mixed cryoglobulinemia and hepatitis C virus infection. *Minerva Med* 2001; **92**:35–42.

472. Gharagozloo S, Khoshnoodi J, Shokri F. Hepatitis C virus infection in patients with essential mixed cryoglobulinemia, multiple myeloma and chronic lymphocytic leukemia. *Pathol Oncol Res* 2001;**7**:135–139.

473. Abrahamian GA, Cosimi AB, Farrell ML, Schoenfeld DA, Chung RT, Pascual M.

Prevalence of hepatitis C virus-associated mixed cryoglobulinemia after liver transplantation. *Liver Transpl* 2000;**6**:185–190.

474. Schmidt WN, Stapleton JT, LaBrecque DR, et al. Hepatitis C virus (HCV) infection and cryoglobulinemia: analysis of whole blood and plasma HCV-RNA concentrations and correlation with liver histology. *Hepatology* 2000;**31**:737–744.

475. Casato M, de Rosa FG, Pucillo LP, et al. Mixed cryoglobulinemia secondary to visceral Leishmaniasis. *Arthritis Rheum* 1999;**42**: 2007–2011.

476. Perrot H, Thivolet J, Fradin G. Cryoglobulin and rheumatoid factor during primosecondary syphilis. *Presse Med* 1971;**7**:1059–1060.

477. Steere AC, Hardin JA, Ruddy S, Mummaw JG, Malawista SE. Lyme arthritis: correlation of serum and cryoglobulin IgM with activity, and serum IgG with remission. *Arthritis Rheum* 1979;**22**:471–483.

478. Ferri C, Zignego AL, Pileri SA. Cryoglobulins. *J Clin Pathol* 2002;**55**:4–13.

479. Tissot JD, Schifferli JA, Hochstrasser DF, et al. Two-dimensional polyacrylamide gel electrophoresis analysis of cryoglobulins and identification of an IgM-associated peptide. *J Immunol Methods* 1994;**173**:63–75.

480. Levo Y. Nature of cryoglobulinaemia. *Lancet* 1980;**1**:285–287.

481. Lawson EQ, Brandau DT, Trautman PA, Middaugh CR. Electrostatic properties of cryoimmunoglobulins. *J Immunol* 1988;**140**: 1218–1222.

482. Middaugh CR, Litman GW. Atypical glycosylation of an IgG monoclonal cryoimmunoglobulin. *J Biol Chem* 1987;**262**: 3671–3673.

483. Tomana M, Schrohenloher RE, Koopman WJ, Alarcon GS, Paul WA. Abnormal glycosylation of serum IgG from patients with chronic inflammatory diseases. *Arthritis Rheum* 1988;**31**: 333–338.

484. Letendre L, Kyle RA. Monoclonal cryoglobulinemia with high thermal insolubility. *Mayo Clin Proc* 1982;**57**:629–633.

485. Howard TW, Iannini MJ, Burge JJ, Davis JS. Rheumatoid factor, cryoglobulinemia, anti-DNA, and renal disease in patients with systemic lupus erythematosus. *J Rheumatol* 1991;18:826–830.

486. Invernizzi F, Pietrogrande M, Sagramoso B. Classification of the cryoglobulinemic syndrome. *Clin Exp Rheumatol* 1995;13(Suppl 13): S123–128.

487. Keren DF, Di Sante AC, Mervak T, Bordine SL. Problems with transporting serum to the laboratory for cryoglobulin assay: a solution. *Clin Chem* 1985;31:1766–1767.

488. Gorevic PD, Galanakis D. Chapter 10. Cryoglobulins, cryofibrinogenemia, and pyroglobulins. In: Rose NR, Hamilton RG, Detrick B., eds. *Manual of clinical laboratory immunology.* Washington, DC: ASM Press, 2002.

489. Grose MP. Clinical Immunology Tech Sample No. CI-2. *ASCP Tech Sample* 1988;1.

490. Okazaki T, Nagai T, Kanno T. Gel diffusion procedure for the detection of cryoglobulins in serum. *Clin Chem* 1998;44:1558–1559.

491. Gorevic PD, Kassab HJ, Levo Y, et al. Mixed cryoglobulinemia: clinical aspects and long-term follow-up of 40 patients. *Am J Med* 1980;69: 287–308.

492. Meltzer M, Franklin EC. Cryoglobulinemia – a study of twenty-nine patients. I. IgG and IgM cryoglobulins and factors affecting cryoprecipitability. *Am J Med* 1966;40:828–836.

493. Meltzer M, Franklin EC, Elias K, McCluskey RT, Cooper N. Cryoglobulinemia – a clinical and laboratory study. II. Cryoglobulins with rheumatoid factor activity. *Am J Med* 1966;40: 837–856.

494. Markel SF, Janich SL. Complexing of lactate dehydrogenase isoenzymes with immunoglobulin A of the kappa class. *Am J Clin Pathol* 1974;61: 328–332.

495. Backer ET, Harff GA, Beyer C. A patient with an IgG paraprotein and complexes of lactate dehydrogenase and IgG in the serum. *Clin Chem* 1987;33:1937–1938.

496. Pearce CJ, Hine TJ, Peek K. Hypercalcaemia due to calcium binding by a polymeric IgA kappa-paraprotein. *Ann Clin BioChem* 1991;28(Pt 3):229–234.

497. Bakker AJ. Influence of monoclonal immunoglobulins in direct determinations of iron in serum. *Clin Chem* 1991;37:690–694.

498. Bakker AJ, Bosma H, Christen PJ. Influence of monoclonal immunoglobulins in three different methods for inorganic phosphorus. *Ann Clin BioChem* 1990;27(Pt 3):227–231.

499. Dalal BI, Collins SY, Burnie K, Barr RM. Positive direct antiglobulin tests in myeloma patients. Occurrence, characterization, and significance. *Am J Clin Pathol* 1991;96:496–499.

500. Glaspy JA. Hemostatic abnormalities in multiple myeloma and related disorders. *Hematol Oncol Clin N Am* 1992;6:1301–1314.

501. Panzer S, Thaler E. An acquired cryoglobulinemia which inhibits fibrin polymerization in a patient with IgG kappa myeloma. *Haemostasis* 1993;23:69–76.

502. Matsuzaki H, Hata H, Watanabe T, Takeya M, Takatsuki K. Phagocytic multiple myeloma with disseminated intravascular coagulation. *Acta Haematol* 1989;82:91–94.

503. Richard C, Cuadrado MA, Prieto M, et al. Acquired von Willebrand disease in multiple myeloma secondary to absorption of von Willebrand factor by plasma cells. *Am J Hematol* 1990;35:114–117.

504. Schwarzinger I, Stain-Kos M, Bettelheim P, et al. Recurrent, isolated factor X deficiency in myeloma: repeated normalization of factor X levels after cytostatic chemotherapy followed by late treatment failure associated with the development of systemic amyloidosis. *Thromb Haemost* 1992;68:648–651.

505. Redmon B, Pyzdrowski KL, Elson MK, Kay NE, Dalmasso AP, Nuttall FQ. Hypoglycemia due to an insulin-binding monoclonal antibody in multiple myeloma. *N Engl J Med* 1992;326: 994–998.

506. Side L, Fahie-Wilson MN, Mills MJ. Hypercalcaemia due to calcium binding IgM paraprotein in Waldenstrom's macroglobulinaemia. *J Clin Pathol* 1995;48: 961–962.

507. Goodrick MJ, Boon RJ, Bishop RJ, Copplestone JA, Prentice AG. Inaccurate haemoglobin estimation in Waldenstrom's macroglobulinaemia: unusual reaction with monomeric IgM paraprotein. *J Clin Pathol* 1993;46:1138–1139.

508. Fohlen-Walter A, Jacob C, Lecompte T, Lesesve JF. Laboratory identification of cryoglobulinemia from automated blood cell counts, fresh blood samples, and blood films. *Am J Clin Pathol* 2002;117:606–614.

509. Emori HW, Bluestone R, Goldberg LS. Pseudo-leukocytosis associated with cryoglobulinemia. *Am J Clin Pathol* 1973;60:202–204.

510. Taft EG, Grossman J, Abraham GN, Leddy JP, Lichtman MA. Pseudoleukocytosis due to cryoprotein crystals. *Am J Clin Pathol* 1973;60:669–671.

511. Haeney MR. Erroneous values for the total white cell count and ESR in patients with cryoglobulinaemia. *J Clin Pathol* 1976;29:894–897.

512. Hambley H, Vetters JM. Artefacts associated with a cryoglobulin. *Postgrad Med J* 1989;65:241–243.

513. Lesesve JF, Goasguen J. Cryoglobulin detection from a blood smear leading to the diagnosis of multiple myeloma. *Eur J Haematol* 2000;65:77.

7

Examination of urine for proteinuria

URINE PROTEIN COMPOSITION

Evaluation of the protein content of urine samples can provide useful information about the location and degree of damage within the nephron, as well as the presence of monoclonal proteins (Fig. 7.1). The protein composition of urine depends upon both factors intrinsic to the protein itself and on pathophysiological alterations in the patients.

Size and amount of the protein

The glomerulus acts as a barrier to the passage of proteins into the urine. The glomerular basement membrane itself serves as a filter that permits molecules smaller than about 50 Å (about 15 kDa) diameter to pass freely into Bowman's space. Molecules between 15 kDa and 69 kDa (albumin) pass through the glomerulus to a lesser extent and larger molecules are retained in the blood. For albumin itself, its size and charge allow less than 0.1 per cent of its plasma concentration to pass

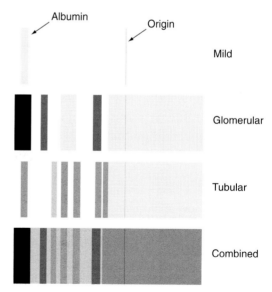

Figure 7.1 Schematic of mild proteinuria with albumin only, glomerular proteinuria, tubular proteinuria and combined glomerular with tubular proteinuria.

into the glomerular ultrafiltrate.[1] Yet, because of its large concentration in the serum, it is often the predominant protein detected in the urine even in cases with only minor glomerular leakage (see below).

Charge of the protein

Charge of the protein is another major influence on glomerular permeability. The glomerulus has a net negative charge because of the presence of negatively charged glycoproteins.[2] Consequently, the poly-anion albumin is repelled by this negatively charged glomerular capillary surface. Neutral dextran of similar molecular weight to albumin will pass much more readily into the glomerular filtrate.[3]

Hydrostatic pressure

Another factor that may alter protein composition of urine is the hydrostatic pressure within the systemic circulation. As blood pressure increases, larger molecules pass through the glomerulus. This results in a glomerular filtrate with a disproportionate number of molecules larger than 100 kDa. Consequently, among patients with primary hypertension and albuminuria, reduction of the blood pressure reduces the albuminuria.[4]

Glomerular filtrate and tubular absorption

Under normal circumstances, the glomerular filtrate contains about 10 mg protein/l.[5] This is composed of numerous low molecular weight proteins and polypeptides, along with albumin. As mentioned above, despite its size and negative charge, its large concentration in serum accounts for the presence of albumin in the glomerular filtrate. The most common small proteins in the glomerular filtrate include α_1-acid glycoprotein (orosomucoid, 40 kDa), α_1-microglobulin (27 kDa), β_2-microglobulin (12 kDa),

γ trace protein (12 kDa), and retinol-binding protein (20 kDa). More than 90 per cent of the low molecular weight proteins that pass into the glomerular filtrate under physiological conditions are reabsorbed by binding to specific receptors on the proximal convoluted tubules, where they are catabolized. Absorption is also influenced by the charge on the protein. Proteins that are relatively anionic are more readily absorbed than cationic proteins. Despite the reabsorption, a small amount of protein (less than 150 mg/24 h), mainly albumin, normally finds its way into the urine.

Protein in normal urine

In most urine samples, insufficient protein is present for detection by standard electrophoretic techniques. Consequently, normal urine must be concentrated to facilitate examination of the various protein fractions. In concentrating the urine, one must be concerned with the possible loss of low molecular weight proteins that are useful to determine the location of damage within the nephron.

Normally, adults excrete about 100–150 mg/24 h and children excrete 60–100 mg/24 h of protein (mainly albumin) in the urine. Often, a 24-h collection is difficult to obtain. Because of this, the ratio of total protein to creatinine on random samples of urine has been employed as an alternative method to study urine protein. For adults, the normal protein/creatinine ratio is < 0.2, for children age 6–24 months it is < 0.5, and for children older than 24 months it is 0.2–0.25.[6]

A faintly staining albumin band may be seen by using electrophoresis on concentrated urine. The albumin band in urine samples migrates closer to the anode than it does in the serum and there is often anodal slurring that smears albumin toward the positive electrode. This migration of albumin is attributed to the binding of anions to its surface.[7] In addition to albumin, a small amount of Tamm–Horsfall protein (80 kDa), the heavily glycosylated (about 30 per cent carbohydrate) secretory product from the cells in the thick ascending limb of the loop of Henle,[8] makes up most of the other protein found in normal urine. Too little is present for detection by electrophoresis.

Electrophoresis on the typical 50–100 times concentrated sample of normal urine does not disclose staining in the γ-region. However, small amounts of polyclonal free light chains are usually present in concentrated urine from patients with mild glomerular or tubular disease. It is important to distinguish these polyclonal FLC from the monoclonal free light chains (MFLCs, formerly Bence Jones proteins; discussed below). Heavy chains are normally secreted by plasma cells only as part of intact immunoglobulin molecules. Uncoupled heavy chains remain in the endoplasmic reticulum and are degraded.[9,10] Unlike normal heavy chains, however, normal light chains may be secreted as part of the intact immunoglobulin molecule, or as free light chains.[10,11] The excess free light chains that occur as monomers (mainly κ) with a molecular mass of about 25 kDa or as non-covalent dimers (mainly λ) with a mass of 50 kDa are small enough to pass through the glomerulus and are reabsorbed by the proximal convoluted tubules where they are catabolized.

When excessive amounts of free light chains are present because of an increase in immunoglobulin production, or glomerular or tubular damage, they are not completely reabsorbed by the tubules, resulting in a light chain proteinuria. If the excreted light chains are polyclonal, they migrate broadly in the β- and γ-regions. Plasma cells in the bone marrow of patients with polyclonal increases in the serum gamma globulins often produce large amounts of free polyclonal light chains. If one is unaware that polyclonal free light chains can be found under these conditions, an immunoelectrophoretic or immunofixation pattern could be mistaken for MFLC because of the relatively restricted heterogeneity of free polyclonal light chains (Fig. 7.2).

MEASUREMENT OF TOTAL URINE PROTEIN

The most common test for proteinuria is the dipstick. However a dipstick should never be used when screening for MFLC; dipsticks are adequate only as general screens for proteinuria. Further, in assessing urine for the presence of MFLC it is important to discard the use of the classic heat precipitation test of the urine. Most MFLC that are in the urine in large quantities will precipitate when heated to 56°C and upon further heating will dissolve; unfortunately, so will some polyclonal light chains. Further, the heat test misses at least 30–50 per cent of true monoclonal free light chains because some MFLC do not precipitate and dissolve under these conditions at any concentration and others are present in too small an amount to be detected by this insensitive method.[12] Although sulfosalicylic acid (SSA) is more sensitive than Albusticks (Bayer AG, Leverkusen, Germany), I have seen many cases of obvious MFLC proteinuria detected by urine protein electrophoresis and immunofixation that gave negative SSA tests. Therefore, when screening urine for MFLC, I recommend performing urine protein electrophoresis and immunofixation on concentrated urine (see below). Examination of the urine sediment for the presence of cells, casts, and crystals can be helpful in establishing etiology and prognosis. It is worth repeating since a negative dipstick does not rule out MFLC.

There are several methods currently in use in clinical laboratories to measure urinary proteins. A 24 h collection of urine provides a good sample on which to base the measurement of total protein. The classic biuret method is still used in a few laboratories to measure total urine protein. It reacts with peptide bonds and provides an equal measurement of both MFLC and the other urine proteins.[13] Other techniques do not detect MFLC to the same extent as they do other proteins. Although it does measure a variety of proteins with the same sensitivity, the biuret method is not amenable to automation.[14] Therefore, in recent years, a wide variety of other methods have been employed in clinical laboratories.

Because of their ease of automation, more common assays now used to quantify urine protein are turbidimetry by trichloroacetic acid (TCA),[15] TCA–Ponceau S,[16] and several dye-binding

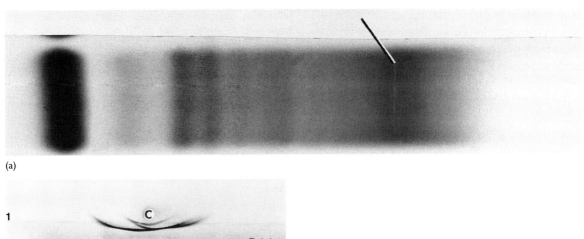

(a)

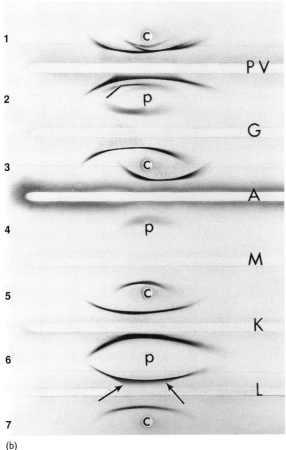

(b)

Figure 7.2 (a) Urine from a patient with chronic osteomyelitis and renal tubular damage. The pattern shows a combined glomerular and tubular proteinuria. Around the origin, a hazy density (indicated) may resemble a monoclonal free light chain. (Urine concentrated 100-fold; Panagel system stained with Coomassie Blue.) (b) Immunoelectrophoresis of urine from a patient with chronic osteomyelitis and renal tubular damage. Patient urine (P) alternates with control serum (C). Reaction with pentavalent (PV) antisera demonstrates a large arc, which extends into the trough, and two smaller arcs (the larger of the small arcs is indicated; the smallest arc is barely visible just above the P). The larger of the two small arcs corresponds with an arc that reacts with anti-IgG (G) and the other with anti-IgA (A). Typically, there is no reaction with anti-IgM (M), since this large molecule finds its way into the urine only with extreme glomerular damage. Note that both the anti-κ (K) and anti-λ (L) precipitin lines extend into the trough, indicating the great concentration of both of these light chains in the urine despite the rather tiny amounts of IgG and IgA heavy chain present. Therefore, they are free polyclonal light chains (a relatively common finding). The fact that the λ arc extends into the trough and looks a bit 'restricted' (arrows) means that it has been mistakenly interpreted as a monoclonal free light chain.

methods.[17,18] Unfortunately, the dye-binding methods differ in their ability to react with proteins of different charge. Therefore, the reaction varies with the albumin to globulin ratio. One aid to consistency in performance of dye-binding assays is standardization of the calibrator. Marshall and Williams[17] demonstrated that the use of a urinary protein calibrator improved the agreement between the Coomassie Brilliant Blue and the pyrogallol red–molybdate dye-binding methods. Lefevre et al.[18] further noted that there was less than a 15 per cent difference between pyrogallol red–molybdate and pyrocatechol–violet (UPRO Vitros; Ortho-Clinical Diagnostics, Raritan, NJ, USA) when measuring light chain proteinuria. However, in samples with glomerular proteinuria, they noted considerable differences in the total protein obtained.

No single particular automated method is consistently superior to another. Whichever method is chosen in the laboratory should be used for all samples on that patient in order to provide consistent data when following a patient. If two different methods are used on samples taken at two different times, any apparent increase or decrease may be caused by variability in the methods rather than a change in the patient's condition.

CONCENTRATION OF URINE SAMPLES

Urine samples need to be concentrated sufficiently to allow visualization of the protein bands in most techniques used by clinical laboratories for urine protein electrophoresis. Colloidal staining with gold or silver does not require concentration of urine samples.[19] These staining methods are about 200 times more sensitive than protein dye staining methods such as Ponceau S or Amido Black and others that are commonly used to stain urine protein electrophoresis.[20,21] However, colloidal methods are technically more laborious than the dye methods, not readily automated and give grainy results that may obscure subtle bands.

Most laboratories use commercially available differential filtration techniques to concentrate urine. Some are passive diffusion while others use centrifugation to speed the process.[21] The amount of concentrating required will vary depending on the technique used for electrophoresis and on the concentration of protein in a particular sample. For most agarose or cellulose acetate methods when stained by protein dyes, a 50–100 times concentration is usually sufficient for urine specimens that have low concentrations of protein. While some have suggested concentrating urine as much as 300–600 times to detect tiny quantities of MFLC,[22] such concentrating is technically laborious and can result in highly viscous samples that are difficult to process.[21] In samples with a large amount of proteinuria, too much concentration may overload the

gel. Yet, Kaplan and Levinson[13] caution against using only 10–20 times concentration, claiming it to be insufficient to detect some cases with MFLC. I have not found concentrating urine more than 50–100 times to be necessary to detect the pattern of proteinuria, or, as discussed below, clinically relevant amounts of MFLC by immunofixation.

The speed at which the stationary filtration devices concentrate the urine is inversely proportional to the concentration of proteins in the sample. Therefore, samples with large protein concentrations, such as those from patients with a non-selective proteinuria, will concentrate slowly. There is no reason to concentrate such samples even 50 times. Our laboratory places the urine samples into the concentrators and allows them to concentrate for up to 4 h. We monitor them to prevent ones that concentrate quickly from evaporating. Thus, one may concentrate samples for a standard period of time as long as one records the final concentration.

Concentration of urine is especially important to detect small, but clinically relevant quantities of MFLC by immunofixation. Detection of even small quantities of MFLC may have importance in cases of amyloid AL (amyloid associated with immunoglobulin light chain) or B-cell lymphoproliferative disorders (see Chapter 6). To rule out amyloidosis, all adult patients with nephrotic syndrome should have an immunofixation of both serum and urine.[23] Unfortunately, with current techniques, not all cases of amyloid AL will have positive urine samples for the presence of MFLC by immunofixation,[24,25] and biopsy is still the main method of diagnosing these cases. However, the advent of new, highly sensitive and specific nephelometric methods to quantify free light chains in both serum and urine (see below) has provided a method that is capable of detecting many cases of amyloid AL that formerly were not detected by urine protein electrophoresis or even by immunofixation.[26]

The practical limit of detection of MFLC by immunofixation is related to the background of polyclonal free light chains that occur in the urine.[27] The limited heterogeneity of these molecules results in a banding pattern that makes reliable distinction of MFLC at concentrations similar to those of the bands found with polyclonal free light chains unreliable with currently available techniques. Because we readily see the ladder pattern on 50× concentrations of urine with even the small quantities of protein present in cases of mild tubular proteinuria, our laboratory does not concentrate samples more than that.

PROTEINURIA AFTER MINOR INJURY

Relatively minor injury has been associated with transient proteinuria. Vigorous physical exercise may cause significant increases in the concentration of albumin in the urine.[28] Congestive heart failure, normal pregnancy, heavy alcohol consumption and reactions to drugs are all associated with minor, usually transient proteinuria.[29-33] Patients with non-renal febrile illnesses also may have proteinuria.[34-36] During such episodes, usually only an albumin band is seen on electrophoresis. However, cases with a combined tubular and glomerular proteinuria have been reported.[37] The etiology of the proteinuria in these conditions is unclear. Within 2 days after such illness, most of the proteinuria has resolved. Pre-eclampsia and gestational hypertension are associated with relatively large amounts of proteinuria that may exceed 25 g/l. These patients may have glomerular, glomerular with tubular leakage and even nonselective glomerular patterns (see below).[38-41]

OVERFLOW PROTEINURIA

When large amounts of relatively low molecular weight proteins are present in the serum, they are cleared through the glomeruli and overwhelm the ability of the tubules to reabsorb them. Therefore, this condition is termed overflow proteinuria.[1,42,43] It is commonly seen in urine from patients with multiple myeloma, where large amounts of MFLC are produced. However, other conditions can result in the production of large amounts of protein that may mimic MFLC in the urine.

Massive hemolysis and crush injuries will result in the release of large quantities of hemoglobin and myoglobin, respectively, which will produce a single large band in the urine (Fig. 7.3).[44] Among other small molecules, β_2-microglobulin, eosinophil-derived neurotoxin and lysozyme have been reported to cause a band in protein electrophoresis that could be mistaken for MFLC (Fig. 7.4).[45-47] This is why immunofixation or measurement of free light chains should be used to evaluate unexplained bands seen on urine protein electrophoresis to avoid an erroneous interpretation of MFLC.

GLOMERULAR PROTEINURIA

Severe renal disease results in a profound proteinuria, the characteristics of which can be defined by urine protein electrophoresis in many cases. In general, patients with glomerular disease excrete considerably more protein than do patients with tubular disease. This reflects the pathophysiology in which tubular dysfunction prevents absorption

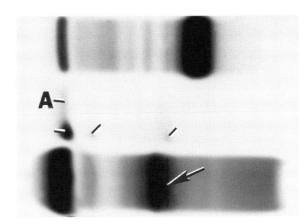

Figure 7.3 The top urine sample has a tubular proteinuria pattern with a massive monoclonal free light chain (MFLC) in the β–γ region. The middle urine sample contains a small amount of albumin (A), but has been contaminated by the sample below. This results in prominent staining in the albumin, α₁- and β-regions only in the lower portions of the second lane (indicated). The bottom sample is from a urine with considerable hemolysis and has a prominent hemoglobin band (arrow) that cannot be distinguished from an MFLC without an immunofixation. (Paragon SPE2 system stained with Paragon Violet.)

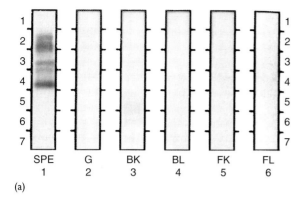

(a)

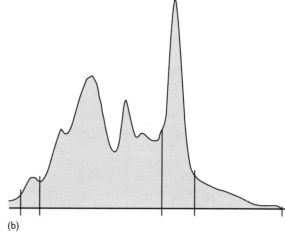

(b)

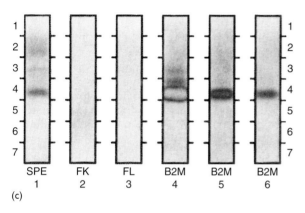

(c)

Figure 7.4 (a) Immunofixation of urine demonstrates a prominent β-region band in the serum protein electrophoresis (SPE) (acid-fixed) lane. However, no immunoglobulins are demonstrated by anti-IgG (G), anti-bound κ (BK), anti-bound λ (BL), anti-free κ (FK) or anti-free λ (FL). (b) The densitometric scan indicates the size of the β-region band. (c) Immunofixation performed with antibodies against β_2-microglobulin with the urine at decreasing concentrations in lane 4 (note the prominent antigen excess effect at this higher concentration of urine); lanes 5, and 6 demonstrate the identity of this band. Figure contributed by Beverly C. Handy,[47] and used with permission from *Archives of Pathology and Laboratory Medicine*.

of the few small proteins that normally pass through the glomerulus, whereas glomerular disease literally opens the floodgates to serum proteins of 50 kDa or more.

A wide variety of conditions result in sufficient damage of the glomerular capillary wall to permit large molecules to pass into the glomerular filtrate. Diabetes mellitus and immunological renal diseases make up the vast majority of the cases with glomerular proteinuria.[2] When glomerular damage occurs, large molecules pass into the glomerular filtrate along with smaller proteins. The smaller proteins and some of the larger molecular weight proteins can be reabsorbed by the renal tubules. However, the absorptive capacity of the tubules is limited to approximately 1 g of protein over 24 h.

Urine protein electrophoresis can distinguish glomerular from tubular damage.[48,49] In glomerular proteinuria patterns, large molecules such as albumin, α_1-antitrypsin, and transferrin predominate

(Fig. 7.5).[50] When sufficient quantities of protein are lost into the urine, the patient suffers from the nephrotic syndrome and displays the severe protein loss pattern on serum protein electrophoresis (see Chapter 5).

All glomerular damage is not the same. The amount of proteinuria and the size of the molecules permitted to pass through the glomerular capillary walls will depend on the degree of injury to the glomeruli. A useful measure of glomerular selectivity (the degree of damage) is the selectivity index.

The concept of 'selectivity' refers to the ability of the glomerulus to permit only smaller proteins for passage through the glomerular basement membrane. If very large molecules, such as IgG, IgM or α_2-macroglobulin leak through the glomerulus in similar proportions to smaller molecules such as albumin or transferrin, then the proteinuria is termed 'non-selective'; that is, the glomerulus is not able to discriminate between these two sizes. The

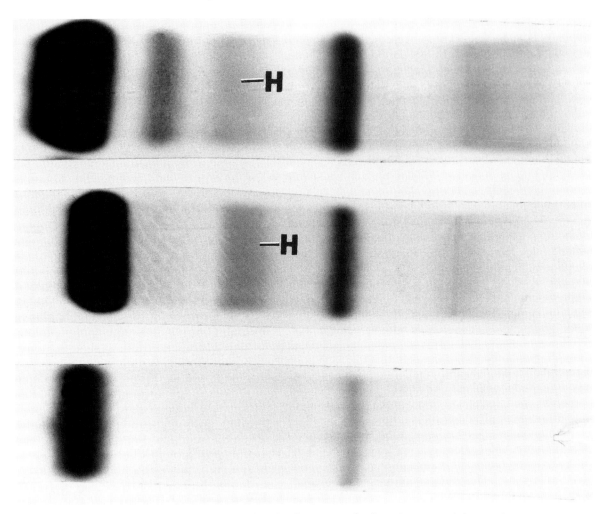

Figure 7.5 Three examples of glomerular proteinuria. Note that the urine samples do not line up exactly because they were electrophoresed on three separate runs. In all three, the main protein excreted is albumin with varying amounts of α_1-antitrypsin, transferrin and other globulins. The top and middle samples have moderate amounts of haptoglobin (H), indicating that the glomeruli have lost some degree of selectivity. The discrete smaller bands which are seen in the α- and β-regions of patients with tubular proteinuria are not seen. The middle sample has a small restriction around the origin that needs an immunofixation. (Top two urine samples concentrated 25-fold; bottom urine concentrated 100-fold. Panagel system stained with Coomassie Blue.)

greater the glomerular damage, the less selective is the proteinuria. The degree of glomerular damage correlates with the level of the selectivity index (SI).[51]

The SI is estimated from the ratio of the clearance of one of these large molecules to that of albumin. Several studies have examined the implications of the selectivity index as a marker for patients with a variety of glomerular diseases.[52–54] Tencer et al.[52,53] reported that by using SI with IgM or α_2-macroglob-

ulin they were able to distinguish cases with minimal change nephropathy and other transient conditions from those with crescentic glomerulonephritis. The selectivity index based on IgG was not as useful because of its lower molecular weight.[52] The formula for the selectivity index using a specific high molecular weight (MW) analyte is:

$$\text{Selectivity index} = \frac{(\text{high MW analyte ur/albumin ur})}{(\text{high MW analyte ser/albumin ser})}$$

where: high MW ur = high molecular weight analyte concentration in urine; albumin ur = albumin concentration in urine; high MW ser = high molecular weight analyte concentration in serum; and albumin ser = albumin concentration in serum.

The total amount of proteinuria increases with decreasing selectivity of the glomerulus. Typically, mild glomerular damage (mainly albumin) has less than 1500 mg/24 h, moderate damage (selective glomerular proteinuria) has from 1500–3000 mg/24 h and non-selective patterns have greater than 3000 mg/24 h. When a random urine sample has been obtained, the ratio of total protein to creatinine for low-grade proteinuria (mainly albumin) is 0.2–1.0, moderate proteinuria (selective glomerular proteinuria) is from 1.0 to 5.0 and values > 5.0 are seen in non-selective patterns (nephrotic).[6] In a case of minor glomerular damage where selectivity is maintained, albumin, α_1-antitrypsin, occasionally α_1-antichymotrypsin and transferrin are found on electrophoresis of concentrated urine (Fig. 7.5). In the most severe cases of glomerular damage, selectivity is lost and the urine protein electrophoresis resembles the normal serum protein pattern. A non-selective pattern, then, will also display the large haptoglobin, α_2-macroglobulin, and intact immunoglobulin molecules in addition to the molecules seen in selective glomerular proteinuria.

Some laboratories use sodium dodecyl sulfate–polyacrylamide gel electrophoresis (SDS-PAGE) to evaluate urine protein excretion.[55] It delineates the molecular weight of proteins escaping from the damaged nephron, providing information about the specific small proteins in the urine from patients with tubular damage.[56–62] As such, it allows one to distinguish individuals that have combined selective glomerular and tubular proteinuria, from those that have non-selective proteinuria. Using both SDS-PAGE and fractional excretion of α_1-microglobulin to divide glomerular selectivity into high (minimal damage), moderate or low selectivity, Bazzi et al.[57] reported that the amount of glomerular selectivity correlates with final outcome and with response to therapy. However, SDS-PAGE remains a more laborious technique than routine agarose or acetate electrophoresis and lacks a ready method for automation. A relatively simpler technique employing SDS-agarose gel electrophoresis to separate the urine proteins by molecular weight is now available for the same purpose.[63] This technique has the advantage of dispensing with the concentration step required for routine urine protein electrophoresis on agarose or cellulose acetate. However, this technique is not as sensitive to detect MFLC as immunofixation on concentrated urine.[64]

TUBULAR PROTEINURIA

When there is damage to the tubule reabsorptive function, small proteins in the glomerular filtrate (α_1-acid glycoprotein, α_1-microglobulin, β_2-microglobulin, γ-trace protein, and retinol-binding protein) pass into the urine.[65,66] Because of their low concentration in serum compared with other proteins these small molecules are undetectable when the serum is examined by electrophoretic techniques used in clinical laboratories. However, when they pass into the urine unencumbered by the more plentiful larger molecules, they can be visualized on concentrated specimens studied by electrophoresis (Fig. 7.1).

Acute tubular necrosis, the most common cause of acute renal failure, can result from a wide variety of injuries.[67–69] Some forms are reversible, such as those caused by transient ischemia, exposure to heavy metals, toxins, or radiocontrast dyes.[68,70–74] Chronic causes include congenital Fanconi syndrome, Dent disease (an X-linked condition caused by inactivation of the renal chloride channel gene), and acquired Fanconi syndrome (the most common cause of which is myeloma kidney).[75,76] Light chain Fanconi syndrome is often associated with intracytoplasmic crystals formed from MFLC in the proximal convoluted tubules,[77] although some cases of Fanconi syndrome caused by light chain deposition do not form crystals, perhaps owing to the absence of side-chains in the CDR-L3 loop which are needed for dimer formation.[78]

Patients with tubular disease have a normal

glomerular selectivity because only small molecules pass into the glomerular filtrate. These molecules bind to receptors on the proximal convoluted tubules. After they are reabsorbed, the proteins are catabolized by the tubular epithelium into amino acids that are reused by the body.[79] When the tubules are damaged, these small molecules pass into the urine.

The amount of proteinuria that results from tubular damage is only a fraction of that seen in glomerular disease. In tubular damage, traces of albumin accompany the smaller molecules into the urine (Fig. 7.6). Because of the relatively small amounts of proteinuria resulting from tubular disease (1–3 g/24 h), the urine must be concentrated 50× for the protein to be visualized by electrophoresis. The albumin band is considerably fainter than that seen in glomerular proteinuria. Typically, it will be similar in intensity, to slightly more intense, than the α_1-microglobulin and β_2-microglobulin bands (Fig. 7.7).

When examining electrophoretic patterns from patients with tubular disease, one finds considerable variation because the amount of damage differs from one patient to another. However, a

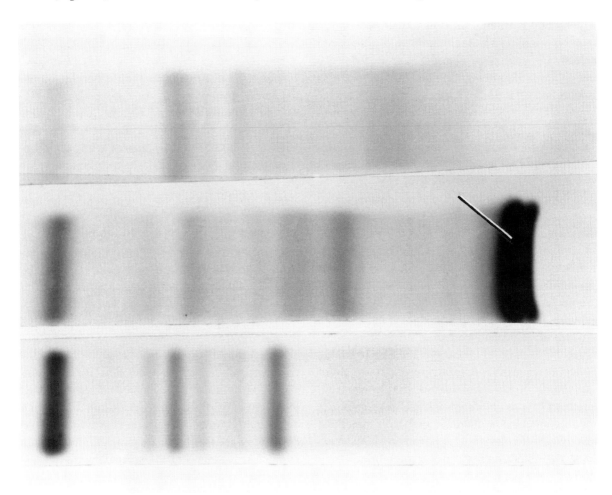

Figure 7.6 Three examples of tubular proteinuria. Note again, that these three samples were performed on three different runs and have slightly different migrations. With tubular proteinuria, the glomeruli are relatively intact and do not allow large amounts of albumin and other large molecules to pass into the glomerular filtrate. Therefore, albumin is a relatively minor component, or slightly greater than the many smaller molecules seen in the α- and β-regions. Note that the middle sample is from a patient with a κ monoclonal free light chain (indicated). This sample has two fused bands perhaps due to the formation of both monomer and dimer free monoclonal κ chains. (Urine concentrated 100-fold. Panagel system stained with Coomassie Blue.)

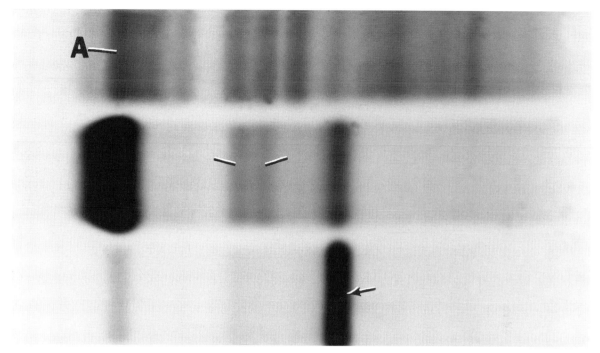

Figure 7.7 Three urine samples with different patterns for comparison on the same gel. The top sample shows a striking tubular proteinuria pattern with an albumin band (A) that stains no darker than the multiple bands in the α-, β- and γ-regions. The middle sample is a predominantly glomerular pattern of proteinuria, but the two bands in the α_2-region (indicated) are consistent with a coexisting tubular leakage, because they correspond to the type of staining seen with the α_2 microglobulins rather than the broad haptoglobin pattern shown in Fig. 7.5. The bottom sample has barely discernable albumin and α_2-region bands (just below the indicated bands in the middle sample). The strong band which lines up with the transferrin region (arrow) is an monoclonal free light chain. It would be inconsistent to have such a large transferrin band with such a minor tubular proteinuria and almost no albumin. (Urine concentrated 100-fold. Paragon SPE2 system stained with Paragon Violet.)

consistent finding is that albumin is much less dominant than in glomerular proteinuria. Individuals with chronic renal disease will often have a combined glomerular and tubular pattern indicative of widespread damage to the nephron (Fig. 7.8). In these patients, the albumin content resembles that of glomerular proteinuria, but the smaller bands in the α- and β-region can also be observed. The occurrence of combined glomerular and tubular proteinuria may be obscured in urine from individuals that have a non-selective glomerular leakage pattern. In those individuals, either the use of SDS-PAGE (see above) or measurement of specific small molecular weight proteins (see below) may be needed to detect the coexistence of tubular dysfunction. Since SDS-PAGE provides a separation of the molecules on the basis of molecular weight, it can be useful to make this distinction in samples with nonselective pattern.

Rarely, one may see analbuminuria in cases of tubular proteinuria. Sun et al.[80] reported such a case in a diabetic, hypertensive patient with tubular proteinuria. They hypothesized that the patient may have had end-stage kidney disease such that the glomerular filtration rate was so low as to allow practically no protein through the glomeruli. Yet, the tubules may have leaked the smaller proteins, possibly as a result of analgesic abuse in that case (Fig. 7.9).

Quantification of low molecular weight proteins may be used as an indication of tubular proteinuria. This may be especially important as an assessment of renal function in patients with multiple myeloma.[81] The urinary protein most often used to evaluate this is β_2-microglobulin, a molecule that

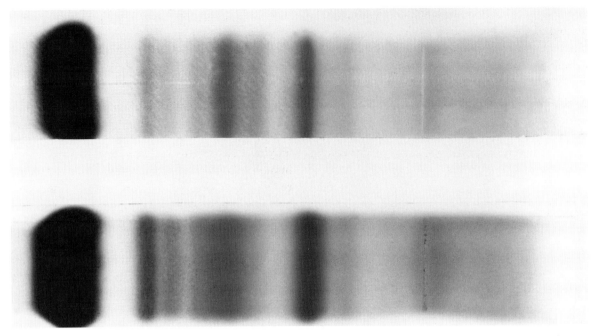

Figure 7.8 Two samples from patients with combined glomerular and tubular proteinuria. Whereas albumin predominates in both samples, many discrete bands are present in the α- and β-regions. (Urine concentrated 25-fold. Panagel system stained with Coomassie Blue.)

has homology with immunoglobulins.[82–86] β_2-Microglobulin migrates between transferrin and C3 on electrophoresis. While it is elevated in tubular proteinuria, it is unstable at pH < 6.5 and, therefore, reliance solely upon quantification of urine β_2-microglobulin to detect tubular damage or dysfunction may result in a false negative.[87,88] α_1-Microglobulin is another plasma glycoprotein that passes freely into the urine. Because it is relatively stable at the acid pH of urine some have suggested that it may serve as a better indicator for renal tubular dysfunction than β_2-microglobulin.[89–91]

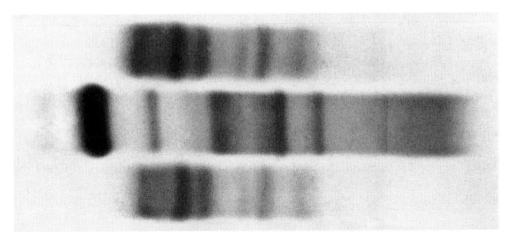

Figure 7.9 Urine (top and bottom lanes) and serum (middle lane) are from a patient with diabetes, hypertension, and renal impairment. Despite the presence of considerable tubular proteinuria, an albumin band is not present in this patient's urine. Figure contributed by Tsieh Sun.[80]

However, Donaldson found that although β$_2$-microglobulin is more susceptible to denaturation under acidic conditions, other marker proteins, including α$_1$-microglobulin, also deteriorate below pH 6.0 and suggested that routine alkalinization of urine upon voiding be employed to enhance detection of these proteins.[88] In addition to these two molecules, other suggested markers for detecting tubular proteinuria include retinol-binding protein, lysozyme, and N-acetyl-β D-glucosaminidase.[65,83,92,93]

FACTITIOUS PROTEINURIA

The protein bands identified can also give information about factitious proteinuria.[94–97] For example, in cases reported by Tojo et al.[95] and Sutcliffe et al.,[96] the electrophoresis of urine demonstrated unusual protein bands. The albumin migrated almost in the α$_1$-antitrypsin region and the other major bands also did not line up. The problem was species: immunofixation confirmed that the protein was egg albumin. Therefore, be suspicious when protein bands do not migrate correctly; ask for a new sample under controlled collection conditions. Identification of the abnormal band is the key to correct diagnosis of factitious proteinuria. However, before suggesting this diagnosis, recall that denaturation also may produce unusual bands. Denaturation of proteins in urine may occur because of deterioration of a sample with time, bacterial overgrowth, the presence of proteases from leukocytes, or laboratory errors such as the accidental addition of a biuret reagent prior to electrophoresis. When abnormal bands are seen that are not monoclonal gammopathies, hemoglobin, myoglobin, lysozyme, eosinophil-derived neurotoxin, or β$_2$-microglobulin, a repeat fresh urine sample is often the best approach.

MONOCLONAL FREE LIGHT CHAINS

In this book, I am encouraging the use of the term MFLC instead of the traditional term, Bence Jones

proteins. This reflects confusion that I have observed when some physicians equate the presence of a monoclonal intact molecule, such as IgG, in the urine with a Bence Jones protein. It is not. Further, it does not have the same significance for the patients. Monoclonal free light chains are associated with a greater likelihood that the patient has a dire condition such as multiple myeloma or amyloid AL than does an intact monoclonal immunoglobulin. They are more likely to damage renal tubules and deposit in glomeruli, tubules and other locations.

Monoclonal free light chains have been referred to as 'Bence Jones proteins' to recognize the contribution made by Henry Bence Jones. His description of the first reported case of multiple myeloma recorded the characteristic of MFLC to precipitate when weakly acidified urine that contained them was warmed to 40–58°C, only to redissolve upon heating to 100°C.[98] He called the protein a 'hydrated deutoxide of albumen' in that paper. Dr Thomas Watson referred the patient, a 44-year-old man who complained of chest pains after a fall, to Dr William MacIntyre. Doctor MacIntyre studied the urine, noting its peculiar thermal characteristics and sent the specimen to Dr Jones who reported his studies in 1848. Two years later, MacIntyre reported the urine characteristics in detail.[99,100] Doctor John Dalrymple examined the slides of the bone marrow from the autopsy on this patient. He noted a large number of nucleated cells that varied in size and shape, also noting irregular cells with two or three nuclei (likely the malignant plasma cells).[101,102]

Doctor Jones was a prominent physician and during his lifetime was described by Florence Nightingale as being 'the best chemical doctor in London'.[103] Another odd fact concerns the hyphen that is occasionally placed between his middle and last names. Although one often sees his name hyphenated, he did not hyphenate his name in any of the 40 papers and books that he wrote.[102] Further, in his contemporary reference books, he was listed under Jones.[102] The hyphen was added by some of his descendants.[102] I presume they were members of the 'Bence' clan. The final point of note is that the test he and Dr MacIntyre used was quite insensitive and should never be used today to detect MFLC.

In the 1960s, the recognition that Bence Jones proteins in the urine were homogeneous populations of immunoglobulin fragments played an important role in the delineation of antibody structure.[104] Although the normal product of plasma cells is an intact immunoglobulin, these cells also produce excess free light chains.[11] This differs from heavy chains that normally are secreted only as part of intact immunoglobulin molecules. When light chains are not present, the normal heavy chains remain in the endoplasmic reticulum and are degraded.[9,10] Unlike normal heavy chains, however, normal light chains may be secreted as part of the intact immunoglobulin molecule, or as free light chains.[10] Most of the κ free light chains are secreted as monomers (25 kDa), although dimerization certainly also occurs, whereas most λ light chains exist as dimers (50 kDa).[10,105] Both monomeric and dimeric light chains pass through the glomerulus and are reabsorbed by the proximal convoluted tubules where they are catabolized. Because of their different sizes, however, free κ chains have a clearance rate of approximately three times that of λ light chains.[105]

When there is a polyclonal increase in immunoglobulin production, such as occurs in chronic inflammation and some autoimmune diseases, the amount of polyclonal free light chains produced also increases and, together with other low molecular weight proteins, they may surpass the ability of renal tubules to reabsorb them. This results in the presence of overflow proteinuria of polyclonal free light chains in the urine, which may produce a 'ladder pattern' on examination of the urine by immunofixation (see below).[27]

There are many disorders of plasma cell and lymphoplasmacytic cells where proliferation of a single clone occurs, termed plasma cell dyscrasias. The variety of conditions parallels the broad spectrum of their clinical significance. At the one end of the spectrum is multiple myeloma and at the other end is monoclonal gammopathy of undetermined significance (MGUS). In between are Waldenström macroglobulinemia, amyloid AL, B-cell non-Hodgkin lymphoma and B-cell leukemia. These conditions may result in production of excessive amounts of MFLC from the clone of plasma cells (or lymphoplasmacytic cells in the case of lymphoma, leukemia and Waldenström macroglobulinemia). The free light chain products may exist as monomers, dimers, tetramers, or light chain fragments. Occasionally, they may bind to other serum components.[21,106] Patients suspected of having amyloid AL should have urine immunofixation to identify MFLC.[23]

In myeloma, when large amounts of the MFLC are present, they overwhelm the capacity of the proximal convoluted tubules and can be detected in the urine (overflow proteinuria). It is important to separate polyclonal free light chains from MFLC for diagnostic purposes. The presence of MFLC in urine samples often portends a poorer prognosis, as this protein is much more frequently seen in association with myeloma (50–60 per cent) and amyloidosis (60 per cent) than it is with 'benign monoclonal gammopathy' (14 per cent) that today would be termed MGUS.[107–109]

RENAL DAMAGE CAUSED BY MFLC

The pathology resulting from such proliferation of plasma cells relates to the lesions created by the plasma cells themselves, interference with the production of normal bone marrow elements, and damage caused by the monoclonal immunoglobulin products of the clone on various organ systems. Because the free light chains and their fragments readily pass through the glomerulus and are reabsorbed by the renal tubules, a major portion of the pathology resides in the kidney.[110] Monoclonal free light chains may damage renal tubules directly or indirectly by facilitating release of intracellular lysozymes.[26,111] This results in dysfunction of the proximal convoluted tubules. In addition, light chain cast nephropathy or deposits of light chains in the glomeruli or around the tubules as light chain or as amyloid AL also result in renal dysfunction.[106,110,112] Almost all patients who have

> 1 g/24 h of MFLC in the urine suffer from renal tubular dysfunction.[26,113]

When the proximal convoluted tubules are damaged, the patients suffer Fanconi syndrome.[114] This manifests as a loss of amino acids, glucose, phosphate and bicarbonate in the urine.[114–116] The MFLC-mediated damage to the proximal convoluted tubules of patients with multiple myeloma is the most common cause of acquired Fanconi syndrome.[76] In some of these cases, crystals can be found in the renal tubules (proximal tubules, distal tubules and collecting ducts).[117,118] These patients can have a paradoxically low serum calcium due to loss into the urine.[119] After therapy for myeloma and decline of MFLC, some cases of Fanconi syndrome have shown improvement in tubular function.[120]

There have been several studies indicating the potential toxic effects of MFLC on kidney tubules. Batuman et al.[116] used MFLC purified from patients with multiple myeloma to study their effect on transport of phosphate and glucose by cultured rat proximal tubule cells. They reported that both κ and λ MFLC were able to inhibit the uptake of these analytes in a dose-dependent manner, whereas albumin had no such effect. This group later demonstrated that Na–K-ATPase activity of primary cell cultures from rat proximal convoluted tubule cells also was inhibited by monoclonal free λ light chain.[114] Na–K-ATPase is an important part of the physiological mechanism for the sodium gradient in these cells. Further, Pote et al.[112] used cultures of human proximal convoluted tubules to demonstrate a direct toxic effect of MFLC on the ability of these cells to proliferate as well as inducing apoptosis within 2 days of exposure to the light chains *in vitro*. These studies indicate that a direct toxic effect by the MFLC on proximal convoluted tubules may explain the occurrence of Fanconi syndrome in some patients with plasma cell dyscrasias.

Both the distal tubules and collecting ducts are involved in myeloma cast nephropathy. They contain prominent renal casts and renal function suffers from loss of concentrating ability and acidification.[112] The third complementarity-determining region (CDR3) of both κ and λ MFLC binds to Tamm–Horsfall proteins resulting in the formation of the dense cast material present in the distal tubules and collecting ducts in myeloma kidney.[8,112,121,122] The presence of these casts in the distal tubules is associated with dilatation of the tubular lumen, atrophy of the tubule epithelium, and significantly worsens the survival compared with patients with pure light chain deposition disease.[106,110] In addition to the light chain–Tamm–Horsfall protein complex, casts contain monocytes, lymphocytes, multinucleated giant cells, occasionally tubular epithelium, and neutrophils.[106] Unfortunately, these consequences may recur in patients that receive renal transplantation. Short et al.[123] suggested that the histological pattern of damage in the patient's kidneys helps to predict the outcome of subsequent renal allograft. Individuals that had light chain deposition disease with a proliferative glomerulonephritis in their native kidneys had worse graft survival than those with cast nephropathy.[123] The presence of light chain nephropathy is a serious complication that should cause the clinician to consider the implementation of early dialysis.[124]

Deposits of the MFLC as either fibrillar amyloid AL or non-fibrillar light chain deposition disease (LCDD) may occur in glomeruli or around tubules in extracellular spaces.[106,110] The same distribution is found in cases of heavy chain deposition disease (HCDD). Amyloid AL deposits are irregular in their distribution, Congo Red-positive and display the characteristic apple-green birefringence under conditions of polarized light (see Chapter 6). In contrast, LCDD deposits usually follow basement membranes, are not fibrillar and are Congo Red-negative.[106]

DETECTION AND MEASUREMENT OF MFLC

Guidelines for clinical and laboratory evaluation of patients with monoclonal gammopathy recommend that detection of MFLC is best achieved by immunofixation of concentrated urine.[125,126] The measurement of MFLC once demonstrated by immunofixation currently requires quantification of the MFLC in a 24-h collection of urine.[125,126] However, these guidelines were written before sensitive and specific automated assays for detecting free light chains (FLC) in serum were readily available. The availability of these assays (see below) makes it possible to follow MFLC by convenient and consistent serum samples. Because most laboratories are still using the urine electrophoresis and immunofixation tests at the time of this writing, I will review evaluation of urine in some detail. However, all laboratories should be following the emerging literature on automated measurement of FLC in serum and consider incorporating them into their evaluation of patients with monoclonal gammopathies.

DETECTION OF MFLC IN THE URINE AND SERUM BY ELECTROPHORESIS AND IMMUNOFIXATION

There has been some controversy regarding the optimum specimen to use for detection of MFLC. Although many authorities recommend a 24-h urine collection for the initial detection of MFLC,[125–127] others accept random urine samples for analysis.[128] Brigden and coworkers[128] reported that an early morning voided sample was as good as, and perhaps superior to, a 24-h sample. They noted, however, that random samples collected at times other than the early morning void were clearly inferior for the initial detection of MFLC (Table 7.1). The early morning voided sample has

been concentrated naturally and will provide excellent material for study. Note, however, that if a patient has a random sample taken at another time, a positive result is useful, but a negative result will not rule out MFLC. Indeed, even a negative result on an early morning void may engender a repeat analysis on a second sample from a clinician that has a high index of suspicion. Alternatively, the laboratory now may suggest the use of FLC assays on serum in cases with a high index of suspicion (see below). The new automated serum FLC assays have been reported to detect abnormal κ/λ ratios in a large number of patients that had been thought to have non-secretory myeloma.[129]

After the initial detection of a MFLC on the early morning urine, I recommend a 24-h sample to quantify the amount of MFLC. This is especially important in patients with light chain disease, as the amount of urine MFLC is a key indicator of tumor burden and response to therapy. Once patients have been informed that a monoclonal protein is present in their urine, they should be sufficiently motivated to provide a reliable 24-h collection. Once again, the new automated serum FLC assays (discussed below) have been shown to provide good information to follow patients with MFLC and may be more convenient and perhaps more accurate than the urine assays.[26] Recently, Salomo et al.[130] reported the use of high-resolution Sebia Hydragel HR (Sebia, Issy-les-Moulineaux, France) agarose electrophoresis on unconcentrated urine to measure the amount of MFLC. They found that they could estimate the concentration of MFLC in urine by comparing the densitometric scans of staining intensities of the MFLC bands relative to the staining intensities of albumin solutions.

Table 7.1 Detection of monoclonal free light chains in urine samples from patients with multiple myeloma[a]

24-h Sample	+ Early morning	– Early morning	+ Random sample	– Random sample
Positive = 17	17	0	14	3
Negative = 3	2	1	0	3

[a]Data from Brigden et al.[128]

For initial detection of the MFLC in urine, we first review our files on that patient to determine if we have a previous serum protein electrophoresis or immunofixation study demonstrating an M-protein. If so, we use that information to determine the set-up for the immunofixation. For example, if the patient is known to have an IgGκ M-protein, we perform immunofixation for IgG, κ and λ on the initial urine study. If there is no history, we perform electrophoresis on concentrated urine and immunofixation for κ and λ. I prefer to use antisera against total (bound and free) κ and against total λ for these studies rather than antisera that reacts solely with FLC. An MFLC will usually have different electrophoretic migration and/or concentration from the intact immunoglobulin molecule. Therefore, as shown in Fig. 7.10, we report the MFLC when the light chain band has a distinctly different electrophoretic migration than the heavy chain band. In cases where the intact immunoglobulin and the light chain overlap in migration, they typically differ sufficiently in concentration that it

is not necessary to use antisera against FLC (Fig. 7.11). In instances where the intact immunoglobulin and the light chain have both the same migration and the same concentration, it is not possible to rule out MFLC by this method. In those cases, laboratories may wish to perform immunofixation with antisera specific to FLC.[21] We prefer the antisera to total light chains because we have found them to be stronger than the anti-FLC antisera and more sensitive in detecting small MFLC. Some of my evaluations of antisera against FLC have found their specificity to be disappointing.

If no suspicious bands are present, the sample is reported to be negative for MFLC. However, if one or more M-protein bands are seen, the analysis is repeated with antisera against the most common heavy chains found in the urine (i.e. IgG and IgA). If this is a 24-h urine sample, we perform a total protein measurement (see above) and use a densitometric scan of the known location of the MFLC (correlating it with the immunofixation) to quantify the 24-h MFLC content of the urine (Fig. 7.12). If

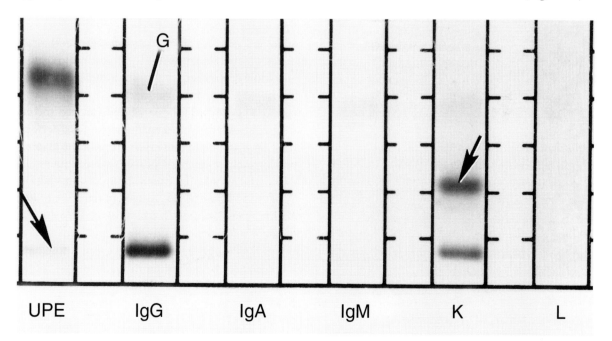

Figure 7.10 Immunofixation on 50-fold concentrated urine. The urine protein electrophoresis (UPE) lane (fixed in acid) demonstrates an albumin band and a tiny band in the slow γ-region (arrow). This band is identified as IgG and κ by reactivity in those two lanes. In addition, however, an anodal band is present in the κ (K) lane that does not react with any heavy chain (arrow). This is a monoclonal free light chain (MFLC). Note also at the anodal end of the IgG lane is a partial fragment of IgG (G). (Beckman Paragon system stained with Paragon Violet; anode at the top.)

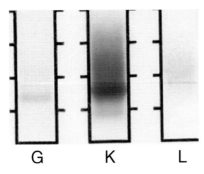

Figure 7.11 Immunofixation on 50-fold concentrated urine. The IgG lane shows a faint band. The κ (K) lane has a monoclonal free light chain (MFLC) that migrates only slightly faster than the band in the IgG region, but is broader and darker in staining intensity. Although they have similar migrations, I consider this to be MFLC rather than intact IgG κ. The λ (L) lane shows diffuse staining. (Sebia immunofixation.)

it is a random sample, we report the type of MFLC and intact immunoglobulin and recommend that a 24-h sample be obtained to quantify the M-protein.

All urine samples suspected of having MFLC should be evaluated by immunofixation.[125,126,131] We

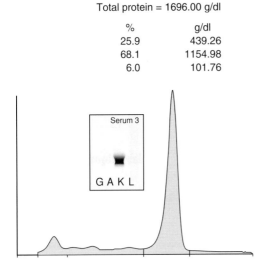

Figure 7.12 Densitometric scan of 50-fold concentrated urine with a small albumin band to the left and a large monoclonal free light chain (MFLC) to the right. The immunofixation of the urine (the label serum 3 is from the manufacturer's kit, but this is obviously not serum since there is no staining at all in the IgG lane) demonstrates a large κ MFLC. The protein concentration circled is measured in mg/dl. (Sebia β$_{1,2}$ gel and Sebia immunofixation.)

have found urine from patients with multiple myeloma who had MFLC detected only by immunofixation of the urine. Even more sensitive techniques than immunofixation have been used to detect MFLC in clinical laboratories. Immunoblotting has been recommended as an alternative to immunofixation.[132] Using this technique, Pezzoli and Pascali[22] concentrated urine up to 300-fold and found that they could identify B-cell lymphoproliferative disorders with small amounts of MFLC. However, as discussed above, our laboratory does not concentrate urine greater than 50–100 times.

Whereas MFLC in the urine are usually associated with multiple myeloma, MGUS or other B-cell lymphoproliferative processes, Pascali and Pezzoli have reported their occurrence in the urine from 32 per cent of the 28 patients they studied who had multiple sclerosis.[133,134] Similarly, Mehta et al.[135] found increased quantities of free light chains both in patients with multiple sclerosis and in individuals with other neurological diseases. Perhaps the new automated immunoassay for FLC (see below) will be able to shed more light on these interesting findings.

Monoclonal free light chains may be seen in the *serum* protein electrophoresis in four circumstances. (1) Most commonly, small amounts of MFLC accompany the intact monoclonal protein in the serum (Fig. 7.13). These may be overlooked and are often hidden by the other β-region bands. Therefore, the hypogammaglobulinemia that often accompanies light chain myeloma should prompt investigation of the urine and serum by immunofixation (Fig. 7.14). (2) When MFLC occur as tetrameric light chains, they are too large to pass through the glomerular basement membrane and usually produce a spike in the serum. (3) As a result of renal damage with sufficient loss of nephrons to reduce clearance of MFLC in the serum. (4) The MFLC may bind to other serum proteins including transthyretin, albumin, α$_1$-antitrypsin, and transferrin (Fig. 7.15).[136,137] The last can usually be demonstrated by finding numerous bands by immunofixation that resolve upon treatment with 2-mercaptoethanol (as described in Chapter 6).

Occasional cases of cryo-Bence Jones proteins

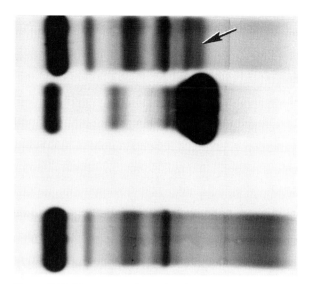

Figure 7.13 The serum in the top lane has a small band just cathodal to the C3 band. This subtle band (arrow) could have been overlooked. The second sample is the urine concentrated 100-fold from the same patient. An obvious monoclonal free light chain is present. The third sample is from a normal urine concentrated 100-fold. The lane appears empty. The bottom sample is a serum with a polyclonal increase in the γ-region. (Paragon SPE2 system stained with Paragon Violet.)

have been reported in the serum.[138–145] These usually give an artifact at the origin. Of course, when a monoclonal light chain is identified in the serum, the laboratory should rule out the possibility of IgD or IgE heavy chain.

Although only a handful of patients with pure tetrameric light chain disease has been reported,[146–152] Solling et al.[148] found that 25 per cent of patients with κ-secreting myelomas have detectable tetrameric kappa chains in their serum (coexisting with dimeric and monomeric forms). The reported cases of tetrameric light chain disease are dramatic because MFLC is *only* in the serum unless considerable renal damage has occurred, which often sends the laboratorian off on a fruitless effort to identify the non-existent heavy chain. Tetrameric MFLC have been described in patients with multiple myeloma, primary amyloidosis and angioimmunoblastic lymphadenopathy.[153] Polymerization of κ light chains may result in a hyperviscosity syndrome.[154,155] Interestingly, polymerization of light chains does not increase the nephrotoxicity of MFLC.[148]

False positive MFLC in urine by electrophoresis

Some urine samples are sent to evaluate the pattern of proteinuria, and not to screen for MFLC (see above for patterns of proteinuria). In those cases, an immunofixation of the urine is not included as part of the routine test in our laboratory. Just as with serum, a variety of proteins can produce bands in the urine that may be mistaken for monoclonal gammopathies on urine protein electrophoresis. The most common protein that may produce a large band in urine is hemoglobin. Depending on the characteristics of the system used (gel type, ionic strength, and pH – see chapter 1 for details), it may migrate from the α_2- to the β-globulin region (Fig. 7.3). Usually, when large quantities of hemoglobin are present, the red color of the urine is good evidence for the identity of the band. However, when smaller quantities of hemoglobin are present in the urine, the color may not serve as a sufficient guide. The naked eye can detect amounts >300 mg/l in urine.[156] In those cases where there is uncertainty, the performance of immunofixation for κ and λ will rule out the presence of a MFLC. Hobbs and Levinson suggest performing a 415 nm scan on an unstained gel sample of the patient's urine.[109] This wavelength is near the Soret band (414 nm) where hemoglobin shows maximal absorbance.[157] Hobbs and Levinson[109] found that they could detect as little as 0.2 g/l of hemoglobin by this technique. Although rarely commercial antisera may not always react appropriately with monoclonal proteins,[21,158–160] our laboratory has not found this to be a problem in distinguishing between a hemoglobin band and an M-protein on urine. The location of hemoglobin on our gels plus the negative κ and λ immunofixation is sufficient evidence for us to rule out an MFLC. If heavy chain disease is part of the differential diagnosis (see Chapter 6), immunofixation for IgG, IgA and IgM are needed.

Other molecules will also produce bands in urine that may be confusing. In crush injuries and some myopathies a myoglobin band may be seen.[44] The distinction between myoglobin and hemoglobin is useful for the clinician. Hamilton et al.[156] point out that myoglobin usually appears as a light brown color whereas hemoglobin is usually red, especially in a fresh sample. However, aged samples, unstable hemoglobin and methemoglobin will also appear a brownish color.[156] The best solution to this issue is to perform an immunoassay to detect the myoglobin.

As mentioned above, β_2-microglobulin, eosinophil-derived neurotoxin, and lysozyme, when present in sufficient quantity, will produce a band in protein electrophoresis that could be mistaken for MFLC.[45,46,47] Proteins added to the urine in cases of factitious proteinuria (see above) may produce peculiar bands (usually in the albumin and α_1-region).

A new source of unusual banding in the urine is found in samples from patients with pancreas and pancreas–renal transplants. Since 1987, the preferred drainage of their exocrine pancreas secretions is into the urinary bladder.[161] It is well documented that their urine contains discrete forms of the pancreatic enzymes.[162–164] Indeed, evaluation of these enzymes has been suggested as a means to follow rejection in these patients.[162–164] Nonetheless, the bands may be confusing to laboratories. I have seen

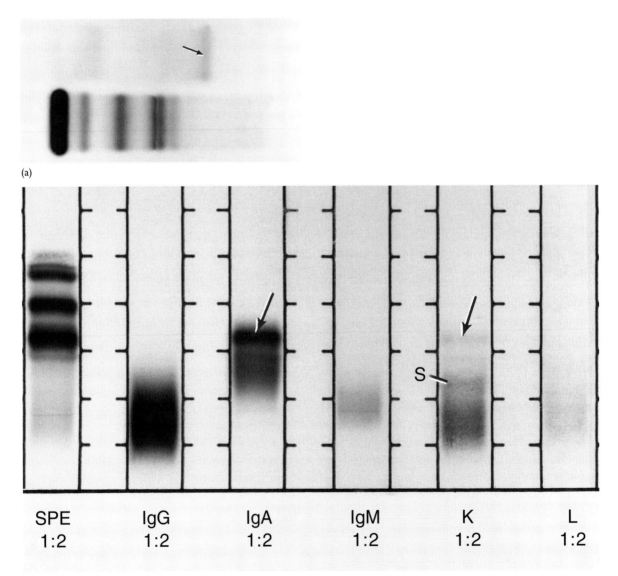

(a)

SPE   IgG   IgA   IgM   K   L
1:2   1:2   1:2   1:2   1:2   1:2

(b)

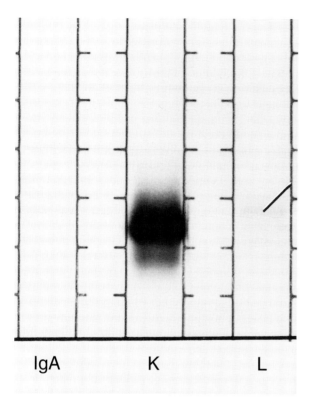

IgA   K   L

(c)

Figure 7.14 (a) A urine concentrated 100-fold (top) and a serum diluted 1:3 from the same patient (bottom) are shown. The serum shows a hypogammaglobulinemia and the urine has a suspicious band in the fast γ-region (arrow). (Paragon SPE2 system stained with Paragon Violet.) (b) Immunofixation of the serum from (a) shows a small IgA κ monoclonal gammopathy (arrows) which had been hidden in the β-region beneath the prominent β₁-lipoprotein and transferrin bands. In addition, there is a small, but suspicious band in the κ region (S). Dilutions of the serum used for immunofixation of serum are shown below the immunoglobulin antisera used. SPE, serum protein electrophoresis for comparison. (Beckman Paragon system stained with Paragon Violet.) (c) Immunofixation of 100-fold concentrated urine of the urine from (a) shows a massive κ monoclonal free light chain protein band, no IgA band and a tiny λ-band (indicated) thought to be part of the ladder pattern. (Beckman Paragon system stained with Paragon Violet; anode at the top.)

two such cases (Fig. 7.16), one of which has been recently documented as being due to pancreatic enzymes by Song et al.[165] Although this is a somewhat unusual source of confusion it re-emphasizes the two key things that the laboratorian should do when confronted by unusual bands that may be MFLC in urine: perform the immunofixation for κ and λ, then if they are negative call the clinician to discuss the case.

There is always the rare possibility of a false negative immunofixation due to an antigen excess effect (see Chapter 3). When a large band seen on urine protein electrophoresis fails to yield a reaction with κ and λ (yet is not consistent with the sources mentioned above); one may wish to consider a repeat immunofixation on 10- and 100-fold diluted urine to help rule out antigen excess problems.[166] It is unusual that these dilutions yield a result that was missed with the original sample. Antigen excess effects are usually easy to recognize.

One should also consider including antisera against heavy chains to rule out the possibility of heavy chain disease. This should give a distinct band that matches the one in the serum. In performing immunofixation for heavy chains in urine, however, one must be aware that the α₂-region of urine may demonstrate the presence of fragments of free polyclonal γ-chain that may be present in urine samples (Fig. 7.10).[167] The presence of these fragments is not uncommon in urine, although the exact cause is unclear. The fragments may result from breakdown of IgG by many factors such as proteases or bacterial enzymes.

FALSE NEGATIVE MFLC IN URINE BY ELECTROPHORESIS

Roach et al.[168] provided an excellent example of a potential source of false-negative urine for MFLC. They presented a pattern in urine protein electrophoresis that resembled the 'ladder pattern'

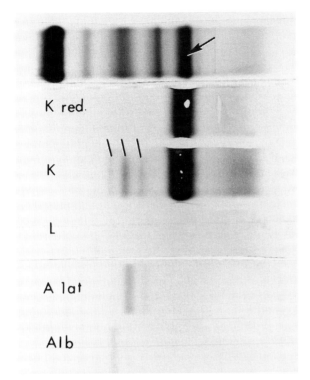

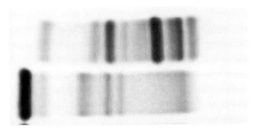

Figure 7.16 A urine protein electrophoresis is in the top lane and serum from the same patient is in the bottom lane. Several discrete bands are present throughout the β- and γ-regions. This is not a tubular pattern of proteinuria. Those bands would be much smaller and migrate in the α-and β-regions, not the γ-region. This urine is from a patient with a pancreas transplant. The exocrine duct of the pancreas is drained through the urinary bladder and the pancreatic enzymes may be found in the urine. Case contributed by Lu Song.

Figure 7.15 This serum has a densely staining band in the C3 area (arrow). When immunofixation was performed with anti-κ (K) and anti-λ (L), the dense κ-band was seen indicating that this was a κ monoclonal gammopathy. No reaction was seen with the other heavy chain antisera (not shown). Also, in the κ reaction are three other more anodal bands (indicated). When the patient's serum was reduced with 2-mercaptoethanol (K red), these extra bands disappeared. They are identified by performing immunofixation of the purified κ chain with antisera to α₁-antitrypsin (A-1at) and albumin (Alb). (Panagel system stained with Coomassie Blue; anode to the left.)

We have seen similar cases in our laboratory. The key lesson is that the 'ladder pattern' refers only to the multiple faint bands seen on immunofixation of urine, not on electrophoresis of urine. Immunofixation is several times more sensitive than urine protein electrophoresis. When the bands are distinct enough to be seen by urine protein electrophoresis, they deserve an immunofixation to rule out MFLC.

FALSE POSITIVE MFLC IN URINE BY IMMUNOFIXATION

Immunofixation on concentrated urine often shows multiple, small and somewhat indistinct bands in the κ and λ lanes that have been termed the 'ladder pattern' (Fig. 7.18).[27,169] The ladder

seen on immunofixation (Fig. 7.17). Their immunology studies, however, demonstrated that this reflected multimer formation of a MFLC.[168]

Figure 7.17 Urine protein electrophoresis on a sample concentrated 50-fold. The γ-region shows numerous closely spaced bands that resemble a ladder pattern on immunofixation. Immunoelectrophoresis demonstrated that this was a case of λ monoclonal free light chain. When unusual patterns are seen on urine protein electrophoresis, immunological studies are needed. Case contributed by Drs Adrian O. Vladutiu and Barbara M. Roach.

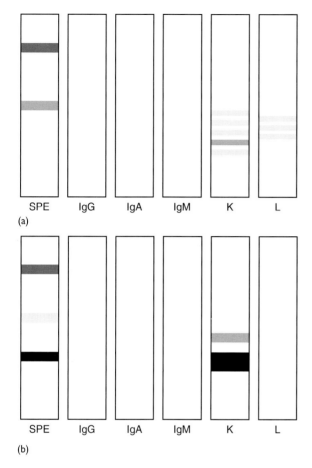

(a)

(b)

Figure 7.18 (a) Schematic view of the ladder pattern in urine. Several evenly spaced bands with variable weak staining are often seen in the κ (K) reaction and occasionally seen in the λ (L) reaction. (b) Schematic view of monoclonal free light chain to contrast with (a). In this case, a smaller second band is seen just anodal to the large MFLC. This is often due to monomer and dimer MFLC and would not be mistaken for a ladder pattern. SPE, serum protein electrophoresis.

pattern now is well recognized and with experience, will not cause problems in interpreting the vast majority of cases of urine immunofixation. The typical ladder pattern demonstrates five or six faint, regular and somewhat diffuse bands with a notable hazy background staining between the bands. These bands may be seen in any urine sample with a polyclonal increase in immunoglobulins.[23,27,169–171] The ladder pattern likely represents the relatively limited heterogeneity of normal polyclonal free light chain molecules. They are more often seen in the κ lane than in the λ lane of an immunofixation gel, but may be seen in both. They are usually regularly spaced which is why Harrison refers to them as a 'ladder' pattern.[169,170] They may be seen when there is considerable polyclonal IgG and IgA present (Fig. 7.19), or when there is little intact immunoglobulin present (Fig. 7.20).

Bailey et al.[27] and Hess et al.[166] noted that when these bands are relatively dense, they may be especially confusing with small MFLC. Some laboratories use descriptions of these patterns as 'oligoclonal' or 'pseudo-oligoclonal' in their reports. These terms are inaccurate and confusing to clinicians. Since these are not MFLC proteins, I sign urine samples with ladder patterns as 'negative for monoclonal free light chains (Bence Jones proteins)'. However, as noted by Hess et al.,[166] in some cases one of the bands may be more prominent than the others or spaced in an unusual pattern (Fig. 7.21), such that one cannot unequivocally rule out an MFLC. In such cases, a repeat immunofixation using larger dilutions may help; however, occasionally I am not certain whether or not the band is a true MFLC. In those cases, I report that 'a small restriction is seen in the κ (or λ if appropriate) chain reactivity. The clinical significance of detection of MFLC that are present in amounts so small that their existence is obscured by the presence of a ladder pattern is not known. A repeat urine sample in 3–6 months will help to determine if the process regresses'. This allows the clinician to know what we know: the case is unusual, a small restriction is seen which we cannot confidently classify as a MFLC protein, a follow-up is needed to be sure this is not a process that will progress. I avoid using the term 'MFLC or Bence Jones' proteins until I am certain of that diagnosis. As always, it is useful to obtain more clinical history on such cases. Another alternative that is now available in such situations is to recommend that a serum FLC assay be performed. Patients with even very small quantities of true MFLC in urine will almost always have an altered κ/λ ratio in serum.[26]

Finally, a modest, but straightforward case of MFLC is shown in Fig. 7.22. Note that the K lane contains several small ladder rungs and there is some diffuse staining in the L lane. Nonetheless, this case is clearly MFLC.

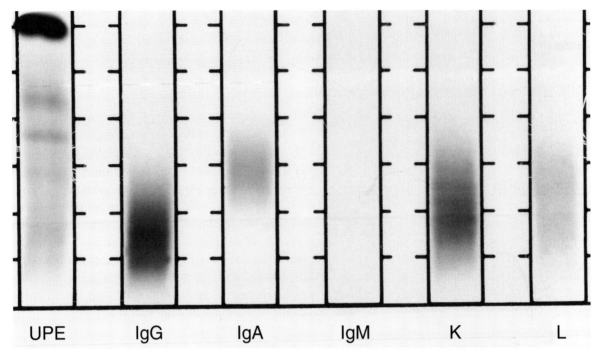

Figure 7.19 Urine containing polyclonal IgG and IgA with a ladder pattern with the κ antiserum (K) and a diffuse weak staining with λ antiserum (L). The interpretation is 'Negative for MFLC (monoclonal free light chain).' (Beckman Paragon system stained with Paragon Violet; anode at the top.)

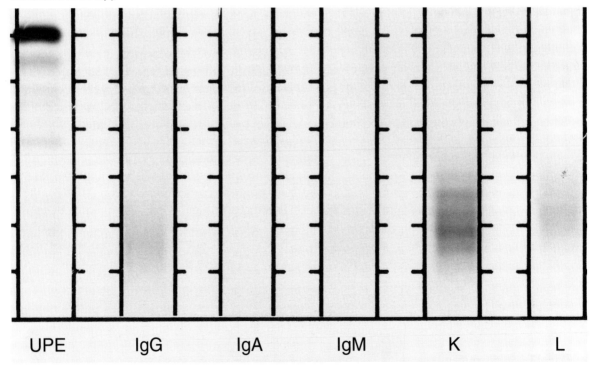

Figure 7.20 Urine containing very little polyclonal IgG and no detectable polyclonal IgA or IgM. The ladder pattern is quite prominent in the reaction with the κ antiserum (K) and also shows up very faintly with the λ antiserum (L). (Beckman Paragon system stained with Paragon Violet; anode at the top.)

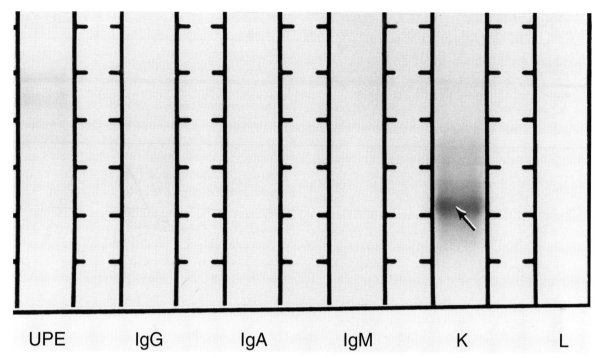

Figure 7.21 A faint diffuse pattern with a hint of a ladder is seen with the κ antisera (K). One band is particularly prominent (arrow), although it is a very small band. The significance of such small bands is not known. I do not report this as a monoclonal free light chain. I note that a small (tiny) κ-restriction is seen and recommend repeating the urine immunofixation at 3–6 months to see if the process evolves or regresses. (Paragon system stained with Paragon Violet; anode at the top.)

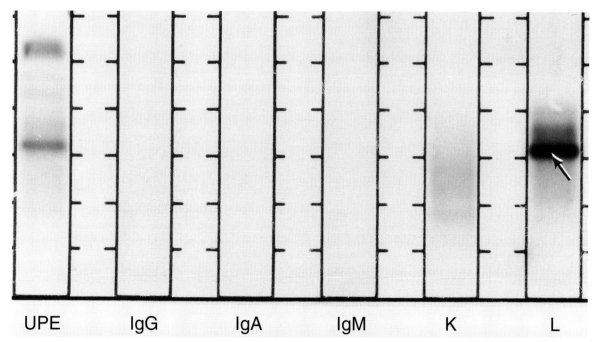

Figure 7.22 Typical immunofixation of a urine with a prominent λ monoclonal free light chain (MFLC) (arrow) and a faint ladder pattern with the κ antisera (K).(Beckman Paragon system stained with Paragon Violet; anode at the top.)

NEPHELOMETRY TO MEASURE TOTAL KAPPA AND TOTAL λ LIGHT CHAINS IN URINE

When the medical literature refers to the measurement of κ and λ chains it usually refers to the total κ and total λ, rather than the free κ and free λ. Total κ is the κ bound to heavy chain (intact immunoglobulin κ) plus the free κ. Total λ refers to the λ bound to heavy chain (intact immunoglobulin λ) plus the free λ. In serum, the amounts of free κ and free λ are insignificant when compared with the κ and λ that are bound to intact immunoglobulin molecules. When I refer to assays specific only for free κ and free λ I will use those terms. Caution should be exercised when reviewing articles in the literature that merely refer to the measurement of κ and λ to check the specificity of the assay being described.

Although there was hope that assays for total κ and total λ might replace the need for immunofixation to detect MFLC, this has not been realized in practice. Some studies reported that laboratories might use nephelometric measurements of total κ and total λ in the urine as a more cost-effective way to screen urine samples for the presence of MFLC than by immunofixation. Levinson[172] compared rate nephelometry using antisera that reacted with both total κ and total λ chains. In normal urine, the poor sensitivity of the nephelometric method used often gave a result of 'too low to detect' for κ and/or λ. With the recent advent of automated FLC assays, this approach has been abandoned.

TECHNIQUES TO MEASURE FREE κ AND FREE λ LIGHT CHAINS IN SERUM AND/OR URINE

Many assays have been devised in the past 20 years that attempted to measure free κ and free λ light chains in serum and/or urine.[35,105,129,173–177] Because there are orders of magnitude between the concentration of bound versus free light chains, some means to distinguish them needed to be employed. Physically separating the free light chains from intact immunoglobulins was used in early studies to facilitate the measurement of free light chains by polyclonal antibodies directed against total κ chains.[176] However, physical separation techniques are problematic for clinical laboratories that depend upon automated immunoassays for efficiency.

In order to use immunological techniques to distinguish between free and bound light chain in the same serum sample special antisera that are highly specific for antigenic determinants only expressed on free light chains needed to be made.[105] The antisera needed to react only with determinants are unavailable to react when the light chains are bound to the heavy chains. Both polyclonal and monoclonal antibodies have been developed for use in these assays.

Polyclonal antibodies are prepared by immunizing an animal (typically a rabbit, goat, or sheep) with purified MFLC, often from the urine of a patient with multiple myeloma (Fig. 7.23). The serum from this animal is extensively adsorbed with intact immunoglobulin molecules. Unfortunately, the adsorption may not remove all antibodies that react with bound light chains. Further, the adsorption also dilutes the concentration of antibodies reactive with the antigenic determinants expressed only on unbound light chains. Because of this, more recent studies have used further purification by reacting the adsorbed sera onto free light chains attached to Sepharose 4B (Pharmacia, Uppsala, Sweden).[105] After extensive washing of the affinity column, elution of the antibodies provides a highly purified, specific antisera against free light chain. To facilitate the automation, Bradwell et al.[105] produced F(ab′)2 fragments that were attached to latex beads. Such antibody-coated beads (The Binding Site Limited, Birmingham, UK) are readily adapted for use on currently available instrumentation: Beckman IMMAGE (Beckman Coulter, Fullerton, CA, USA), and Dade Behring BNII nephelometer (Dade Behring Inc., Deerfield, IL, USA).

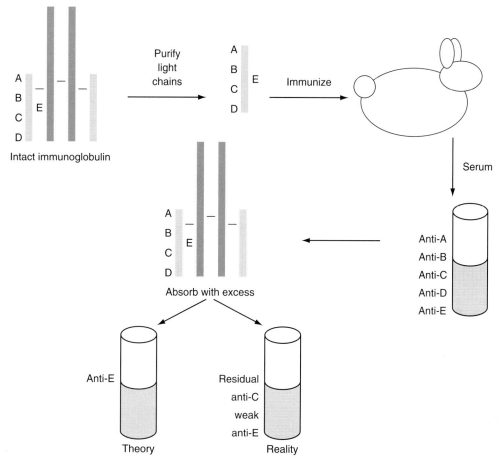

Figure 7.23 Commercial polyclonal sera with reactivities to free light chain can be created because some light chain antigenic determinants are 'hidden' in intact molecules. In the example shown, determinant E in the intact molecule is not available to react with antisera. When the light chains are separated from the heavy chains, this determinant is now expressed along with the many antigenic determinants (A, B, C, D) which are also expressed in the intact molecule. When these light chains are used to immunize an animal, antibodies against all of these determinants can result. By absorbing these antisera with intact molecules, *in theory*, only the antisera against free light chain determinants will remain. However, these antisera are often very weak, and often crossreact with intact molecules. This is why recent methods required further purification on Sepharose 4B columns coated with purified light chain.

Monoclonal antibodies have also been produced against the antigenic determinants of free light chains that are hidden when bound to heavy chains.[173,174,178] This provides highly specific antisera; however, monoclonal proteins may not react with the entire spectrum of light chains produced. Eventually, reagents composed of cocktails of monoclonal antibodies may provide the breadth and specificity of reaction required for optimal results. However, at present, the highly purified polyclonal products seem to have an edge.

The serum ratio of total κ to total λ (which repre-

sents overwhelmingly intact immunoglobulins that contain κ or λ light chains respectively) is 2:1. However, the most recent studies of serum using highly specific polyclonal or monoclonal antibodies that are able to distinguish bound from unbound light chains have demonstrated a free κ/free λ ratio of approximately 1:2.[105,174] Abe et al.,[174] who used monoclonal antibodies in an enzyme-linked immunosorbent assay hypothesized that the disparity in free versus intact light chain ratios may be due to a higher rate of production by λ plasma cells than κ plasma cells. They further

Table 7.2 Free light chain measurements in serum and urine[a]

Reference	Serum free κ (mean mg/l ± SD)	Serum free λ (mean mg/l ± SD)	Urine free κ (mean mg/24 h ± SD)	Urine free λ (mean mg/24 h ± SD)
Abe et al.[174]	16.6 ± 6.1	33.8 ± 4.8	2.96 ± 1.84	1.07 ± 0.69
Bradwell et al.[105]	8.4 ± 2.66	14.5 ± 4.4	5.5 ± 4.95	3.17 ± 3.3

[a]Table modified from Bradwell et al., Table 2, with permission.[105]

suggested that the quaternary structural differences between κ and λ free light chains also may play a role. Bradwell et al.,[105] who used polyclonal antibodies in an automated immunoassay suggested that since κ chains exist predominately as monomers (25 kDa) they will be cleared more quickly through the glomeruli than the mainly dimeric λ free light chains (50 kDa). The normal serum and urine free κ and λ concentrations are noted in Table 7.2. Whereas the serum free light chain ratios are the opposite of serum intact light chain ratios, in urine, the free κ/free λ ratios in the two studies are 2.8:1 and 1.7:1; both similar to the total κ/total λ ratio usually quoted for serum. This reflects the molecular sieve properties of the glomerulus to favor passage of the smaller κ molecules into the urine.

Bradwell et al. reported a free light chain κ/λ ratio in serum to be 1:1.62 with a 95 per cent confidence interval of 1:2.75–1:0.99.[105] This free light chain κ/λ ratio in serum provided consistent discrimination between individuals with myeloma and those with polyclonal increases in immunoglobulins.[26,105] Sera from patients with multiple myeloma or Waldenström macroglobulinemia contain increased concentrations of the free light chain type associated with the M-protein in all 27 cases examined (Fig. 7.24).[105] They also

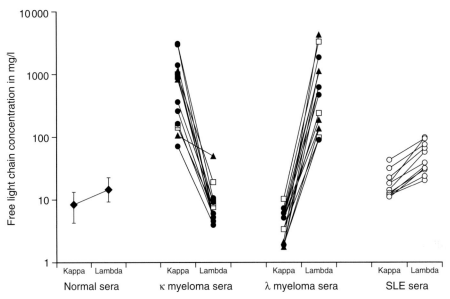

Figure 7.24 Comparison of free light chain concentrations (means and 95% confidence interval shown) in 100 normal sera, sera from 27 patients with monoclonal gammopathies (IgG, closed circle; IgA, closed triangle; IgM, open square), and sera from 12 patients with systemic lupus erythematosus (SLE). Figure contributed by Dr Arthur Bradwell, and used with permission.[105]

Table 7.3 Comparison of initial serum free light chain (FLC) and 24-h urine M-protein in patients with light chain multiple myeloma

Patients	24-h urine M-protein g/24 h	Serum κ FLC (normal 2–11.2 mg/l)	Serum λ FLC (normal 6.8–25.2 mg/l)	Serum κ/λ (normal 0.09–0.89)
κ Patients				
1	0.33	22.6	12.3	1.837
2	2.25	4370	10.1	432.673
3	1.77	699	14.3	48.881
4	1.36	1390	5.6	248.214
5	1.81	693	1.2	577.500
6	0.10	276	1.1	250.909
7	0.03	2820	8.1	348.148
8	0.30	1230	9.8	125.510
9	1.31	351	5.1	68.824
λ Patients				
10	0.29	10.4	235	0.044
11	1.53	1.0	1030	0.001
12	3.04	3.3	15 900	0.000
13	6.32	4.8	116	0.041
14	3.31	7.0	11 000	0.001
15	0.11	1.1	46.5	0.024
16	4.11	0.9	545	0.002
17	5.02	6.8	1390	0.005
18	1.23	7.2	1430	0.005
19	4.20	9.2	4690	0.002
20	1.46	3.1	71.8	0.043
21	7.14	1.2	65.4	0.018
22	10.81	6.9	2700	0.003
23	4.01	16.9	2360	0.007
24	6.48	10.5	1420	0.007
25	0.79	46.7	10 000	0.005
26	0.10	5.9	81.6	0.072
27	2.99	1.00	326	0.003
28	5.58	4.5	3570	0.001

ªTable modified from Abraham et al., Table 1, with permission.[26]

investigated sera from 12 patients with systemic lupus erythematosus as an example of the effect of polyclonal proliferation of immunoglobulins on the κ/λ ratio. While the level of circulating free light chain was increased for both κ and λ, the free κ/free λ ratios were in the reference range they established.[105] As would be expected in situations of decreased renal clearance, polyclonal free light chains also occur in patients receiving chronic hemodialysis.[179]

In their studies, Bradwell et al.[105] reported two patients with a negative urine study for MFLC using radial immunodiffusion (RID) screening with a sensitivity of 40 mg/l, where the serum nephelometric immunoassay demonstrated increased free light chains of the monoclonal type.

Abraham et al.[26] compared the urine and serum levels for free κ and free λ light chains using the same nephelometric immunoassay with a normal free κ/λ of 0.09–0.89.[105] Their data is shown in Table 7.3. The selected population all had monoclonal free light chains in the urine, but only three of nine patients with κ-secreting monoclonal gammopathies and 13 of 19 patients with λ-secreting monoclonal gammopathies had a monoclonal peak on the serum protein electrophoresis. Nonetheless, all of the patients had abnormal κ/λ ratios (Table 7.3). As a control group, they evaluated seven patients with lupus glomerulonephritis. None of the seven patients had an abnormal κ/λ ratio, although one had an elevation of the free λ light chains in serum. This indicates that serum measurements of free κ and free λ light chains by automated immunoassays may provide a viable alternative to the current 24-h urine collections to follow the clinical response of patients with multiple myeloma.[26]

The serum free light chain assays may be of considerable assistance with another difficult problem, the 'non-secretory' cases of multiple myeloma. Drayson et al.[129] have reported the utility of automated immunoassays for free light chains in serum from patients with the diagnosis of 'non-secretory' multiple myeloma where no monoclonal protein in either serum or urine could be demonstrated by traditional methods of immunofixation and electrophoresis. They detected increased concentrations of one free light chain along with an abnormal free κ/λ ratio in sera from 19 of the 28 patients with the diagnosis of non-secretory multiple myeloma. Of course, these assays do not prove the existence of a monoclonal protein. A polyclonal increase in one or the other light chain type could result in an abnormal free κ/λ ratio. However, Drayson et al.'s finding that the clinical changes during follow-up of six patients correlated with the changes in the free light chain concentration during that time period suggest that this assay will be of use in many patients with non-secretory myeloma (a name we may need to change – 'pauci-secretory'?).[129]

The technique of automated immunoassays for free light chain measurement in serum and urine is evolving rapidly. Currently, reagents are available from The Binding Site, Ltd, and assays have been performed on the Beckman IMMAGE and on the Dade Behring BNII. The reader is encouraged to seek the most recent information about this technique.

REFERENCES

1. Waller KV, Ward KM, Mahan JD, Wismatt DK. Current concepts in proteinuria. *Clin Chem* 1989;35:755–765.

2. Kaysen GA, Myers BD, Couser WG, Rabkin R, Felts JM. Mechanisms and consequences of proteinuria. *Lab Invest* 1986;54:479–498.

3. Cooper EH. Proteinuria. *Am Assn Clin Chem Specific Protein Analysis* 1984;1:1–11.

4. Ritz E, Nowicki M, Fliser D, Horner D, Klimm HP. Proteinuria and hypertension. *Kidney Int Suppl* 1994;47:S76–80.

5. Guder WG, Hofmann W. Markers for the diagnosis and monitoring of renal tubular lesions. *Clin Nephrol* 1992;38:S3–7.

6. Wallach J. *Interpretation of diagnostic tests.* Philadelphia: Lippincott, Williams & Wilkins, 2000.

7. Wieme R. *Agar gel electrophoresis.* Amsterdam: Elsevier, 1965.

8. Ying WZ, Sanders PW. Mapping the binding domain of immunoglobulin light chains for Tamm–Horsfall protein. *Am J Pathol* 2001;158: 1859–1866.

9. Bole DG, Hendershot LM, Kearney JF. Posttranslational association of immunoglobulin heavy chain binding protein with nascent heavy chains in nonsecreting and secreting hybridomas. *J Cell Biol* 1986;102:1558–1566.

10. Dul JL, Aviel S, Melnick J, Argon Y. Ig light chains are secreted predominantly as monomers. *J Immunol* 1996;157:2969–2975.

11. Shapiro AL, Scharff MD, Maizel JV, Uhr JW. Synthesis of excess light chains of gamma globulin by rabbit lymph node cells. *Nature* 1966;211:243–245.

12. Kyle RA. Diagnosis and management of multiple myeloma and related disorders. *Prog Hematol* 1986;14:257–282.

13. Kaplan IV, Levinson SS. Misleading urinary protein pattern in a patient with hypogammaglobulinemia: effects of mechanical concentration of urine. *Clin Chem* 1999;45: 417–419.

14. Rice EW. Improved biuret procedure for routine determination of urinary total proteins in clinical proteinuria. *Clin Chem* 1975;21: 398–401.

15. Shahangian S, Brown PI, Ash KO. Turbidimetric measurement of total urinary proteins: a revised method. *Am J Clin Pathol* 1984;81:651–654.

16. Pesce MA, Strande CS. A new micromethod for determination of protein in cerebrospinal fluid and urine. *Clin Chem* 1973;19:1265–1267.

17. Marshall T, Williams KM. Total protein determination in urine: elimination of a differential response between the Coomassie blue and pyrogallol red protein dye-binding assays. *Clin Chem* 2000;46:392–398.

18. Lefevre G, Bloch S, Le Bricon T, Billier S, Arien S, Capeau J. Influence of protein composition on total urinary protein determined by pyrocatechol-violet (UPRO vitros) and pyrogallol red dye binding methods. *J Clin Lab Anal* 2001;15:40–42.

19. Cesati R, Dolci A. Simple gold overstaining enhances sensitivity of automated electrophoresis of unconcentrated urine. *Clin Chem* 1996;42: 1293–1294.

20. Matsuda K, Hiratsuka N, Koyama T, et al. Sensitive method for detection and semiquantification of Bence Jones protein by cellulose acetate membrane electrophoresis using colloidal silver staining. *Clin Chem* 2001;47: 763–766.

21. Levinson SS, Keren DF. Free light chains of immunoglobulins: clinical laboratory analysis. *Clin Chem* 1994;40:1869–1878.

22. Pezzoli A, Pascali E. Urine collection for the detection of Bence Jones proteinuria. *Am J Clin Pathol* 1991;95:266–268.

23. Gertz MA, Lacy MQ, Dispenzieri A. Immunoglobulin light chain amyloidosis and the kidney. *Kidney Int* 2002;61:1–9.

24. Hidaka H, Ikeda K, Oshima T, Ohtani H, Suzuki H, Takasaka T. A case of extramedullary plasmacytoma arising from the nasal septum. *J Laryngol Otol* 2000;114:53–55.

25. Lau CF, Fok KO, Hui PK, et al. Intestinal obstruction and gastrointestinal bleeding due to systemic amyloidosis in a woman with occult plasma cell dyscrasia. *Eur J Gastroenterol Hepatol* 1999;11:681–685.

26. Abraham RS, Clark RJ, Bryant SC, et al. Correlation of serum immunoglobulin free light chain quantification with urinary Bence Jones protein in light chain myeloma. *Clin Chem* 2002;48:655–657.

27. Bailey EM, McDermott TJ, Bloch KJ. The urinary light-chain ladder pattern. A product of improved methodology that may complicate the recognition of Bence Jones proteinuria. *Arch Pathol Lab Med* 1993;117:707–710.

28. Newman DJ, Pugia MJ, Lott JA, Wallace JF, Hiar AM. Urinary protein and albumin excretion corrected by creatinine and specific gravity. *Clin Chim Acta* 2000;294:139–155.

29. Ambuhl PA, Muller V, Binswanger U. Transient proteinuria after infusion of a gelatin plasma volume expander. *Clin Nephrol* 1999;52: 399–400.

30. Ito S, Ueno M, Izumi T, Arakawa M. Induction of transient proteinuria, hematuria, and glucosuria by ethanol consumption in Japanese alcoholics. *Nephron* 1999;**82**:246–253.

31. Lipatov IS, Kupaev IA, Kozupitsa GS. Pregnancy outcomes in women with a pathological weight gain, vascular dysfunction, transitory edemas and transient proteinuria. *Akush Ginekol* 1995;**6**: 16–18.

32. Nicot GS, Merle LJ, Charmes JP, et al. Transient glomerular proteinuria, enzymuria, and nephrotoxic reaction induced by radiocontrast media. *JAMA* 1984;**252**:2432–2434.

33. Reuben DB, Wachtel TJ, Brown PC, Driscoll JL. Transient proteinuria in emergency medical admissions. *N Engl J Med* 1982;**306**: 1031–1033.

34. Alpert HC, Lohavichan C, Presser JI, Papper S. 'Febrile' proteinuria. *South Med J* 1974;**67**: 552–554.

35. Hemmingsen L, Skaarup P. Urinary excretion of 10 plasma proteins in patients with febrile diseases. *Acta Med Scand* 1977;**201**:359–364.

36. Mori M. Febrile proteinuria. *Rinsho Byori* 1979;(Suppl 36):85–93.

37. Jensen H, Henriksen K. Proteinuria in non-renal infectious diseases. *Acta Med Scand* 1974;**196**: 75–82.

38. Derfler K, Hauser C, Endler M, Nowotny C, Lapin A, Balcke P. Proteinuria in normal pregnancy and in EPH gestosis. *Acta Med Austriaca.* 1989;**16**:13–18.

39. Kaltenbach FJ, Wilhelm C. Selectivity of glomerular proteinuria and liver function in gestosis. *Zentralbl Gynakol* 1994;**116**: 340–343.

40. Sifuentes Alvarez A. Electrophoretic profile of proteinuria in normal pregnancy and in gestational hypertension. *Ginecol Obstet Mex* 1995;**63**:147–151.

41. Dendorfer U, Anders HJ, Schlondorff D. Urine diagnosis: proteinuria. *Dtsch Med Wochenschr* 2001;**126**:1310–1313.

42. Carroll MF, Temte JL. Proteinuria in adults: a diagnostic approach. *Am Fam Physician* 2000;**62**:1333–1340.

43. Voiculescu M. Advances in the study of proteinurias. I. The pathogenetic mechanisms. *Med Interne* 1990;**28**:265–277.

44. Lillehoj EP, Poulik MD. Normal and abnormal aspects of proteinuria. Part I: Mechanisms, characteristics and analyses of urinary protein. Part II: clinical considerations. *Exp Pathol* 1986;**29**:1–28.

45. Tardy F, Bulle C, Prin L, Cordier JF, Deviller P. High concentrations of eosinophil-derived neurotoxin in patients' urine mimic lysozyme far-cathodic bands in agarose gel electrophoresis. *Clin Chem* 1993;**39**:919–920.

46. Sexton C, Buss D, Powell B, O'Connor M, Rainer R, Woodruff R, Cruz J, Pettenati M, Rao PN, Case LD. Usefulness and limitations of serum and urine lysozyme levels in the classification of acute myeloid leukemia: an analysis of 208 cases. *Leuk Res* 1996;**20**:467–472.

47. Handy BC. Urinary beta2-microglobulin masquerading as a Bence Jones protein. *Arch Pathol Lab Med* 2001;**125**:555–557.

48. Bottini PV, Ribeiro Alves MA, Garlipp CR. Electrophoretic pattern of concentrated urine: Comparison between 24-hour collection and random samples. *Am J Kidney Dis.* 2002;**39**:E2.

49. Levinson SS. Urine protein electrophoresis and immunofixation electrophoresis supplement one another in characterizing proteinuria. *Ann Clin Lab Sci* 2000;**30**:79–84.

50. Killingsworth LM, Tyllia MM. Protein analysis finding clues to disease in urine. *Diagn Med* 1980;May–June:69–75.

51. Myers B. *In vivo* evaluation of glomerular permoselectivity in normal and nephrotic man. In: M A, ed. *Proteinuria*. New York: Plenum, 1985.

52. Tencer J, Bakoush O, Torffvit O. Diagnostic and prognostic significance of proteinuria selectivity index in glomerular diseases. *Clin Chim Acta* 2000;**297**:73–83.

53. Tencer J, Torffvit O, Thysell H, Rippe B, Grubb A. Proteinuria selectivity index based upon alpha 2-macroglobulin or IgM is superior to the IgG based index in differentiating glomerular diseases. Technical note. *Kidney Int* 1998;**54**:2098–2105.

54. Kouri T, Harmoinen A, Laurila K, Ala-Houhala I, Koivula T, Pasternack A. Reference intervals for the markers of proteinuria with a standardized bed-rest collection of urine. *Clin Chem Lab Med* 2001;**39**:418–425.

55. Marshall T, Williams KM. Electrophoretic analysis of Bence Jones proteinuria. *Electrophoresis* 1999;**20**:1307–1324.

56. Bazzi C, Petrini C, Rizza V, et al. Characterization of proteinuria in primary glomerulonephritis. SDS-PAGE patterns: clinical significance and prognostic value of low molecular weight ('tubular') proteins. *Am J Kidney Dis* 1997;**29**:27–35.

57. Bazzi C, Petrini C, Rizza V, Arrigo G, D'Amico G. A modern approach to selectivity of proteinuria and tubulointerstitial damage in nephrotic syndrome. *Kidney Int* 2000;**58**: 1732–1741.

58. Bruning T, Thier R, Mann H, et al. Pathological excretion patterns of urinary proteins in miners highly exposed to dinitrotoluene. *J Occup Environ Med* 2001;**43**:610–615.

59. Ikonomov V, Melzer H, Nenov V, Stoicheva A, Stiller S, Mann H. Importance of sodium dodecyl sulfate pore-graduated polyacrylamide gel electrophoresis in the differential diagnostic of Balkan nephropathy. *Artif Organs* 1999;**23**: 75–80.

60. Koliakos G, Papachristou F, Papadopoulou M, Trachana V, Gaitatzi M, Sotiriou I. Electrophoretic analysis of urinary proteins in diabetic adolescents. *J Clin Lab Anal* 2001;**15**: 178–183.

61. Schreiber S, Hamling J, Zehnter E, et al. Renal tubular dysfunction in patients with inflammatory bowel disease treated with aminosalicylate. *Gut* 1997;**40**:761–766.

62. Woo KT, Lau YK. Pattern of proteinuria in tubular injury and glomerular hyperfiltration. *Ann Acad Med Singapore* 1997;**26**:465–470.

63. Le Bricon T, Erlich D, Bengoufa D, Dussaucy M, Garnier JP, Bousquet B. Sodium dodecyl sulfate–agarose gel electrophoresis of urinary proteins: application to multiple myeloma. *Clin Chem* 1998;**44**:1191–1197.

64. Keren DF. Detection and characterization of monoclonal components in serum and urine. *Clin Chem* 1998;**44**:1143–1145.

65. Bang LE, Holm J, Svendsen TL. Retinol-binding protein and transferrin in urine. New markers of renal function in essential hypertension and white coat hypertension? *Am J Hypertens* 1996;**9**:1024–1028.

66. Ikeda M, Moon CS, Zhang ZW, et al. Urinary alpha1-microglobulin, beta2-microglobulin, and retinol-binding protein levels in general populations in Japan with references to cadmium in urine, blood, and 24-hour food duplicates. *Environ Res* 1995;**70**:35–46.

67. Stark J. Acute renal failure. Focus on advances in acute tubular necrosis. *Crit Care Nurs Clin North Am* 1998;**10**:159–170.

68. Lameire N, Vanholder R. Pathophysiologic features and prevention of human and experimental acute tubular necrosis. *J Am Soc Nephrol* 2001;**12**(Suppl 17):S20–32.

69. Weisberg LS, Allgren RL, Genter FC, Kurnik BR. Cause of acute tubular necrosis affects its prognosis. The Auriculin Anaritide Acute Renal Failure Study Group. *Arch Intern Med* 1997;**157**: 1833–1838.

70. Spoto S, Galluzzo S, De Galasso L, Zobel B, Navajas MF. Prevention of acute tubular necrosis caused by the administration of non-ionic radiologic contrast media. *Clin Ter* 2000;**151**: 323–327.

71. Dussol B, Reynaud-Gaubert M, Saingra Y, Daniel L, Berland Y. Acute tubular necrosis induced by high level of cyclosporine A in a lung transplant. *Transplantation* 2000;**70**:1234–1236.

72. Weisberg LS, Allgren RL, Kurnik BR. Acute tubular necrosis in patients with diabetes mellitus. *Am J Kidney Dis.* 1999;**34**:1010–1015.

73. Yanagisawa H, Nodera M, Wada O. Inducible nitric oxide synthase expression in mercury chloride-induced acute tubular necrosis. *Ind Health* 1998;**36**:324–330.

74. Lo RS, Chan JC, Cockram CS, Lai FM. Acute tubular necrosis following endosulphan insecticide poisoning. *J Toxicol Clin Toxicol* 1995;**33**:67–69.

75. Igarashi T, Inatomi J, Ohara T, Kuwahara T, Shimadzu M, Thakker RV. Clinical and genetic studies of CLCN5 mutations in Japanese families with Dent's disease. *Kidney Int* 2000;**58**: 520–527.

76. Minemura K, Ichikawa K, Itoh N, et al. IgA-kappa type multiple myeloma affecting proximal and distal renal tubules. *Intern Med* 2001;**40**:931–935.

77. Markowitz GS, Flis RS, Kambham N, D'Agati VD. Fanconi syndrome with free kappa light chains in the urine. *Am J Kidney Dis* 2000;**35**: 777–781.

78. Deret S, Denoroy L, Lamarine M, et al. Kappa light chain-associated Fanconi's syndrome: molecular analysis of monoclonal immunoglobulin light chains from patients with and without intracellular crystals. *Protein Eng* 1999;**12**:363–369.

79. Maker J. Tubular proteinuria: clinical implications. In: Avram M, ed. *Proteinuria*. New York: Plenum, 1985.

80. Sun T, Lien Y, Mailloux L. A case of proteinuria with analbuminuria. *Clin Chem* 1985;**31**: 1905–1906.

81. Corso A, Serricchio G, Zappasodi P, et al. Assessment of renal function in patients with multiple myeloma: the role of urinary proteins. *Ann Hematol* 1999;**78**:371–375.

82. Saatci U, Ozdemir S, Ozen S, Bakkaloglu A. Serum concentration and urinary excretion of beta 2-microglobulin and microalbuminuria in familial Mediterranean fever. *Arch Dis Child* 1994;**70**:27–29.

83. Nomiyama K, Liu SJ, Nomiyama H. Critical levels of blood and urinary cadmium, urinary beta 2-microglobulin and retinol-binding protein for monitoring cadmium health effects. *IARC Sci Publ.* 1992;**118**:325–340.

84. Boesken WH, Effenberger C, Krieger HP, Jammers W, Stierle HE. Fractional clearance of beta-2-microglobulin in the diagnostic and prognostic assessment of kidney diseases. *Klin Wochenschr* 1989;**67**:31–36.

85. Tsukahara H, Fujii Y, Tsuchida S, et al. Renal handling of albumin and beta-2-microglobulin in neonates. *Nephron* 1994;**68**:212–216.

86. Valles P, Peralta M, Carrizo L, et al. Follow-up of steroid-resistant nephrotic syndrome: tubular proteinuria and enzymuria. *Pediatr Nephrol* 2000;**15**:252–258.

87. Blumsohn A, Morris BW, Griffiths H, Ramsey CF. Stability of beta 2-microglobulin and retinol binding protein at different values of pH and temperature in normal and pathological urine. *Clin Chim Acta* 1991;**195**:133–137.

88. Donaldson MD, Chambers RE, Woolridge MW, Whicher JT. Stability of alpha 1-microglobulin, beta 2-microglobulin and retinol binding protein in urine. *Clin Chim Acta* 1989;**179**:73–77.

89. Yu H, Yanagisawa Y, Forbes MA, Cooper EH, Crockson RA, MacLennan IC. Alpha-1-microglobulin: an indicator protein for renal tubular function. *J Clin Pathol* 1983;**36**: 253–259.

90. Bernard A, Vyskocyl A, Mahieu P, Lauwerys R. Effect of renal insufficiency on the concentration of free retinol-binding protein in urine and serum. *Clin Chim Acta* 1988;**171**:85–93.

91. Bernard AM, Lauwerys RR. Retinol binding protein in urine: a more practical index than urinary beta 2-microglobulin for the routine screening of renal tubular function. *Clin Chem* 1981;**27**:1781–1782.

92. Ginevri F, Piccotti E, Alinovi R, et al. Reversible tubular proteinuria precedes microalbuminuria and correlates with the metabolic status in diabetic children. *Pediatr Nephrol* 1993;**7**:23–26.

93. Mengoli C, Lechi A, Arosio E, et al. Contribution of four markers of tubular proteinuria in detecting upper urinary tract infections. A multivariate analysis. *Nephron* 1982;**32**:234–238.

94. Korneti P, Tasic V, Cakalaroski K, Korneti B. Factitious proteinuria in a diabetic patient. *Am J Nephrol* 2001;**21**:512–513.

95. Tojo A, Nanba S, Kimura K, et al. Factitious proteinuria in a young girl. *Clin Nephrol* 1990;**33**:299–302.

96. Sutcliffe H, Rawlinson PS, Thakker B, Neary R, Mallick N. Factitious proteinuria: diagnosis and protein identification by use of isoelectric focusing. *Clin Chem* 1988;**34**:1653–1655.

97. Mitas JA 2nd. Exogenous protein as the cause of nephrotic-range proteinuria. *Am J Med* 1985;**79**: 115–118.

98. Jones H. On a new substance occurring in the urine of a patient with mollities ossium. *Phil Trans R Soc London* 1848;**138**:55–62.

99. Clamp JR. Some aspects of the first recorded case of multiple myeloma. *Lancet* 1967;**ii**: 1354–1356.

100. MacIntyre W. Case of mollities and fragilitas ossium, accompanied with urine strongly charged with animal matter. *Med-Chir Trans* 1850;**33**: 211–232.

101. Dalrymple J. On the microscopic character of mollities ossium. *Dublin Q J Med Sci* 1846;**2**: 85–95.

102. Rosenfeld L. Henry Bence Jones (1813–1873): the best 'chemical doctor' in London. *Clin Chem* 1987;**33**:1687–1692.

103. Putnam FW. Henry Bence Jones: the best chemical doctor in London. *Perspect Biol Med* 1993;**36**:565–579.

104. Boffa GA, Zakin MM, Faure A, Fine JM. Contribution to the study of the relationships between the chains of gamma-G immunoglobulins and Bence-Jones proteins. *Transfusion* 1967;**10**:169–181.

105. Bradwell AR, Carr-Smith HD, Mead GP, et al. Highly sensitive, automated immunoassay for immunoglobulin free light chains in serum and urine. *Clin Chem* 2001;**47**:673–680.

106. Picken MM, Shen S. Immunoglobulin light chains and the kidney: an overview. *Ultrastruct Pathol* 1994;**18**:105–112.

107. Pick AI, Shoenfeld Y, Skvaril F, et al. Asymptomatic (benign) monoclonal gammopathy – a study of 100 patients. *Ann Clin Lab Sci* 1977;**7**:335–343.

108. Kyle RA, Therneau TM, Rajkumar SV, et al. A long-term study of prognosis in monoclonal gammopathy of undetermined significance. *N Engl J Med* 2002;**346**:564–569.

109. Hobbs GA, Levinson SS. Hemoglobin interference with urinary Bence Jones protein analysis on electrophoresis. *Ann Clin Lab Sci.* 1996;**26**:71–75.

110. Lin J, Markowitz GS, Valeri AM, et al. Renal monoclonal immunoglobulin deposition disease: the disease spectrum. *J Am Soc Nephrol* 2001;**12**:1482–1492.

111. Winearls CG. Acute myeloma kidney. *Kidney Int* 1995;**48**:1347–1361.

112. Pote A, Zwizinski C, Simon EE, Meleg-Smith S, Batuman V. Cytotoxicity of myeloma light chains in cultured human kidney proximal tubule cells. *Am J Kidney Dis* 2000;**36**:735–744.

113. Cooper EH, Forbes MA, Crockson RA, MacLennan IC. Proximal renal tubular function in myelomatosis: observations in the fourth Medical Research Council trial. *J Clin Pathol* 1984;**37**:852–858.

114. Guan S, el-Dahr S, Dipp S, Batuman V. Inhibition of Na-K-ATPase activity and gene expression by a myeloma light chain in proximal tubule cells. *J Invest Med* 1999;**47**:496–501.

115. Batuman V, Sastrasinh M, Sastrasinh S. Light chain effects on alanine and glucose uptake by renal brush border membranes. *Kidney Int* 1986;**30**:662–665.

116. Batuman V, Guan S, O'Donovan R, Puschett JB. Effect of myeloma light chains on phosphate and glucose transport in renal proximal tubule cells. *Ren Physiol Biochem* 1994;**17**: 294–300.

117. Truong LD, Mawad J, Cagle P, Mattioli C. Cytoplasmic crystals in multiple myeloma-associated Fanconi's syndrome. A morphological study including immunoelectron microscopy. *Arch Pathol Lab Med* 1989;**113**:781–785.

118. Schillinger F, Hopfner C, Montagnac R, Milcent T. IgG kappa myeloma with Fanconi's syndrome and crystalline inclusions. Immunohistochemical and ultrastructural study. *Presse Med* 1993;**22**: 675–679.

119. Horn ME, Knapp MS, Page FT, Walker WH. Adult Fanconi syndrome and multiple myelomatosis. *J Clin Pathol* 1969;**22**:414–416.

120. Uchida S, Matsuda O, Yokota T, et al. Adult Fanconi syndrome secondary to kappa-light chain myeloma: improvement of tubular functions after treatment for myeloma. *Nephron* 1990;**55**:332–335.

121. Huang ZQ, Sanders PW. Localization of a single binding site for immunoglobulin light chains on human Tamm–Horsfall glycoprotein. *J Clin Invest* 1997;**99**:732–736.

122. Sanders PW, Booker BB. Pathobiology of cast nephropathy from human Bence Jones proteins. *J Clin Invest* 1992;**89**:630–639.

123. Short AK, O'Donoghue DJ, Riad HN, Short CD, Roberts IS. Recurrence of light chain nephropathy in a renal allograft. A case report and review of the literature. *Am J Nephrol* 2001;**21**:237–240.

124. Abbott KC, Agodoa LY. Multiple myeloma and light chain-associated nephropathy at end-stage renal disease in the United States: patient characteristics and survival. *Clin Nephrol* 2001;**56**:207–210.

125. Keren DF. Procedures for the evaluation of monoclonal immunoglobulins. *Arch Pathol Lab Med* 1999;**123**:126–132.

126. Keren DF, Alexanian R, Goeken JA, Gorevic PD, Kyle RA, Tomar RH. Guidelines for clinical and laboratory evaluation patients with monoclonal gammopathies. *Arch Pathol Lab Med* 1999;**123**:106–107.

127. Kyle RA. Sequence of testing for monoclonal gammopathies. *Arch Pathol Lab Med* 1999;**123**:114–118.

128. Brigden ML, Neal ED, McNeely MD, Hoag GN. The optimum urine collections for the detection and monitoring of Bence Jones proteinuria. *Am J Clin Pathol* 1990;**93**:689–693.

129. Drayson M, Tang LX, Drew R, Mead GP, Carr-Smith H, Bradwell AR. Serum free light-chain measurements for identifying and monitoring patients with nonsecretory multiple myeloma. *Blood* 2001;**97**:2900–2902.

130. Salomo M, Gimsing P, Nielsen LB. Simple method for quantification of Bence Jones proteins. *Clin Chem* 2002;**48**:2202–2207.

131. DelBuono L, Keren DF. Detection of Bence Jones proteinuria by high-resolution electrophoresis and immunofixation. *Am J Clin Pathol* 1989;**92**:541(Abstr).

132. Graziani MS, Righetti G. Immunoblotting for detecting Bence Jones proteinuria. *Clin Chem* 1987;**33**:1079–1080.

133. Pascali E, Pezzoli A. Bence Jones proteins in the urine of patients with multiple sclerosis. *Clin Chem* 1989;**35**:1550–1551.

134. Pezzoli A, Pascali E. Bence Jones proteinuria in multiple sclerosis. *Clin Chem* 1987;**33**:1923–1924.

135. Mehta PD, Cook SD, Troiano RA, Coyle PK. Increased free light chains in the urine from patients with multiple sclerosis. *Neurology* 1991;**41**:540–544.

136. Hobbs J. Bence Jones Proteins. *Essays Med Biochem* 1975;**1**:105–131.

137. Laurell CB. Complexes formed *in vivo* between immunoglobulin light chain kappa, prealbumin and/or alpha-1-antitrypsin in myeloma sera. *Immunochemistry* 1970;**7**:461–465.

138. Nabeshima Y, Ikenaka T. Primary structure of cryo Bence-Jones protein (Tog) from the urine of a patient with IgD myeloma. *Mol Immunol* 1979;**16**:439–444.

139. Kanoh T, Niki T, Murata Y, Ohta M. Multiple myeloma associated with cryo-Bence Jones protein: report of a case and review of the literature. *Nippon Naika Gakkai Zasshi* 1978;**67**:160–165.

140. Kojima M, Kobayashi Y, Murakawa E. Crystallizable, cryo-precipitable lambda Bence Jones protein in a case of IgD multiple myeloma. *Nippon Ketsueki Gakkai Zasshi* 1978;**41**:81–89.

141. Finazzi Agro A, Crifo C, Natali PG, Chersi A. Differential denaturation of a crystalline Bence-Jones type cryoprotein as monitored by fluorescence. *Ital J Biochem* 1978;**27**:36–42.

142. Harris RI, Kohn J. A urinary cryo-Bence Jones protein gelling at room temperature. *Clin Chim Acta* 1974;**53**:233–237.

143. Hirai H, Doi I, Kawai T. Autopsy case of multiple myeloma with cryo-Bence Jones protein. *Naika* 1969;**23**:585–591.

144. Kawai T, Tadano J. Cryo-Bence Jones protein. First case in Japan. *Igaku To Seibutsugaku* 1967;**74**:251–257.

145. Alper CA. Cryoglobulinuria: studies of a cryo-Bence Jones protein. *Acta Med Scand Suppl* 1966;**445**:200–205.

146. Inoue N, Togawa A, Yawata Y. Tetrameric Bence Jones protein – case report and review of the literature. *Nippon Ketsueki Gakkai Zasshi.* 1984;47:1456–1459.

147. Hom BL. Polymeric (presumed tetrameric) lambda Bence Jones proteinemia without proteinuria in a patient with multiple myeloma. *Am J Clin Pathol* 1984;82:627–629.

148. Solling K, Solling J, Lanng Nielsen J. Polymeric Bence Jones proteins in serum in myeloma patients with renal insufficiency. *Acta Med Scand* 1984;216:495–502.

149. Inoue S, Nagata H, Yozawa H, Terai T, Hasegawa H, Murao M. A case of IgG myeloma with tetrameric Bence Jones proteinemia and abnormal fibrin polymerization. *Rinsho Ketsueki* 1980;21:200–207.

150. Kozuru M, Benoki H, Sugimoto H, Sakai K, Ibayashi H. A case of lambda type tetramer Bence-Jones proteinemia. *Acta Haematol* 1977;57:359–365.

151. Togawa A, Imamura Y. Study on the Bence-Jones protein – study on paraproteins detected in myeloma patients with tetrameric Bence-Jones proteins (type kappa) (author's transl). *Nippon Ketsueki Gakkai Zasshi.* 1975;38:571–581.

152. Caggiano V, Dominguez C, Opfell RW, Kochwa S, Wasserman LR. IgG myeloma with closed tetrameric Bence Jones proteinemia. *Am J Med* 1969;47:978–985.

153. Kosaka M, Iishi Y, Okagawa K, Saito S, Sugihara J, Muto Y. Tetramer Bence Jones protein in the immunoproliferative diseases. Angioimmunoblastic lymphadenopathy, primary amyloidosis, and multiple myeloma. *Am J Clin Pathol* 1989;91:639–646.

154. Carter PW, Cohen HJ, Crawford J. Hyperviscosity syndrome in association with kappa light chain myeloma. *Am J Med* 1989;86:591–595.

155. Khan P, Roth MS, Keren DF, Foon KA. Light chain disease associated with the hyperviscosity syndrome. *Cancer* 1987;60:2267–2268.

156. Hamilton RW, Hopkins MB 3rd, Shihabi ZK. Myoglobinuria, hemoglobinuria, and acute renal failure. *Clin Chem* 1989;35:1713–1720.

157. Ross D, Prenant M, Bessis M. On the proper use of the Soret band for hemoglobin detection in erythrocytic cells. *Blood Cells* 1978;4:361–367.

158. Bush D, Keren DF. Over- and underestimation of monoclonal gammopathies by quantification of kappa- and lambda-containing immunoglobulins in serum. *Clin Chem* 1992;38:315–316.

159. Su L, Keren DF, Warren JS. Failure of anti-lambda immunofixation reagent mimics alpha heavy-chain disease. *Clin Chem* 1995;41:121–123.

160. Levinson SS. Studies of Bence Jones proteins by immunonephelometry. *Ann Clin Lab Sci* 1992;22:100–109.

161. Prieto M, Sutherland DE, Goetz FC, Rosenberg ME, Najarian JS. Pancreas transplant results according to the technique of duct management: bladder versus enteric drainage. *Surgery.* 1987;102:680–691.

162. See WA, Smith JL. Urinary levels of activated trypsin in whole-organ pancreas transplant patients with duodenocystostomies. *Transplantation.* 1991;52:630–633.

163. See WA, Smith JL. Urinary trypsin levels observed in pancreas transplant patients with duodenocystostomies promote *in vitro* fibrinolysis and *in vivo* bacterial adherence to urothelial surfaces. *Urol Res* 1992;20:409–413.

164. Zheng T, Lu Z, Merideth N, Lanza RP, Soon-Shiong P. Early markers of pancreas transplant rejection. *Am Surg* 1992;58:630–633.

165. Song L, Allison N, Lorah S, Seiple J. An abnormal urine protein electrophoresis pattern associated with a patient who received simultaneous pancreas kidney transplant. *Clin Chem* 2002;48:A29.

166. Hess PP, Mastropaolo W, Thompson GD, Levinson SS. Interference of polyclonal free light chains with identification of Bence Jones proteins. *Clin Chem* 1993;39:1734–1738.

167. Charles EZ, Valdes AJ. Free fragments of gamma chain in the urine. A possible source of confusion with gamma heavy-chain disease. *Am J Clin Pathol* 1994;101:462–464.

168. Roach BM, Meinke JS, Sridhar N, Vladutiu AO. Multiple narrow bands in urine protein electrophoresis. *Clin Chem* 1999;45:716–718.

169. Harrison HH. The 'ladder light chain' or 'pseudo-oligoclonal' pattern in urinary immunofixation electrophoresis (IFE) studies: a distinctive IFE pattern and an explanatory hypothesis relating it to free polyclonal light chains. *Clin Chem* 1991;37:1559–1564.

170. Harrison HH. Fine structure of 'light-chain ladders' in urinary immunofixation studies revealed by ISO-DALT two-dimensional electrophoresis. *Clin Chem* 1990;36:1526–1527.

171. MacNamara EM, Aguzzi F, Petrini C, et al. Restricted electrophoretic heterogeneity of immunoglobulin light chains in urine: a cause for confusion with Bence Jones protein. *Clin Chem* 1991;37:1570–1574.

172. Levinson SS. An algorithmic approach using kappa/lambda ratios to improve the diagnostic accuracy of urine protein electrophoresis and to reduce the volume required for immunoelectrophoresis. *Clin Chim Acta* 1997;262:121–130.

173. Axiak SM, Krishnamoorthy L, Guinan J, Raison RL. Quantitation of free kappa light chains in serum and urine using a monoclonal antibody based inhibition enzyme-linked immunoassay. *J Immunol Methods* 1987;99:141–147.

174. Abe M, Goto T, Kosaka M, Wolfenbarger D, Weiss DT, Solomon A. Differences in kappa to lambda (kappa:lambda) ratios of serum and urinary free light chains. *Clin Exp Immunol* 1998;111:457–462.

175. Brouwer J, Otting-van de Ruit M, Busking-van der Lely H. Estimation of free light chains of immunoglobulins by enzyme immunoassay. *Clin Chim Acta* 1985;150:267–274.

176. Solling K. Free light chains of immunoglobulins in normal serum and urine determined by radioimmunoassay. *Scand J Clin Lab Invest* 1975;35:407–412.

177. Waldmann TA, Strober W, Mogielnicki RP. The renal handling of low molecular weight proteins. II. Disorders of serum protein catabolism in patients with tubular proteinuria, the nephrotic syndrome, or uremia. *J Clin Invest* 1972;51:2162–2174.

178. Nelson M, Brown RD, Gibson J, Joshua DE. Measurement of free kappa and lambda chains in serum and the significance of their ratio in patients with multiple myeloma. *Br J Haematol* 1992;81:223–230.

179. Wakasugi K, Sasaki M, Suzuki M, Azuma N, Nobuto T. Increased concentrations of free light chain lambda in sera from chronic hemodialysis patients. *Biomater Artif Cells Immobilization Biotechnol* 1991;19:97–109.

APPENDIX

Capillary zone electrophoresis (CZE) has dramatically improved the efficiency and quality of serum protein electrophoresis in many laboratories. Recently, techniques to perform this technique in the urine have become available in both experimental and clinical settings. In this Appendix, I show examples contributed by Margaret A. Jenkins and Cynthia Blessum.

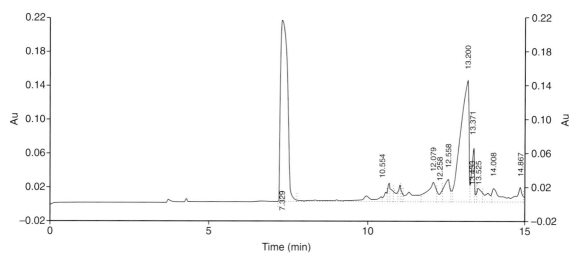

A7.1 This is a capillary zone electropherogram of unconcentrated human serum showing predominantly albumin proteinuria. Total urine protein is 0.17 g/l. The peak with the highest absorbance is at approximately 7.3 min. This is the urea/creatinine peak. Albumin is the largest protein peak and occurs at 13.2 min. The smaller peaks before and after albumin are small amounts of other molecules that absorb at 200 nm. The technique was performed on a Beckman MDQ CE instrument; assay buffer was 150 mM boric acid pH 9.7 containing calcium lactate. Separation voltage was 18 kV and detection was at an absorbance of 200 nm. This figure was contributed by Margaret Jenkins, Austin and Repatriation Medical Centre, Heidelberg, Australia.

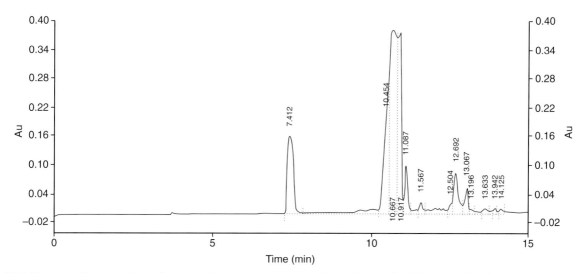

A7.2 This is a capillary zone electropherogram of unconcentrated urine with a predominantly tubular pattern. Total urine protein is 1.40 g/l. Once again, the largest peak is due to urea/creatinine at approximately 7.3 min. The albumin peak at approximately 13.2 min is difficult to see between the numerous tubular protein peaks. Conditions as in A7.1. This figure was contributed by Margaret Jenkins, Austin and Repatriation Medical Centre, Heidelberg, Australia.

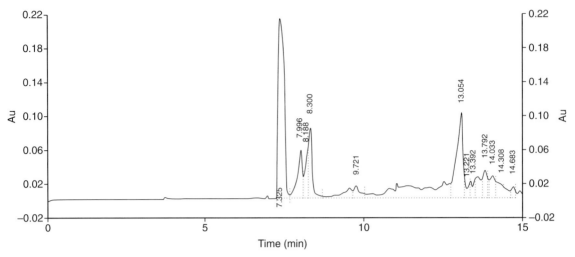

A7.3 This is a capillary zone electropherogram of unconcentrated urine with monoclonal free light chain (MFLC) and albumin. Total urine protein is 3.59 g/l. The urea/creatinine peak is at approximately 7.3 min. Immediately following this peak are two sharp peaks that merge at their base. They are at 7.9 min and 8.3 min and represent a double free κ MFLC. The albumin peak is prominent at approximately 13.1 min. Conditions as in A7.1. This figure was contributed by Margaret Jenkins, Austin and Repatriation Medical Centre, Heidelberg, Australia.

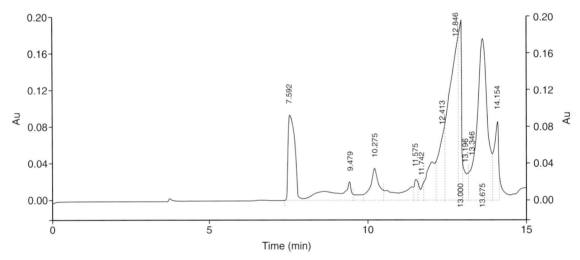

A7.4 This is a capillary zone electropherogram of unconcentrated urine with combined glomerular and tubular proteinuria. Total protein is 5.2 g/l. Again, the urea/creatinine is a good marker peak at approximately 7.6 min. The albumin peak is the second largest and occurs at 13.7 min. Conditions as in A7.1. This figure was contributed by Margaret Jenkins, Austin and Repatriation Medical Centre, Heidelberg, Australia.

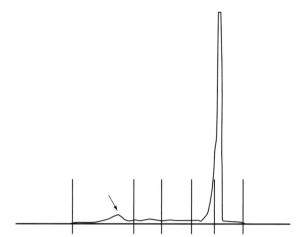

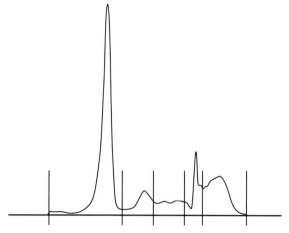

A7.5 This is a capillary zone electropherogram of unconcentrated urine with a prominent monoclonal free light chain (MFLC). The total protein is > 1.5 g/l. No urea/creatinine peak is present on the patterns performed on the Paragon CZE 2000; neither are the precise elution times noted. The albumin peak (arrow) is the small peak toward the anode. A large peak due to the κ MFLC is seen at the β–γ region interface. The technique was performed on a Paragon CZE 2000, borate buffer, pH 10.0. Electrophoresis at 10.5 kV, 24°C for 4 min. This figure is courtesy of Cynthia Blessum, Beckman Coulter, Inc.

A7.6 This is a capillary zone electropherogram of unconcentrated urine with a combined glomerular and tubular pattern. The total protein is > 1.5 g/l. No urea/creatinine peak is present on the patterns performed on the Paragon CZE 2000. This pattern closely resembles a normal serum protein electrophoresis pattern with a few exceptions. The α_2-region has the appearance of an irregular mesa and drops off sharply just before the transferrin band. There is some β–γ bridging and irregularity in the γ-region. The technique was performed on a Paragon CZE 2000, borate buffer, pH 10.0. Electrophoresis at 10.5 kV, 24°C for 4 min. This figure is courtesy of Cynthia Blessum, Beckman Coulter, Inc.

A7.7 This is a capillary zone electropherogram of unconcentrated urine with a glomerular pattern. The total protein is > 1.5 g/l. No urea/creatinine peak is present on the patterns performed on the Paragon CZE 2000. Basically, two peaks are seen. The prominent albumin peak and a small β_1-region peak due to the presence of transferrin. The technique was performed on a Paragon CZE 2000, borate buffer, pH 10.0. Electrophoresis at 10.5 kV, 24°C for 4 min. This figure is courtesy of Cynthia Blessum, Beckman Coulter, Inc.

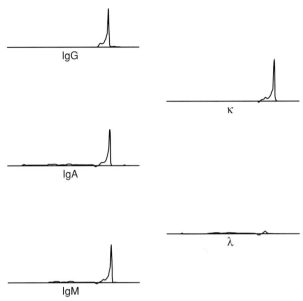

A7.8 This is a capillary zone electropherogram of immunosubtraction on an unconcentrated urine with λ monoclonal free light chain. Each square shows the pattern of migration after the urine was treated with beads coated with antibodies against the specific immunoglobulin class noted in each square. The prominent γ-region spike is present in all the sectors, except the one pretreated with anti-λ. The technique was performed on a Paragon CZE 2000, borate buffer, pH 10.0. Electrophoresis at 10.5 kV, 24°C for 4 min. This figure is courtesy of Cynthia Blessum, Beckman Coulter, Inc.

8

Approach to pattern interpretation in cerebrospinal fluid

EARLY ELECTROPHORETIC STUDIES

Early studies by Kabat et al.[1,2] reported that cerebrospinal fluid (CSF) from control individuals contained relatively little γ-globulin compared with serum, whereas CSF from patients with a variety of neurological conditions had an elevation of the total CSF protein with a decreased ratio of albumin/globulin. They pointed out that in patients with either multiple sclerosis or neurosyphilis, there was a consistent increase in both the γ and the transthyretin (prealbumin) fractions.[1,2] Currently (as discussed below), the most specific laboratory test for multiple sclerosis is the demonstration of oligoclonal bands in the CSF that are not present in a corresponding serum.[3,4]

CEREBROSPINAL FLUID PROTEIN COMPOSITION

Source of CSF proteins

Cerebrospinal fluid is an ultrafiltrate of plasma that is continuously produced at the rate of about 500 ml/day in the choroid plexus.[5] Since the total volume of CSF is only 135 ml, it must turn over every 6 h.[6] Reabsorption occurs into the bloodstream at the superior sagittal sinus by the arachnoid granulations.[7] The vast majority of CSF proteins are serum proteins that pass through the blood–CSF barrier at the choroid plexus. Only about 20 per cent of CSF proteins are synthesized locally.[8] Proteins passing through the choroid plexus are limited by molecular size, charge, their concentration in plasma and the integrity of the blood–CSF barrier.[6,9] Whereas, their plasma concentration and the integrity of the blood–CSF barrier may change dramatically in some disease processes, the size and the charge of the proteins are relatively constant factors (with some exceptions in charge and genetic structural variants). The sieve effect of the blood–CSF barrier is not as clear-cut as that of the glomerular basement membrane in the kidney; however, small molecules such as transthyretin (prealbumin) preferentially pass into the CSF, and the larger α_2-macroglobulin and haptoglobin are greatly restricted.[10,11]

While most proteins are passively transferred into the CSF, some, such as transferrin, have an active mechanism. Transferrin binds to specific receptors on the endothelium of cerebral capillaries and neurons.[12,13] Once within the cytoplasm, transferrin releases its attached iron and some of the transferrin molecules also lose their terminal sialic acid residues from its carbohydrate side-chain. This forms the desialated transferrin (τ protein also called 'CSF-specific slow transferrin') that exists in CSF along with the usual sialated form of transferrin.[6] As discussed below, the presence of this desialated transferrin can be used as a marker of CSF leakage into nasal and aural fluids as a result of damage to the cranial vault.[14] Although some of the desialated transferrin finds its way into the blood, most of it is quickly taken up by receptors on reticulo-endothelial cells that do not bind transferrin containing the terminal sialic acid residues.[6] The presence of desialated transferrin (also termed carbohydrate-deficient transferrin) in the blood has become a convenient marker for the presence of alcoholism.[15-18] However, even in alcoholics, the concentration of desialated transferrin is too low to interfere with electrophoretic techniques that have been employed to detect CSF leakage.

Electrophoretic pattern of normal CSF

Because of the molecular sieve action of the choroid plexus and the presence of proteins unique to CSF, the protein electrophoresis pattern seen with CSF differs considerably from the pattern seen with serum (Fig. 8.1). Normally, the transthyretin (prealbumin) band in serum is barely visible, whereas this band is increased relative to the other protein bands in concentrated CSF (Fig. 8.2). This results from both a preferential transport of transthyretin because of its size and charge characteristics as well as its local synthesis by the epithelium of the choroid plexus.[19,20] Because of this increase in its relative concentration, transthyretin had been used to detect CSF leakage into nasal and aural fluids.[6] However, since the advent of both immunofixation to detect desialated transferrin and measurement of prostaglandin D synthase (formerly β-trace protein) (discussed below), I no longer recommend studies for transthyretin to detect CSF leakage.[21]

Isoforms of albumin, transferrin and immunoglobulins comprise the vast majority of CSF proteins.[5] Albumin is a major protein band in CSF, as in serum, but it usually migrates more toward the anode in CSF than in serum. α_1-Lipoprotein tends to overlap albumin or migrates anodally to it

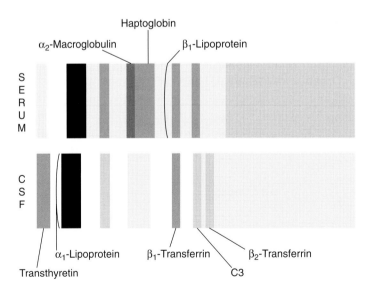

Figure 8.1 Schematic comparison of serum versus concentrated cerebrospinal fluid (CSF). The CSF transthyretin band is much stronger than in the corresponding serum. In contrast the α_2-region is considerably weaker in CSF because the high molecular weight haptoglobin and α_2-macroglobulin are restricted from passing across the blood–CSF barrier under normal circumstances. Increased protein in this region is an indication of a disturbed blood–CSF barrier. In the β_2-region of CSF an extra band is present, β_2-transferrin that is not normally present in serum. The γ-region of CSF normally stains considerably lighter than the γ-region in the corresponding serum.

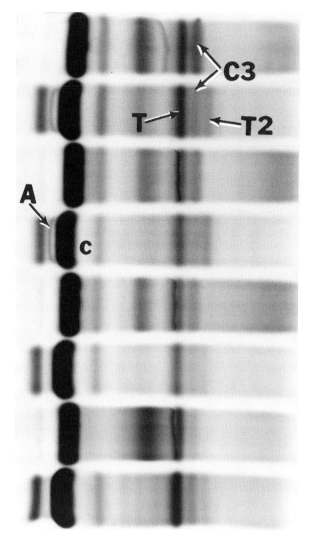

Figure 8.2 *Comparison of several pairs of serum diluted 1:3 and CSF concentrated 80-fold (serum above and CSF below from each patient for comparison). The transthyretin (prealbumin) band anodal to the albumin is considerably more prominent in the CSF. The α₁-lipoprotein band (A) migrates more anodally in CSF than in the corresponding serum. It is often hidden by the albumin band, but occasionally can be seen between transthyretin and albumin. The area between albumin and α₁-antitrypsin is, therefore, much clearer (C) in the CSF than in the adjacent serum. The α₂-region bands are always much denser in serum, unless there is a damaged blood–CSF barrier. Note that the C3 band (C3) is more prominent in the serum than in the corresponding CSF below it. Cerebrospinal fluid contains two transferrin bands, the β₁-region transferrin band (T) corresponds to that of the adjacent serum, but a β₂-region transferrin band (T2) is normally present only in CSF. (Paragon SPE2 system stained with Paragon Violet.)*

in CSF, therefore, the region between albumin and α_1-antitrypsin stains more weakly with protein dyes such as Amido Black in CSF than in serum (Fig. 8.2). The α_1-antitrypsin band may be slightly more diffuse in the CSF than in the serum due to desialation of this protease inhibitor within the CSF.[22] As stated earlier, the large α_2-macroglobulin and haptoglobin molecules do not pass readily into the CSF and, therefore, the α_2-region is weakly stained in normal CSF compared to serum. When the α_2-region stains strongly, likely causes include a traumatic tap or damage to the blood–CSF barrier.

The β-region has two major bands; however, unlike serum, the bands reflect different forms of transferrin. The more anodal band (Figs 8.1 and 8.3) is transferrin, which is structurally the same as its serum counterpart. There is relatively little C3 in the CSF. As mentioned above, the second major β-region band is the desialated form of transferrin.[23] The small amount of C3 found in CSF is located just anodal to the desialated transferrin band (Fig. 8.3). IgM is also part of the slow β-region in serum, but, because of its size, IgM does not readily pass into the CSF, although it may be formed locally and as such may serve as a marker for early intrathecal immune response.[24]

As with the serum, the most clinically significant region is γ-globulin. Normally, this region contains little protein, even after concentration. The IgG that is present in the CSF tends to show less heterogeneity than serum IgG. In the γ-region, a 2.3 kDa protein called γ-trace protein, not an immunoglobulin, may be seen (depending on the method of concentration – some concentrators exclude molecules this small).[25,26]

Because of the significance of oligoclonal bands in the diagnosis of multiple sclerosis, it is important to be aware of the location of γ-trace protein (pI 9.5) or artifacts that may occur in some electrophoretic systems.[27,28] The location of these non-immunoglobulin bands should become obvious with experience. I batch several CSF samples on the same gel and this allows for comparison from one sample to the next. True oligoclonal bands differ in their location in the γ-region from one case to another. Therefore, if one finds bands in two or more CSF samples from

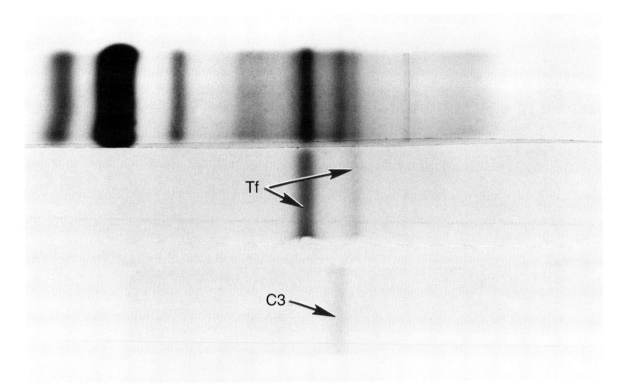

Figure 8.3 The top sample is CSF concentrated 80-fold. To demonstrate the constituents of the β-region bands in the CSF, immunofixation with anti-transferrin (Tf) and anti-C3 are shown immediately below. Note the two transferrin bands normally present in the CSF. (Panagel system stained with Coomassie Blue.)

different individuals on the same gel, one should consider the presence of some artifact either in that run or inherent to the system one is using. In the mid-γ-region of some electrophoretic systems, one may find a slight sharpening of the γ-band owing to restricted migration of CSF IgG, which could be confused with oligoclonal bands (Fig. 8.4).[29] In some electrophoretic methods, I have noticed two faint bands in the fast γ-region that have been mistaken for oligoclonal bands (Fig. 8.5). They do not stain with anti-immunoglobulin reagents. More recent techniques use specific identification of oligoclonal bands by immunostaining on all samples. Immunofixation of CSF has the advantage of requiring much less CSF (since a concentration step can be eliminated) and also allows one to rule out non-specific bands caused by γ-trace protein (Fig. 8.6).

The tally of total CSF proteins reveals a content about 1/350th of that in plasma. In adults up to about 50 years of age, the concentration is 12–60 mg/dl (0.12–0.60 g/l).[30] However, in children, there is a considerable age-dependent change that must be taken into account when looking for increased protein concentrations.[31,32] Biou et al.[33] reported that during the first 6 months of life, there is a dramatic decline in the total protein content of CSF. This difference reflects the immature blood–CSF barrier of the newborn (especially of premature infants) that permits larger amounts of protein to transfer into the CSF than does the blood–CSF barrier of older children (Table 8.1). Similarly, a gradual increase in CSF occurs, perhaps because of a less stable blood–CSF barrier, in individuals over the age of 45 years.[10,11]

Damaged blood–CSF barrier

Alterations of CSF protein patterns occur in a wide variety of conditions. However, there is little clini-

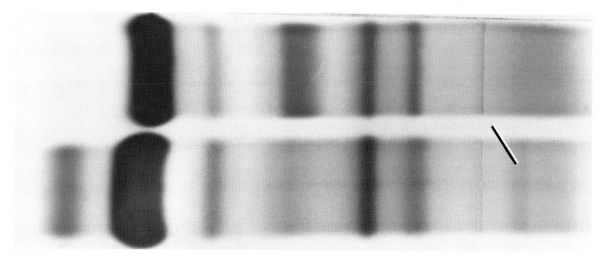

Figure 8.4 Serum diluted 1:4 (top) and cerebrospinal fluid (CSF) concentrated 80-fold (bottom) from the same patient are shown. Note the single band (indicated) in the γ-region of the CSF, which is not present in the corresponding serum. This is not an immunoglobulin by immunofixation and should not be confused with oligoclonal bands seen in patients with multiple sclerosis. The presence of these bands emphasizes the importance of using immunological identification of the bands. (Panagel system stained with Coomassie Blue.)

cal diagnostic significance for alterations other than those in the γ-region. Meningitis results in an elevated CSF total protein because of increased blood–CSF barrier permeability. With increased permeability the total protein content of the CSF increases, as does the proportion of larger proteins such as those in the α_2-region. Yet, there are better laboratory methods to support the diagnosis of meningitis, such as CSF differential cell counts and serum C-reactive protein levels to distinguish between bacterial and aseptic meningitis.[6,34] An

elevated CSF albumin or CSF total protein can be helpful in confirming the diagnosis of Guillain–Barré syndrome in the face of a normal CSF differential cell count.[6] Thompson and Keir[6] also point out the useful finding of decreased CSF transthyretin concentration relative to the other CSF proteins as an indicator of obstructed CSF flow within the spinal cord.

In addition to meningitis, a damaged blood–CSF barrier can result from other sources of inflammation (such as encephalitis from a variety of infec-

Table 8.1 Cerebrospinal fluid (CSF) total protein in children[a]

Age	n	50th Percentile	5–95% Interval
1–8 days	26	71 (0.71)	33–108 (0.33–1.08)
8–30 days	76	59 (0.59)	31–90 (0.31–0.90)
1–2 months	155	47 (0.47)	27–77 (0.27–0.77)
2–3 months	115	35 (0.35)	18–60 (0.18–0.60)
3–6 months	66	23 (0.23)	10–40 (0.10–0.40)
6 months–10 years	599	18 (0.18)	10–32 (0.10–0.32)
10–16 years	37	22 (0.22)	10–41 (0.10–0.41)

[a]Data from Biou et al. expressed as mg/dl (g/l).[33]

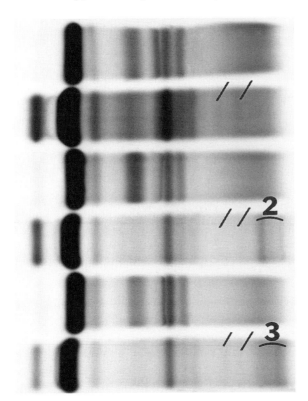

Figure 8.5 Several sera diluted 1:3 and their corresponding CSF concentrated 80-fold immediately below each serum is shown. Note the two faint bands that are indicated in each CSF sample. These are not immunoglobulins and should be ignored when examining samples for oligoclonal bands. Note that true oligoclonal bands (seen in the slow γ-regions of CSF specimens 2 and 3 stain darker than these bands. Note also that the two bands indicated in the top three samples vary in staining intensity, roughly correlating with the amount of protein in the CSF. (Paragon SPE2 system stained with Paragon Violet.)

tious agents), cerebrovascular accidents, metastatic or primary tumors of the central nervous system (CNS), hydrocephalus or herniated intervertebral discs.[35-38] The disturbed blood–CSF barrier permits the larger α_2 molecules to penetrate into the CSF and enhances the staining in this region (Fig. 8.7). Also, the total CSF protein is increased because of a proportionately larger amount of other proteins passing into the CSF. On the electrophoretic strip itself, it is difficult to distinguish this pattern from that of a traumatic tap, where some whole blood or plasma is mixed with the CSF. Often, with a traumatic tap, the sample will have some hemoglobin, giving it a red or pink tinge.

Because albumin is formed only in the liver, the ratio of CSF albumin to serum albumin is a standard index (albumin quotient, Q Alb) for the integrity of the blood–CSF barrier.[10,11] This technique was used to document the increased permeability of the blood–CSF barrier with age.[39] For ages 18–44 years, Q Alb has a mean of 4.7 with a standard deviation of 1.2, whereas from ages 45–88 years, the mean increases to 5.9 with a standard deviation of 2.1.[10,11] The Q Alb is useful not only in confirming leakage of the blood–CSF barrier, but also in ruling it out.[40]

ELECTROPHORETIC METHODS TO STUDY CSF

A wide variety of methods have been used for the electrophoretic evaluation of CSF. I currently recommend the use of methods that enhance the sensitivity and specificity by using immunological identification of bands, such as isoelectric focusing or immunofixation methods. The least sensitive are methods where routine gel electrophoresis is used.[41] A recent study of a commercial agarose gel electrophoresis method to detect oligoclonal bands recorded a disappointing 53 per cent positive among individuals with clinically unambiguous multiple sclerosis.[42] Furthermore, in addition to relatively poor sensitivity, even high-resolution agarose and cellulose methods require that the sample be concentrated (typically 80-fold on a commercial ultrafiltration device) before staining with a protein dye such as Coomassie Brilliant Blue. Unfortunately, this increases the volume of CSF required for analysis. The use of silver stains on unconcentrated CSF has been advocated as a means to decrease the volume requirements while preserving the sensitivity of the assay.[43,44] Recently, CSF has also been studied by capillary zone electrophoresis (CZE).[45] This method is able to detect oligoclonal bands, has the advantages of not requiring concentration or staining and has a shorter analysis time. However, none of the available CZE procedures is currently approved by the US Food and Drug Administration (FDA) for this type of analysis.

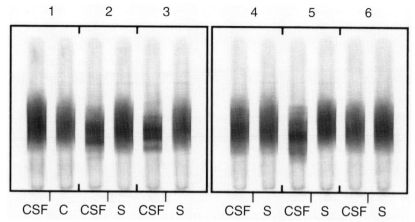

Figure 8.6 Immunofixation performed on unconcentrated cerebrospinal fluid (CSF). In this technique, CSF samples are alternated with serum. Note that in sample 1, each lane contains CSF from a separate patient because neither had serum available at the time of the assay. In the remainder of the cases, they are paired CSF and serum from the same patient. Note the diffuse staining of the two negative CSF samples in the sample 1 pair. In contrast, oligoclonal bands are present in the CSF lanes of samples 2, 3, and 5. The sera for all of these samples are stained diffusely with no evidence of oligoclonal bands. Samples 4 and 6 are negative in both the CSF and serum lanes. (Sebia CSF immunofixation gel.)

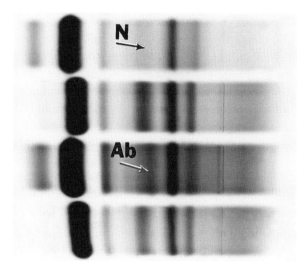

Figure 8.7 The top sample is a normal cerebrospinal fluid (CSF) concentrated 80-fold with its corresponding serum diluted 1:3 directly below it. Note the relatively light staining α_2-region in the normal CSF (N). The CSF sample below the normal serum is an 80-fold concentrated sample from a patient with a compromised blood–CSF barrier. All of the bands stain darker in this CSF than usual. The abnormal α_2-region (Ab) shows much denser staining than is present in normal CSF. The presence of these large molecules (α_2-macroglobulin and haptoglobin) indicates either a damaged blood–CSF barrier or a traumatic tap. The corresponding serum is in the bottom lane. (Paragon SPE2 system stained with Paragon Violet).

Sensitive and specific methods are available for the detection of oligoclonal bands.[6,23,27,41,43,44,46] Isoelectric focusing is a more challenging technique than routine agarose or cellulose acetate, but provides crisp oligoclonal bands that have been demonstrated with silver or immunoenzymatic staining to enhance band visibility (Fig. 8.8).[4,31,32,41,47] Lunding et al.[41] were able to detect oligoclonal bands in all 20 cases of multiple sclerosis examined by their isoelectrofocusing technique which was enhanced by immunofixation, while only nine of the 20 patients had oligoclonal bands on agarose gel electrophoresis and only nine had an IgG Index above the cut-off of 0.72.

An alternative to isoelectric focusing is the use of immunofixation on agarose-based systems to improve both the sensitivity of the detection of oligoclonal bands as well as specificity by documenting their immunoglobulin nature (Fig. 8.6).[48–50] Although this method was first suggested by Cawley et al.[49] in 1976, the recent development of automated commercial methods to perform these assays has provided a practical means for clinical laboratories to take advantage of the improved specificity and sensitivity that this method provides. Richard et al.[51] demonstrated a

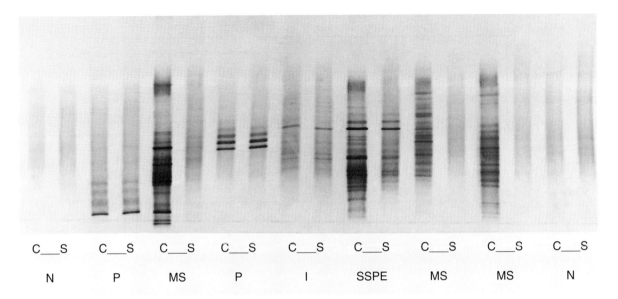

C__S   C__S   C__S   C__S   C__S   C__S   C__S   C__S   C__S

N   P   MS   P   I   SSPE   MS   MS   N

Figure 8.8 Isoelectric focusing (anode at the top) followed by nitrocellulose blotting and immunofixation with antiserum against IgG is one of the most sensitive methods to detect oligoclonal bands. This photograph compares oligoclonal banding of cerebrospinal fluid (C) and serum (S) pairs from the following situations: Normal (N), paraproteinemia (P), multiple sclerosis (MS), subacute sclerosing panencephalitis (SSPE), and a peripheral inflammatory response – not within the central nervous system (I). Photograph provided by E. J. Thompson.[6]

sensitivity and specificity of 83 per cent and 79 per cent respectively for clinically definite multiple sclerosis using an immunofixation peroxidase method on unconcentrated CSF. With immunofixation on agarose gels, not as many bands are seen as on the isoelectric focusing methods and they are broader in their migration. The number of bands and their electrophoretic migration tend to remain constant during active and inactive disease over a period of years.[52,53] However, the number of bands *per se* is not recommended to be used in clinical decision-making.[54]

I prefer batching several samples on the same gel to facilitate comparison of positive and negative samples. Aside from the improvement in efficiency that this offers, it makes artifacts due to specific gel preparation relatively obvious. I recommend that serum from the patient accompany the CSF sample. My laboratory will report a negative study for CSF oligoclonal bands when serum is not provided. However, when a CSF sample contains one or more bands in the CSF and no serum is provided, no final interpretation can be made with confidence. The report is always appended with

'serum must be run with CSF sample to improve the specificity of this information.' Our laboratory accepts a serum up to 2 weeks after the CSF sample was run to assay as the control serum. If the serum had oligoclonal bands, they will still be present in a large enough concentration to be detected. This cut-off is arbitrary and has not been rigorously investigated.

Although the blood–CSF barrier excludes most immunoglobulins, some do cross the blood–CSF barrier. For example, in Fig. 8.9, a patient with an obvious monoclonal gammopathy in the serum has had some of it transfer into the CSF. When CSF samples contain one prominent band and we are not sent a corresponding serum, we recommend that the clinician examine the serum and urine for the presence of a monoclonal gammopathy. In patients with prominent oligoclonal banding in the serum, these immunoglobulins will also find their way into the CSF (Fig. 8.10). Typically, they stain more strongly in the serum than in the corresponding CSF, whereas, if the bands had originated in the CSF (due to local synthesis in patients with multiple sclerosis), they would not be detectable at

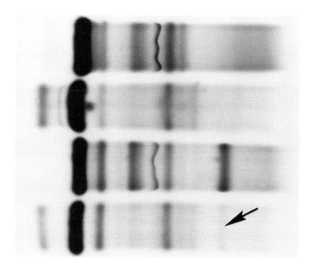

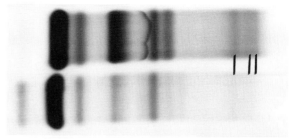

Figure 8.10 A serum sample diluted 1:3 in the top lane shows prominent α_1- and α_2-globulins along with three distinct oligoclonal bands. The cerebrospinal fluid (CSF) below is concentrated 80-fold and has the same three oligoclonal bands barely visible. Because of the density of staining of the bands in the serum and their faint staining in the CSF, I report that since both CSF and serum have the same oligoclonal bands that it is considered negative for CSF oligoclonal bands. The patient had a systemic inflammatory condition, not multiple sclerosis. (Paragon SPE2 system stained with Paragon Violet.)

Figure 8.9 The bottom cerebrospinal fluid (CSF) is concentrated 80-fold and has a small monoclonal band (arrow). This has likely diffused across the blood–CSF barrier from the serum, because the corresponding serum (diluted 1:3) immediately above has a much denser band in the same region. Whereas one cannot rule out involvement of the central nervous system by the monoclonal proliferation of plasma cells, one would expect that the staining in the CSF would be stronger or at least equal to that seen in the serum (relative to the density of other bands). A normal serum and its corresponding CSF are in the top two lanes for comparison. (Paragon SPE2 system stained with Paragon Violet.)

all in the serum. Under conditions of damage to the blood–brain barrier, one must be very suspicious of bands present in both the CSF and in the corresponding serum. By using the lack of oligoclonal bands in the corresponding serum as a criterion for reporting the presence of oligoclonal bands in the CSF, one improves the specificity of the assay with only a slight decrease in sensitivity. In one study using high-resolution agarose, the sensitivity of the CSF oligoclonal band test was 84.9 per cent with a specificity of 78.9 per cent when the lack of these bands in the corresponding serum sample was required, however, when the requirement for no corresponding oligoclonal bands in the serum was withdrawn, the specificity dropped to 64.8 per cent.[55]

It is necessary to define what will be included as an 'oligoclonal band pattern.' When using the less specific agarose gel-based methods, it is important to exclude other non-immunoglobulins that may be seen as bands in the γ-region. For example, γ-trace protein band or the slight mid-γ restriction normally seen with γ CSF proteins can be problematic in cases with only a few bands. I also excluded artifacts such as the two fast γ restrictions mentioned above seen on the Paragon SPE2 system (Figs 8.4 and 8.5). Once these artifacts are known, these methods will demonstrate the more prominent oligoclonal bands (Figs 8.11 and 8.12).

The literature supports switching from high-resolution agarose gel electrophoresis to isoelectric focus or other techniques that use unconcentrated CSF and identify the bands as IgG. In the 2002 College of American Pathologists (CAP) Survey M-B, 218 laboratories (93 per cent) were listed as using electrophoresis to perform analysis of oligoclonal bands, whereas only 17 (7 per cent) were listed as performing isoelectric focusing.[56] At that time our laboratory was performing the Sebia Hydragel CSF (Sebia, Issy-les-Moulineaux, France) by immunofixation, there was no category for this type of testing and we listed the method as 'other'. These figures indicate that while most investigators now recommend isoelectric focusing and/or immunofixation methods to enhance the sensitivity and specificity of the examination of CSF for oligoclonal bands, the vast majority of clinical

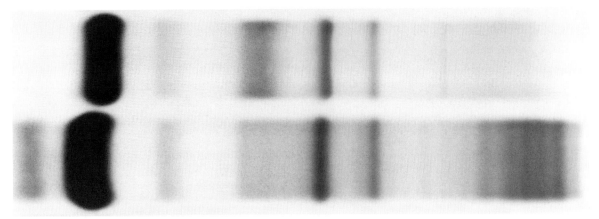

Figure 8.11 The bottom lane contains cerebrospinal fluid (CSF) concentrated 80-fold from a patient with multiple sclerosis and the corresponding serum diluted 1:4 is immediately above. Several oligoclonal bands are evident in the CSF. (Panagel system stained with Coomassie Blue.)

laboratories participating in this survey do not use these more sensitive and specific methods.

Some authors note that IgM oligoclonal bands may be useful to document the onset of multiple sclerosis or to document acute relapse.[4,24,57–60] However, the clinical evidence of relapse itself may

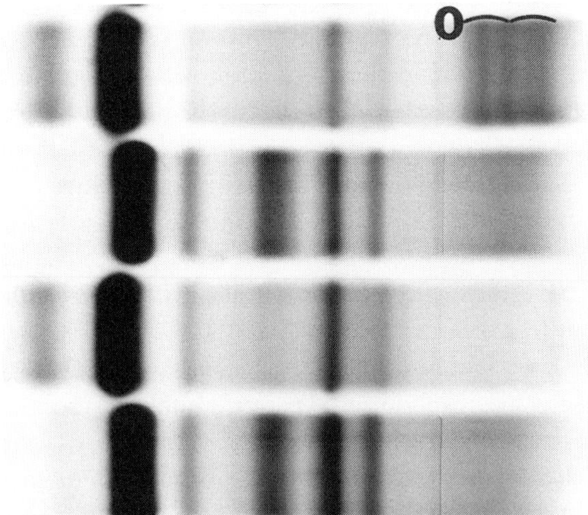

Figure 8.12 The top sample is a cerebrospinal fluid (CSF) concentrated 80-fold from a patient with multiple sclerosis and the corresponding serum diluted 1:3 is immediately below. The γ-region of the CSF from this patient has several densely staining oligoclonal bands (O) which are not in the corresponding serum. The CSF is in the third lane and has a faint, slow γ-band not seen in its corresponding serum below. This single band is insufficient for an interpretation of oligoclonal bands. (Paragon SPE2 system stained with Paragon Violet).

provide sufficient evidence of its occurrence, or, if objective information is needed, immunoassay for myelin basic protein in CSF is an excellent indicator of disease activity.[61]

MULTIPLE SCLEROSIS AND OLIGOCLONAL BANDS

The most common reason for performing electrophoretic analysis of CSF is to help in the evaluation of a patient suspected of having multiple sclerosis.[4] Although detection of oligoclonal bands in the γ-region is not specific for multiple sclerosis, examination of the CSF for the presence of oligoclonal bands is helpful because the clinical diagnosis of multiple sclerosis can be difficult and *supportive* laboratory data is useful to the clinician.[3]

Multiple sclerosis is a disease predominately of young adults (beginning 20–40 years of age) and is more frequent in women (2:1). It occurs in about 100 per 100 000 individuals among Caucasian populations.[62] The incidence of multiple sclerosis is, however, much lower in Asian populations. In Japan, the incidence is 0.7–3.8 cases per 100 000.[63,64] Further, the Western type of multiple sclerosis tends to diffusely involve the CNS, whereas the Asian type more selectively involves the optic nerves and spinal cord.[64] The epidemiological differences are paralleled by a difference in the occurrence of oligoclonal bands. In Western countries, over 90 per cent of individuals with diffuse multiple sclerosis have oligoclonal IgG bands in their CSF. Among Japanese patients with diffuse multiple sclerosis, only about half have oligoclonal bands and they are present in only about 10 per cent of individuals with the more selective form of multiple sclerosis.[64,65]

In multiple sclerosis, localized destruction of myelin occurs in the CNS.[66] The immune system has been implicated by the demonstration of myelin-reactive T lymphocytes and the oligoclonal bands in CSF reflecting local synthesis of immunoglobulins.[62] Presenting clinical symptoms and signs of multiple sclerosis are highly variable and include: weakness, diplopia, optic neuritis, paresthesias, numbness, poor vibration sensation leading to difficulty with coordinated movements, absence of abdominal reflexes, trigeminal neuralgia (in young adults), vertigo, and easy fatigability.[67,68] The diagnosis of multiple sclerosis depends on recurrent episodes of the above phenomena involving at least two anatomic sites. Magnetic resonance imaging (MRI) is both a diagnostic tool, and a marker to monitor the progress of the disease. It serves as a measurable baseline for therapeutic trials.[69,70] Because early clinical signs and symptoms are non-specific, the diagnosis of multiple sclerosis at this stage can be quite difficult. The clinical laboratory provides useful information to support the diagnosis in many of these patients.

The etiology of multiple sclerosis is unknown, but genetic factors have been implicated mainly by family studies. The fact that 26 per cent of monozygotic twins both develop multiple sclerosis compared with only 2.3 per cent of dizygotic twins is strong evidence for a genetic basis for multiple sclerosis.[71] Further support for this idea comes from the observation of a higher risk for developing multiple sclerosis in offspring of affected individuals than in spouses of affected individuals.[72] Alterations of the immune system have been implicated in this process. For example, patients with multiple sclerosis have a restricted immune response with predominately IgG1 subclass.[73] Genetic linkage studies have indicated a variety of associations, but the strongest is with HLA-DR2 (human leukocyte antigen) genes.[62,74] A genetic predisposition would be consistent with findings of distinctive idiotypes and a proclivity toward development of autoantibody-secreting cells in multiple sclerosis.[75,76] Evidence that the IgG heavy chain repertoire in plaques from patients with multiple sclerosis differs from that of their peripheral blood lymphocytes implies a specific CNS antigen-driven targeting that may have a genetic basis.[77]

Several tests can be performed by clinical laboratories to aid in the diagnosis of multiple sclerosis: oligoclonal bands in CSF, CSF IgG synthesis (IgG Index), CSF myelin basic protein, and serum antibodies against myeline oligodendrocyte glycoprotein (MOG) and against myelin basic protein (MBP).[78,79] Detection of oligoclonal bands is the most sensitive for supporting the initial diagnosis.[80] Seres et al.[36] reported that, whereas 76 per cent of the 37 patients with clinically documented multiple sclerosis had an elevated IgG Index, 91 per cent had oligoclonal banding as demonstrated by agarose gel electrophoresis. Anti-MOG and anti-MBP in serum are strong predictors for early conversion of early to clinically definite multiple sclerosis.[78]

Patients with multiple sclerosis have an increased local (CSF) synthesis of immunoglobulins. This is demonstrated by the fact that oligoclonal bands are found in the CSF, but not in the corresponding serum in over 90 per cent of these patients.[3,36,41,55,79,81] Multiple sclerosis patients lacking oligoclonal bands in the CSF have fewer plasma cells within the meninges and fewer plaques at time of autopsy than patients whose CSF contains oligoclonal bands. This suggests that the oligoclonal bands are a reflection of the local synthesis of immunoglobulin by plasma cells in the diseased tissue.[82] Although there have been many studies on a wide variety of possible antigens (many viral), the specific antigen(s) against which most these antibodies are being made has not been identified. It is likely that many antigens are responsible.

Myelin basic protein is elevated in CSF during acute episodes of multiple sclerosis.[83,84] As such, it is useful in following disease activity.[85] However, since myelin basic protein may not be detectable during periods of quiescence, it may give a false negative result in these patients.[86,87] Also, myelin basic protein will be elevated in other conditions involving damage to myelin in the CNS.[81]

Calculation of CSF IgG index

Patients with multiple sclerosis do not usually have elevated total protein in the CSF, as indicated by a normal CSF albumin quotient.[36] The ratio of CSF IgG to serum IgG (there are several formulas for this that use various corrections for serum proteins) provides an estimate of local production of IgG within the CNS. Normally, the immunoglobulin synthesized within the CSF is a relatively small amount compared with the other CSF proteins. However, in multiple sclerosis, there is an increased local synthesis of IgG that results in a decrease in the albumin/globulin ratio of the CSF. An IgG index is used to correct for decreases in albumin/globulin ratio associated with diseases that merely increase the permeability of the blood brain barrier.[11] The upper limit of normal is 0.72.[41]

$$\text{CSF (IgG index)} = \frac{(\text{CSF IgG/serum IgG})}{(\text{CSF albumin/serum albumin})}$$

Reports on the sensitivity of an elevated IgG index for multiple sclerosis vary widely from slightly less than 50 per cent to as high as 80 per cent.[41,88] Although an elevated IgG index is useful with a patient in the appropriate clinical setting, elevated local production of IgG is also seen in patients with viral encephalitis, bacterial meningitis, neurosyphilis, subacute sclerosing panencephalitis, acute poliomyelitis, and Guillain–Barré syndrome.[89] Elevated IgG index has also been reported in patients with neurosarcoidosis, systemic lupus erythematosus and other conditions, but these cases do not usually have oligoclonal bands.[90–93] Patients whose CSF IgG elevation is caused by the presence of a systemic polyclonal gammopathy, with immunoglobulin passively diffusing into the CSF, will have a normal IgG index because the serum IgG value will negate this factor in the equation. A combination of the IgG index and oligoclonal banding yields confirmatory evidence of multiple sclerosis in more than 90 per cent of cases. Papadopoulos et al.[94] caution that to minimize errors in laboratory methodology, the same immunochemical method (e.g. nephelometry) should be used to calculate the IgG and the albumin of both the CSF and the serum.

Interpretation of O-band studies

In multiple sclerosis, the oligoclonal IgG is synthesized locally in the CNS. Therefore, the oligoclonal bands in multiple sclerosis patients are present in the CSF and not in their serum.[4,71,73,95] My standard laboratory sign-outs for CSF are provided in Table 8.2. Please feel free to use them. Occasionally, there are unusual features to a case that require me to depart from the standard sign-out and write a unique interpretation.

The presence of diffusely staining immunoglobulin in both CSF and serum is a negative pattern. I encourage the clinicians to submit a serum with the CSF, however, a negative CSF can be interpreted without a serum. The serum may be helpful, however, because the presence of a monoclonal gammopathy in the serum may be related to neurological symptoms.

A positive result requires at least two immunoglobulin bands in the CSF with no matching bands in the serum (Fig. 8.13). I do not report the number of oligoclonal bands in the interpretation. Avasarala et al.[54] documented that the number of oligoclonal bands is an insensitive prognostic indicator and recommended that it not be used to influence decisions about therapy. The bands persist, unchanged in pattern in most patients. The clones of B lymphocytes in the CNS seem to be quite robust, because even following

Table 8.2 Interpretations of cerebrospinal fluid (CSF) protein electrophoresis

CSF is negative for oligoclonal bands.

CSF is positive for oligoclonal bands. No bands are seen in the corresponding serum.

CSF contains oligoclonal bands. Since the corresponding serum contains the same oligoclonal bands, this is not
specific enough to be considered supportive evidence for multiple sclerosis.

Oligoclonal bands are seen in the CSF. Without a corresponding serum sample, we are uncertain as to the
significance of these findings. Recommend: serum for comparison with the CSF pattern. *Repeat CSF is not needed if
the serum is sent within the next 2 weeks.*

One γ-band is seen in the CSF. This is insufficient to be supportive evidence for multiple sclerosis.

A monoclonal band is present in the serum. Recommend serum evaluation of monoclonal gammopathy.

autologous hematopoietic stem cell transplantation the oligoclonal bands were found to persist despite magnetic resonance imaging (MRI) evidence of reduction in some lesions.[96]

When I observe oligoclonal bands in both the CSF and serum, I consider this to be a result of diffusion of the serum oligoclonal bands into the CSF. This

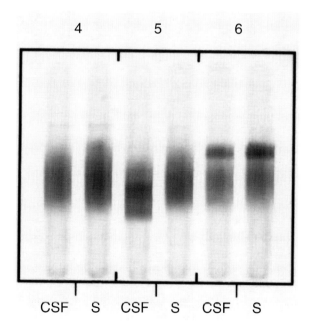

Figure 8.13 Immunofixation performed on unconcentrated cerebrospinal fluid (CSF). Sample 4 has a negative CSF and negative serum (S). Sample 5 has several oligoclonal bands in the CSF and none in the corresponding serum. Sample 6 has a monoclonal band in both the CSF and the serum. (Sebia CSF immunofixation gel.)

has been referred to as the 'mirror pattern'.[4] An occasional case of multiple sclerosis with oligoclonal bands in both locations does occur, but this is not considered useful in confirming the presence of multiple sclerosis.[95]

The presence of oligoclonal bands in the CSF with no serum sample submitted is of questionable significance. When this occurs, I recommend that a serum be sent within the next 2 weeks for comparison with this CSF pattern. The 2 weeks is an arbitrary period, which reflects the fact that if the oligoclonal bands came from a systemic process, the IgG from those systemic clones should have a half-life of about 2–3 weeks. Therefore, at least half of the amount should be present. Obviously, the ideal is to have the serum accompany the CSF sample. But the patient can avoid a needless repeat lumbar puncture if the serum can be obtained relatively soon after the first sample.

When there is one small band in the CSF but none in the serum, I note its presence but advise the clinician that this is not sufficiently strong evidence to support the diagnosis of multiple sclerosis.

Finally, occasionally I have observed a relatively large monoclonal band in the CSF; these cases also have the band in the serum. When I observe this, I recommend an evaluation of the patient for the monoclonal gammopathy (which may be accounting for the neurological symptoms).

OTHER CONDITIONS WITH CSF OLIGOCLONAL BANDS

One early sign of multiple sclerosis is optic neuritis. Examination of the CSF for the presence of oligoclonal bands and/or an elevated IgG index increases the risk that these individuals will go on to develop multiple sclerosis. However, a normal CSF study in this group cannot rule out that possibility.[97]

A positive CSF oligoclonal band test is not pathognomonic for multiple sclerosis. Oligoclonal bands can be found in a wide variety of neurological diseases, including inflammation, neoplasia, cerebrovascular accidents, structural CNS lesions, demyelinating diseases, and some peripheral neuropathies (Table 8.3).[98–107] Except for those patients with subacute sclerosing panencephalitis, the percentage of patients with these conditions who have oligoclonal bands is considerably less than the approximately 90 per cent of multiple sclerosis patients with oligoclonal bands. Fortunately, multiple sclerosis is not part of the differential diagnosis in most of these clinical conditions.

Because the oligoclonal band test is non-specific, it should be used in defined situations, such as in the case of a patient that has had few clinical episodes suggestive of multiple sclerosis, but the diagnosis is not yet secure. The presence of oligoclonal bands in those patients is useful supportive information. A negative test will cause the clinician to review the clinical features, as 80–90 per cent of patients with multiple sclerosis should have oligoclonal bands in the CSF. The laboratory test should never be used as the sole evidence of multiple sclerosis. Using a sensitive polyacrylamide gel electrophoresis technique, Coret et al.[108] reported the diagnoses associated with CSF oligoclonal bands in a study of 488 patients suffering from neurological disease (Table 8.4). While multiple sclerosis represented the vast majority of cases, almost half of their patients with infectious diseases had oligoclonal bands, as did 11 per cent of patients with vascular malignancies.

Table 8.3 Conditions in which oligoclonal bands may be found in cerebrospinal fluid (CSF)

Multiple sclerosis[a]
Subacute sclerosing panencephalitis
Creutzfeldt–Jakob disease
Meningoencephalitis
Spinal cord compression
Guillain–Barré syndrome
Syphilis
Peripheral neuropathy
Optic neuritis
Hydrocephalus
Cerebrovascular accident
Immune complex vasculitis
Systemic lupus erythematosus
Diabetes
Whipple's disease
Neoplasms
Acquired immune deficiency syndrome (AIDS)
Lyme disease
Fever of unknown origin

[a]In multiple sclerosis about 90% of patients will have oligoclonal bands in the CSF. In most of the other conditions listed, such bands are uncommon but *may* be seen.

Central nervous system systemic lupus erythematosus (CNS lupus)

Some patients with systemic lupus erythematosus (SLE) develop central nervous system (CNS) involvement, that manifests a variety of symptoms: psychosis, cranial nerve palsy, seizures, cerebrovascular accidents, and transverse myelopathy.[109,110] The incidence of CNS manifestations in patients with SLE varies widely from 25 per cent, reported in a large clinical series, to as high as 75 per cent in a retrospective postmortem series.[111] Central nervous system involvement can be the cause of death in as many as 13 per cent of these patients.[112] Unfortunately, symptoms of CNS lupus can be

Table 8.4 Occurrence of cerebrospinal fluid (CSF) oligoclonal bands in patients with neurological disease

Diagnosis	Per cent of patients with CSF O-bands
Definite multiple sclerosis	84
Probable multiple sclerosis	46
Inflammatory infectious diseases	43
Possible multiple sclerosis	7
Vascular malignancies	11
Other neurological diseases	4
Control CSF	0

ªData from Coret et al.[108]

mimicked by steroid psychosis, and the clinician is occasionally faced with a patient with known SLE who is receiving steroid therapy and demonstrating psychotic symptoms. Should the steroids be increased (for CNS lupus) or should they be tapered (for steroid psychosis)? Over the years, there have been many attempts at establishing laboratory tests that would help with this differential diagnosis. Largely, they have failed. Levels of C3, C4, anti-DNA, oligoclonal bands and, more recently, interleukin-1 and interleukin-6 levels in CSF are, at best, only partly helpful.[113] One recent improvement in the use of these assays is the adoption of the ratios (Q) of CSF C3 to serum C3, and CSF C4 to serum C4 rather than the absolute values. The CSFQ3 and Q4 are increased in some patients with CNS lupus.[114] Nonetheless, the clinical picture remains the gold standard for whether the patient has CNS lupus.[109,110,115]

The presence of oligoclonal bands in the CSF also has been suggested as an aid in the diagnosis of CNS lupus. Unfortunately, this is a poor marker because only a minority of patients with CNS lupus have such bands.[116] Further, a negative oligoclonal band test in a patient with clinically suspected CNS lupus will not cause the clinician to withhold therapy. I recommend use of the CSF

oligoclonal band test only as a confirmatory test for patients suspected of having multiple sclerosis. The increased IgG levels in CSF of some patients with CNS lupus result primarily from an impaired blood–CSF barrier.[111]

Serum antibody against ribosomal P has been suggested as a marker for patients with lupus psychosis.[117-120] These antibodies give a pattern on the fluorescent antinuclear antibody (ANA) test that shows both cytoplasmic and nuclear staining similar to that seen with mitochondrial antibody. Previously, confirmation of this antibody required either Western blot or a specific immunoassay for ribosomal P.[121] However, a recently reported enzyme-linked immunosorbent assay (ELISA) that used a 22 amino acid peptide that corresponds to a common epitope on ribosomal P0, P1, and P2 had an 83 per cent concordance with Western blot.[122] When this ELISA was used to test sera from 178 consecutive patients with SLE and 28 others with CNS lupus, the presence of anti-ribosomal P was associated with clinically active disease, high levels of anti-dsDNA and decreased C4 levels.[122] Unfortunately, only 11 of the 28 patients with CNS lupus were positive for anti-ribosomal P. While this is significantly greater than in unselected SLE patients, it leaves more than 60 per cent of individuals with CNS lupus as false negatives. Therefore, in its present form, a positive suggests more clinically active disease and is supportive, but not diagnostic evidence for CNS lupus. Also, the absence of the antibody does not rule out CNS lupus.[123]

DETECTION OF CSF LEAKAGE IN NASAL AND AURAL FLUID FOLLOWING HEAD TRAUMA

Leakage of CSF into nasal or aural cavities is most often caused by trauma, but also results from intracranial surgical procedures, infection, hydrocephalus, congenital malformations, and neoplasms.[124] In order to prevent the development of meningitis, the CSF leakage must be differentiated from allergic rhinitis or infectious rhinosinusitis as soon as possible.[125,126] There are two highly specific methods available to detect CSF leakage: β₂-transferrin demonstration by immunofixation or immunoblotting and measurement of β-trace protein (prostaglandin D synthase). At present, the detection of β₂-transferrin is more commonly employed. However, it requires non-standard immunofixation or immunoblotting techniques, as described below. In contrast, prostaglandin D synthase may be measured by either nephelometry or immunofixation and will likely become the preferred method for this situation.[125,127–131]

As mentioned earlier in this chapter, the presence of neuraminidase in the CNS causes desialation of some transferrin molecules in the CNS. The loss of these negatively charged sialic acid groups from some transferrin molecules results in two transferrin bands by electrophoresis: the β₁ fraction (the same as that found in serum), and a more cathodal β₂-transferrin band (τ fraction, also called CSF-specific transferrin).[124] β₂-Transferrin is not normally present in serum, tears, saliva, sputum, nasal or aural fluid, perilymph or endolymph, although it is present in aqueous and vitreous humor.[132] However, one must be cautious to

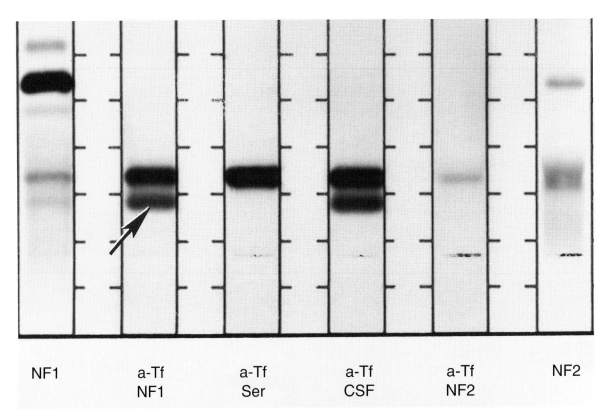

| NF1 | a-Tf NF1 | a-Tf Ser | a-Tf CSF | a-Tf NF2 | NF2 |

Figure 8.14 Examination of two nasal fluids for the presence of β₂-transferrin. The electrophoresis of nasal fluid 1 (NF1) fixed in acid is shown in the far left lane, and its immunofixation with anti-transferrin (a-Tf NF1) is in the second lane. We had a control serum from NF1 patient which is in the third lane (a-Tf Ser). A control cerebrospinal fluid (CSF) (not from either patient) reacted with anti-transferrin is in the fourth lane (a-Tf CSF). The immunofixation with anti-transferrin for nasal fluid 2 (NF2) is in the fifth lane (a-Tf NF2) and the electrophoresis of NF2 fixed in acid is in the far right lane. A β₂-region band is present in both the NF1 sample (arrow) and in the control CSF. No such band is present in the NF2 sample. Note, however, that the NF2 sample stains more weakly and has less protein. Therefore, one cannot exclude a sensitivity problem. Also, a serum was not sent with the NF2 sample, so if a β₂-transferrin band had been seen, we would have had to obtain a serum to exclude a genetic variant.

control for genetic variants of transferrin. Immunological identification of β_2-transferrin provides a sensitive and specific tool to distinguish CSF leakage from serous nasal or aural fluid. Immunofixation, or more sensitive immuno-blotting procedures on these fluids have proven useful to make this distinction.[14,124,133]

My laboratory performs immunofixation on 10× concentrated samples of the nasal or aural fluid. After concentration, 3–5 ml of unknown fluid are applied onto the gel in two lanes with the patient's serum diluted 1:3 in an adjacent lane to control for genetic variants of transferrin. A CSF control is applied to two other lanes as a positive control for β_2-transferrin (Fig. 8.14). After a 5-min diffusion time, the gels are gently blotted and the samples are electrophoresed at 100 V for 30 min. The gels are then overlaid with 80 μl of antiserum against human transferrin. Following a 35 min incubation at room temperature in a moisture chamber, the gels are washed twice in isotonic saline, dried, stained with Paragon Violet for 5 min, and destained in two washes of 10 per cent glacial acetic acid.[14]

Western blotting may also be performed to detect some of this leakage as well as genetic variants of transferrin.[134–136] This involves performing serum protein electrophoresis and then blotting the proteins onto nitrocellulose paper. The paper is then incubated with anti-transferrin (Fig. 8.15). By using immunoenzyme conjugates, this technique may provide greater sensitivity and lower background than immunofixation. Normansell et al.[137] report a sensitivity of 1 mg/ml and their procedure has the further advantage of not requiring a concentration step. By using a combination of isoelectric focusing on polyacrylamide gel, direct immunofixation and silver staining, Roelandse et al.[138] have further improved the sensitivity of this procedure.

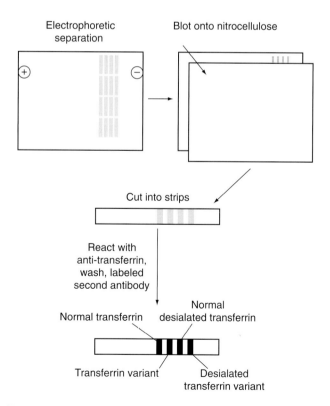

Figure 8.15 Schematic view of Western blot to detect transferrin variants.

When interpreting these qualitative electrophoretic assays for β_2-transferrin, one must exclude false positives that may result from genetic protein variants that migrate in the β_2-region (Figs 8.16 and 8.17).[134] Fortunately, the most common transferrin variant migrates anodal to β_1-transferrin. Rare cases have been reported, however, with a variant transferrin band at the β_2 position.[134,139] If the serum control shows such a band, then this technique will not provide a definitive answer for that patient. If one could obtain CSF from that patient, however, it might show that the desialated form of that transferrin variant has an even greater cathodal migration that could be recognized. Normally, however, one does not need to have a CSF sample from the patient if the serum does not show a β_2-migrating transferrin band. Finally, a negative result does not rule out CSF leakage; it may represent a sensitivity problem. A remote possibility is a patient with congenital atransferrinemia, but this would be obvious from the patient's serum control lane.[140] Therefore, when a negative result is obtained, a cautionary note about the limitations of the procedure is added.

The other method to detect CSF leakage involves detection of prostaglandin D synthase, formerly known as β-trace protein. There are two isoforms

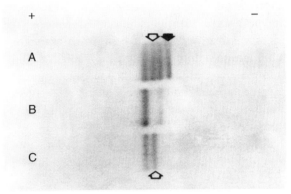

Figure 8.17 Western blot of the sample from Fig. 8.16. For the Western blot, the proteins were separated by agarose electrophoresis and then transferred to nitrocellulose. Strips of nitrocellulose were reacted with anti-transferrin and then with peroxidase labeled anti-globulin. This produces a result similar to immunofixation, but is more sensitive and may have less background. The patient's cerebrospinal fluid (CSF) is in lane A, a normal CSF is in lane B and the patient's serum is in lane C. The patient's CSF (lane A) shows four bands. The band with the open arrow is the variant TfD which corresponds to the variant in the patient's serum (C), also labeled with an open arrow. The desialated variant is seen in the patient's CSF as the most cathodal band (filled arrow, lane A). The normal CSF in lane B shows the usual β_1-transferrin band and the slower, usually weaker-staining β_2-transferrin band. This photograph was provided by Drs Arthur J. Sloman and Robert H. Kelly.[134]

of this enzyme. The CNS isoform of prostaglandin D synthase catalyses the conversion of prostaglandin H2 to prostaglandin D2A.[141] It is involved in the rapid eye movement during sleep.[142] Felgenhauer et al. first established an Ouchterlony method to detect this protein as an indication of CSF leakage.[143] This method was too insensitive, however, and others suggested the use of immunoelectrophoresis.[128] Of 59 cases of CSF leakage examined in this manner, 57 were detected, two were missed and 73 true negatives were ruled out for leakage, resulting in a sensitivity of 97.3 per cent and specificity of 100 per cent.[128] Kleine et al.[21] used nephelometry to detect prostaglandin D synthase and compared the results to detection of β_2-transferrin by isoelectric focusing with transferrin specific immunofixation. Although they found the prostaglandin D synthase method to be more sensitive and less prone to interference, they recommended complementary use of the two assays to

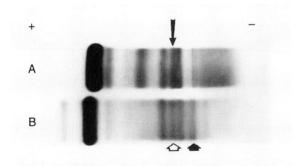

Figure 8.16 Agarose gel of serum (lane A) and cerebrospinal fluid (CSF; lane B) from a patient with a slow-moving transferrin variant. The serum contains the normal transferrin band and a cathodal variant (long arrow) migrates between it and C3. The CSF contains two abnormal bands. One (short open arrow) has a similar mobility to the variant band in the serum and the second (short filled arrow) migrates more toward the γ-region. The latter band is due to the desialated transferrin variant. This photograph was provided by Drs Arthur J. Sloman and Robert H. Kelly.[134]

detect nasal and aural leakage of CSF. More recent nephelometric studies substantiate the claimed sensitivity of greater than 90 per cent with a concentration of prostaglandin D synthase (β-trace protein) of 6 mg/l or higher and a specificity of 100 per cent. Detection of this enzyme may soon become the method of choice to identify CSF leakage.[21,125,128–131,142–145]

REFERENCES

1. Kabat EA, Landow H, Moore DH. Electrophoretic patterns of concentrated cerebrospinal fluid. *Proc Soc Ex Biol Med* 1942;**49**:260–263.

2. Kabat EA, Moore DH, Landow H. An electrophoretic study of the protein components in cerebrospinal fluid and their relationship to the serum proteins. *J Clin Invest* 1942;**21**:571–577.

3. Falip M, Tintore M, Jardi R, Duran I, Link H, Montalban X. Clinical usefulness of oligoclonal bands. *Rev Neurol* 2001;**32**:1120–1124.

4. Sindic CJ. CSF analysis in multiple sclerosis. *Acta Neurol Belg* 1994;**94**:103–111.

5. Sickmann A, Dormeyer W, Wortelkamp S, Woitalla D, Kuhn W, Meyer HE. Towards a high resolution separation of human cerebrospinal fluid. *J Chromatogr B Analyt Technol Biomed Life Sci* 2002;**771**:167–196.

6. Thompson EJ, Keir G. Laboratory investigation of cerebrospinal fluid proteins. *Ann Clin Biochem* 1990;**27**(Pt 5):425–435.

7. Fishman RA. *Cerebrospinal fluid in diseases of the nervous system.* Philadelphia: WB Saunders, 1992.

8. Jeppson JO, Laurell CB, Franzen B. Agarose gel electrophoresis. *Clin Chem* 1979;**25**:629–638.

9. Griffin DE, Giffels J. Study of protein characteristics that influence entry into the cerebrospinal fluid of normal mice and mice with encephalitis. *J Clin Invest* 1982;**70**:289–295.

10. Blennow K, Fredman P, Wallin A, Gottfries CG, Skoog I, Wikkelso C, Svennerholm L. Protein analysis in cerebrospinal fluid. III. Relation to blood–cerebrospinal fluid barrier function for formulas for quantitative determination of intrathecal IgG production. *Eur Neurol* 1993;**33**:134–142.

11. Blennow K, Fredman P, Wallin A, et al. Protein analysis in cerebrospinal fluid. II. Reference values derived from healthy individuals 18–88 years of age. *Eur Neurol* 1993;**33**:129–133.

12. Moos T. Immunohistochemical localization of intraneuronal transferrin receptor immunoreactivity in the adult mouse central nervous system. *J Comp Neurol* 1996;**375**:675–692.

13. Kissel K, Hamm S, Schulz M, Vecchi A, Garlanda C, Engelhardt B. Immunohistochemical localization of the murine transferrin receptor (TfR) on blood-tissue barriers using a novel anti-TfR monoclonal antibody. *Histochem Cell Biol* 1998;**110**:63–72.

14. Zaret DL, Morrison N, Gulbranson R, Keren DF. Immunofixation to quantify beta 2-transferrin in cerebrospinal fluid to detect leakage of cerebrospinal fluid from skull injury. *Clin Chem* 1992;**38**:1908–1912.

15. Rukstalis MR, Lynch KG, Oslin DW, et al. Carbohydrate-deficient transferrin levels reflect heavy drinking in alcohol-dependent women seeking treatment. *Alcohol Clin Exp Res* 2002;**26**:1539–1544.

16. Arndt T, Kropf J. Alcohol abuse and carbohydrate-deficient transferrin analysis: are screening and confirmatory analysis required? *Clin Chem* 2002;**48**:2072–2074.

17. Suzuki Y, Saito H, Suzuki M, Hosoki Y, Sakurai S, Fujimoto Y, Kohgo Y. Up-regulation of transferrin receptor expression in hepatocytes by habitual alcohol drinking is implicated in hepatic iron overload in alcoholic liver disease. *Alcohol Clin Exp Res* 2002;**26**:26S-31S.

18. Hermansson U, Helander A, Brandt L, Huss A, Ronnberg S. The Alcohol Use Disorders Identification Test and carbohydrate-deficient transferrin in alcohol-related sickness absence. *Alcohol Clin Exp Res* 2002;**26**:28–35.

19. Duan W, Cole T, Schreiber G. Cloning and nucleotide sequencing of transthyretin (prealbumin) cDNA from rat choroid plexus and liver. *Nucl Acids Res* 1989;**17**:3979.

20. Harms PJ, Tu GF, Richardson SJ, Aldred AR, Jaworowski A, Schreiber G. Transthyretin (prealbumin) gene expression in choroid plexus is strongly conserved during evolution of vertebrates. *Comp Biochem Physiol B* 1991;**99**:239–249.

21. Kleine TO, Damm T, Althaus H. Quantification of beta-trace protein and detection of transferrin isoforms in mixtures of cerebrospinal fluid and blood serum as models of rhinorrhea and otorrhea diagnosis. *Fresenius J Anal Chem* 2000;**366**:382–386.

22. Whicher JT. The interpretation of electrophoresis. *Br J Hosp Med* 1980;**24**:348–356.

23. Blennow K, Fredman P. Detection of cerebrospinal fluid leakage by isoelectric focusing on polyacrylamide gels with silver staining using the PhastSystem. *Acta Neurochir (Wien)* 1995;**136**:135–139.

24. Rijcken CA, Thompson EJ, Teelken AW. An improved, ultrasensitive method for the detection of IgM oligoclonal bands in cerebrospinal fluid. *J Immunol Methods* 1997;**203**:167–169.

25. Hiraoka A, Arato T, Tominaga I, Eguchi N, Oda H, Urade Y. Analysis of low-molecular-mass proteins in cerebrospinal fluid by sodium dodecyl sulfate capillary gel electrophoresis. *J Chromatogr B Biomed Sci Appl* 1997;**697**: 141–147.

26. Link H. Isolation and partial characterization of 'trace' proteins and immunoglobulin G from cerebrospinal fluid. *J Neurol Neurosurg Psychiatry* 1965;**28**:552–559.

27. Stibler H. The normal cerebrospinal fluid proteins identified by means of thin-layer isoelectric focusing and crossed immunoelectrofocusing. *J Neurol Sci* 1978;**36**:273–288.

28. Laurenzi MA, Link H. Localization of the immunoglobulins G, A and M, beta-trace protein and gamma-trace protein on isoelectric focusing of serum and cerebrospinal fluid by immunofixation. *Acta Neurol Scand* 1978;**58**:141–147.

29. Laurell CB. Composition and variation of the gel electrophoretic fractions of plasma, cerebrospinal fluid and urine. *Scand J Clin Lab Invest Suppl* 1972;**124**:71–82.

30. Lott JA, Warren P. Estimation of reference intervals for total protein in cerebrospinal fluid. *Clin Chem* 1989;**35**:1766–1770.

31. Tibbling G, Link H, Ohman S. Principles of albumin and IgG analyses in neurological disorders. I. Establishment of reference values. *Scand J Clin Lab Invest* 1977;**37**:385–390.

32. Link H, Tibbling G. Principles of albumin and IgG analyses in neurological disorders. II. Relation of the concentration of the proteins in serum and cerebrospinal fluid. *Scand J Clin Lab Invest* 1977;**37**:391–396.

33. Biou D, Benoist JF, Nguyen-Thi C, Huong X, Morel P, Marchand M. Cerebrospinal fluid protein concentrations in children: age-related values in patients without disorders of the central nervous system. *Clin Chem* 2000;**46**:399–403.

34. Tatara R, Imai H. Serum C-reactive protein in the differential diagnosis of childhood meningitis. *Pediatr Int* 2000;**42**:541–546.

35. Chamoun V, Zeman A, Blennow K, Fredman P, Wallin A, Keir G, Giovannoni G, Thompson EJ. Haptoglobins as markers of blood – CSF barrier dysfunction: the findings in normal CSF. *J Neurol Sci* 2001;**182**:117–121.

36. Seres E, Bencsik K, Rajda C, Vecsei L. Diagnostic studies of cerebrospinal fluid in patients with multiple sclerosis. *Orv Hetil* 1998;**139**:1905–1908.

37. Qin D, Ma J, Xiao J, Tang Z. Effect of brain irradiation on blood–CSF barrier permeability of chemotherapeutic agents. *Am J Clin Oncol* 1997;**20**:263–265.

38. Hallgren R, Terent A, Wide L, Bergstrom K, Birgegard G. Cerebrospinal fluid ferritin in patients with cerebral infarction or bleeding. *Acta Neurol Scand* 1980;**61**:384–392.

39. Pakulski C, Drobnik L, Millo B. Age and sex as factors modifying the function of the blood-cerebrospinal fluid barrier. *Med Sci Monit* 2000;**6**:314–318.

40. Haussermann P, Kuhn W, Przuntek H, Muller T. Integrity of the blood–cerebrospinal fluid barrier in early Parkinson's disease. *Neurosci Lett* 2001; **300**:182–184.

41. Lunding J, Midgard R, Vedeler CA. Oligoclonal bands in cerebrospinal fluid: a comparative study of isoelectric focusing, agarose gel electrophoresis and IgG index. *Acta Neurol Scand* 2000;**102**: 322–325.

42. Niederwieser G, Bonelli RM, Brunner D, et al. Diagnostic accuracy of an agarose gel electrophoretic method in multiple sclerosis. *Clin Chem* 2001;**47**:144.

43. Lane JR, Bowles KJ, Normansell DE. The detection of oligoclonal IgG bands in unconcentrated cerebrospinal fluid by isoelectric focusing on thin-layer agarose gels and silver staining. *Arch Pathol Lab Med* 1986;**110**: 26–29.

44. Wybo I, van Blerk M, Malfait R, Goubert P, Gorus F. Oligoclonal bands in cerebrospinal fluid detected by PhastSystem isoelectric focusing. *Clin Chem* 1990;**36**:123–125.

45. Ivanova M, Tzvetanova E, Jetcheva V, Kilar F. Abnormal protein patterns in blood serum and cerebrospinal fluid detected by capillary electrophoresis. *J Biochem Biophys Methods* 2002;**53**:141–150.

46. Verbeek MM, de Reus HP, Weykamp CW. Comparison of methods for the detection of oligoclonal IgG bands in cerebrospinal fluid and serum: results of the Dutch Quality Control survey. *Clin Chem* 2002;**48**:1578–1580.

47. Hansson LO, Link H, Sandlund L, Einarsson R. Oligoclonal IgG in cerebrospinal fluid detected by isoelectric focusing using PhastSystem. *Scand J Clin Lab Invest* 1993;**53**:487–492.

48. Caudie C, Allausen O, Bancel J. Detection of oligoclonal IgG bands in cerebrospinal fluid by immunofixation after electrophoretic migration in the automated Hydrasys sebia system. *Ann Biol Clin (Paris)* 2000;**58**:376–379.

49. Cawley LP, Minard BJ, Tourtellotte WW, Ma BI, Chelle C. Immunofixation electrophoretic techniques applied to identification of proteins in serum and cerebrospinal fluid. *Clin Chem* 1976;**22**:1262–1268.

50. Cavuoti D, Baskin L, Jialal I. Detection of oligoclonal bands in cerebrospinal fluid by immunofixation electrophoresis. *Am J Clin Pathol* 1998;**109**:585–588.

51. Richard S, Miossec V, Moreau JF, Taupin JL. Detection of oligoclonal immunoglobulins in cerebrospinal fluid by an immunofixation–peroxidase method. *Clin Chem* 2002;**48**:167–173.

52. Nilsson K, Olsson JE. Analysis for cerebrospinal fluid proteins by isoelectric focusing on polyacrylamide gel: methodological aspects and normal values, with special reference to the alkaline region. *Clin Chem* 1978;**24**: 1134–1139.

53. Olsson JE, Link H, Muller R. Immunoglobulin abnormalities in multiple sclerosis. Relation to clinical parameters: disability, duration and age of onset. *J Neurol Sci* 1976;**27**:233–245.

54. Avasarala JR, Cross AH, Trotter JL. Oligoclonal band number as a marker for prognosis in multiple sclerosis. *Arch Neurol* 2001;**58**: 2044–2045.

55. Davenport RD, Keren DF. Oligoclonal bands in cerebrospinal fluids: significance of corresponding bands in serum for diagnosis of multiple sclerosis. *Clin Chem* 1988;**34**: 764–765.

56. College of American Pathologists (CAP). Cerebrospinal Fluid Survey M-B. *Participant Summary Rep* 2002:6.

57. Villar LM, Gonzalez-Porque P, Masjuan J, Alvarez-Cermeno JC, Bootello A, Keir G. A sensitive and reproducible method for the detection of oligoclonal IgM bands. *J Immunol Methods* 2001;**258**:151–155.

58. Nelson D, Fredman P, Borjeson J. A high-sensitivity immuno-chemiluminescence technique for detection of oligoclonal IgG and IgM in unconcentrated cerebrospinal fluid. *Scand J Clin Lab Invest* 1994;**54**:51–54.

59. Sharief MK, Thompson EJ. Intrathecal immunoglobulin M synthesis in multiple sclerosis. Relationship with clinical and cerebrospinal fluid parameters. *Brain* 1991;114(Pt 1A):181–195.

60. Sindic CJ, Monteyne P, Laterre EC. Occurrence of oligoclonal IgM bands in the cerebrospinal fluid of neurological patients: an immunoaffinity-mediated capillary blot study. *J Neurol Sci* 1994;124:215–219.

61. Ohta M, Ohta K, Ma J, et al. Clinical and analytical evaluation of an enzyme immunoassay for myelin basic protein in cerebrospinal fluid. *Clin Chem* 2000;46:1326–1330.

62. Hellings N, Raus J, Stinissen P. Insights into the immunopathogenesis of multiple sclerosis. *Immunol Res* 2002;25:27–51.

63. Kuroiwa Y, Igata A, Itahara K, Koshijima S, Tsubaki T. Nationwide survey of multiple sclerosis in Japan. Clinical analysis of 1084 cases. *Neurology* 1975;25:845–851.

64. Nakashima I, Fujihara K, Itoyama Y. Oligoclonal IgG bands in Japanese multiple sclerosis patients. *J Neuroimmunol* 1999;101:205–206.

65. Fukazawa T, Kikuchi S, Sasaki H, et al. The significance of oligoclonal bands in multiple sclerosis in Japan: relevance of immunogenetic backgrounds. *J Neurol Sci* 1998;158:209–214.

66. Zipp F. Apoptosis in multiple sclerosis. *Cell Tissue Res* 2000;301:163–171.

67. Lublin FD. The diagnosis of multiple sclerosis. *Curr Opin Neurol* 2002;15:253–256.

68. Bernard CC, Kerlero de Rosbo N. Multiple sclerosis: an autoimmune disease of multifactorial etiology. *Curr Opin Immunol* 1992;4:760–765.

69. Arnold DL, Matthews PM. MRI in the diagnosis and management of multiple sclerosis. *Neurology* 2002;58:S23–31.

70. McFarland HF. The emerging role of MRI in multiple sclerosis and the new diagnostic criteria. *Mult Scler* 2002;8:71–72.

71. Ebers GC, Bulman DE, Sadovnick AD, et al. A population-based study of multiple sclerosis in twins. *N Engl J Med* 1986;315:1638–1642.

72. Ebers GC, Yee IM, Sadovnick AD, Duquette P. Conjugal multiple sclerosis: population-based prevalence and recurrence risks in offspring. Canadian Collaborative Study Group. *Ann Neurol* 2000;48:927–931.

73. Vandvik B, Natvig JB, Wiger D. IgG1 subclass restriction of oligoclonal IgG from cerebrospinal fluids and brain extracts in patients with multiple sclerosis and subacute encephalitides. *Scand J Immunol* 1976;5:427–436.

74. Fogdell A, Olerup O, Fredrikson S, Vrethem M, Hillert J. Linkage analysis of HLA class II genes in Swedish multiplex families with multiple sclerosis. *Neurology* 1997;48:758–762.

75. LaGanke CC, Freeman DW, Whitaker JN. Cross-reactive idiotypy in cerebrospinal fluid immunoglobulins in multiple sclerosis. *Ann Neurol* 2000;47:87–92.

76. Sellebjerg F, Jensen CV, Christiansen M. Intrathecal IgG synthesis and autoantibody-secreting cells in multiple sclerosis. *J Neuroimmunol* 2000;108:207–215.

77. Owens GP, Burgoon MP, Anthony J, Kleinschmidt-DeMasters BK, Gilden DH. The immunoglobulin G heavy chain repertoire in multiple sclerosis plaques is distinct from the heavy chain repertoire in peripheral blood lymphocytes. *Clin Immunol* 2001;98:258–263.

78. Berger T, Rubner P, Schautzer F et al. Antimyelin antibodies as a predictor of clinically definite multiple sclerosis after a first demyelinating event. *N Engl J Med* 2003;349:139–145.

79. Gerson B, Cohen SR, Gerson IM, Guest GH. Myelin basic protein, oligoclonal bands, and IgG in cerebrospinal fluid as indicators of multiple sclerosis. *Clin Chem* 1981;27:1974–1977.

80. Bloomer LC, Bray PF. Relative value of three laboratory methods in the diagnosis of multiple sclerosis. *Clin Chem* 1981;27:2011–2013.

81. Correale J, de Los Milagros Bassani Molinas M. Oligoclonal bands and antibody responses in multiple sclerosis. *J Neurol* 2002;249:375–389.

82. Farrell MA, Kaufmann JC, Gilbert JJ, Noseworthy JH, Armstrong HA, Ebers GC.

Oligoclonal bands in multiple sclerosis: clinical-pathologic correlation. *Neurology* 1985;35: 212–218.

83. Cohen SR, Herndon RM, McKhann GM. Radioimmunoassay of myelin basic protein in spinal fluid. An index of active demyelination. *N Engl J Med* 1976;295:1455–1457.

84. Whitaker JN, Lisak RP, Bashir RM, et al. Immunoreactive myelin basic protein in the cerebrospinal fluid in neurological disorders. *Ann Neurol* 1980;7:58–64.

85. Lamers KJ, de Reus HP, Jongen PJ. Myelin basic protein in CSF as indicator of disease activity in multiple sclerosis. *Mult Scler* 1998;4: 124–126.

86. Martin-Mondiere C, Jacque C, Delassalle A, Cesaro P, Carydakis C, Degos JD. Cerebrospinal myelin basic protein in multiple sclerosis. Identification of two groups of patients with acute exacerbation. *Arch Neurol* 1987;44: 276–278.

87. Warren KG, Catz I. Diagnostic value of cerebrospinal fluid anti-myelin basic protein in patients with multiple sclerosis. *Ann Neurol* 1986;20:20–25.

88. Hershey LA, Trotter JL. The use and abuse of the cerebrospinal fluid IgG profile in the adult: a practical evaluation. *Ann Neurol* 1980;8: 426–434.

89. Salazar-Grueso EF, Grimaldi LM, Roos RP, Variakojis R, Jubelt B, Cashman NR. Isoelectric focusing studies of serum and cerebrospinal fluid in patients with antecedent poliomyelitis. *Ann Neurol* 1989;26:709–713.

90. Hirohata S, Hirose S, Miyamoto T. Cerebrospinal fluid IgM, IgA, and IgG indexes in systemic lupus erythematosus. Their use as estimates of central nervous system disease activity. *Arch Intern Med* 1985;145: 1843–1846.

91. Hung KL, Chen WC, Huang CS. Diagnostic value of cerebrospinal fluid immunoglobulin G (IgG) in pediatric neurological diseases. *J Formos Med Assoc* 1991;90:1055–1059.

92. Weller M, Stevens A, Sommer N, Wietholter H, Dichgans J. Cerebrospinal fluid interleukins,

immunoglobulins, and fibronectin in neuroborreliosis. *Arch Neurol* 1991;48:837–841.

93. Borucki SJ, Nguyen BV, Ladoulis CT, McKendall RR. Cerebrospinal fluid immunoglobulin abnormalities in neurosarcoidosis. *Arch Neurol* 1989;46:270–273.

94. Papadopoulos NM, Costello R, Kay AD, Cutler NR, Rapoport SI. Combined immunochemical and electrophoretic determinations of proteins in paired serum and cerebrospinal fluid samples. *Clin Chem* 1984;30:1814–1816.

95. Keshgegian AA, Coblentz J, Lisak RP. (University of Pennsylvania Case Conference): oligoclonal immunoglobulins in cerebrospinal fluid in multiple sclerosis. *Clin Chem* 1980;26: 1340–1345.

96. Saiz A, Carreras E, Berenguer J, et al. MRI and CSF oligoclonal bands after autologous hematopoietic stem cell transplantation in MS. *Neurology* 2001;56:1084–1089.

97. Nikoskelainen E, Frey H, Salmi A. Prognosis of optic neuritis with special reference to cerebrospinal fluid immunoglobulins and measles virus antibodies. *Ann Neurol* 1981;9:545–550.

98. Hampel H, Kotter HU, Padberg F, Korschenhausen DA, Moller HJ. Oligoclonal bands and blood–cerebrospinal-fluid barrier dysfunction in a subset of patients with Alzheimer disease: comparison with vascular dementia, major depression, and multiple sclerosis. *Alzheimer Dis Assoc Disord* 1999;13:9–19.

99. Grimaldi LM, Castagna A, Lazzarin A, et al. Oligoclonal IgG bands in cerebrospinal fluid and serum during asymptomatic human immunodeficiency virus infection. *Ann Neurol* 1988;24:277–279.

100. Sandberg-Wollheim M. Optic neuritis: studies on the cerebrospinal fluid in relation to clinical course in 61 patients. *Acta Neurol Scand* 1975;52:167–178.

101. Cohen O, Biran I, Steiner I. Cerebrospinal fluid oligoclonal IgG bands in patients with spinal arteriovenous malformation and structural central nervous system lesions. *Arch Neurol* 2000;57:553–557.

102. Mehta PD, Kulczycki J, Patrick BA, Sobczyk W,

Wisniewski HM. Effect of treatment on oligoclonal IgG bands and intrathecal IgG synthesis in sequential cerebrospinal fluid and serum from patients with subacute sclerosing panencephalitis. *J Neurol Sci* 1992;**109**:64–68.

103. Izquierdo G, Aguilar J, Angulo S, Giron JM. Oligoclonal bands in serum and cerebrospinal fluid in patients with suspected neuroborreliosis. *Med Clin (Barc)* 1992;**98**:516–517.

104. Dalakas M. Oligoclonal bands in the cerebrospinal fluid of post-poliomyelitis muscular atrophy. *Ann Neurol* 1990;**28**:196–197.

105. Skotzek B, Sander T, Zimmermann J, Kolmel HW. Oligoclonal bands in serum and cerebrospinal fluid of patients with HIV infection. *J Neuroimmunol* 1988;**20**:151–152.

106. Gessain A, Caudie C, Gout O, Vernant JC, Maurs L, Giordano C, Malone G, Tournier-Lasserve E, Essex M, de-The G. Intrathecal synthesis of antibodies to human T lymphotropic virus type I and the presence of IgG oligoclonal bands in the cerebrospinal fluid of patients with endemic tropical spastic paraparesis. *J Infect Dis* 1988;**157**:1226–1234.

107. Jamieson DG, Mehta PD, Lavi E. Oligoclonal immunoglobulin bands in cerebrospinal fluid of a patient with lymphocytic choriomeningitis. *Ann Neurol* 1986;**19**:386–388.

108. Coret F, Vilchez JJ, Enguidanos MJ, Lopez-Arlandis J, Fernandez-Izquierdo S. The presence of oligoclonal bands in the cerebrospinal fluid in various neurologic diseases. *Rev Clin Esp* 1989;**185**:231–234.

109. Small P, Mass MF, Kohler PF, Harbeck RJ. Central nervous system involvement in SLE. Diagnostic profile and clinical features. *Arthritis Rheum* 1977;**20**:869–878.

110. Jennekens FG, Kater L. The central nervous system in systemic lupus erythematosus. Part 2. Pathogenetic mechanisms of clinical syndromes: a literature investigation. *Rheumatology (Oxford)* 2002;**41**:619–630.

111. Johnson RT, Richardson EP. The neurological manifestations of systemic lupus erythematosus. *Medicine (Baltimore)* 1968;**47**:337–369.

112. Dubois EL, Wierzchowiecki M, Cox MB, Weiner JM. Duration and death in systemic lupus erythematosus. An analysis of 249 cases. *JAMA* 1974;**227**:1399–1402.

113. Alcocer-Varela J, Aleman-Hoey D, Alarcon-Segovia D. Interleukin-1 and interleukin-6 activities are increased in the cerebrospinal fluid of patients with CNS lupus erythematosus and correlate with local late T-cell activation markers. *Lupus* 1992;**1**:111–117.

114. Jongen PJ, Doesburg WH, Ibrahim-Stappers JL, Lemmens WA, Hommes OR, Lamers KJ. Cerebrospinal fluid C3 and C4 indexes in immunological disorders of the central nervous system. *Acta Neurol Scand* 2000;**101**:116–121.

115. van Dam AP. Diagnosis and pathogenesis of CNS lupus. *Rheumatol Int* 1991;**11**:1–11.

116. Winfield JB, Shaw M, Silverman LM, Eisenberg RA, Wilson HA 3rd, Koffler D. Intrathecal IgG synthesis and blood–brain barrier impairment in patients with systemic lupus erythematosus and central nervous system dysfunction. *Am J Med* 1983;**74**:837–844.

117. Weiner SM, Otte A, Schumacher M, et al. Diagnosis and monitoring of central nervous system involvement in systemic lupus erythematosus: value of F-18 fluorodeoxyglucose PET. *Ann Rheum Dis* 2000;**59**:377–385.

118. Isshi K, Hirohata S. Differential roles of the anti-ribosomal P antibody and antineuronal antibody in the pathogenesis of central nervous system involvement in systemic lupus erythematosus. *Arthritis Rheum* 1998;**41**:1819–1827.

119. Georgescu L, Mevorach D, Arnett FC, Reveille JD, Elkon KB. Anti-P antibodies and neuropsychiatric lupus erythematosus. *Ann N Y Acad Sci* 1997;**823**:263–269.

120. Bonfa E, Elkon KB. Clinical and serologic associations of the antiribosomal P protein antibody. *Arthritis Rheum* 1986;**29**:981–985.

121. Teh LS, Bedwell AE, Isenberg DA, Gordon C, Emery P, Charles PJ, Harper M, Amos N, Williams BD. Antibodies to protein P in systemic lupus erythematosus. *Ann Rheum Dis* 1992;**51**:489–494.

122. Tzioufas AG, Tzortzakis NG, Panou-Pomonis E, et al. The clinical relevance of antibodies to ribosomal-P common epitope in two targeted systemic lupus erythematosus populations: a large cohort of consecutive patients and patients with active central nervous system disease. *Ann Rheum Dis* 2000;**59**:99–104.

123. Hay EM, Isenberg DA. Autoantibodies in central nervous system lupus. *Br J Rheumatol* 1993;**32**:329–332.

124. Meurman OH, Irjala K, Suonpaa J, Laurent B. A new method for the identification of cerebrospinal fluid leakage. *Acta Otolaryngol* 1979;**87**:366–369.

125. Arrer E, Meco C, Oberascher G, Piotrowski W, Albegger K, Patsch W. beta-Trace protein as a marker for cerebrospinal fluid rhinorrhea. *Clin Chem* 2002;**48**:939–941.

126. Oberascher G. Cerebrospinal fluid otorrhea – new trends in diagnosis. *Am J Otol* 1988;**9**: 102–108.

127. Arrer E, Gibitz HJ. Detection of beta 2-transferrin with agarose gel electrophoresis, immunofixation and silver staining in cerebrospinal fluid, secretions and other body fluids. *J Clin Chem Clin Biochem* 1987;**25**:113–116.

128. Bachmann G, Achtelik R, Nekic M, Michel O. Beta-trace protein in diagnosis of cerebrospinal fluid fistula. *HNO* 2000;**48**:496–500.

129. Bachmann G, Nekic M, Michel O. Clinical experience with beta-trace protein as a marker for cerebrospinal fluid. *Ann Otol Rhinol Laryngol* 2000;**109**:1099–1102.

130. Bachmann G, Petereit H, Djenabi U, Michel O. Predictive values of beta-trace protein (prostaglandin D synthase) by use of laser-nephelometry assay for the identification of cerebrospinal fluid. *Neurosurgery* 2002;**50**: 571–577.

131. Petereit HF, Bachmann G, Nekic M, Althaus H, Pukrop R. A new nephelometric assay for beta-trace protein (prostaglandin D synthase) as an indicator of liquorrhoea. *J Neurol Neurosurg Psychiatry* 2001;**71**:347–351.

132. Tripathi RC, Millard CB, Tripathi BJ, Noronha A. Tau fraction of transferrin is present in human aqueous humor and is not unique to cerebrospinal fluid. *Exp Eye Res* 1990;**50**: 541–547.

133. Rouah E, Rogers BB, Buffone GJ. Transferrin analysis by immunofixation as an aid in the diagnosis of cerebrospinal fluid otorrhea. *Arch Pathol Lab Med* 1987;**111**:756–757.

134. Sloman AJ, Kelly RH. Transferrin allelic variants may cause false positives in the detection of cerebrospinal fluid fistulae. *Clin Chem* 1993;**39**:1444–1445.

135. Porter MJ, Brookes GB, Zeman AZ, Keir G. Use of protein electrophoresis in the diagnosis of cerebrospinal fluid rhinorrhoea. *J Laryngol Otol* 1992;**106**:504–506.

136. Keir G, Zeman A, Brookes G, Porter M, Thompson EJ. Immunoblotting of transferrin in the identification of cerebrospinal fluid otorrhoea and rhinorrhoea. *Ann Clin Biochem* 1992;**29**(Pt 2):210–213.

137. Normansell DE, Stacy EK, Booker CF, Butler TZ. Detection of beta-2 transferrin in otorrhea and rhinorrhea in a routine clinical laboratory setting. *Clin Diagn Lab Immunol* 1994;**1**:68–70.

138. Roelandse FW, van der Zwart N, Didden JH, van Loon J, Souverijn JH. Detection of CSF leakage by isoelectric focusing on polyacrylamide gel, direct immunofixation of transferrins, and silver staining. *Clin Chem* 1998;**44**:351–353.

139. Verheecke P. On the tau-protein in cerebrospinal fluid. *J Neurol Sci* 1975;**26**:277–281.

140. Hamill RL, Woods JC, Cook BA. Congenital atransferrinemia. A case report and review of the literature. *Am J Clin Pathol* 1991;**96**: 215–218.

141. Kanaoka Y, Ago H, Inagaki E, et al. Cloning and crystal structure of hematopoietic prostaglandin D synthase. *Cell* 1997;**90**:1085–1095.

142. Sri Kantha S. Prostaglandin D synthase (beta-trace protein): a molecular clock to trace the origin of REM sleep? *Med Hypotheses* 1997;**48**:411–412.

143. Felgenhauer K, Schadlich HJ, Nekic M. Beta trace-protein as marker for cerebrospinal fluid fistula. *Klin Wochenschr* 1987;**65**:764–768.

144. Tumani H, Nau R, Felgenhauer K. Beta-trace protein in cerebrospinal fluid: a blood–CSF barrier-related evaluation in neurological diseases. *Ann Neurol* 1998;44:882–889.

145. Bachmann G, Petereit H, Djenabi U, Michel O. Measuring beta-trace protein for detection of perilymph fistulas. *HNO* 2002;50:129–133.

9

Laboratory strategies for diagnosing monoclonal gammopathies

In the past decade there have been several changes to the testing available for the clinical evaluation of patients suspected of harboring a monoclonal gammopathy. As detailed in Chapters 2 and 3, improved resolution on gels and capillary zone electrophoresis, together with automated and semi-automated systems of electrophoresis and immunofixation, have provided more efficient and sensitive methods for these studies. Recent immunochemical methods to measure free light chains in serum and urine promise to enhance further our ability to detect and follow monoclonal proteins in these patients. Because of the sensitivity and efficiency of the new methodologies, our laboratory has changed our method of evaluating monoclonal gammopathies in the past few years. These methods allow for the efficient detection and immunochemical characterization of most mono-clonal gammopathies in 1 day.

There is no perfect strategy to detect monoclonal gammopathies. A laboratory with a relatively large volume of testing may prefer to use automated screening methods, such as capillary zone electro-phoresis, a laboratory with a more modest volume may choose a semi-automated gel-based method and one with a smaller volume may prefer to use a manual method. These decisions need to be based on the technical demands, available skills and economic realities of the individual laboratory. When reviewing the alternatives presented in this chapter, consider how each strategy would work in your specific laboratory situation.

GUIDELINES FOR CLINICAL AND LABORATORY EVALUATION OF MONOCLONAL GAMMOPATHIES

The major reason for incorporating serum protein electrophoresis and immunofixation into the clinical laboratory is to improve the detection of monoclonal gammopathies. In 1998, the College of American Pathologists Conference XXXII convened a panel of experts to provide recommendations for the clinical and laboratory evaluation of patients suspected of having a monoclonal gammopathy. As a result of that conference and several subsequent teleconferences, guidelines were agreed upon and reported in the *Archives of Pathology and Laboratory Medicine*.[1] The findings of the expert panel were reported as nine guidelines and published with detailed articles to provide a basis for selection of the best available strategy to detect, characterize and follow patients with monoclonal gammopathies. Since the publication of these guidelines, I have received suggestions from several sources to improve the information presented. I will include some of the suggestions, together with a review of the guidelines in the following discussion. The reader is encouraged to review the guidelines document and the articles that accompanied them to flesh out the details involved in selecting patients to be tested, optimal methods for following those patients, and evaluation of unusual circumstances, such as cryoglobulins.[1-6]

Guideline 1 recommends that electrophoretic techniques capable of high-resolution be used to evaluate serum and urine on samples suspected of containing a monoclonal gammopathy. Because there are many techniques available for electrophoresis at various levels of resolution (see Chapter 2), the practical definition offered in one of the papers that accompanied the guidelines was that the method provide crisp separation of the transferrin (β_1) and C3 (β_2) bands.[5] Direct examination of the gel itself was encouraged. However, at the time of the conference, capillary zone electrophoresis (CZE) with high-resolution electropherograms, such as those presented in this book, were not widely used in clinical laboratories. Although capillary zone electrophoresis provides virtual gel images, the basic information is the electropherogram. This is the opposite of the situation with gels, where the gels are the basic information and the densitometric scans provide useful adjunctive information about the quantity of protein in specific regions of the gel. Use of methods providing low resolution was discouraged. Guideline 1 recommended use of densitometry (I currently include electropherograms) to quantify the M-protein peak thereby providing an estimate of tumor burden and a reproducible way to follow the patient's course.

Guideline 1 also noted that these recommendations apply most commonly to clinical disorders that suggest the presence of malignant B-cell/plasma-cell lymphoproliferative disorders, as detailed in Chapter 6 of this book. Although not included in these guidelines, it is important to recognize that individuals with immunodeficiency diseases (congenital, acquired or iatrogenic) also may present with monoclonal gammopathies and/or prominent oligoclonal banding that may require sensitive techniques such as immunoblotting for detection (see Chapter 6).[7-11]

Guideline 2 recommends the use of immunofixation to define the abnormal protein type. It also noted that in cases where serum protein electrophoresis screen is negative, but there is a high clinical suspicion that the patient may harbor a plasma-cell dyscrasia, immunofixation with antisera against κ and λ may be useful to detect more subtle M-proteins. Since publication of these guidelines, the semi-automated immunofixation technique with pentavalent antisera (Penta: one reagent antiserum that detects IgG, IgA, IgM, κ and λ) has become available and provides the advantage of being able to detect heavy chain diseases as well as monoclonal gammopathies in one lane while providing a control serum protein electrophoresis in the adjacent lane (Fig. 9.1). Guideline 2 also recommended that immunofixation may be useful to investigate subtle bands that cause asymmetry in the γ-region or distortions of the β-region bands. Finally, Guideline 2 discouraged the use of immunoelectrophoresis because it is less sensitive than immunofixation, often more difficult to interpret and slower. The advantages and disadvantages of those techniques are detailed in Chapter 3. Immunoblotting was not mentioned in the Guidelines because it is not commonly used in clinical laboratories of the individuals at the conference. However, it is a sensitive method to detect subtle M-proteins and oligoclonal bands. It is especially useful for evaluating patients with monoclonal proteins as a result of immunodeficiency problems, detecting monoclonal free light chains (MFLC) in urine and the more subtle bands that occur in patients with lymphoma and leukemia.[12-18]

Guideline 3 recommends that after an M-protein is identified, it should be followed by measurement of the spike on densitometry (electropherogram measurements are the equivalent) in preference to immunonephelometric, immunoturbidimetric or radial immunodiffusion techniques that are standardized against polyclonal rather than monoclonal immunoglobulins (see later). The only exception is when a small M-protein is obscured by a serum protein band, such as C3, in which case measurement of the immunoglobulin type may be more accurate. This guideline also noted that there is no reason to repeat immunofixation on a previously characterized M-protein unless there has been a change in the electrophoretic migration, development of an additional M-protein, or for confirmation of remission after treatment. Otherwise, repeating immunofixation on a previously characterized M-protein is wasteful. When I receive requests for a redundant immunofixation I append a note to the report that repeat immunofixation on previously characterized M-proteins is not indicated.

Guideline 4 recommends that clinicians order measurements of serum immunoglobulins at the time of detection of the M-protein to determine the level of the uninvolved immunoglobulins. This Guideline also cautions that immunoglobulin measurement should not be the sole primary screening technique for M-proteins. Nephelometry and turbidimetry were preferred to radial immunodiffusion as methods to measure the total immunoglobulins (IgG, IgA, and IgM).

Guideline 5 recommends that all patients suspected of having plasma-cell dyscrasias have a 24-h urine sample studied as well as the serum. As discussed in Chapter 7, some studies indicate that an early morning void may be adequate for the initial screening if a 24-h urine sample is not obtained.[19] However, once a MFLC has been detected, a 24-h collection is currently the standard to use as a baseline to follow these patients. Guideline 5 also noted that the urine should not be tested with dipsticks, sulfosalicylic acid or the ancient acidified heat precipitation tests as screens. Urine should have the total protein measured by one of the techniques discussed in Chapter 7 and then have both immunofixation and urine protein electrophoresis performed on concentrated urine. With the recent availability of immunoassays for

free light chains (FLCs) in the serum and the urine, it may soon be possible to follow the MFLC by merely studying the FLC in a serum sample.[20–23] However, at present, there is relatively little information in the literature to acquaint clinicians with the possibilities of these techniques. One of our tasks in the laboratory is to encourage the use of efficient sensitive new techniques. This seems to fit that description. By the end of 2003 an article will be published in our laboratory publication (The Warde Report) that will be sent to clinicians who use our laboratory. Please feel free to review material from the Warde Report (www.Wardelab.com).

Guideline 6 recommends intervals that are useful to follow patients with previously identified monoclonal gammopathies in serum or urine. The Guidelines recognized that the follow-up time would vary depending on the clinical circumstances. For individuals that are actively being treated for a lymphoplasmacytic neoplasm, examination of both serum and urine are recommended at 1- to 2-month intervals. At the other end of the spectrum are patients that have been diagnosed after careful clinical and laboratory evaluation to have monoclonal gammopathy of undetermined significance (MGUS). These individuals only require annual evaluation of serum and urine along with their physical examination, unless there has been a change in their clinical condition.

Guideline 7 addresses the issue of hyperviscosity syndrome. On occasion, hyperviscosity requires intervention by emergency plasma exchange. By performing serum viscosity and serum protein electrophoresis prior to the first plasma exchange, one may be able to correlate the level of the M-protein with symptoms. Then by following the M-protein as described in Guidelines 3 and 6, the clinician may anticipate the need for plasma exchange by a rise of the M-protein toward the level where the patient experiences viscosity problems.

Guideline 8 recognizes that cryoglobulins are unique problems in evaluating patients with monoclonal gammopathies, as well as some autoimmune and infectious conditions. Not all patients with M-proteins require study for cryoglobulins. This

Guideline discouraged screening for cryoglobulins in individuals with vague, non-specific symptoms. It is recommended that evaluation for cryoglobulins be performed on individuals who have specific clinical features reflecting cold sensitivity (detailed in Chapter 6). For the evaluation, the initial collection and transport of the specimen are known to be critical.[2,24,25] A 10 ml sample of blood should be collected in a prewarmed tube (37°C) and transported to the laboratory at that temperature (unless the venepuncture may be performed in the laboratory). After separating the serum from the clot, the sample should be split into two tubes, one kept at 37°C the other at 4°C or up to 7 days. A precipitate that forms in the 4°C tube, but not in the 37°C tube should be measured and characterized as described in Chapter 6.

Guideline 9 reaffirms the importance of the techniques used in evaluating M-proteins. It recognized that either gel- or capillary-based electrophoretic techniques of high-resolution were preferred. Guideline 9 also recommended immunofixation and immunoselection (for documenting cases of heavy chain disease). Immunoselection, however, is a highly specialized technique and should only be performed by laboratories that have experience with its use (see Chapter 3). The guidelines also recognize the potential value for immunosubtraction.

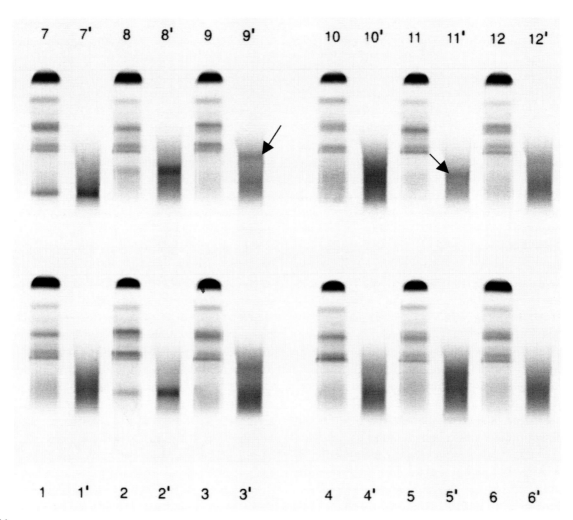

(a)

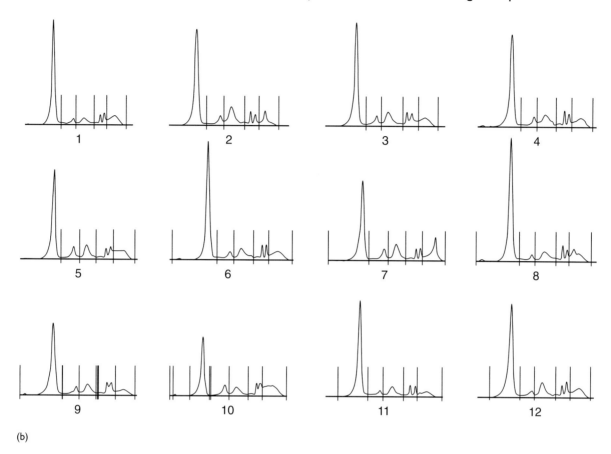

(b)

Figure 9.1 (a) Twelve serum samples are assayed on this Penta (pentavalent) screening immunofixation gel. Each sample is placed in two lanes. The first lane of each sample provides an acid fixation of the serum protein electrophoresis for orientation, whereas the second lane is the immunofixation with the Penta reagent (Sebia, Penta immunofixation). (b) Electropherograms by capillary zone electrophoresis for the same twelve samples shown in (a). Note in specimen 1 a normal electropherogram is present. This corresponds to the normal pattern of sample 1 in (a) and the Penta immunofixation in 1′ shows a diffuse pattern. In contrast, in specimen 2 of the electropherogram, there is a small, but obvious M-protein in the mid-γ-region. In (a) this corresponds to the restriction seen in 2 and confirmation in 2′ that it is an immunoglobulin. A more difficult sample is number 9. On the electropherogram is shown only a bridging between β_1- and β_2-globulins. However, in (a), the arrow in 9′ demonstrates that this is a small M-protein that requires further characterization. Similarly, in the electropherogram for specimen 11 a tiny restriction is seen in the fast γ-region where we commonly see a fibrinogen band in samples with inadequate clotting. However, the Penta of sample 11′ in (a) shows that the restriction is caused by an immunoglobulin (arrow) and deserves further evaluation. (Paragon CZE 2000.)

These guidelines form the basis for developing the strategy that a laboratory could adopt in evaluating serum and urine for monoclonal gammopathies. The use of serum protein electrophoresis with immunofixation or even more sensitive techniques such as immunoblotting increases the detection of small monoclonal gammopathies. Is there a screening technique that will detect clinically relevant bands and distinguish them from bands that are not clinically relevant? Unfortunately, there is no such test. Small monoclonal gammopathies are occasionally part of infectious conditions, often part of MGUS, which may be transient.[26–31] All monoclonal gammopathies must be followed for the remainder of the patient's life because some are a reflection of a malignant, potentially malignant process or a dysregulation of the immune system.

INITIAL SCREEN BY SERUM PROTEIN ELECTROPHORESIS AND PENTA IMMUNOFIXATION

Regardless of which other technique is used, the initial screen should be performed by electrophoresis on both serum and urine. As mentioned in the guidelines, methods that provide crisp resolution of the β-region bands are recommended. Currently, when serum is sent to our laboratory to evaluate for a possible monoclonal gammopathy by immunofixation, we first perform serum protein electrophoresis using a high-resolution capillary zone method. If no suspicious band is seen, we perform a Penta immunofixation that will detect IgG, IgA, IgM, k and λ (Fig. 9.1). The combination of a normal serum protein electrophoresis and a normal Penta immunofixation should detect virtually all monoclonal proteins in serum. When a band suggesting an M-protein is seen on the initial serum protein electrophoresis, we go straight to immunosubtraction or immunofixation to identify the heavy and light chain type, foregoing the Penta screen. If the band is quite small, immunofixation is chosen. Immunosubstraction by the capillary zone method is highly reliable, requires very little technologist time and can be performed in under an hour.[32] However, with small M-proteins it is sometimes difficult to detect the effect of the immunosubtraction, and for those I prefer immunofixation. It is unusual that a case requires measurement of immunoglobulins to identify the monoclonal protein.

The availability of semi-automated screening techniques such as the Penta is helpful in dealing with the occasional subtle restrictions or distortion of β-region bands. I use the Penta immunofixation to demonstrate that the cause of the restriction is or is not an immunoglobulin. If it is an immunoglobulin, an immunofixation or immunosubtraction will identify it. The most common cause for non-immunoglobulin bands when using CZE is the presence of radiocontrast dyes.[33] The proteins that cause non-immunoglobulin restrictions are discussed in Chapter 6.

When using automated immunosubtraction, all evaluations are for IgG, IgA, IgM, κ and λ. However, with immunofixation, we can tailor the reaction depending on the electrophoretic migration of the M-protein. A slow γ-migrating restriction results in immunofixation with antisera against IgG, κ and λ. If the monoclonal protein is not identified by this, immunofixation is repeated with IgA, and IgM antisera; IgD and IgE are only evaluated when the other antisera fail to identify the monoclonal protein.

A fast γ- or β-restriction is evaluated by immunofixation with antisera against IgG, IgA, IgM, κ and λ. Monoclonal restrictions of κ or λ in the serum with no corresponding restriction of IgG, IgA or IgM by immunofixation are usually caused by light chain disease. However, these cases require evaluation for possible IgD or IgE monoclonal proteins. One may perform immunofixation for IgD and IgE or one may quantify the levels of these isotypes. A urine study for MFLC by immunofixation is always recommended when a monoclonal protein is suspected, regardless of the serum immunofixation findings. The early morning void has been shown to be as sensitive as a 24-h sample (although the latter is needed to quantify the MFLC once it has been demonstrated).[19]

κ/λ TOTAL (NOT FREE) QUANTIFICATION AND THE DIAGNOSIS OF MONOCLONAL GAMMOPATHIES

Before the availability of automated high-resolution screening, CZE-based automated immunosubtraction and the automated Penta immunofixation, our laboratory used a combination of serum protein electrophoresis (using a high-resolution method) and quantification of IgG, IgA, IgM with measurement of total κ and λ in serum to identify many of the M-proteins encountered. I no longer find this to be a useful method to identify the M-protein. The method took advantage of the normal ratio κ/λ in the serum in our population, 1.2–2.6 (± 2 SD).[34] When patients have a chronic infectious disease with a marked elevation of immunoglobulins, there is a polyclonal expansion of B-cell clones. In this situation, although the total amounts of κ and λ are elevated, the κ/λ ratio usually remains in the normal range. In contrast, when a monoclonal gammopathy is present owing to a malignant process such as multiple myeloma, there *usually* is a marked alteration in the κ/λ ratio.[34] This results from the combined effect of the increase in the monoclonal protein of the single light chain type and the suppression of normal polyclonal immunoglobulin-secreting clones that occur in myeloma. However, a κ/λ ratio cannot be used to predict whether one is dealing with MGUS or multiple myeloma.

Unfortunately, in some circumstances, serum protein electrophoresis, immunoglobulin quantification and the κ/λ ratio are inadequate to characterize a monoclonal gammopathy. For example, in a double (biclonal) gammopathy, where one neoplastic clone secretes a κ-containing monoclonal protein and the other clone secretes a λ-containing monoclonal protein, the κ/λ ratio may give misleading information. If one does not have a serum protein electrophoresis pattern to demonstrate an obvious double gammopathy, one could seriously underestimate the process. Because of the presence of normal polyclonal immunoglobulins, the κ/λ ratio is too insensitive to detect relatively small monoclonal gammopathies. The κ/λ ratio has great difficulty detecting monoclonal gammopathies that are smaller than 400 mg/dl.[35] Immunoglobulin measurements should not be used without a serum protein electrophoresis to include or rule out the presence of a monoclonal gammopathy.

When there are massive increases in the amount of a monoclonal protein, nephelometric determinations of a particular immunoglobulin component can grossly overestimate or underestimate the amount present.[36, 37] Ideally, the amount of the serum M-protein measured by densitometry or

electropherogram, the heavy chain isotype of the monoclonal protein and the light chain type of the monoclonal protein would be very close in concentration (unless a considerable amount of MFLC is present). This is often not the case. For example, in Fig. 9.2, compare IgG, IgA, IgM, κ, and λ concentrations and the total protein of the spike (by densitometry) in selected patients with prominent monoclonal gammopathies. Note that in some individuals, there is a very poor correlation between the concentration of the light chain and the heavy chain. In patient RK, the concentration of the spike is similar to that of the κ light chain, but the IgM concentration is twice as high as the light chain concentration. In patient FC, the IgA concentration is twice as high as κ and the spike is intermediate in amount. The concentration of λ in patient SP is half again greater than that of the spike, whereas the IgA heavy chain has an intermediate concentration (this could be explained by a large MFLC, but this was not the case in patient SP). Yet, in contrast to these examples, patients NG, MS, and EK have reasonably good correlation between the densitometric scan

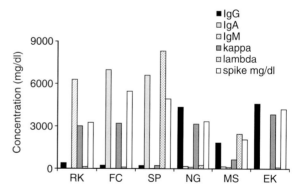

Figure 9.2 Immunoglobulin and monoclonal spike (from densitometry) values on selected patients with monoclonal gammopathies. Note that in the last serum protein electrophoresis patients (NG, MS, and EK) there is reasonably good correlation of the spike and the monoclonal proteins as measured by nephelometry. However, in the first serum protein electrophoresis patients (RK, FC, and SP), there is disparity between these values. Therefore, in following those patients, if one were to use the densitometric information one time and the light chain or heavy chain number another time, one would not have consistent results.

and the heavy and light chains involved in the monoclonal gammopathy.

If these differences between the nephelometric amounts and densitometry were merely due to antigen excess, then adjustment of the dilution of patients' serum should have changed the antigen concentration and achieved a different result. The overestimations or underestimations of immunoglobulins by nephelometry are not accounted for, however, by dilution studies to bring constituents into the equivalence region (Fig. 9.3).[36, 37] The problem is caused by variation in the specificities of the reagent antisera used.

Comparison of quantification of κ and λ monoclonal gammopathies by two manufacturer's reagents demonstrated poor correlation of the κ-containing monoclonal gammopathies (Fig. 9.4), despite excellent correlation with polyclonal samples (Fig. 9.5).[37] As the concentration of the monoclonal proteins increased, when examined by these reagents, the values were less likely to agree

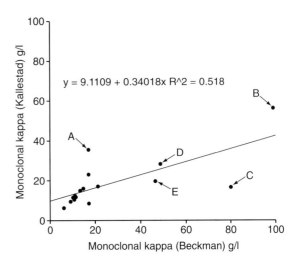

Figure 9.4 A poor correlation was found with comparison of κ monoclonal gammopathies in serum quantified by reagents from two different manufacturers. (Data from Bush and Keren.)[37]

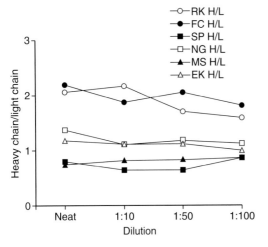

Figure 9.3 Dilution of the sera from Fig. 9.1 with re-assay by nephelometry did not correct for the differences between the heavy and light chains (ideally the ratio should be 1:1). Although light chains weigh about half as much as heavy chains, the measurement is of light chains attached to the heavy chain and is expressed as κ-containing immunoglobulin. The fact that the ratio of heavy/light chains (H/L) does not correct to 1.0 with increasing dilution of at least two of the six samples indicates that the higher ratio in those two samples was not due only to simple antigen excess effect.

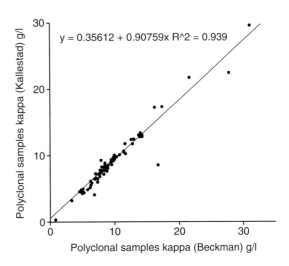

Figure 9.5 A good correlation was found with comparison of normal and polyclonal increases in κ in serum quantified by reagents from two different manufacturers. (Data from Bush and Keren.)[37]

between reagents. The samples labeled A, B, C, D, and E are compared with the densitometric information in Table 9.1. If the densitometric information is regarded as the gold standard, neither manufacturer's reagent consistently gave 'correct' results with these samples. Neither was there a consistent trend. With sample A, the Beckman reagent

Table 9.1 Comparison of five non-linear κ monoclonal samples

	Sample (g/l)				
Method	A	B	C	D	E
Kallestad	35.40	57.00	16.70	28.20	19.70
Beckman	16.80	98.60	79.80	48.60	46.50
Densitometry[b]	16.10	57.00	82.90	25.20	22.00

[a]Data from Bush and Keren.[37]
[b]Gamma optical density (OD) = densitometric scans of the γ-region of serum protein electrophoresis gels.

agreed closely with the densitometry measurement while the Kallestad reagent gave a value twice as high. In contrast, with sample B the Kallestad reagent agreed with the densitometric measurement, with the Beckman reagent twice as high. These are both good reagents for measuring polyclonal immunoglobulins. However, in extreme concentrations of monoclonal immunoglobulins and suppression of the normal polyclonal population, nephelometric techniques may give, at best, rough estimates of the amount of monoclonal protein present.

The differences between results obtained on the same monoclonal proteins with different reagent antisera may result from the lack of standardization of polyclonal reagents for nephelometry. Take the example of determinants A, B, C, D, and E on the polyclonal immunoglobulins shown in the solution in Fig. 9.6. The reagent antibody shown has strong reactivity against determinants A, C and E. It produces an excellent precipitate with polyclonal immunoglobulins that have an abundance of all sites represented. However, with monoclonal proteins that express mainly B and D, this antiserum has poor reactivity, considerably underestimating the concentration of the monoclonal protein present. Of course, the typical reactions using reagent antisera are much more complex, with varying affinities of antibodies against a vast number of epitopes. However, these are standardized against polyclonal serum in which these epitopes are roughly evenly distributed from one

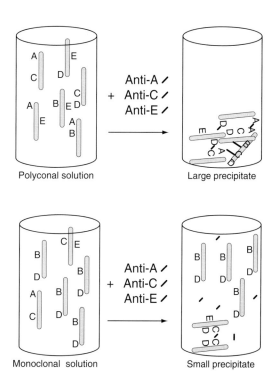

Figure 9.6 Schematic representation of problems involving the limited antigen expression of some monoclonal proteins and the inferior precipitates that may occur. With a polyclonal preparation of immunoglobulin (many epitopes expressed: A, B, C, D, and E) on top, the reagent antisera that reacts with epitopes A, C, and E can form a precipitate with all of the molecules present (disregarding equilibrium for this illustration). However, in the monoclonal expansion of molecules that express mainly determinants B and D, only a weak precipitate is formed. The amount of precipitate formed depends upon the affinity and amount of antibodies directed against the major epitopes of the monoclonal immunoglobulins.

individual to another. In monoclonal proteins, there is a distortion of the type of epitopes expressed, and the standardization may be considerably off.

Thus, κ/λ ratios and immunoglobulin quantifications were useful to identify M-proteins when used along with serum protein electrophoresis. However, their potential to provide misinformation is a key factor in my no longer using this strategy. The availability of better electrophoretic screening techniques, semi-automated techniques, immunofixation electrophoresis and immunosubtraction by CZE provides our laboratory with a more efficient and reliable way to perform the electrophoresis with immunofixation, as described above.

Despite the problems discussed here, if one wishes to use immunoglobulin measurements along with serum protein electrophoresis to avoid excessive performance of immunofixation, one should take a logical approach. To this end, algorithms have been proposed for deciding whether a serum examined by serum protein electrophoresis and immunoglobulins quantification has sufficient likelihood of a monoclonal gammopathy to require further study by immunofixation. In simpler algorithm, Liu et al.[38,39] recommend the use of five independent criteria (any one abnormality results in immunofixation) (Table 9.2). In this algorithm, any abnormal potential monoclonal band seen on electrophoresis should result in an immunofixation. They also recognize the importance of hypogammaglobulinemia as a sign of monoclonal gammopathies. To avoid redundant performance of immunofixation in the many polyclonal increases in γ-globulins, they relate the concentrations of IgA and IgM to that of IgG. If there is a polyclonal increase in immunoglobulins, IgG is almost always involved. Therefore, if IgA or IgM increase disproportionately greater than the increase in IgG (which they define as IgG/IgA < 3.0 and IgG/IgM < 4.0), a monoclonal gammopathy should be suspected. Finally, the presence of a low or high κ/λ ratio should be evaluated for a monoclonal gammopathy in this system. Using these criteria to determine whether they will perform further studies such as immunofixation, they achieved a sensitivity of detection of monoclonal gammopathies of 98 per cent.[38,39] A multinational study by Jones and colleagues[40,41] used computer-based algorithms to fine tune this process, yet they misassigned 2.5 per cent of monoclonal gammopathies when they were present in concentrations greater than 1000 mg/dl (Table 9.3).

The algorithms may be useful for evaluating the immunochemical information in the context of serum protein electrophoresis information. Regardless which system is used, however, occasional cases of monoclonal gammopathy will not be detected. Therefore, clinicians should be encouraged to send repeat serum and urine samples on cases where suspicion of a monoclonal gammopathy persists after a negative analysis (which may include an immunofixation). False negatives do occur. Some cases may be non-secretory myeloma. Those may benefit from the FLC immunoassays as described in Chapter 7. About half of the cases of non-secretory myeloma were found to have abnormal κ/λ ratios in serum.[42] Just as anywhere else in the laboratory, the occasional sample may be switched with another. Alternatively, a cryoglobulin may have precipitated because of routine use of room-temperature processing.

Table 9.2 Criteria for further workup of probable monoclonal gammopathy[a]

Independent criteria for monoclonal gammopathy
Abnormal potential monoclonal band on electrophoresis
Hypogammaglobulinemia (IgG < 700 mg/dl)
Disproportionate increase in IgA (IgG/IgA < 3.0)
Disproportionate increase in IgM (IgG/IgM <4.0)
Abnormal κ/λ (< 1.0 or > 3.0)

[a]These five criteria are independent. The presence of a single criterion is sufficient to indicate a possible monoclonal gammopathy.[38,39]

Table 9.3 Decision pathway used to classify monoclonal gammopathies

Step 1. Analyse serum protein electrophoresis and IgG, IgA, IgM, κ, and λ

Step 2. Decide whether classification can be attempted:

if (a) single band and κ/λ > 0.75[a] and < 2.30[a]

or (b) multiple bands are present

or (c) no bands are present and κ/λ > 0.75[a] and < 2.30[a]

then **report immunofixation is necessary**

Step 3. Assign light chain class:

if κ/λ gt; 2.30[a], then light chain class = κ

if κ/λ < 0.75[a], then light chain class = λ

Step 4. Assign heavy chain class:

heavy chain = immunoglobulin in highest concentration (expressed as multiples

of the standard deviation of the reference range to correct for the normally higher

presence of IgG than IgA or IgM)

Step 5. Calculate monoclonal light chain concentration (LC:MC):

if LC = κ, then

[k:MC] = [k:total] − {[l:total] × 2.20[a]}

(accounting for the polyclonal κ which is assumed to be present)

if LC = λ, then [l:MC] = [l:total] − [k:total]/0.95[a] (accounting for

the polyclonal λ, which is assumed to be present)

Step 6. Detect free light chains:

if [light chain:total] > [heavy chain:total] × 0.8,[a]

then free light chains are present

Step 7. Check confirmatory criteria:

if [heavy chain] > [light chain:monoclonal component] × 0.75[a]

and [heavy chain]$_{SD}$[b] > − 0.35[a]

then **report heavy chain:light chain (e.g. IgG κ) and free light chain (e.g. κ), if present**

Step 8. Check for possible IgD or free light chain disease:

if [light chain:total] > [heavy chain:total] × 0.80[a]

and [IgG]$_{SD}$ < − 0.80[a]

and [IgA]$_{SD}$ < − 0.80[a]

and [IgM]$_{SD}$ < − 0.80[a]

then **report free light chain disease or IgD is possible and immunofixation is**

necessary for classification

Step 9. Unable to classify:

report immunofixation is necessary for classification

[a]Number derived from iterative procedure described in Jones et al.[40, 41]

[b][Heavy chain]$_{SD}$ = concentration expressed as multiple of the standard deviation of the reference range.

FOLLOWING MONOCLONAL GAMMOPATHIES

Because of the variation in nephelometric information, I follow the recommendation of the Guidelines described above to use densitometric scan or electropherogram pattern (with CZE) to follow monoclonal gammopathies once they have been characterized. Previously, I followed β-migrating monoclonal proteins with nephelometric information, but was wary when values differed markedly from the densitometric information on very large monoclonal proteins (> 3.0 g/dl). I now establish a baseline even with interference by the transferrin, C3 or β_1-lipoprotein bands. By keeping a record of the location of previous measurements, one can provide a reproducible estimate of the M-protein. I use nephelometric or turbidimetric measurements to follow IgD monoclonal gammopathies when they are present in small quantities.

CLUES TO DETECTING MONOCLONAL GAMMOPATHIES

Recognizing the various appearances that monoclonal gammopathies can display on the serum protein electrophoresis is key. For example hypogammaglobulinemia is an important finding that is highly suggestive of a monoclonal gammopathy or a B-cell lymphoproliferative disorder (Table 9.4). It is not a normal finding in older

Table 9.4 Conditions associated with isolated[a] hypogammaglobulinemia

Multiple myeloma (especially light chain disease)
Chronic lymphocytic leukemia
Well-differentiated lymphocytic lymphoma
Immunodeficiency
Amyloidosis
Chemotherapy

[a]Isolated means no other serum abnormalities.

individuals. Older individuals usually have normal immunoglobulin concentrations. Total B-cell number and immunoglobulin content are the same in older individuals as in young. This erroneous impression has a grain of truth because some specific antibody levels do differ in older patients from those in younger patients, while the total gravimetric quantity of IgG does not change significantly. Because of decreased suppressor T-cell function and altered helper function, more autoantibodies are seen as individuals age, whereas the response elicited to foreign antigens is often weaker in the elderly than in the younger patients. Certainly older individuals have a much higher incidence of monoclonal gammopathies that may be related to deterioration of their immune regulatory functions. Hypogammaglobulinemia (which we define by densitometric scan or electropherogram of the γ-region) is an important clue that the patient may have light chain disease or one of the B-cell neoplasms described in Chapter 6, or the effect of chemotherapy. When hypogammaglobulinemia is present, I recommend a urine and serum immunofixation to rule out MFLC.

Other abnormalities in the serum must be carefully scrutinized. Monoclonal proteins may occasionally bind to normal serum components altering their migration. Consequently, any abnormal band from the α_1- to the γ-region is regarded with suspicion. Most can be interpreted by understanding the pattern diagnoses outlined in Chapters 4 and 5. A nephrotic patient will have a markedly elevated α_2-macroglobulin, and an iron-deficient patient will have an elevated transferrin band. Any unexplained band should cause the interpreter to inform the clinician about the abnormal pattern and perform immunofixation.

When one is consulted about the results from the general chemistry laboratory information of a patient, low albumin, elevated calcium, elevated total protein, elevated sedimentation rate, or decreased albumin/globulin ratio are sufficient abnormalities to recommend evaluation of serum and urine for the presence of a monoclonal gammopathy. Information about these chemical abnormalities should raise one's level of suspicion when viewing a serum protein electrophoresis pattern. Similarly, routine hematological screening tests that demonstrate plasma cells on the differential, rouleaux formation on the blood smear, or a bone marrow with > 10 per cent plasma cells are highly suspicious. These relevant laboratory findings should result in evaluation of serum and urine for a monoclonal gammopathy.

MAINTAINING AN ACTIVE FILE OF ALL MONOCLONAL PROTEINS

To conserve time and important laboratory resources, one should maintain a file on all patients with monoclonal proteins. When a sample is received with a request to evaluate for monoclonal protein, the old file is checked for previous findings.[43] Serum protein electrophoresis is performed, and the monoclonal band (if in the γ-region) is quantified by densitometry.[44] If there is no change in pattern, there is no reason to perform immunofixation because the monoclonal protein was characterized on the original sample. However, any change in the pattern of migration, or the appearance of a second band should result in an immunofixation. By comparing the present results with previous ones, the laboratory helps clinicians follow the patient by determining the change (if any) of the monoclonal protein quantity. The same quantification method should be used to follow a particular patient. For example, if the densitometric scan of the γ-region was used to estimate the concentration of a γ-migrating monoclonal gammopathy, continue to use this on subsequent samples, unless there is a change in the pattern (such as the development of a second monoclonal band).

SCREENING AND FOLLOW-UP OF MFLC

If the patient has MFLC, 24-h urine should be collected, total protein determined, and densitometry used to establish the percentage of MFLC.

Tetrameric light chain disease must be followed by serum samples, as the molecules are too large to pass into the urine.[45] As with the serum, any change in the electrophoretic migration or the development of other suspicious bands should trigger a reinvestigation, complete with immunofixation to determine if the patient is developing a double gammopathy, or if the course of the condition has altered.

When the clinician sends urine to be evaluated for MFLC, we perform serum protein electrophoresis on 50-fold concentrated urine. When the urine has considerable protein, it will not readily concentrate to this level. As the guidelines recommend, the acidified heat test, sulfosalicylic acid test and urine dipsticks are inadequate screens for MFLC.[46] Urine immunofixation has replaced immunoelectrophoresis on all samples in our laboratory. With immunofixation, a false negative may occur because the dilution used may be inappropriate; therefore, occasionally, an additional dilution may be useful to be certain of the final result when an antigen excess effect is seen. In practice, we have found that such dilutions are rarely needed. Most commercial antibodies are sufficiently strong that even when an antigen excess situation occurs, it is usually obvious. We end up repeating less than 1 per cent of our urine immunofixation with additional dilutions. When using immunofixation on urine, one must be aware of the ladder patterns that occur mainly with κ and to a lesser extent with λ (Chapter 7). As always, clinicians should be encouraged to send a second sample (preferably the early morning void or a 24-h urine) if the first is negative and myeloma or amyloidosis is still part of their differential diagnosis. I also recommend serum assays for FLC in these situations because of its proven sensitivity in detecting light chain myeloma and some cases of non-secretory myeloma.[42,47]

MONOCLONALS WHICH MAY BE DIFFICULT TO DIAGNOSE

Monoclonal proteins can create diagnostic dilemmas for the clinical laboratory. Tetrameric light

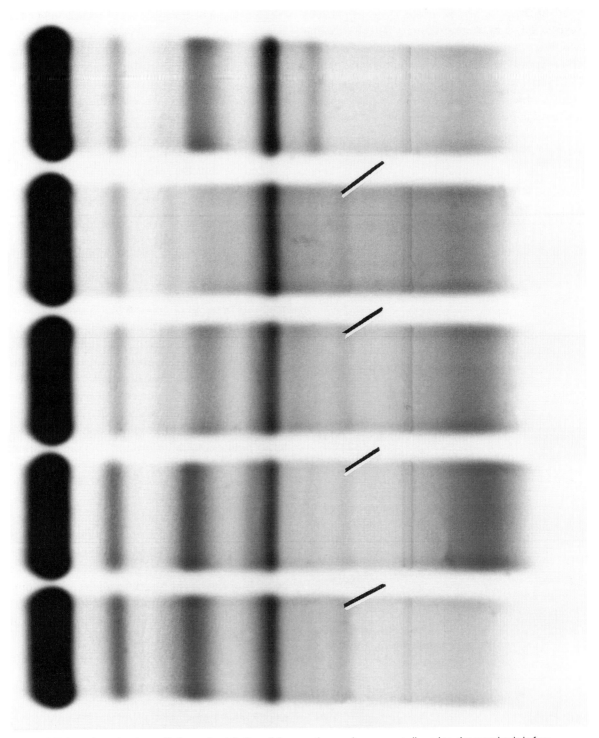

Figure 9.7 Several samples show a fibrinogen band (indicated) because the samples were not allowed to clot completely before separation of the serum. (Panagel system stained with Coomassie Blue; anode at the left.)

chain disease can be missed because the serum spike is often in the β-region, where it may be confused with other proteins, and because of molecular size the light chains in this situation do not appear in the urine as MFLC. Fibrinogen in an incompletely clotted specimen may be mistaken for a monoclonal protein (Fig. 9.7). This error will be avoided by checking the serum for a clot (if a small one is present, it indicates that some fibrinogen was left), and repeating the sample (always the best choice when one is not certain of the diagnosis). If still uncertain, immunofixation with Penta, or anti-light chain antisera or even antisera against fibrinogen will reveal the true nature of the protein. A possible trap, however, lies in the non-specific reactivity of some commercial antisera. We have found that commercial antisera against immunoglobulins occasionally will react with fibrinogen, C3, C4 and transferrin. For example, in the case shown in Fig. 9.8, a fibrinogen band reacted with antisera against IgM. When this antisera was reacted against normal plasma, it gave the same line. Therefore, when evaluating new lots of reagents used for immunofixation in the laboratory, it is a good idea to test it against a sample of plasma to be sure that this will not be a problem, for example, in specimens from patients on anticoagulants.

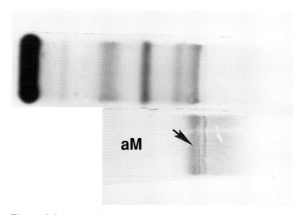

Figure 9.8 Immunofixation of a sample with a prominent fibrinogen band gave a false positive reaction (arrow) with antisera against IgM. (Panagel system stained with Coomassie Blue; anode at the left).

Polyclonal free light chains in the urine can occasionally obscure the presence of a monoclonal gammopathy. When the immunofixation pattern of urine gives dense staining, a repeat immunofixation should be performed in order to rule out a monoclonal process.[48] Uncommonly, it may not be possible to distinguish between polyclonal and monoclonal free light chains migrating in a ladder-banding pattern (see Chapter 7). In those instances, I advise the clinician to follow both the urine and serum after 2–3 months to see if the process evolves.

FINAL WORDS

The strategies reviewed in this chapter provide efficient processing of specimens, benefiting the patient, the clinician, and the laboratory. They help prevent inappropriate ordering and over-utilization of the laboratory. With the incorporation of newer methods such as semi-automated gel-based and CZE, immunosubtraction and Penta immunofixation screens, more rapid turnaround times are possible, which have been appreciated by clinicians. In general, when I discuss unusual cases with clinicians, they appear to like being involved in the evaluation of challenging specimens. An occasional clinician has objected to the report of a small monoclonal protein or hypogammaglobulinemia in patients that were clinically well. However, a few of these prove to be associated with lymphoproliferative processes. We are not able to determine the meaning of each abnormality detected by these sensitive methods, but we do know (as has been discussed) the important conditions that need to be ruled in or ruled out. When we have a very small monoclonal gammopathy, about which we are uncertain after talking to the clinician and studying the urine, we recommend repeating the evaluation in a few months. If it represents an early neoplastic monoclonal process, it will still be there or may have progressed to the point where it will be readily detectable. Myeloma is not treated in the early clinical stages, therefore, there has been no problem with delay in diagnosis. If it was merely an oligoclonal expansion due to some infection or other process, it will likely have resolved by this time.

It is important to view our diagnostic process in context of the causes for monoclonal gammo-pathies. For example, when classifying the etiology of monoclonal gammopathies as caused by B-cell malignancies and B-cell benign processes, Radl and colleagues[14] estimate that the newer, more sensitive techniques detect one true malignancy for 100 benign processes. Typically, the monoclonal product is much larger in malignant than in benign processes. However, size of the gammopathy alone is inadequate to distinguish MGUS from malignant or potentially malignant processes. When dealing with small monoclonal bands in a serum sample, I recommend that the clinicians follow both the serum and the urine.

Occasionally, we receive requests for immunofix-ation on serum samples from children. These are relevant in the evaluation of children with a wide variety of immunodeficiency diseases. Myeloma in children is vanishingly rare. The normal κ/λ ratio in children increases with age from about 1.0 at 4 months to the adult values of about 2.0 by mid-adolescence (15 years).[49,50] When I receive a request for an immunofixation on a child, I con-tact the clinician to find out what they are consid-ering in the differential diagnosis of the case. Invariably, in my experience, the clinicians are concerned about the possibility of an immunodefi-ciency disease such as Wiskott–Aldrich syndrome, Bruton's X-linked agammaglobulinemia, an immunoglobulin subclass deficiency, or a comple-ment deficiency because the child has had recur-rent pyogenic infections. This is helpful information because individuals with various con-genital and acquired immune disorders can have monoclonal or oligoclonal gammopathies with abnormal κ/λ ratios.[51,52]

Finally, a cooperative relationship between the laboratory and clinician is one of the critical points to success in dealing with difficult cases. Contacting the ordering physician for more clinical information can be helpful. When uncertain about a result, repeat the procedure, speak to the clini-cian, perform further studies, occasionally send for a fresh sample, or send the sample off for reference work.

REFERENCES

1. Keren DF, Alexanian R, Goeken JA, Gorevic PD, Kyle RA, Tomar RH. Guidelines for clinical and laboratory evaluation patients with monoclonal gammopathies. *Arch Pathol Lab Med* 1999;**123**; 106–107.

2. Kallemuchikkal U, Gorevic PD. Evaluation of cryoglobulins. *Arch Pathol Lab Med* 1999;**123**; 119–125.

3. Alexanian R, Weber D, Liu F. Differential diagnosis of monoclonal gammopathies. *Arch Pathol Lab Med* 1999;**123**;108–113.

4. Goeken JA, Keren DF. Introduction to the report of the consensus conference on monoclonal gammopathies. *Arch Pathol Lab Med* 1999;**123**; 104–105.

5. Keren DF. Procedures for the evaluation of monoclonal immunoglobulins. *Arch Pathol Lab Med* 1999;**123**;126–132.

6. Kyle RA. Sequence of testing for monoclonal gammopathies. *Arch Pathol Lab Med* 1999;**123**; 114–118.

7. Radl J. Monoclonal gammapathies. An attempt at a new classification. *Neth J Med* 1985;**28**; 134–137.

8. Radl J. Differences among the three major categories of paraproteinaemias in aging man and the mouse. A minireview. *Mech Ageing Dev* 1984;**28**;167–170.

9. Radl J, Liu M, Hoogeveen CM, et al. Monoclonal gammopathies in long-term surviving rhesus monkeys after lethal irradiation and bone marrow transplantation. *Clin Immunol Immunopathol* 1991;**60**; 305–309.

10. Radl J, Valentijn RM, Haaijman JJ, Paul LC. Monoclonal gammapathies in patients undergoing immunosuppressive treatment after

renal transplantation. *Clin Immunol Immunopathol* 1985;37;98–102.

11. Radl J. Light chain typing of immunoglobulins in small samples of biological material. *Immunology* 1970;19;137–149.

12. Norden AG, Fulcher LM, Heys AD. Rapid typing of serum paraproteins by immunoblotting without antigen-excess artifacts. *Clin Chem* 1987;33;1433–1436.

13. Nooij FJ, van der Sluijs-Gelling AJ, Jol-van der Zijde CM, van Tol MJ, Haas H, Radl J. Immunoblotting techniques for the detection of low level homogeneous immunoglobulin components in serum. *J Immunol Methods* 1990;134;273–281.

14. Radl J, Wels J, Hoogeveen CM. Immunoblotting with (sub)class-specific antibodies reveals a high frequency of monoclonal gammopathies in persons thought to be immunodeficient. *Clin Chem* 1988;34;1839–1842.

15. Beaume A, Brizard A, Dreyfus B, Preud'homme JL. High incidence of serum monoclonal Igs detected by a sensitive immunoblotting technique in B-cell chronic lymphocytic leukemia. *Blood* 1994;84;1216–1219.

16. Withold W, Rick W. An immunoblotting procedure following agarose gel electrophoresis for subclass typing of IgG paraproteins in human sera. *Eur J Clin Chem Clin Biochem* 1993;31; 17–21.

17. Withold W, Reinauer H. An immunoblotting procedure following agarose gel electrophoresis for detection of Bence Jones proteinuria compared with immunofixation and quantitative light chain determination. *Eur J Clin Chem Clin Biochem* 1995;33;135–138.

18. Koide N, Suehira S. Clinical application of immunoblotting to the detection of monoclonal immunoglobulins and Bence Jones protein. *Rinsho Byori* 1987;35;638–643.

19. Brigden ML, Neal ED, McNeely MD, Hoag GN. The optimum urine collections for the detection and monitoring of Bence Jones proteinuria. *Am J Clin Pathol* 1990;93;689–693.

20. Bradwell AR, Carr-Smith HD, Mead GP, et al. Highly sensitive, automated immunoassay for immunoglobulin free light chains in serum and urine. *Clin Chem* 2001;47;673–680.

21. Abraham RS, Clark RJ, Bryant SC, et al. Correlation of serum immunoglobulin free light chain quantification with urinary Bence Jones protein in light chain myeloma. *Clin Chem* 2002;48;655–657.

22. Abraham RS, Charlesworth MC, Owen BA, et al. Trimolecular complexes of lambda light chain dimers in serum of a patient with multiple myeloma. *Clin Chem* 2002;48; 1805–1811.

23. Katzmann JA, Clark RJ, Abraham RS, et al. Serum reference intervals and diagnostic ranges for free kappa and free lambda immunoglobulin light chains: relative sensitivity for detection of monoclonal light chains. *Clin Chem* 2002;48; 1437–1444.

24. Gorevic PD, Galanakis D. Chapter 10. Cryoglobulins, cryofibrinogenemia, and pyroglobulins. In: Rose NR, Hamilton RG, Detrick B, eds. *Manual of clinical laboratory immunology*. Washington, DC: ASM Press, 2002.

25. Keren DF, Di Sante AC, Mervak T, Bordine SL. Problems with transporting serum to the laboratory for cryoglobulin assay: a solution. *Clin Chem* 1985;31;1766–1767.

26. Godeau P, Herson S, De Treglode D, Herreman G. Benign monoclonal immunoglobulins during subacute infectious endocarditis. *Coeur Med Interne* 1979;18;3–12.

27. Herreman G, Godeau P, Cabane J, Digeon M, Laver M, Bach JF. Immunologic study of subacute infectious endocarditis through the search for circulating immune complexes. Preliminary results apropos of 13 cases. *Nouv Presse Med* 1975;4;2311–2314.

28. Keren DF, Morrison N, Gulbranson R. Evolution of a monoclonal gammopathy (MG) documented by high-resolution electrophoresis (HRE) and immunofixation (IFE). *Lab Med* 1994;25: 313–317.

29. Kanoh T. Fluctuating M-component level in relation to infection. *Eur J Haematol* 1989;42; 503–504.

30. Larrain C. Transient monoclonal gammopathies associated with infectious endocarditis. *Rev Med Chil* 1986;114;771–776.

31. Crapper RM, Deam DR, Mackay IR. Paraproteinemias in homosexual men with HIV infection. Lack of association with abnormal clinical or immunologic findings. *Am J Clin Pathol* 1987;88;348–351.

32. Katzmann JA, Clark R, Wiegert E, et al. Identification of monoclonal proteins in serum: a quantitative comparison of acetate, agarose gel, and capillary electrophoresis. *Electrophoresis* 1997;18;1775–1780.

33. Blessum CR, Khatter N, Alter SC. Technique to remove interference caused by radio-opaque agents in clinical capillary zone electrophoresis. *Clin Chem* 1999;45;1313.

34. Keren DF, Warren JS, Lowe JB. Strategy to diagnose monoclonal gammopathies in serum: high-resolution electrophoresis, immunofixation, and kappa/lambda quantification. *Clin Chem* 1988;34;2196–2201.

35. Laine ST, Soppi ET, Morsky PJ. Critical evaluation of the serum kappa/lambda light-chain ratio in the detection of M proteins. *Clin Chim Acta* 1992;207;143–149.

36. Riches PG, Sheldon J, Smith AM, Hobbs JR. Overestimation of monoclonal immunoglobulin by immunochemical methods. *Ann Clin Biochem* 1991;28(Pt 3):253–259.

37. Bush D, Keren DF. Over- and underestimation of monoclonal gammopathies by quantification of kappa- and lambda-containing immunoglobulins in serum. *Clin Chem* 1992;38;315–316.

38. Liu Y-C, Valenzuela R, Weick J, Slaughter S. Verification of monoclonality criteria for initial serum screening. *Am J Clin Pathol* 1991;96;417 (abstract).

39. Liu Y-C, Valenzuela R, Slaughter S. Sensitive and specific immunochemical criteria for characterization of monoclonal gammopathies. *Am J Clin Pathol* 1992;97;458(Abstr).

40. Jones RG, Aguzzi F, Bienvenu J, et al. Use of immunoglobulin heavy-chain and light-chain measurements in a multicenter trial to investigate monoclonal components: I. Detection. *Clin Chem* 1991;37;1917–1921.

41. Jones RG, Aguzzi F, Bienvenu J, et al. Use of immunoglobulin heavy-chain and light-chain measurements in a multicenter trial to investigate monoclonal components: II. Classification by use of computer-based algorithms. *Clin Chem* 1991;37;1922–1926.

42. Drayson M, Tang LX, Drew R, Mead GP, Carr-Smith H, Bradwell AR. Serum free light-chain measurements for identifying and monitoring patients with nonsecretory multiple myeloma. *Blood* 2001;97;2900–2902.

43. Rao KM, Bordine SL, Keren DF. Decision making by pathologists. A strategy for curtailing the number of inappropriate tests. *Arch Pathol Lab Med* 1982;106;55–56.

44. Keren DF, Di Sante AC, Bordine SL. Densitometric scanning of high-resolution electrophoresis of serum: methodology and clinical application. *Am J Clin Pathol* 1986;85;348–352.

45. Hom BL. Polymeric (presumed tetrameric) lambda Bence Jones proteinemia without proteinuria in a patient with multiple myeloma. *Am J Clin Pathol* 1984;82;627–629.

46. Duffy TP. The many pitfalls in the diagnosis of myeloma. *N Engl J Med* 1992;326;394–396.

47. Abraham RS, Clark RJ, Bryant SC, et al. Correlation of serum immunoglobulin free light chain quantification with urinary Bence Jones protein in light chain myeloma. *Clin Chem* 2002;48;655–657.

48. Hess PP, Mastropaolo W, Thompson GD, Levinson SS. Interference of polyclonal free light chains with identification of Bence Jones proteins. *Clin Chem* 1993;39;1734–1738.

49. Herkner KR, Salzer H, Bock A, et al. Pediatric and perinatal reference intervals for immunoglobulin light chains kappa and lambda. *Clin Chem* 1992;38;548–550.

50. Saitta M, Iavarone A, Cappello N, Bergami MR, Fiorucci GC, Aguzzi F. Reference values for immunoglobulin kappa and lambda light chains and the kappa/lambda ratio in children's serum. *Clin Chem* 1992;38;2454–2457.

51. Haraldsson A, Weemaes CM, Kock-Jansen MJ, et al. Immunoglobulin G, A and M light chain ratio in children. *Ann Clin Biochem* 1992;29(Pt 3):271–274.

52. Haraldsson A, Jaminon M, Bakkeren JA, Stoelinga GB, Weemaes CM. Immunoglobulin G, A, and M light chain ratios in some humoral immunological disorders. *Scand J Immunol* 1992;36;57–61.

10

Case studies for interpretation

This chapter allows one to interpret some cases from my files. In most of these cases, I only know the age and sex of the individual at the time of the initial interpretation. I hope that you will review the electrophoretic information, make your interpretation and, where appropriate, any suggestions for the clinicians. The following discussion reviews my interpretations on each case. I have included cases processed with a variety of techniques to give a broad overview of the possible patterns. In the past, some readers have sent me correspondence by mail. I welcome these, but if you want a quicker response (perhaps) you could e-mail me at kerend@wardelab.com. I will get back to you as soon as I can.

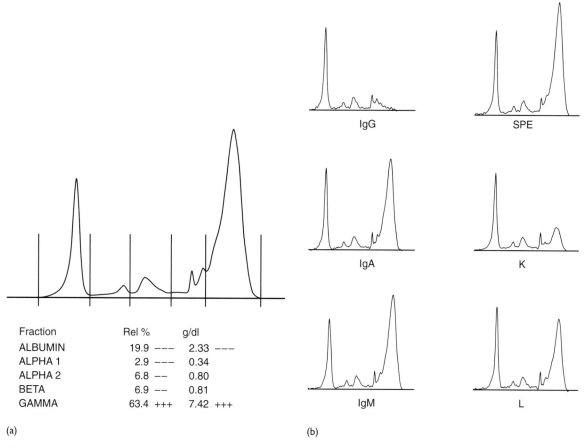

Fraction	Rel %	g/dl
ALBUMIN	19.9 ---	2.33 ---
ALPHA 1	2.9 ---	0.34
ALPHA 2	6.8 --	0.80
BETA	6.9 --	0.81
GAMMA	63.4 +++	7.42 +++

(a)   (b)

Figure 10.1 (a) Capillary zone electropherogram performed on serum from a 42-year-old man. (b) Immunosubtraction of the serum from (a). (Paragon CZE 2000.)

INTERPRETATION FOR FIG. 10.1

The absolute value of albumin is decreased. The α_1-, α_2-, and β-globulins are all decreased along with albumin in their relative percentage of the serum proteins. This results from the massive increase in γ-globulin that has a total concentration of 7.42 g/dl. Such a massive increase in γ-globulin causes one to consider multiple myeloma. The peak of this massive γ-globulin region is relatively sharp, yet the base of the γ-globulin extends throughout the entire γ-globulin region. Nonetheless, such a massive increase deserves an immunosubtraction or immunofixation to rule out monoclonality.

The serum protein electrophoresis (SPE) recapitulates the findings in the capillary zone electropherogram. When serum preincubated with beads coated with anti-IgG was evaluated, the beads subtracted out the massive spike, indicating that it was overwhelmingly due to IgG. Neither preincubation with beads coated with anti-IgA nor with beads coated with anti-IgM decreased the peak. However, what about the light chains? The beads coated with anti-κ removed most of the peak, but not all. How much was removed? We're not given an exact amount but at least three-quarters of the peak seems to be gone. If this was a monoclonal gammopathy, I would expect the peak to be completely removed by the beads. Yet, could one see an antigen excess situation where the amount of antibody on the beads was insufficient to remove all of the antibody? Now consider the results in the L (anti-λ) electropherogram. At first, it may appear that the anti-λ beads have not affected the massive γ-globulin region. It is not readily apparent what has happened unless a frame of reference is established. I use albumin. Note that in the SPE electropherogram the γ-globulin peak is considerably higher than the albumin peak. But, in the anti-λ electropherogram the beads have clearly reduced the peak to well below the height of albumin. This indicates that it has subtracted perhaps one-quarter of the peak. In a polyclonal response the ideal is if the κ-containing IgG accounts for two-thirds of the response and the λ-containing IgG accounts for one-third. Here, the ratio seems to be three-quarters to one-quarter. One can quibble because we do not have

exact quantities. However, if this were a massive IgG κ monoclonal gammopathy, one would expect virtually all of it would be κ-containing and almost none λ. Therefore, the anti-λ beads should have had no noticeable effect. Since it did, this is a massive polyclonal increase in IgG. Such increases are uncommon. The differential includes viral infections such as human immunodeficiency virus, hepatitis C virus, a combination of the two, and Epstein–Barr virus. T-cell immunoblastic lymphoma (angioimmunoblastic lymphadenopathy) may also produce a picture of decreased albumin and polyclonal increase in γ-globulin with β–γ bridging. Fortunately, these conditions can be tested for serologically and with molecular studies. I have also seen an idiopathic case where no underlying cause has been found for a period of over 5 years.

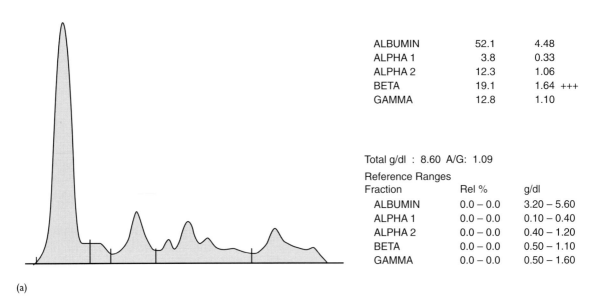

ALBUMIN	52.1	4.48
ALPHA 1	3.8	0.33
ALPHA 2	12.3	1.06
BETA	19.1	1.64 +++
GAMMA	12.8	1.10

Total g/dl : 8.60 A/G: 1.09

Reference Ranges

Fraction	Rel %	g/dl
ALBUMIN	0.0 – 0.0	3.20 – 5.60
ALPHA 1	0.0 – 0.0	0.10 – 0.40
ALPHA 2	0.0 – 0.0	0.40 – 1.20
BETA	0.0 – 0.0	0.50 – 1.10
GAMMA	0.0 – 0.0	0.50 – 1.60

(a)

Figure 10.2 (a) Densitometric scan of serum protein electrophoresis from a 3-year-old girl with a history of a skin rash and anemia. IgA level was elevated, IgM was low and IgG was in the upper limit of normal.

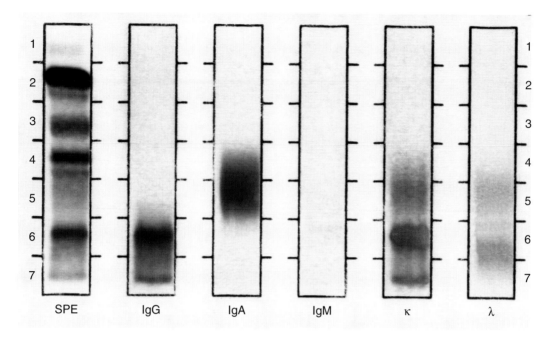

(b)

Figure 10.2 (contd) (b) Immunofixation of this patient's serum (Paragon immunofixation). Case contributed by Joseph M. Lombardo.

INTERPRETATION OF FIG. 10.2

The most remarkable finding is the presence of two distinct, γ-region restrictions. This is an unusual finding in a child. The immunofixation in Fig. 10.2b demonstrates that both the anodal and cathodal restrictions are consistent with IgG κ M-proteins. The immunofixation also confirms the impression of a polyclonal increase in IgA and a very low level (almost absent) IgM. Monoclonal gammopathies, biclonal gammopathies and even prominent oligoclonal banding are unusual in young children. However, they do occur with increased frequency in patients with immunodeficiency syndromes. Although these features bore resemblance to the Wiskott–Aldrich Syndrome, this suggestion was rejected as having been ruled out by the clinician. Wiskott–Aldrich Syndrome is X-linked and has only rarely been reported in girls. The most recent follow-up information from the clinician indicates that this child's immunodeficiency is 'uncharacterized'. What do you think?

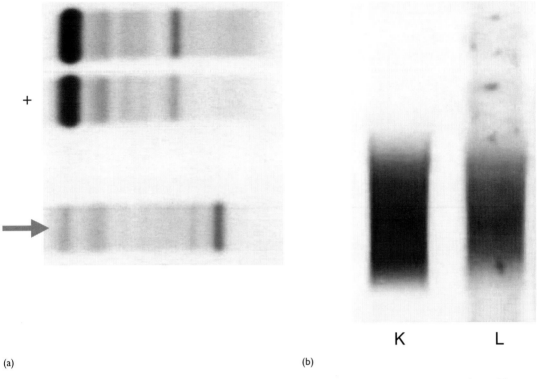

+

→

K   L

(a)   (b)

Figure 10.3 (a) Four urine samples are shown with the anode indicated to the left. The bottom sample came from a 51-year-old man. (Sebia $\beta_{1,2}$ gel). (b) Immunofixation for anti-κ (K) and anti-λ (L) on the bottom sample from (a).

INTERPRETATION OF FIG. 10.3

The bottom urine sample (arrowed) has a tubular proteinuria pattern. The albumin band is small compared with the two samples above it that have glomerular and tubular proteinuria. The numerous small bands and diffuse staining in the α- and β-regions are consistent with the many small proteins that normally pass through the glomerulus and are reabsorbed by the tubules. However, in the β–γ region there is a prominent band. It is suspicious for a monoclonal free light chain (MFLC, or Bence Jones protein). However, it does not produce a band with κ or λ immunofixation. We tested the urine for myoglobin and found that it had a value of 9642 ng/ml (normal is < 90 ng/ml).

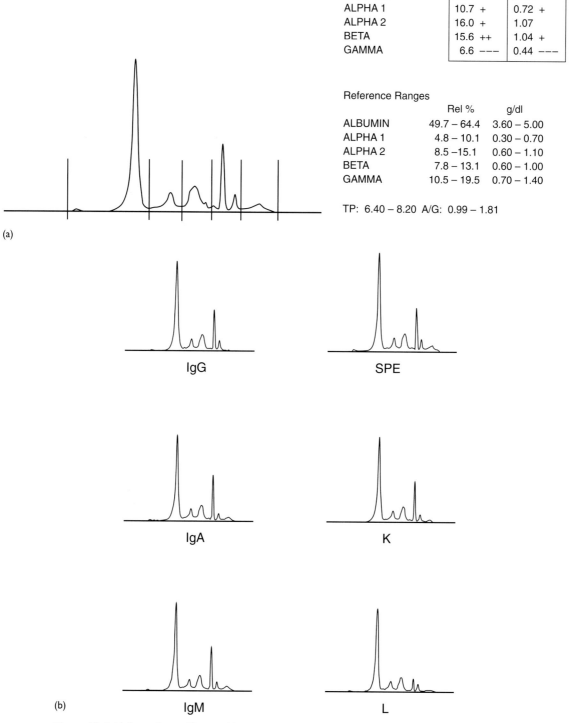

Fraction	Rel %	g/dl
ALBUMIN	51.1	3.42 –
ALPHA 1	10.7 +	0.72 +
ALPHA 2	16.0 +	1.07
BETA	15.6 ++	1.04 +
GAMMA	6.6 – – –	0.44 – – –

Reference Ranges

	Rel %	g/dl
ALBUMIN	49.7 – 64.4	3.60 – 5.00
ALPHA 1	4.8 – 10.1	0.30 – 0.70
ALPHA 2	8.5 –15.1	0.60 – 1.10
BETA	7.8 – 13.1	0.60 – 1.00
GAMMA	10.5 – 19.5	0.70 – 1.40

TP: 6.40 – 8.20 A/G: 0.99 – 1.81

(a)

IgG   SPE

IgA   K

(b)   IgM   L

Figure 10.4 (a) Serum from a 73-year-old woman. (Capillary zone electropherogram, Paragon CZE 2000).
(b) Immunosubtraction of the serum from (a).

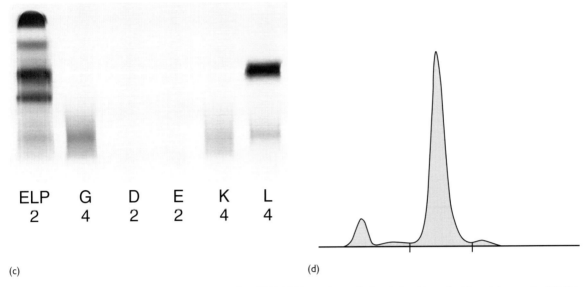

(c)

(d)

Figure 10.4 (*contd*) (c) Immunofixation to rule out IgD and IgE. (d) Urine: the small albumin band is to the left and the massive MFLC is obvious.

INTERPRETATION OF FIG. 10.4

The albumin is slightly decreased with an increase in α_1- and β-globulins. There is also a relative increase in the percentage of α_2-globulin. The β_1-region band (transferrin) is elevated, accounting for the modest increase in the β-region. Note that it barely edges above the upper limit of 1.00 g/dl for that region in our laboratory. The γ-region is decreased considerably and there is a slight mid-γ-region restriction. The combination of decreased albumin and low γ-region are a concern for protein loss. The increase in α_1-region with a low albumin could indicate an acute-phase pattern, but then I would expect the transferrin band to be decreased. The elevated transferrin band could indicate an iron deficiency anemia, or it may be hiding a monoclonal gammopathy. The decrease in the γ-region is out of proportion to the slight decrease in albumin. Because of the small restriction and the concern about the transferrin band, the immuno-subtraction was performed. In the SPE lane, the electropherogram findings are repeated as a frame of reference. Incubation with beads coated with anti-IgG subtracts out the small γ-region restriction. No change is seen in the IgA, IgM, or K lanes. However, when the serum was preincubated with anti-λ, not only was the tiny restriction removed (identifying it as a small IgG λ monoclonal gammopathy), but the 'transferrin' band was cut down to normal size, indicating a likely λ MFLC. Figure 10.4c shows the immunofixation to rule out IgD and IgE. Dilutions of the patient's serum are shown in each lane. It also demonstrates the small IgG λ in the γ-region and, interestingly, it shows the λ MFLC migrating in the α_2-region. Electrophoretic conditions differ between gel-based systems and capillary zone electrophoresis, such that (usually) slight differences in migration of some M-proteins will be seen. The patient's urine in Fig. 10.4d shows the small albumin band on the left and the massive MFLC is obvious. This is another example of how the presence of, in this case, both a very small IgG monoclonal gammopathy and a subtle change in the transferrin peak, helped to make the diagnosis of light chain multiple myeloma.

Fraction	Rel %	g/dl
ALBUMIN	50.4	3.73
ALPHA 1	8.7	0.65
ALPHA 2	16.4 +	1.21 +
BETA	11.7	0.87
GAMMA	12.8	0.95

Reference Ranges

	Rel %	g/dl
ALBUMIN	49.7 – 64.4	3.55 – 5.04
ALPHA 1	4.8 – 10.1	0.25 – 0.74
ALPHA 2	8.5 –15.1	0.55 – 1.14
BETA	7.8 – 13.1	0.55 – 1.04
GAMMA	10.5 – 19.5	0.65 – 1.44

TP: 6.40 – 8.20   A/G: 0.99 – 1.81

(a)

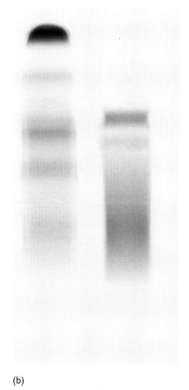

(b)

Figure 10.5 (a) Capillary zone electropherogram for an 80-year-old man (Paragon CZE 2000). (b) Penta (pentavalent) immunofixation of serum from (a) (Sebia Penta Immunofixation).

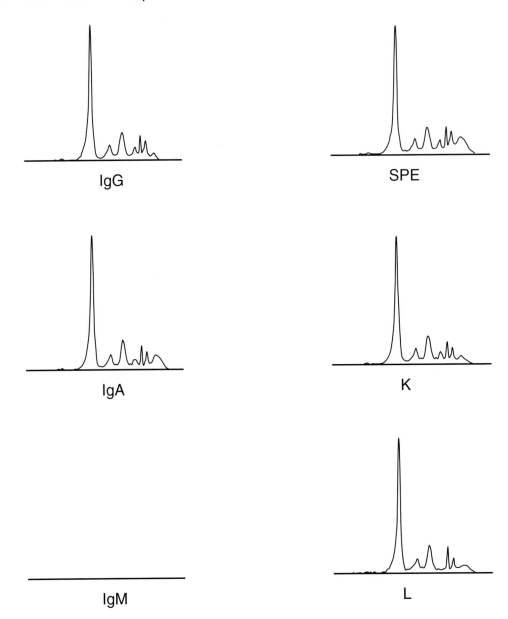

(c)

Figure 10.5 (*contd*) (c) Immunosubtraction of serum from (a).

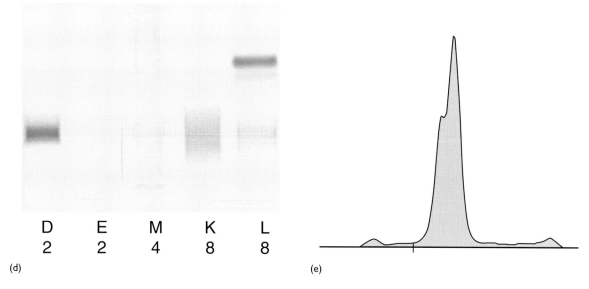

D	E	M	K	L
2	2	4	8	8

(d)

(e)

Figure 10.5 (*contd*) (d) Immunofixation for IgD, IgE, IgM, κ, and λ. (e) Densitometric scan of urine.

INTERPRETATION FOR FIG. 10.5

The electropherogram has a slight increase in the α_2-region, but otherwise the values are normal. However, there is a small band at the interface between the α_2- and β-regions. Sometimes a small hemopexin band may be seen here, but this seems too large. In the γ-region, there is an asymmetry to the usually almost 'normal' distribution of the immuno-globulins here. This asymmetry may represent a poly-clonal increase in a subclass. However, because of these two features, I wanted to see a pentavalent (Penta) immunofixation of this sample. The Penta immunofixation in Fig. 10.5b demonstrates two bands in the α_2-region and a more subtle band in the γ-region. This finding indicates that a monoclonal gammopathy is likely present. Immunosubtraction was performed in Fig. 10.5c, but there was a prob-lem. No record was obtained from the capillary that contained the results of the IgM immunosubtraction, so we just see a flat line. No conclusions may be drawn about IgM from this study. The SPE lane provides the frame of reference for the other areas. In the IgG immunosubtraction, much of the γ-region is removed, but its anodal end (where we saw the asymmetry on the initial electropherogram) now has a well-outlined,

though small band. Also, the small band between α_2 and β is still present. The IgA immunosubtraction has no effect on the band between α_2 and β, but the κ immunosubtraction makes the small band in the fast end of the γ-region more obvious (having removed the κ-containing polyclonal IgG, the non-κ monoclonal gammopathy is more evident). The λ immunosub-traction discloses that both the band between α_2 and β as well as the band at the anodal end of the γ-region has been removed. They are both λ monoclonal gammopathies. But do they have a heavy chain attached? We do not know yet. Thus, an immuno-fixation was performed for IgM, κ, λ, IgD and IgE (Fig. 10.5d). Dilutions of the patient's serum are shown in each lane. Here the band at the anodal end of the γ-region is found to be an IgD λ monoclonal gammopathy and the one between α_2 and β is found to be two bands (the cathodal one quite faint – I hope that you see it), both λ MFLC (possibly a monomer and a dimer). Lastly, the urine densitometric scan in Fig. 10.5e demonstrates the urine contains a large amount of λ MFLC (by immunofixation, the band had no IgD heavy chain attached to it). The weak staining of the λ where it is bound to IgD is not unusual and occasionally leads to an erroneous impression of heavy chain disease.

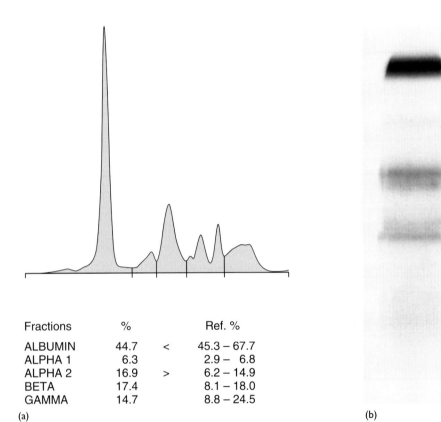

Fractions	%		Ref. %
ALBUMIN	44.7	<	45.3 – 67.7
ALPHA 1	6.3		2.9 – 6.8
ALPHA 2	16.9	>	6.2 – 14.9
BETA	17.4		8.1 – 18.0
GAMMA	14.7		8.8 – 24.5

(a)

(b)

Figure 10.6 (a) Electropherogram for a 72-year-old man (Sebia Capillarys). (b) Penta (pentavalent) immunofixation screen of serum from (a) (Sebia Penta immunofixation).

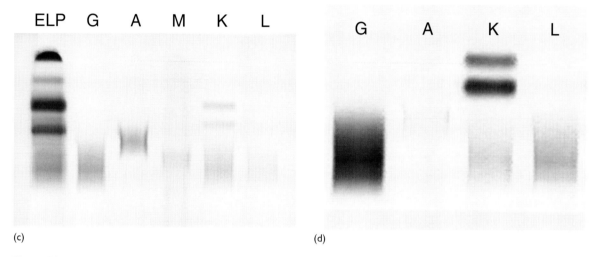

Figure 10.6 *(contd)* (c) Immunofixation of serum from same case. (d) Immunofixation of urine from same case.

INTERPRETATION OF FIG. 10.6

This sample was sent for an immunofixation of the serum. In our laboratory, a request for immunofixation requires both a capillary zone electrophoresis (CZE) and the serum Penta immunofixation. The serum electropherogram has a slightly decreased percentage of albumin with a modest increase in the α_2-globulin. There is a small restriction present at the anodal end of the β_1-region. The γ-region has a few irregularities consistent with oligoclonal bands. The most cathodal band is a little more prominent than the rest. In Fig. 10.6b, the Penta immunofixation screen demonstrates two bands in the α_2- and β-regions. The γ-region of the Penta has a few tiny oligoclonal bands consistent with the electropherogram. Because of the α_2- and β-region bands, the immunofixation in Fig. 10.6c was performed. It demonstrates that the two κ are MFLC. Another immunofixation (not shown) ruled out IgD and IgE. The urine immunofixation in Fig. 10.6d has the same finding as the serum, although the two κ-bands are more prominent. It is likely that the two κ MFLC represent a monomer and a dimer.

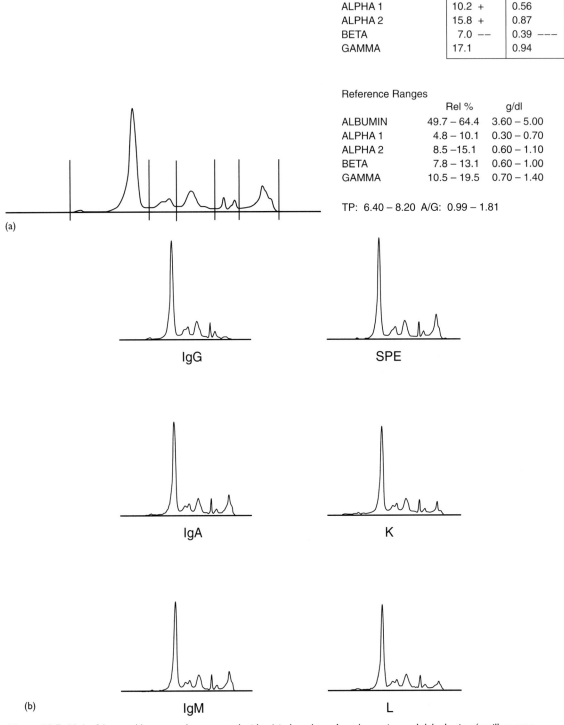

Fraction	Rel %	g/dl	
ALBUMIN	49.9	2.74	---
ALPHA 1	10.2 +	0.56	
ALPHA 2	15.8 +	0.87	
BETA	7.0 --	0.39	---
GAMMA	17.1	0.94	

Reference Ranges

	Rel %	g/dl
ALBUMIN	49.7 – 64.4	3.60 – 5.00
ALPHA 1	4.8 – 10.1	0.30 – 0.70
ALPHA 2	8.5 –15.1	0.60 – 1.10
BETA	7.8 – 13.1	0.60 – 1.00
GAMMA	10.5 – 19.5	0.70 – 1.40

TP: 6.40 – 8.20 A/G: 0.99 – 1.81

(a)

IgG SPE

IgA K

(b) IgM L

Figure 10.7 (a) An 84-year-old woman who presented with a 'viral syndrome' – aches, pains, and dehydration (capillary zone electropherogram, Paragon CZE 2000). (b) Immunosubtraction from same sample.

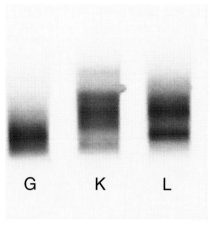

Figure 10.7 (*contd*) (c) Immunofixation for IgG, κ (K) and λ (L) from same sample.

INTERPRETATION OF FIG. 10.7

Albumin is decreased, there is a relative increase in the percentage of α_1- and α_2-globulins. In the α_1-region, there is a prominent α_1-acid glycoprotein (orosomucoid) shoulder just anodal to the α_1-antitrypsin band. The β-region is decreased especially with a decreased transferrin band. These features are consistent an acute-phase reaction. The γ-region has two or maybe three modest-sized restrictions. The immunosubtraction in Fig. 10.7b was performed because of the unusual γ-region. Once again, the SPE lane offers a frame of reference to the immunosubtraction. Removal of IgG eliminates the restrictions, but removal of IgA and IgM by immunosubtraction has no effect on the bands. Some of the bands are lost when κ-containing immunoglobulins are removed (although the most prominent remains) and by removing λ-containing immunoglobulins, the most prominent band is lost, although a few other tiny ones remain. This is a prominent oligoclonal response. The immunofixation also nicely demonstrates the several oligoclonal bands in Fig. 10.7c.

After viewing the immunofixation, I called the clinician and learned that the patient had just died. Her legs had developed petechiae that grew larger and red golf ball-sized bumps broke out on her arms. Her temperature had been 102°F (38.9°C). We performed an anti-neutrophil cytoplasmic antibody (ANCA) test on the serum and it was positive for cytoplasmic ANCA at a titer of 1:320. This antibody was confirmed by a specific enzyme immunoassay that was positive for anti-serine protease 3 (PR3), but negative for anti-myeloperoxidase. These findings are consistent with Wegener's granulomatosis.

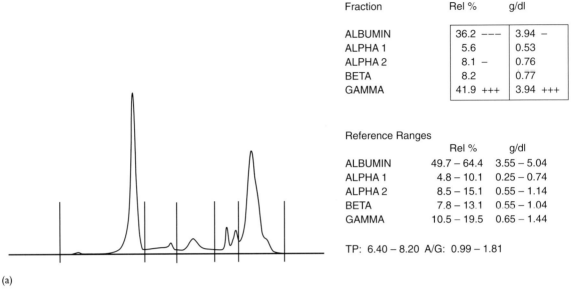

Fraction	Rel %	g/dl
ALBUMIN	36.2 ---	3.94 –
ALPHA 1	5.6	0.53
ALPHA 2	8.1 –	0.76
BETA	8.2	0.77
GAMMA	41.9 +++	3.94 +++

Reference Ranges

	Rel %	g/dl
ALBUMIN	49.7 – 64.4	3.55 – 5.04
ALPHA 1	4.8 – 10.1	0.25 – 0.74
ALPHA 2	8.5 – 15.1	0.55 – 1.14
BETA	7.8 – 13.1	0.55 – 1.04
GAMMA	10.5 – 19.5	0.65 – 1.44

TP: 6.40 – 8.20 A/G: 0.99 – 1.81

(a)

Figure 10.8 (a) Serum from a 64-year-old woman. (Capillary zone electropherogram, Paragon CZE 2000.)

INTERPRETATION OF FIG. 10.8

This serum has a decrease in albumin and a markedly elevated γ-globulin. The elevation is unusual because it has a relatively broad base and a shoulder on the left side. These features are most likely a monoclonal gammopathy. The cathodal shoulder could represent the normal IgG or perhaps a second M-protein. This question is quickly answered by the immunosubtraction pattern in Fig. 10.8b. Using the SPE as the frame of reference, it is clear that subtracting out the IgA removed almost the entire peak, including the shoulder. The residual amount left in the γ-region after subtracting the IgA shows us the patient's normal IgG. It is very low. Similarly, subtracting the λ removes the same amount of the peak leaving a little normal κ-containing IgG as a residual. It would not be a bad idea here to go back and look at the immunosubtraction from Fig. 10.1. There, the massive IgG increase was polyclonal and immunosubtraction of κ-containing immunoglobulins removed about three-quarters of it while immunosubtraction of λ-containing

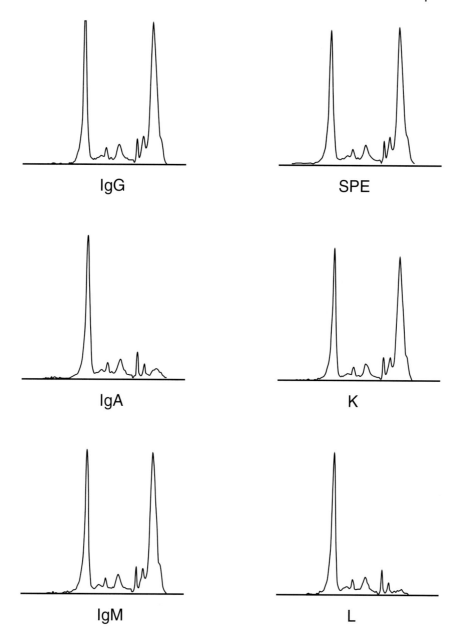

(b)

Figure 10.8 *(contd)* (b) Immunosubtraction on the same sample.

immunoglobulins removed about one-quarter of it. Contrast that with the removal of well over 90–95% of the spike by the relevant immunosubtraction in the present case. Further, the residual γ-region protein left after the κ-immunosubtraction in the present case is normal polyclonal IgG, whereas both the κ- and λ-immunosubtractions in Fig. 10.1 had the same basic shape. The breadth of the present IgA λ as well as the presence of the shoulder likely reflects two features of IgA monoclonal gammopathies. They tend to be heavily glycosylated and thus have a broader migration. They also tend to form multimers that could explain the shoulder.

Fraction	Rel %	g/dl
ALBUMIN	44.2 ––	3.14 ––
ALPHA 1	8.3	0.59
ALPHA 2	15.0	1.07
BETA	10.8	0.77
GAMMA	21.7 ++	1.54 +

Reference Ranges

	Rel %	g/dl
ALBUMIN	49.7 – 64.4	3.55 – 5.04
ALPHA 1	4.8 – 10.1	0.25 – 0.74
ALPHA 2	8.5 –15.1	0.55 – 1.14
BETA	7.8 – 13.1	0.55 – 1.04
GAMMA	10.5 – 19.5	0.65 – 1.44

TP: 6.40 – 8.20 A/G: 0.99 – 1.81

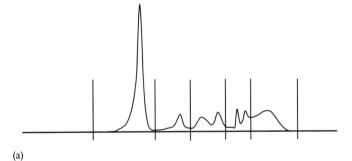

(a)

Figure 10.9 (a) Serum from a 59-year-old man. (Capillary zone electropherogram, Paragon CZE 2000).

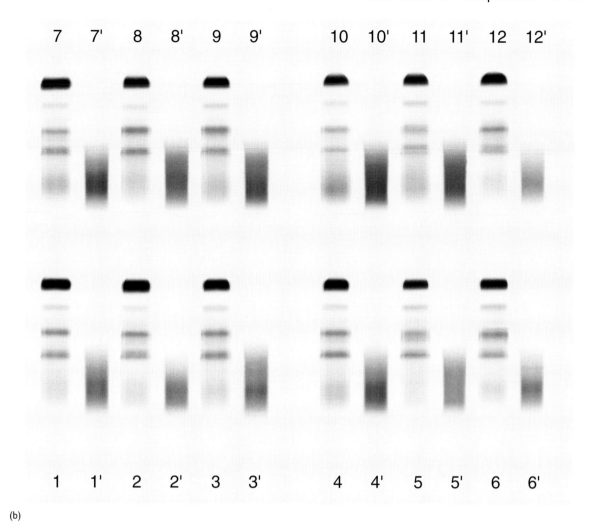

(b)

Figure 10.9 *(contd)* (b) Penta (pentavalent) immunofixation for this case and for Fig. 10.10 (Sebia Penta immunofixation).

INTERPRETATION OF FIG. 10.9

This sample has a decrease in albumin with a modest increase in γ-globulin and possibly a slight β–γ bridge. There is a prominent band in the cathodal end of the α_2-region. In Fig. 10.9b, the Penta immunofixation for this case (number 8 and 8') demonstrates that the α_2-region restriction is not an immunoglobulin. On the gel-based Penta system, a protein dye is used and no such band even appears in the acid-fixed lane (8). In previous cases, it is clear that the Penta immunofixation can detect even quite small M-proteins. So what is this band? I called the clinician and learned that the patient had received a radiocontrast dye during the performance of a stent procedure. This is a problem with the CZE technique because these dyes absorb at the same wavelength that peptide bonds do. As shown on the Penta gel, however, they will not stain with the protein dye. A table of radiocontrast dyes that are known to produce this artifact is present in Table 2.3 (Chapter 2).

Fraction	Rel %	g/dl	
ALBUMIN	56.0	2.52	---
ALPHA 1	12.0 ++	0.54	
ALPHA 2	17.1 ++	0.77	
BETA	10.4	0.47	--
GAMMA	4.5 ---	0.20	---

Reference Ranges

	Rel %	g/dl
ALBUMIN	49.7 – 64.4	3.55 – 5.04
ALPHA 1	4.8 – 10.1	0.25 – 0.74
ALPHA 2	8.5 – 15.1	0.55 – 1.14
BETA	7.8 – 13.1	0.55 – 1.04
GAMMA	10.5 – 19.5	0.65 – 1.44

TP: 6.40 – 8.20 A/G: 0.99 – 1.81

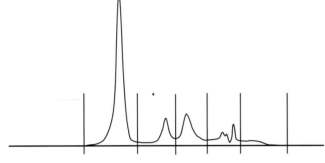

TP g/dl: 4.50 ---   A/G: 1.27   Operator Initials: ALR
Run date: 1/2/03   Edit date: Reviewer Initials:

(a)

Fraction	Rel %	g/dl
ALBUMIN	58.5	4.68
ALPHA 1	6.1	0.49
ALPHA 2	10.6	0.85
BETA	9.1	0.73
GAMMA	15.8	1.26

Reference Ranges

	Rel %	g/dl
ALBUMIN	49.7 – 64.4	3.55 – 5.04
ALPHA 1	4.8 – 10.1	0.25 – 0.74
ALPHA 2	8.5 – 15.1	0.55 – 1.14
BETA	7.8 – 13.1	0.55 – 1.04
GAMMA	10.5 – 19.5	0.65 – 1.44

TP: 6.40 – 8.20 A/G: 0.99 – 1.81

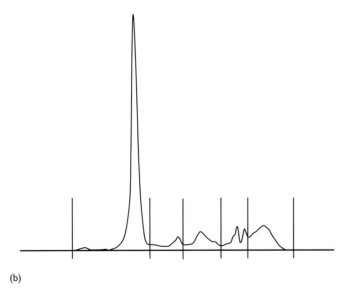

(b)

Figure 10.10 (a) Serum from a 54-year-old woman (capillary zone electropherogram, Paragon CZE 2000). (b) Serum from a 74-year-old man (capillary zone electropherogram, Paragon CZE 2000).

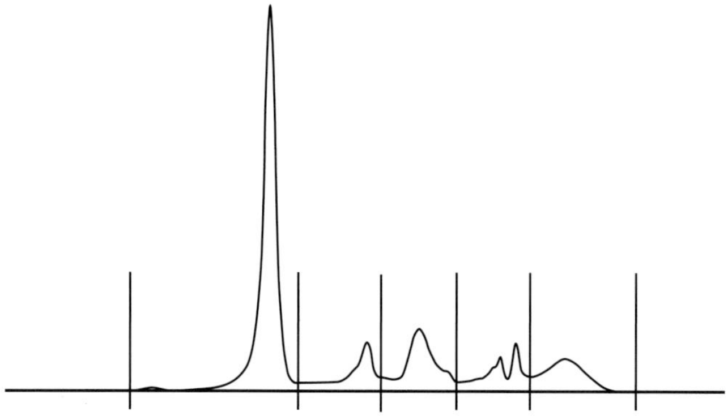

Fraction	Rel %	g/dl
ALBUMIN	51.6	3.41 –
ALPHA 1	10.2 +	0.67
ALPHA 2	15.8 +	1.05
BETA	10.4	0.69
GAMMA	12.0	0.79

Reference Ranges

	Rel %	g/dl
ALBUMIN	49.7 – 64.4	3.55 – 5.04
ALPHA 1	4.8 – 10.1	0.25 – 0.74
ALPHA 2	8.5 – 15.1	0.55 – 1.14
BETA	7.8 – 13.1	0.55 – 1.04
GAMMA	10.5 – 19.5	0.65 – 1.44

TP: 6.40 – 8.20 A/G: 0.99 – 1.81

(c)

Figure 10.10 (*contd*) (c) Serum from a 67-year-old man (capillary zone electropherogram, Paragon CZE 2000).

INTERPRETATION FOR FIG. 10.10

Serum 10.10a is has a decrease in albumin with a relative increase in the percentage of α_1- and α_2-globulins. The γ-globulin is considerably decreased which, together with the low albumin, suggests a protein loss pattern. However, it could also be consistent with immunosuppression due to light chain multiple myeloma. Also, MFLC will cause renal damage possibly resulting in a pattern like this. When I looked at the β-region expecting to see a decrease in transferrin to confirm an impression of an acute-phase pattern, I noticed a discrete band just anodal to transferrin (β_1). The same sample is processed in lanes 5 and 5´ on the Penta gel from Fig. 10.9b. There is no evidence of a monoclonal band. This could represent a transferrin variant.

However, in this case it does not. Look at the same area just anodal to the transferrin band in Figures 10.10b, and 10.10c. Both of these also contain some type of shoulder or suggestion of a double transferrin band. All three samples were run in the same capillary as the first sample. By processing these specimens in other capillaries, the band disappears. It is an artifact. Both the Paragon CZE 2000 and the Sebia Capillarys have several capillaries that serum passes through. Occasionally, if a highly lipemic sample is processed, the capillary may give spurious results for the next few assays. Furthermore, as the capillary ages similar artifacts can occur. Therefore, when I observe an unusual band in a sample, I check other samples that passed through that particular capillary on the same run prior to rendering my interpretation.

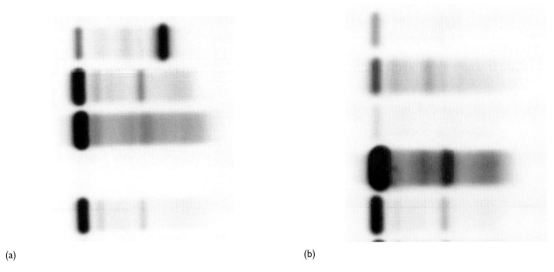

(a)   (b)

Figure 10.11 (a) Electrophoresis of five urine samples concentrated up to 50-fold is shown for interpretation (Sebia $\beta_{1,2}$ gel). (b) Electrophoresis of five more urine samples concentrated up to 50-fold is shown for interpretation (Sebia $\beta_{1,2}$ gel)

INTERPRETATION OF FIG. 10.11

The top lane in Fig. 10.11a contains a small amount of albumin; discrete staining in the α- and β-regions indicate tubular proteinuria. However, the main observation is the presence of a large γ-region band that almost certainly represents MFLC. This was confirmed to be a κ MFLC on immunofixation. The second lane demonstrates a prominent albumin band with an α_1-antitrypsin and transferrin band as main other components. This is predominately glomerular proteinuria. The third sample shows a strong albumin band with both α_1-antitrypsin and transferrin band identifiable; however, there is broad diffuse staining that may reflect some denaturation of the protein. This must have glomerular and tubular leakage, but it needs an immunofixation for further comment.

The fourth sample shows no protein at all. Whenever I see a pattern like this I check the total protein. It was huge (7320 mg/24 h) so electrophoresis was repeated. It appears as the fourth specimen in Fig. 10.11b. This shows a massive proteinuria consistent with a nonselective proteinuria. The patient is a diabetic with glomerular disease. This illustrates the problem of the wick effect. On this automated application system, the samples are placed in a well that is attached to a wick made of paper. The sample needs to 'wick' to the end of the paper for application to the gel. In cases with high protein content, such as this, and ones that contain particulate material (crystals, cellular elements) we have found a problem with adequate sampling. This is discussed further in Chapter 7. The last sample in Fig. 10.11a is also a glomerular pattern.

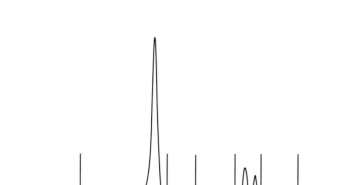

Fraction	Rel %	g/dl
ALBUMIN	50.1	3.75
ALPHA 1	6.5	0.49
ALPHA 2	12.9	0.97
BETA	17.0 +++	1.28 +++
GAMMA	13.4	1.01

Reference Ranges

	Rel %	g/dl
ALBUMIN	49.7 – 64.4	3.60 – 5.00
ALPHA 1	4.8 – 10.1	0.30 – 0.70
ALPHA 2	8.5 – 15.1	0.60 – 1.10
BETA	7.8 – 13.1	0.60 – 1.00
GAMMA	10.5 – 19.5	0.70 – 1.40

TP: 6.40 – 8.20 A/G: 0.99 – 1.81

(a)

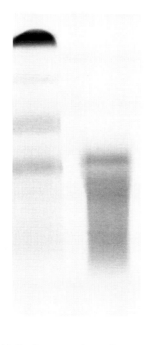

(b)

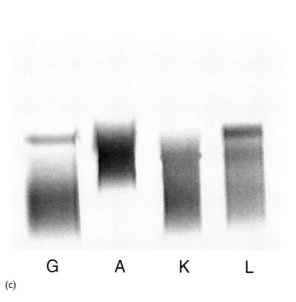

G   A   K   L

(c)

Figure 10.12 (a) Capillary zone electropherogram for a 41-year-old woman (Paragon CZE 2000). (b) Penta (pentavalent) immunofixation on the sample from (a). (c) Immunofixation on the same sample.

INTERPRETATION OF FIG. 10.12

The capillary zone electropherogram has two unusual features. First the transferrin band appears to be broader than usual. Typically, transferrin and C3 bands both are slender and symmetrical. Here, the transferrin band is broad and slightly asymmetrical showing a subtle cathodal shoulder. The second area of concern was the β–γ bridge. A β–γ bridge usually reflects a polyclonal increase in IgA.

It can be seen in a wide variety of circumstances, but one of the more common is cirrhosis and sometimes hepatitis. This pattern looks like neither of those. Further, considering the normal amount of total γ-globulin, the bridge has the unusual appearance of going up into the β-region rather than just being a flat bridge across. Because of these concerns we performed the Penta immunofixation shown in Fig. 10.12b. It demonstrates an immunoglobulin band (likely an M-protein) at the anodal end of the immunofixation (approximately the location of transferrin). There is a second broader area in the β–γ-region that likely reflects the bridging described in the electropherogram. In Fig. 10.12c, the immunofixation demonstrates that the restriction in the transferrin-region was due to the small IgG λ monoclonal gammopathy. The bridge, however, is a polyclonal increase in IgA. This can be determined by looking at the κ and λ lanes. There is a broad increase in both light chain types in the same area of migration that has an increase in the IgA.

Fraction	Rel %		g/dl	
ALBUMIN	37.0	---	2.37	---
ALPHA 1	10.4	+	0.66	
ALPHA 2	18.2	+++	1.17	+
BETA	22.1	+++	1.42	+++
GAMMA	12.3		0.79	

Reference Ranges

	Rel %	g/dl
ALBUMIN	49.7 – 64.4	3.60 – 5.00
ALPHA 1	4.8 – 10.1	0.30 – 0.70
ALPHA 2	8.5 – 15.1	0.60 – 1.10
BETA	7.8 – 13.1	0.60 – 1.00
GAMMA	10.5 – 19.5	0.70 – 1.40

TP: 6.40 – 8.20 A/G: 0.99 – 1.81

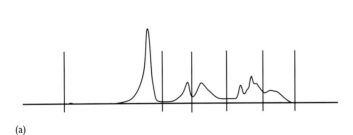

(a)

Figure 10.13 (a) Serum protein electrophoresis from 51-year-old woman (capillary zone electropherogram, Paragon CZE 2000).

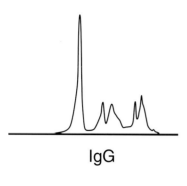

IgG

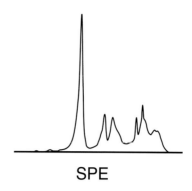

SPE

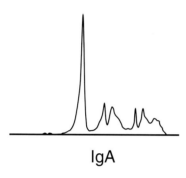

IgA

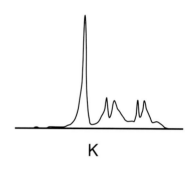

K

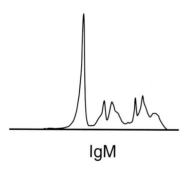

IgM

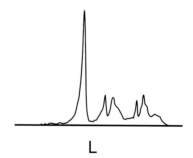

L

(b)

Figure 10.13 (*contd*) (b) Immunosubtraction of serum from Figure (a).

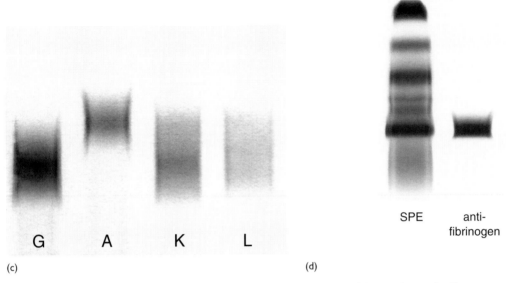

G   A   K   L

SPE   anti-fibrinogen

(c)   (d)

Figure 10.13 (contd) (c) Immunofixation of same serum (Sebia immunofixation). (d) Immunofixation for fibrinogen.

INTERPRETATION OF FIG. 10.13

This serum has a decrease in albumin, a relative increase in the percentage of α_1-globulin, a modest increase in α_2-globulin, and a considerable increase in β-globulin. The transferrin band is decreased and is consistent, together with the above findings, with an acute-phase response. However, forming a cathodal shoulder to the C3 band is a broad, but definite restriction. The γ-region has a slight restriction in its anodal end. But the concentration is in the normal range. Because of the presumption that the slow β-region band was a monoclonal gammopathy, I skipped the Penta step and performed the immunosubtraction shown in Fig. 10.13b. The SPE electropherogram provides the frame of reference for the immunosubtraction study. It demonstrates the unusual cathodal clinging to the C3 band. Unfortunately, it is not removed by any of the immunosubtractions. There is a slight decrease in the β_2-region after IgA has

been subtracted out; however, that is where polyclonal IgA normally migrates and one would assume that this would decrease the overall height of the β_2-region because of that. The shoulder to C3 remains, however, even in the IgA immunosubtraction. Because of this, we performed the immunofixation shown in Fig. 10.13c. Since we already knew that the IgM had no effect on the restriction, we did not include this in the immunofixation. There is a subtle restriction in the middle of the IgG lane that corresponds to a similar subtle restriction in the middle of the κ lane. This may represent a tiny IgG κ monoclonal gammopathy that is most likely part of the acute-phase response, but what is responsible for the shoulder? Finally, I performed an immunofixation for fibrinogen (Fig. 10.13d). The mystery protein is fibrinogen. With the earlier versions of the Paragon CZE 2000 fibrinogen did not show up as a band, but with the new buffer systems it does.

Fraction	Rel %		g/dl	
ALBUMIN	35.1	---	2.70	---
ALPHA 1	8.7		0.67	
ALPHA 2	12.2		0.94	
BETA	19.0	+++	1.46	+++
GAMMA	25.0	+++	1.93	+++

Reference Ranges

	Rel %	g/dl
ALBUMIN	49.7 – 64.4	3.60 – 5.00
ALPHA 1	4.8 – 10.1	0.30 – 0.70
ALPHA 2	8.5 – 15.1	0.60 – 1.10
BETA	7.8 – 13.1	0.60 – 1.00
GAMMA	10.5 – 19.5	0.70 – 1.40

TP: 6.40 – 8.20 A/G: 0.99 – 1.81

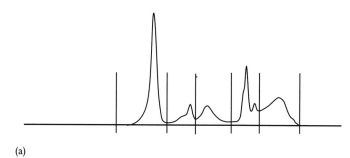

(a)

Fraction	Rel %	g/dl
FRACTION1	35.1	2.70
FRACTION2	8.7	0.67
FRACTION3	12.2	0.94
FRACTION4	1.8	0.14
FRACTION5	12.4	0.95
FRACTION6	4.8	0.37
FRACTION7	25.0	1.93

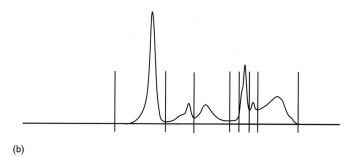

(b)

Figure 10.14 (a) Capillary zone electropherogram of a 56-year-old man (Paragon CZE 2000). (b) Measurement of suspicious band in (a).

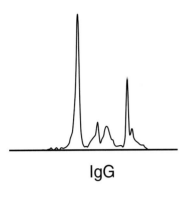

IgG

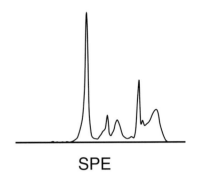

SPE

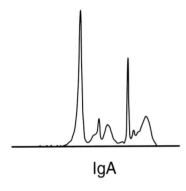

IgA

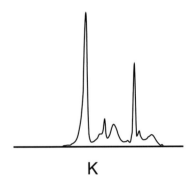

K

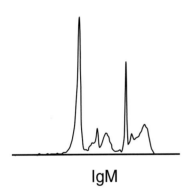

IgM

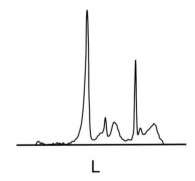

L

(c)

Figure 10.14 (*contd*) (c) Immunosubtraction of same serum.

INTERPRETATION OF FIG. 10.14

The albumin is decreased and there is an increase in both the β- and γ-globulins. The γ-region has a few tiny irregularities, most notably at the cathodal end were it drops off sharply to the baseline. However, the transferrin band is very unusual. It is much broader and taller than normal. Believing this to be a likely M-protein, I measured the band as shown in Fig. 10.14b. I recognize that this measurement includes the residual transferrin, but it provides a baseline that can be used to follow the patient's M-protein over the years. If one prefers, after identifying the isotype, one can measure the total isotype. However, if that turns out to be IgG (although a little improbable in this location, we just saw a case of it in Fig. 10.4) since there is 1.93 g/dl of IgG (assuming the γ-region is all IgG)

this will in fact be an even worse approximation. There are other ways to do this, such as measuring the immunofixation band, but we do not get one with immunosubtraction. Anyway, that is how I measured it. Unfortunately, this is all rather academic because the immunosubtraction in Fig. 10.14c does not demonstrate a monoclonal gammopathy. The large transferrin band remains large in all subtractions. The subtraction of the IgG reduces the γ-region to almost nothing (as expected). Similarly, the subtraction of κ and of λ reduce the γ-region by two-thirds and one-third respectively. The call to the clinician resolved the issue. Yes, it is another radiocontrast dye effect. I should have done a Penta screen and saved the trouble. The alternative would be to use the technique described by Cynthia Blessum to remove the dye (see Chapter 2).

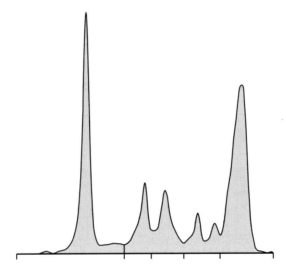

Fractions	%		Ref. %	Conc.	Ref. Conc.
Albumin	35.8	<	45.3 – 67.7	2.4	3.4 – 5.2
Alpha 1	5.7		2.9 – 6.8	0.4	0.2 – 0.4
Alpha 2	12.6		6.2 – 14.9	0.8	0.5 – 1.0
Beta	9.0		8.1 – 18.0	0.6	0.6 – 1.1
Gamma	36.9	>	8.8 – 24.5	2.5	0.6 – 1.6

(a)

Figure 10.15 (a) Capillary zone electropherogram on a 60-year-old woman (Sebia Capillarys).

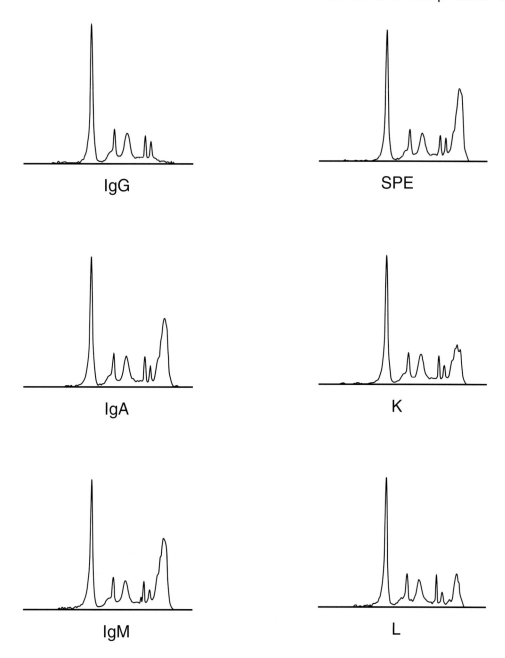

(b)

Figure 10.15 (*contd*) (b) Immunosubtraction of serum from (a) (Paragon CZE 2000).

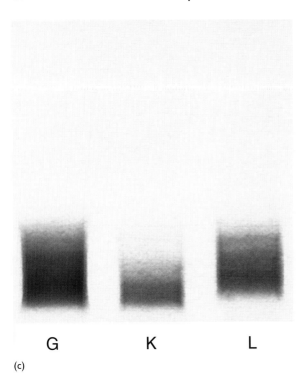

(c)

Figure 10.15 (*contd*) (c) Immunofixation of same serum (Sebia immunofixation).

INTERPRETATION OF FIG. 10.15

The albumin is decreased and there is a large, but relatively broad increase in the γ-globulin. Figure 10.15b is the immunosubtraction on this case. The subtraction of IgG removes virtually all of the γ-region. Immunosubtraction of κ removes about half and of λ removes the other half of the γ-globulin. There is considerable irregularity, however, which shows up in the light chain immunosubtractions that are consistent with an oligoclonal expansion. In Fig. 10.15c is the immunofixation that demonstrates both small κ and small λ bands that confirm this impression. A conversation with the clinician disclosed that this patient has abdominal lymphadenopathy involving adenocarcinoma of unknown primary.

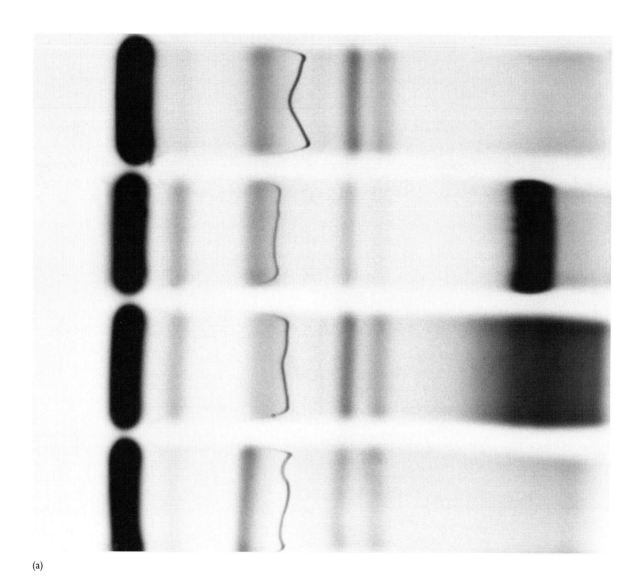

(a)

Figure 10.16 (a) Serum protein electrophoresis on four samples (Paragon SPE2 system stained with Paragon Violet).

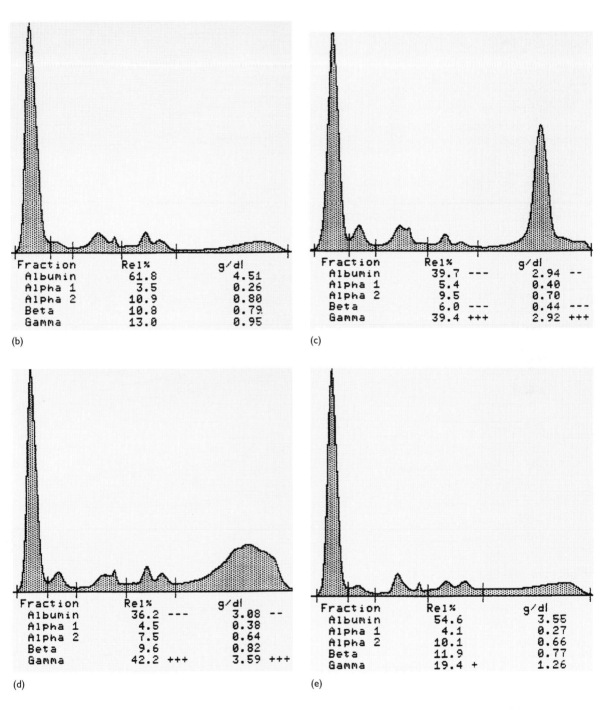

Fraction	Rel%		g/dl	
Albumin	61.8		4.51	
Alpha 1	3.5		0.26	
Alpha 2	10.9		0.80	
Beta	10.8		0.79	
Gamma	13.0		0.95	

(b)

Fraction	Rel%		g/dl	
Albumin	39.7	---	2.94	--
Alpha 1	5.4		0.40	
Alpha 2	9.5		0.70	
Beta	6.0	---	0.44	---
Gamma	39.4	+++	2.92	+++

(c)

Fraction	Rel%		g/dl	
Albumin	36.2	---	3.08	--
Alpha 1	4.5		0.38	
Alpha 2	7.5		0.64	
Beta	9.6		0.82	
Gamma	42.2	+++	3.59	+++

(d)

Fraction	Rel%		g/dl	
Albumin	54.6		3.55	
Alpha 1	4.1		0.27	
Alpha 2	10.1		0.66	
Beta	11.9		0.77	
Gamma	19.4	+	1.26	

(e)

Figure 10.16 (*contd*) (b) Densitometric scan of sample in the top lane of (a), which was from a 61-year-old woman. (c) Densitometric scan of sample in the second lane of (a), which was from an 81-year-old man with dizziness. (d) Densitometric scan of sample in the third lane of (a), which was from a 29-year-old man. (e) Densitometric scan of sample in the bottom lane of (a), which was from a 78-year-old man.

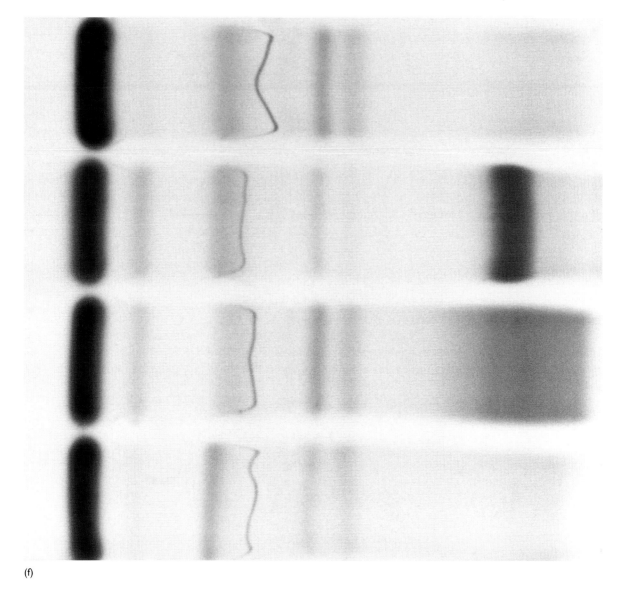

(f)

Figure 10.16 (*contd*) (f) A somewhat underexposed second photograph of the gel in (a).

INTERPRETATION FOR FIG. 10.16

Top lane

The top lane in Fig. 10.16a is a normal serum. The gel stained relatively lightly; in this case α_1-regions do not show to advantage, especially in the top and bottom lanes. However, the densitometric infor-

mation helps here by letting us know they were quantitatively in the normal range.

Second lane

The albumin band is moderately decreased as is the β_2-region. In the γ-region, there is a relatively large spike. By immunofixation, the spike is characterized as an IgM κ monoclonal gammopathy. Clinically, this patient had Waldenström's macroglobulinemia that accounted for her dizziness. I recommended that a urine specimen be evaluated for the presence of MFLC.

Third lane

The albumin band is decreased and there is a marked increase in the γ-globulin. Because of the extreme density of the staining of the γ-globulin region, a second photograph of this gel is given (Fig. 10.16f). This photograph is somewhat underexposed to emphasize the presence of oligoclonal bands in the γ-region. These bands can be appreciated by holding the gel up to a strong light. On the densitometric scan, the irregularities in the γ-region are the counterparts to these oligoclonal bands. The presence of a massive polyclonal increase in γ with oligoclonal bands usually indicates profound infectious disease. Rarely, patterns like this one are seen in patients with angioimmunoblastic

lymphadenopathy, although they often have a β–γ bridging which this sample lacks. This type of pattern may also occur in patients with acquired immunodeficiency syndrome (see Fig. 10.1).

Bottom lane

All of the fractions are normal except γ-globulin. There is a tiny restriction in the slow γ-region. This can be seen directly on the gel and by noting the sharp drop-off at the cathodal end of the densitometric scan (Fig. 10.16e). Compare this to the smoother decrease in the cathodal end of the normal sample in Fig. 10.16b. This very tiny restriction is present in an otherwise normal γ-globulin region. I noted that a tiny γ-globulin restriction was present and that the significance of such tiny restrictions is unclear. I recommended that this serum and a urine have an immunofixation. I also recommended that the serum be re-evaluated in 3–6 months to see if the process resolves.

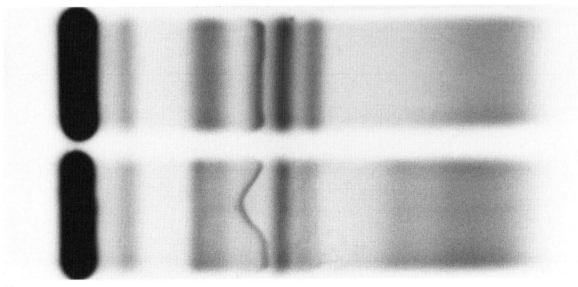

(a)

Figure 10.17 (a) Serum protein electrophoresis on two samples is shown (Paragon SPE2 system stained with Paragon Violet).

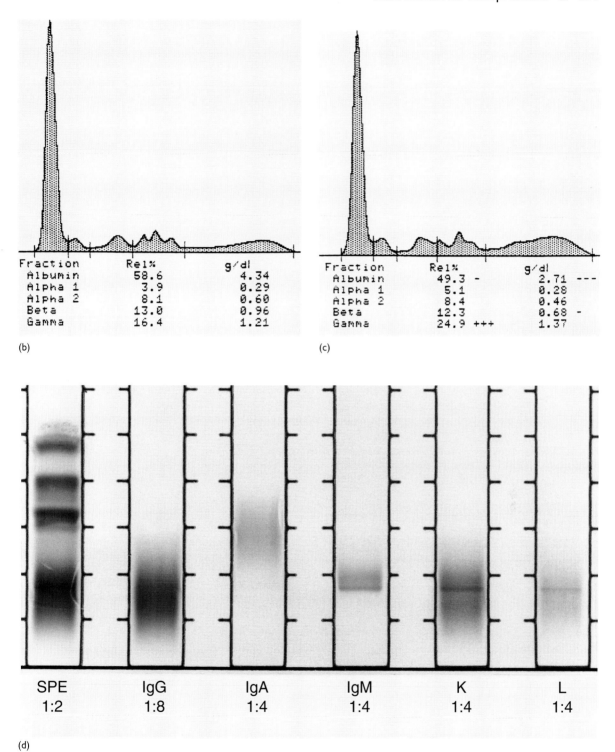

Fraction	Rel%	g/dl
Albumin	58.6	4.34
Alpha 1	3.9	0.29
Alpha 2	8.1	0.60
Beta	13.0	0.96
Gamma	16.4	1.21

(b)

Fraction	Rel%		g/dl	
Albumin	49.3	-	2.71	---
Alpha 1	5.1		0.28	
Alpha 2	8.4		0.46	
Beta	12.3		0.68	-
Gamma	24.9	+++	1.37	

(c)

SPE	IgG	IgA	IgM	K	L
1:2	1:8	1:4	1:4	1:4	1:4

(d)

Figure 10.17 (contd) (b) Densitometric scan of sample in top lane of (a), which was serum from a 59-year-old woman. (c) Densitometric scan of the sample in bottom lane of (a), which was serum from a 28-year-old woman. (d) Immunofixation of serum from sample from bottom lane of (a) (Paragon system stained with Paragon Violet; anode at the top).

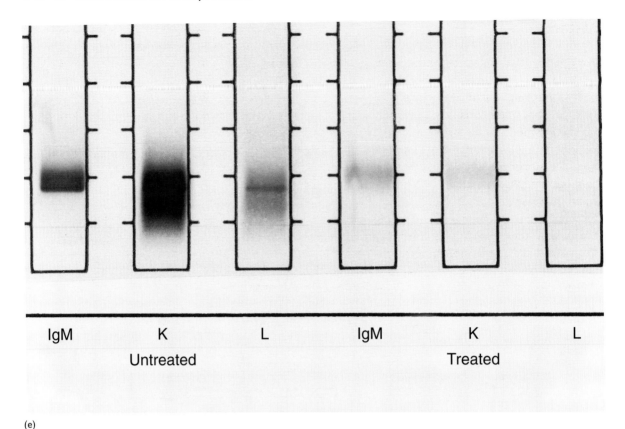

(e)

Figure 10.17 (*contd*) (e) Immunofixation of same sample as in (d) before and after removal of IgG fraction by a commercial column (Paragon system stained with Paragon Violet; anode at the top).

INTERPRETATION FOR FIG. 10.17

Top lane

The top lane in Fig. 10.17a is a normal serum. Note that the densitometric scan in Fig. 10.17b (and the other) barely separates the α_1-region from albumin, yet by looking at the gel, one can clearly see the α_1-region band. Contrast this with the clarity of earlier CZE figures depiction of the α_1-region. With gel-based techniques, it was usually impossible to see the α_1-acid glycoprotein band, but this is commonly seen with CZE, especially in cases with acute-phase reactions (see Chapter 5). Another important feature of these samples is the unusual appearance of the β-regions of the densitometric scans. The β_1-lipoprotein band creates a third band that is confusing when looking at the densitometric scans in the absence of the gel. This is one of the reasons I always show the gel for comparison. With CZE this is not usually a problem. As demonstrated in Chapter 4, β_1-lipoprotein usually migrates in the slow α_2-region and the β-region bands are transferrin and C3. Unlike the capillary zone electropherograms where I demonstrated the ability to detect subtle monoclonal proteins by identifying distortions of the usually sharp and slender transferrin and C3 bands, the muddle of the densitometer β-region patterns makes this very hard. However, the gels themselves provide a reasonably good look at this region.

Bottom lane

There is a considerable decrease in the albumin band, which is best appreciated by the densitometric scan information. In the anodal end of the γ-globulin region there is a faint and somewhat broad restriction. This is also seen as a distortion in the anodal portion of the densitometric scan. This type of distortion may be due to a small monoclonal gammopathy, but may also represent a polyclonal increase in an immunoglobulin subclass. Immunofixation of the serum (Fig. 10.17d) confirmed that there was a restriction in the IgM isotype, but did not demonstrate a corresponding light chain isotype restriction. This is because the patient has a normal quantity of IgG which obscured the light chain restriction. Therefore, a commercial IgG absorbent was used to remove most of the serum IgG and the electrophoresis was repeated. In Fig. 10.17e, the serum treated with the absorbent is compared with untreated serum in the immunofixation reaction. By removing the polyclonal IgG, one can demonstrate that the monoclonal process is due to an IgM κ monoclonal gammopathy. The obscuring of the IgM immunofixation in the present case is the counterpart to the umbrella effect commonly seen on immunoelectrophoresis (see Chapter 3). Owing to the better resolution of immunofixation compared with immunoelectrophoresis, it is very uncommon to have to perform such purifications. In this case, I also recommended that a urine be provided to rule out MFLC.

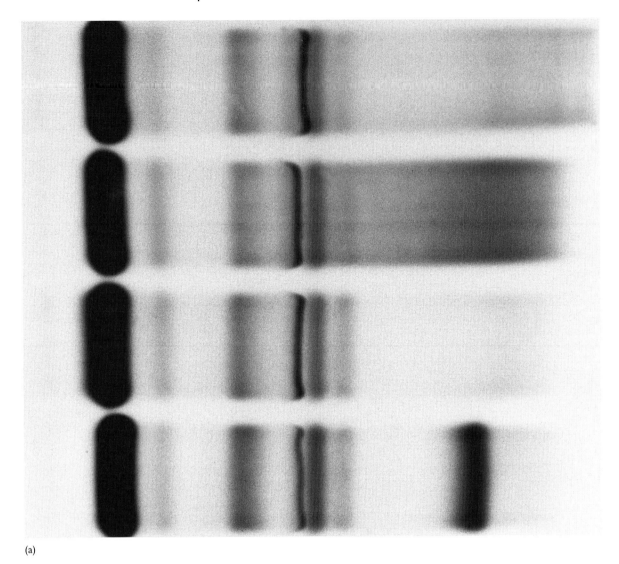

(a)

Figure 10.18 (a) Serum protein electrophoresis on four samples is shown (Paragon SPE2 system stained with Paragon Violet).

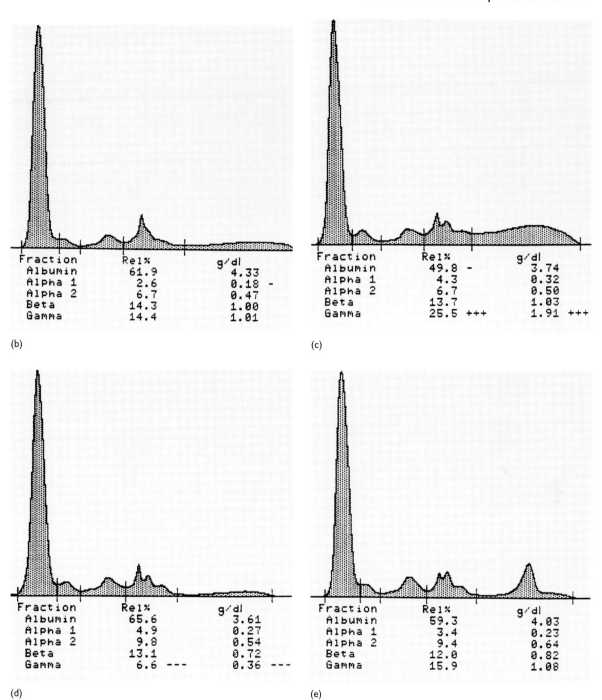

Fraction   Rel%   g/dl
Albumin   61.9   4.33
Alpha 1   2.6   0.18 -
Alpha 2   6.7   0.47
Beta   14.3   1.00
Gamma   14.4   1.01

(b)

Fraction   Rel%   g/dl
Albumin   49.8 -   3.74
Alpha 1   4.3   0.32
Alpha 2   6.7   0.50
Beta   13.7   1.03
Gamma   25.5 +++   1.91 +++

(c)

Fraction   Rel%   g/dl
Albumin   65.6   3.61
Alpha 1   4.9   0.27
Alpha 2   9.8   0.54
Beta   13.1   0.72
Gamma   6.6 ---   0.36 ---

(d)

Fraction   Rel%   g/dl
Albumin   59.3   4.03
Alpha 1   3.4   0.23
Alpha 2   9.4   0.64
Beta   12.0   0.82
Gamma   15.9   1.08

(e)

Figure 10.18 (contd) (b) Densitometric scan of sample in the top lane in (a) (54-year-old woman). (c) Densitometric scan of sample in the second lane in (a) (55-year-old man). (d) Densitometric scan of sample in the third lane in (a) (51-year-old man). (e) Densitometric scan of sample in the bottom lane in (a) (76-year-old woman).

INTERPRETATION FOR FIG. 10.18

Top lane

Although all of the densitometric scan information is normal, note how much better the information is by looking directly at the gel. The gel in the top lane of Fig. 10.18a shows a dense β_1-lipoprotein band in the middle of the β-region with the normal transferrin and smaller normal C3 band also visible. However, the densitometric scan of this sample in Fig. 10.18b indicates the 'spike' due to the β_1-lipoprotein, but it is difficult to perceive the transferrin band and the C3 band does not show up well at all. Look back to the earlier cases that used electropherograms to contrast the consistent crisp β_2-region bands that these electropherograms display. For the present case, direct inspection of the γ-globulin region of the gel discloses the presence of at least two small restrictions. These bands are most likely related to an inflammatory process, but there is no corresponding acute-phase reaction. I recommended a follow-up sample in 3–6 months.

Second lane

There is polyclonal increase in γ-globulins with β–γ bridging. The presence of the β–γ bridge usually corresponds to a polyclonal increase in IgA. Although the β–γ bridge is traditionally associated with cirrhosis, this pattern is also seen in other individuals with chronic inflammatory processes that share an increase in IgA. Autoimmune conditions and infections may have this pattern. This patient has normal albumin with no evidence of anodal slurring (often seen when the bilirubin is elevated), and the α- and β-globulin fractions are normal.

Third lane

When comparing the major bands on this gel, it becomes apparent that the anodal edge of albumin for the case in lane three has a slightly faster anodal migration than the other albumin bands on this gel. In addition, there is a decrease in the γ-globulin region. The other bands are unremarkable. The interpretation of this case notes the presence of the anodal slurring and its possible relation to drug binding (commonly antibiotics or heparin) and emphasizes the isolated hypogammaglobulinemia. I recommend a urine be evaluated for the presence of MFLC in any case of isolated hypogammaglobulinemia.

Bottom lane

Despite the presence of an obvious mid-γ monoclonal gammopathy in case D, there is no increase in the γ-globulin region overall. This patient had been being followed for a known IgG κ monoclonal gammopathy (characterized by a previous immunofixation) for many years. We always examine our records of patients with monoclonal gammopathies when a new sample is sent to us. There had been no significant change in the amount or migration of this monoclonal protein since the examination 12 months previously. This was noted on the report. A repeat was not necessary because there was no change in the migration of the monoclonal protein.

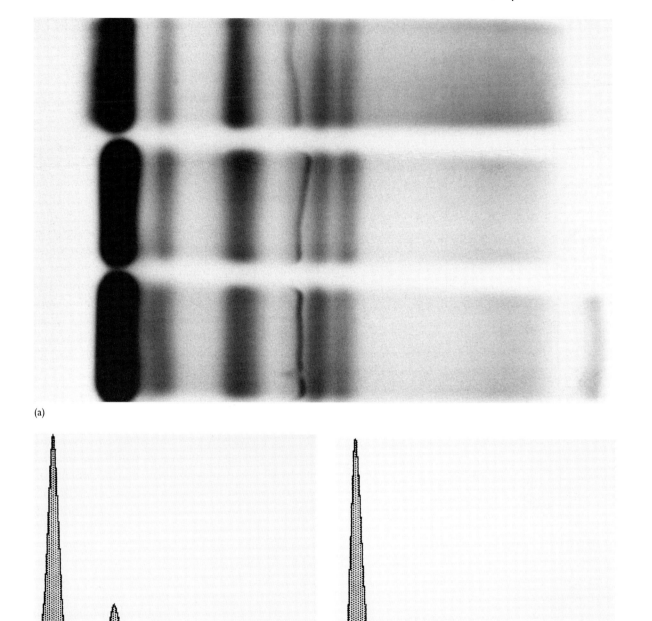

(a)

(b)

Fraction	Rel%		g/dl	
Albumin	43.0	--	2.75	---
Alpha 1	5.6		0.36	
Alpha 2	12.4		0.79	
Beta	15.7		1.00	
Gamma	23.3	+++	1.49	

(c)

Fraction	Rel%		g/dl	
Albumin	55.9		3.58	
Alpha 1	6.8	++	0.44	+
Alpha 2	13.2	+	0.84	
Beta	14.2		0.91	
Gamma	10.0		0.64	

Figure 10.19 (a) Serum protein electrophoresis on three samples (Paragon SPE2 system stained with Paragon Violet). (b) Densitometric scan of sample in the top lane in (a) (76-year-old man). (c) Densitometric scan of sample in the second lane in (a) (61-year-old man).

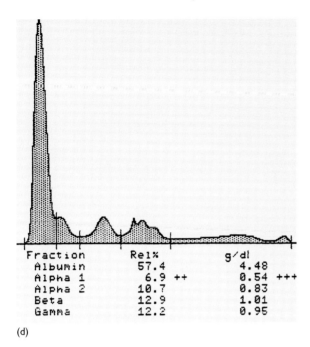

Fraction	Rel%		g/dl	
Albumin	57.4		4.48	
Alpha 1	6.9	++	0.54	+++
Alpha 2	10.7		0.83	
Beta	12.9		1.01	
Gamma	12.2		0.95	

(d)

Figure 10.19 (contd) (d) Densitometric scan of sample in the third lane in (a) (41-year-old woman).

INTERPRETATION FOR FIG. 10.19

Top lane

There is anodal slurring of the albumin band in Fig. 10.19a. Although it is difficult to discern by looking at the photograph of this gel, the densitometric information in Fig. 10.19d demonstrates a decrease in albumin. The other bands are normal, although there is a relative increase in the γ-globulin region. In addition, there is a slight β–γ bridge in this sample. When I spoke to the clinician about this patient, I learned that the patient had hyperbilirubinemia and cirrhosis.

Middle lane

There is a slight increase in the absolute amount of α_1-globulin and a relative increase in both α_1- and α_2 globulins. This is consistent with a mild acute-phase reaction pattern. The transferrin band appears normal, however, and no C-reactive protein band is seen.

Bottom lane

There is a moderate increase in the α_1-globulin, which has not separated as well as I like to see from the albumin band. Once again, looking at the picture of the gel in Fig. 10.19a provides a better view of this than the densitometric scan which does, however, provide quantitative information. This may be due to an increase in the α_1-acid glycoprotein (orosomucoid) which migrates just anodally to α_1-antitrypsin. But unlike the capillary zone electropherograms, even in cases with a marked increase in this band, it is difficult to see on most gel-based systems. Both proteins increase as part of the acute-phase response. Alternatively, there may be an increase in α_1-lipoprotein which may obscure the region between albumin and α_1-antitrypsin. The α_2- and transferrin-bands are both at relatively high normal levels. Therefore, the combination of elevated α_1-lipoprotein, α_1-antitrypsin and transferrin may reflect a hyperestrogen effect rather than an acute-phase response. The most important finding in the case is that of a slow-migrating γ band. A band in this position is almost always due to a monoclonal gammopathy. In urine from a patient with myelogenous leukemia, this band could be due to lysozyme. The previous records from this patient indicated that she had an identical slow γ band 6 months previous to this sample. Immunofixation at that time revealed an IgG κ monoclonal gammopathy. There was no change in migration or amount of the monoclonal protein on the present sample, therefore, immunofixation was not repeated. Urine did not contain a MFLC. Annual follow-up was recommended.

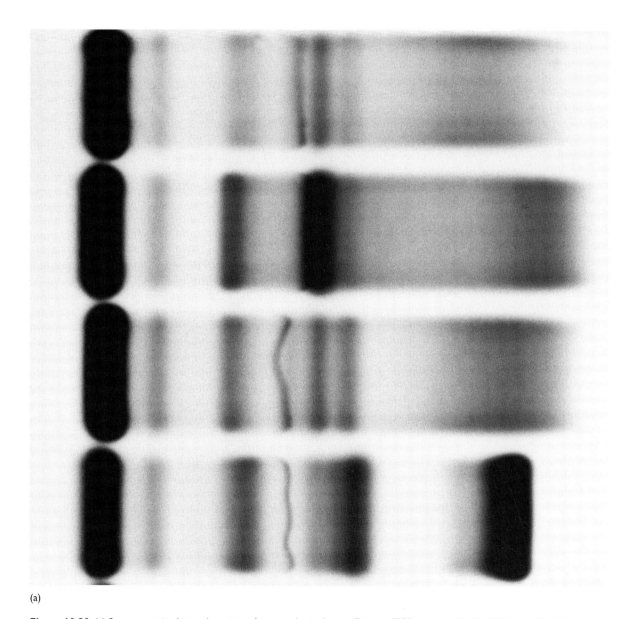

(a)

Figure 10.20 (a) Serum protein electrophoresis on four samples is shown. (Paragon SPE2 system stained with Paragon Violet).

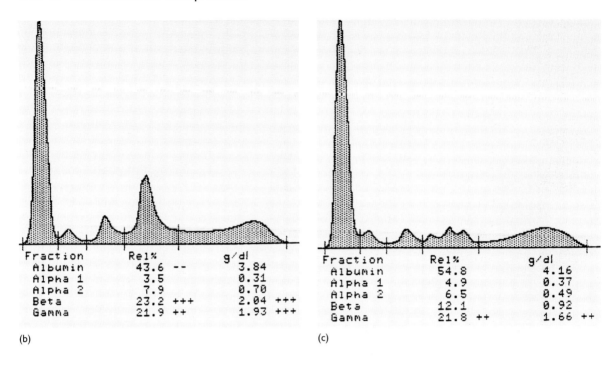

Fraction	Rel%		g/dl	
Albumin	43.6	--	3.84	
Alpha 1	3.5		0.31	
Alpha 2	7.9		0.70	
Beta	23.2	+++	2.04	+++
Gamma	21.9	++	1.93	+++

(b)

Fraction	Rel%		g/dl	
Albumin	54.8		4.16	
Alpha 1	4.9		0.37	
Alpha 2	6.5		0.49	
Beta	12.1		0.92	
Gamma	21.8	++	1.66	++

(c)

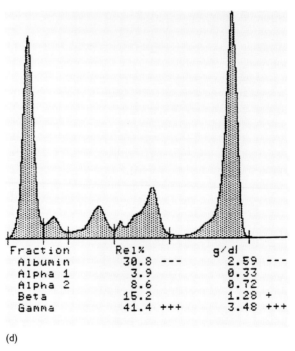

Fraction	Rel%		g/dl	
Albumin	30.8	---	2.59	---
Alpha 1	3.9		0.33	
Alpha 2	8.6		0.72	
Beta	15.2		1.28	+
Gamma	41.4	+++	3.48	+++

(d)

Figure 10.20 (*contd*) (b) Densitometric scan of sample in the second lane in (a) (35-year-old man). The tube was marked as 'grossly hemolyzed'. (c) Densitometric scan of sample in the third lane in (a) (45-year-old man). (d) Densitometric scan of sample in the bottom lane in (a) (75-year-old man).

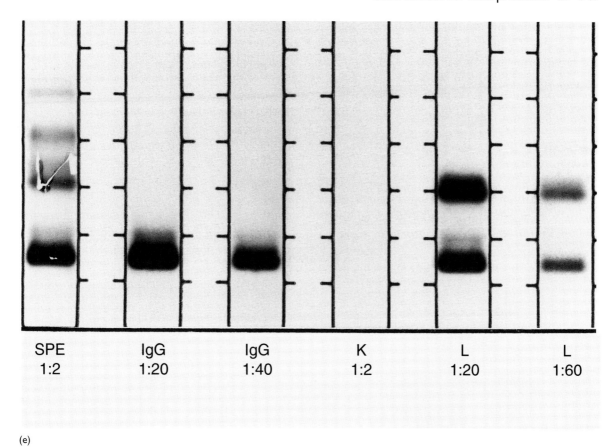

SPE 1:2	IgG 1:20	IgG 1:40	K 1:2	L 1:20	L 1:60

(e)

Figure 10.20 (*contd*) (e) Immunofixation of sample in (d) (Paragon SPE2 system stained with Paragon Violet).

INTERPRETATION FOR FIG. 10.20

Top lane

This is a normal electrophoretic pattern (no densitometry shown).

Second lane

This serum has a marked increase in the β_1-region demonstrated both by the densitometric scan numbers in Fig. 10.20b and by comparing the transferrin bands on the adjacent tracts. The γ-globulin region shows a polyclonal increase and there is β–γ bridging. This is a poorly handled specimen from a patient with a polyclonal increase in γ-globulin. The serum was bright red. The markedly increased β_1-region band is too large for transferrin in an iron-deficient patient. It could represent a monoclonal gammopathy. That possibility, however, is unlikely in the face of the gross hemolysis and the polyclonal increase in γ. If one is uncertain of the nature of such a band, an immunofixation will rule out a monoclonal gammopathy.

Third lane

This serum has a slight increase in the α_1-lipoprotein region between albumin and α_1-antitrypsin. It can be seen best by comparing the region between albumin and α_1-antitrypsin with the serum above and below. There is also a modest polyclonal increase in γ-globulin with no β–γ bridging. This pattern is most consistent with a chronic inflammatory process.

Bottom lane

The serum has a decrease in albumin and an increase in both β- and γ-globulins by the densitometric scan information in Fig. 10.20d. Although one's attention is drawn to the massive γ-globulin spike, one must not ignore the other significant findings on this gel. The β_2-region band is far too large for C3 under any circumstances. On these gels, fibrinogen migrates between the β- and the γ-regions. Therefore, the large band in the C3 region is almost certainly another monoclonal band. In addition to the large γ-region spike mentioned above, there is a decrease in the staining density of the remaining γ-globulin. The immunofixation of this patient's sample (Fig. 10.20e; Paragon SPE2 system stained with Paragon Violet) demonstrates that the β-band is due to λ MFLC and the mid-γ band is due to an IgG λ monoclonal protein. Upon discussing this case with the clinician, I learned that this patient has multiple myeloma. A 24-h urine specimen was recommended to quantify the amount of MFLC.

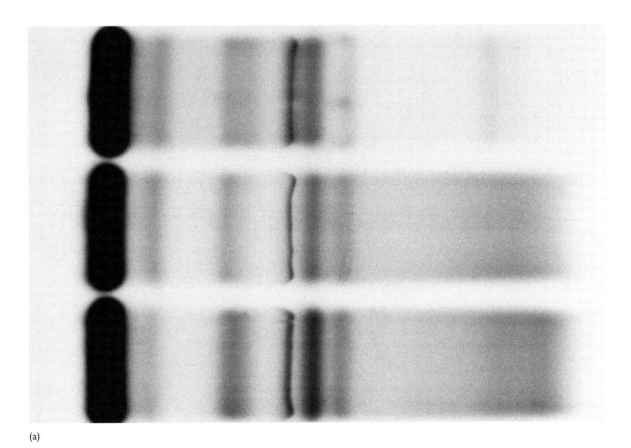

(a)

Figure 10.21 (a) Serum protein electrophoresis on three samples (Paragon SPE2 system stained with Paragon Violet).

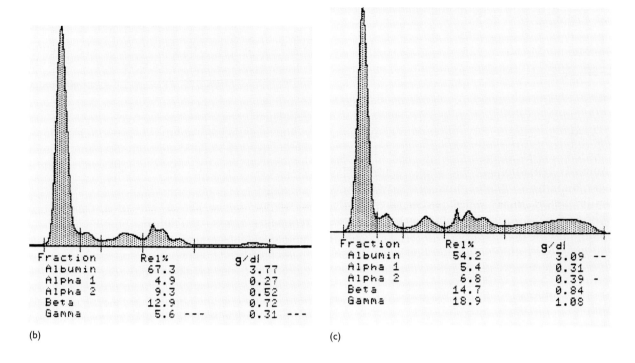

Fraction	Rel%	g/dl	
Albumin	67.3	3.77	
Alpha 1	4.9	0.27	
Alpha 2	9.3	0.52	
Beta	12.9	0.72	
Gamma	5.6 ---	0.31	---

(b)

Fraction	Rel%	g/dl	
Albumin	54.2	3.09	--
Alpha 1	5.4	0.31	
Alpha 2	6.8	0.39	-
Beta	14.7	0.84	
Gamma	18.9	1.09	

(c)

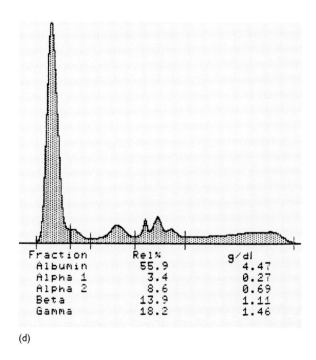

Fraction	Rel%	g/dl
Albumin	55.9	4.47
Alpha 1	3.4	0.27
Alpha 2	8.6	0.69
Beta	13.9	1.11
Gamma	18.2	1.46

(d)

Figure 10.21 (*contd*) (b) Densitometric scan of sample in the top lane in (a) (61-year-old man with lymphocytosis). (c) Densitometric scan of sample in the middle lane in (a) (69-year-old man). (d) Densitometric scan of sample in the bottom lane in (a) (83-year-old woman).

INTERPRETATION FOR FIG. 10.21

Top lane

This serum has a striking hypogammaglobulinemia confirmed by the densitometric scan information in Fig. 10.21b. There is also a small mid-γ restriction. The other major protein bands are unremarkable. Isolated hypogammaglobulinemia is associated with B-cell lymphoproliferative disorders, multiple myeloma (light chain or non-secretory), amyloidosis (AL), humoral immunodeficiency syndromes and chemotherapy. Review of the hematology information on this patient demonstrated that he had chronic lymphocytic leukemia. I also recommended that a 24-h urine be studied by immunofixation and electrophoresis to rule out MFLC.

Middle lane

This serum has a modest decrease in albumin. Otherwise, the serum is unremarkable.

Bottom lane

This woman has an oligoclonal restriction in the mid- to slow γ-region. As discussed before, this is associated with a wide variety of infectious, autoimmune, and immunodeficiency conditions. It represents a restricted clonal proliferation of (in most cases) unknown cause. There is no acute-phase response and the overall quantity of γ-globulin is normal. Although the quantifications of the major protein fractions are normal, the transferrin band seems large compared with the transferrin bands in the two samples above. I recommend an immunofixation on such samples.

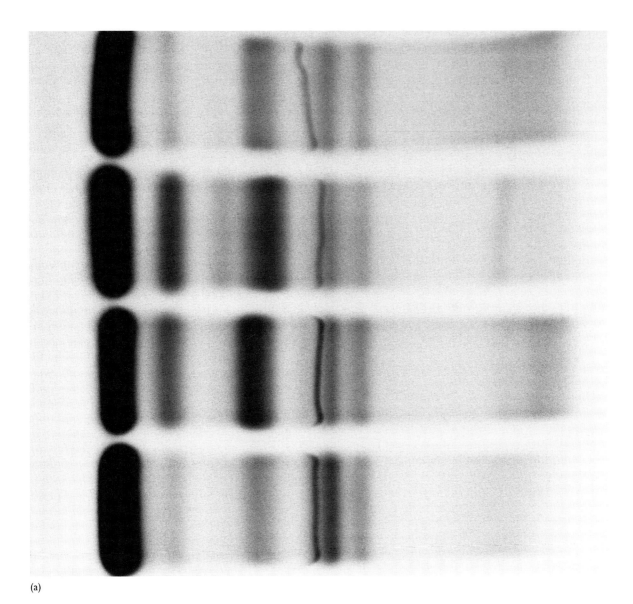

(a)

Figure 10.22 (a) Serum protein electrophoresis on four samples (Paragon SPE2 system stained with Paragon Violet).

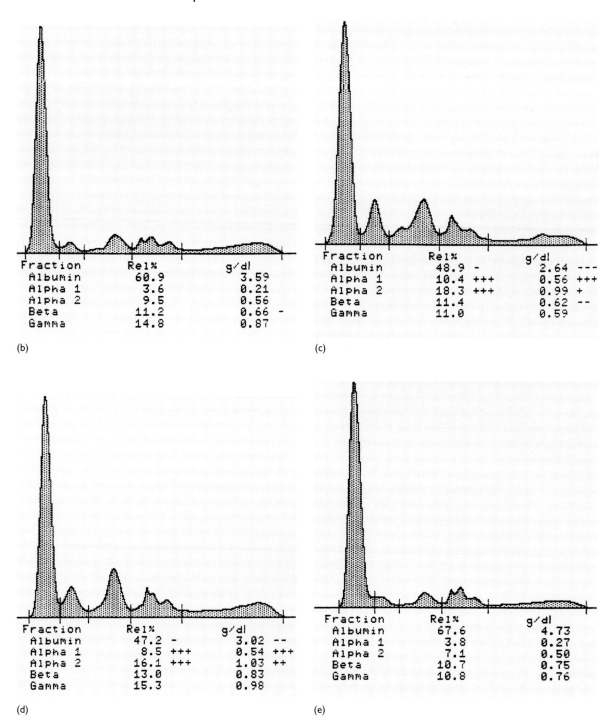

(b)

Fraction	Rel%	g/dl	
Albumin	60.9	3.59	
Alpha 1	3.6	0.21	
Alpha 2	9.5	0.56	
Beta	11.2	0.66	-
Gamma	14.8	0.87	

(c)

Fraction	Rel%		g/dl	
Albumin	48.9	-	2.64	---
Alpha 1	10.4	+++	0.56	+++
Alpha 2	18.3	+++	0.99	+
Beta	11.4		0.62	--
Gamma	11.0		0.59	

(d)

Fraction	Rel%		g/dl	
Albumin	47.2	-	3.02	--
Alpha 1	8.5	+++	0.54	+++
Alpha 2	16.1	+++	1.03	++
Beta	13.0		0.83	
Gamma	15.3		0.98	

(e)

Fraction	Rel%	g/dl
Albumin	67.6	4.73
Alpha 1	3.8	0.27
Alpha 2	7.1	0.50
Beta	10.7	0.75
Gamma	10.8	0.76

Figure 10.22 (*contd*) (b) Densitometric scan of sample in the top lane in (a) (68-year-old man). (c) Densitometric scan of sample in the second lane in (a) (67-year-old man). (d) Densitometric scan of sample in the third lane in (a) (55-year-old man). (e) Densitometric scan of sample in the bottom lane in (a) (43-year-old man).

INTERPRETATION FOR FIG. 10.22

Top lane

Although there is a slightly decreased β-region by the densitometric scan in Fig. 10.22b, the remainder of this sample is unremarkable. I interpret this as one value slightly abnormal. No repeat is recommended.

Second lane

There is a decrease in the albumin (Fig. 10.22c) with a very slight accentuation of its anodal migration compared with the anodal edge of the albumin bands above and below it. The α_1-globulin is markedly increased. There is a prominent inter α_1–α_2-region band (associated with other acute-phase reactants; see Chapters 4 and 5) and a modest increase in haptoglobin (hp2-2). Transferrin is slightly decreased compared with the other β_1-region bands on this gel and the β-region is decreased quantitatively by the densitometric information. There is a small mid-γ-band consistent with C-reactive protein. To be certain, an immunofixation is needed to rule out a small M-protein. These features define a classic acute-phase reaction in a patient likely receiving antibiotics (anodal slurring of albumin).

Third lane

The densitometric information in Fig. 10.22d demonstrates a decrease in albumin, although anodal slurring is not seen. Both α_1- and α_2-globulins are increased. A tiny mid-γ-region band is seen just below the C-reactive protein band in the sample above. In addition, there is an increased staining in the slow γ-region. These features are consistent with an acute-phase reaction with an oligoclonal response.

Bottom lane

This is a normal serum electrophoretic pattern.

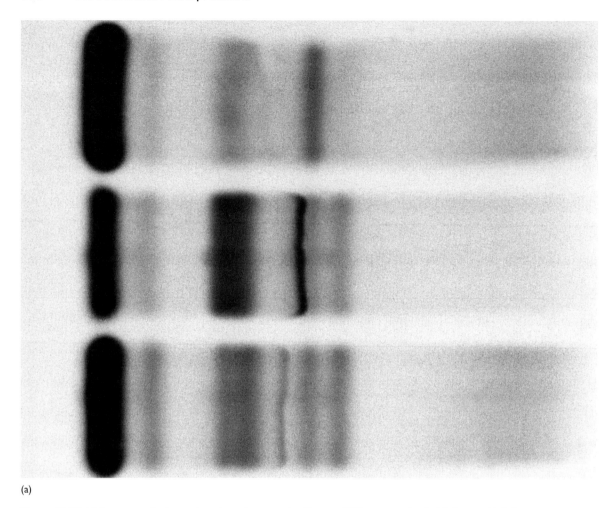

(a)

Figure 10.23 (a) Serum protein electrophoresis on three samples (Paragon SPE2 system stained with Paragon Violet).

INTERPRETATION FOR FIG. 10.23

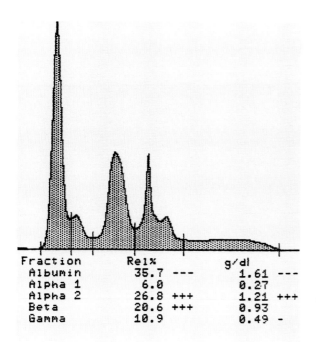

Fraction	Rel%		g/dl	
Albumin	35.7	---	1.61	---
Alpha 1	6.0		0.27	
Alpha 2	26.8	+++	1.21	+++
Beta	20.6	+++	0.93	
Gamma	10.9		0.49	-

(b)

Figure 10.23 (*contd*) (b) Densitometric scan of sample in the middle lane in (a) (74-year-old man).

Top lane

Typical for lyophilized controls, the C3 (β_2-region) band stains very weakly. No densitometric scan is shown.

Middle lane

This serum has a markedly decreased serum albumin and a moderate decrease in γ-globulin. In addition, there is a marked increase in the α_2-globulin region and a prominent β_1-lipoprotein band. These features are typical of nephrotic syndrome.

Bottom lane

Normal electrophoretic pattern. No densitometric pattern is shown.

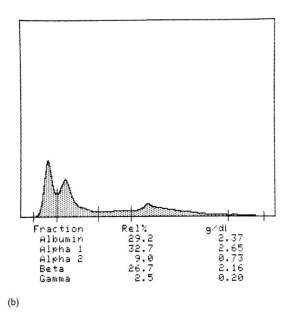

(a)

(b)

Fraction	Rel%	g/dl
Albumin	29.2	2.37
Alpha 1	32.7	2.65
Alpha 2	9.0	0.73
Beta	26.7	2.16
Gamma	2.5	0.20

Figure 10.24 (a) Serum protein electrophoresis on three samples is shown. Case X is from a 46-year-old man with liver disease. (Panagel system stained with Amido Black). (b) Densitometric scan of case X.

INTERPRETATION FOR FIG. 10.24

The top and bottom serum sample in this figure are normal. The electrophoretic pattern (X) is markedly abnormal. Two dense bands are seen in the albumin region. The densitometric scan in Fig. 10.24b demonstrates that the two albumin bands are unequal in height. Further, the cathodal albumin band has a broad shoulder that covers up the α_1-antitrypsin band. There is a diffuse haze between the second albumin region band and transferrin. No distinct α_2 band is seen. C3 is absent although a diffuse haze extends from the transferrin band to about the origin. No staining is found in the γ-globulin region. One interpreter thought that this was a bisalbuminemia in a patient with severe liver disease. When I first saw this pattern, I doubted that it came from a human; I know of no disease that could cause such a pattern.

This is an artifact due to partial denaturation of the serum proteins. On questioning the technologist, it became clear that at the same time he was monitoring this serum protein electrophoresis gel, he used adjacent tubes for the total protein determination (biuret technique). We suspected that a drop of biuret reagent may have fallen into the patient's sample. We were able to reproduce this artifact by placing a drop of biuret reagent into the serum and performing electrophoresis. When unusual samples like this one are found and the interpretation is unclear, a repeat analysis should be done. In this case, the repeat revealed a normal electrophoretic pattern. If the unusual pattern repeats, one should call the clinician and ask for a new sample.

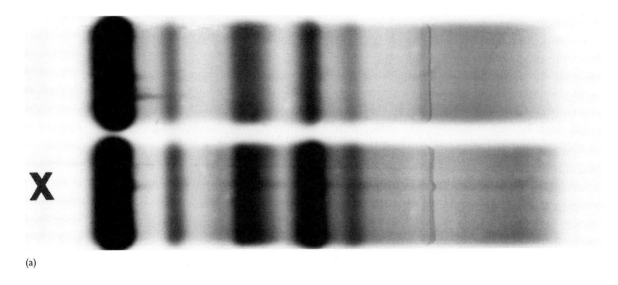

(a)

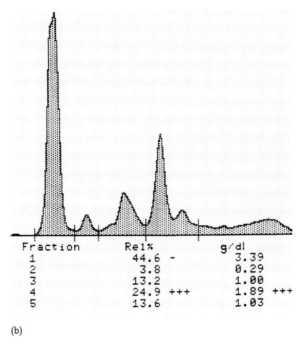

(b)

Figure 10.25 (a) Serum protein electrophoresis on two samples. Case X is from a 72-year-old woman (Paragon SPE2 stained with Paragon Violet). (b) Densitometric scan of serum from case X.

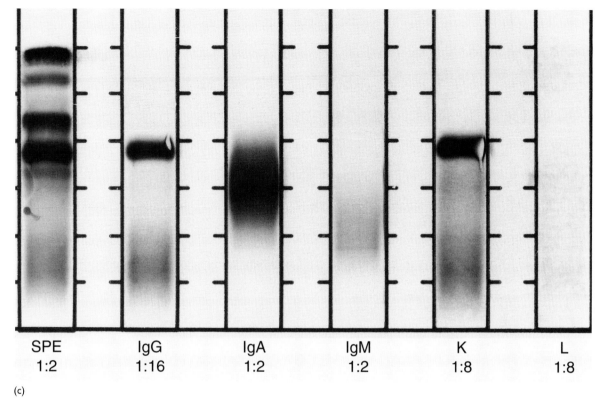

SPE	IgG	IgA	IgM	K	L
1:2	1:16	1:2	1:2	1:8	1:8

(c)

Figure 10.25 (*contd*) (c) Immunofixation of serum from case X (Paragon system stained with Paragon Violet; anode at the top).

INTERPRETATION FOR FIG. 10.25

The electrophoretic pattern for both samples demonstrates irregularities (best seen at the cathodal end of the albumin bands) that reflect inadequate blotting of the gel prior to application of the sample. The top sample is normal. The bottom sample (X) has an enormous band in the transferrin region. The densitometric scan information in Fig. 10.25b confirms the visual impression of the large β_1-region band. A transferrin band would never be this large, even in the presence of iron deficiency. The immunofixation in Fig. 10.25c reveals an IgG κ monoclonal gammopathy. This is an unusual location for an IgG monoclonal gammopathy. I had expected to find an IgA monoclonal protein in this location. Note that at the dilution used, the broad polyclonal nature of the IgA protein is obvious. Only faint staining is seen in the λ reaction. This is too dilute for optimal conditions with the reagent we were using at the time. It should have been diluted to about half the concentration of κ for a better reaction (1:4 or 1:5). I also recommended that urine be evaluated for MFLC.

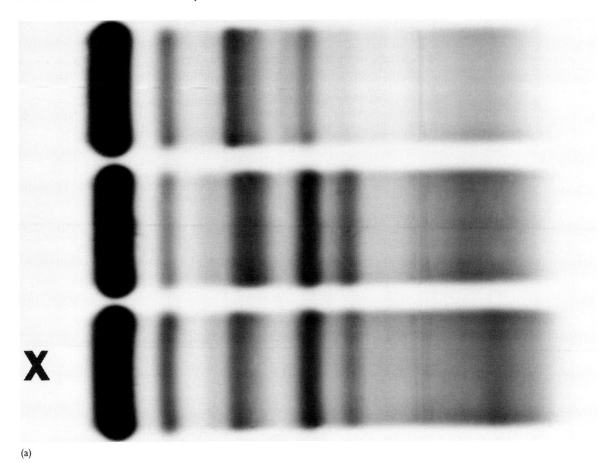

(a)

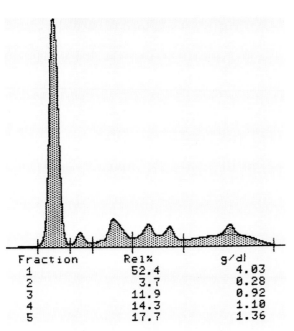

Fraction	Rel%	g/dl
1	52.4	4.03
2	3.7	0.28
3	11.9	0.92
4	14.3	1.10
5	17.7	1.36

(b)

Figure 10.26 (a) Serum protein electrophoresis on three samples. Case X is from a 71-year-old man (Paragon SPE2 stained with Paragon Violet). (b) Densitometric scan of serum from case X.

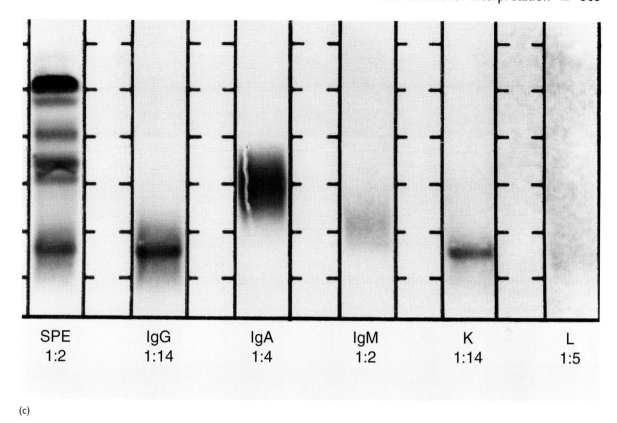

SPE   IgG   IgA   IgM   K   L
1:2   1:14   1:4   1:2   1:14   1:5

(c)

Figure 10.26 (*contd*) (c) Immunofixation of serum from case X (Paragon system stained with Paragon Violet; anode at the top).

INTERPRETATION OF FIG. 10.26

The bottom sample (X) has a slight mid-γ band. This is also obvious on the densitometric scan in Fig. 10.26b. The immunofixation in Fig. 10.26c identifies this band as an IgG κ monoclonal gammopathy. One could argue that with such tiny restrictions, a follow-up serum and a urine to rule out MFLC would be sufficient. Certainly the presence of a normal amount of γ-globulin other than the restriction is evidence in favor of this process being a monoclonal gammopathy of undetermined significance. In this case, the urine turned out to be negative for MFLC. Note how much more darkly the γ-globulin regions appear on this gel stained with Paragon Violet compared with Fig. 10.24 that used the Panagel system stained with Amido Black. This is why one should know which system is being used when interpreting gel patterns.

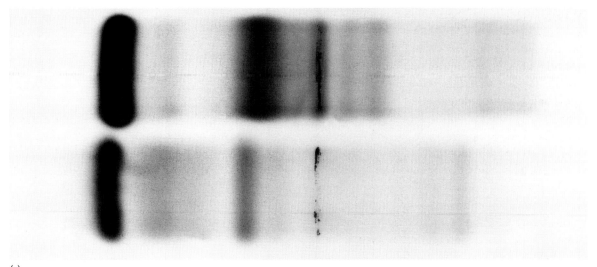

(a)

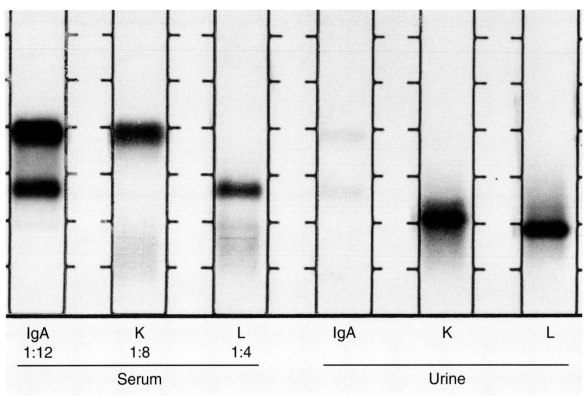

IgA	K	L	IgA	K	L
1:12	1:8	1:4			
	Serum			Urine	

(b)

Figure 10.27 (a) Serum (top lane) and urine (bottom lane) protein electrophoresis from a 76-year-old man. (Paragon SPE2 stained with Paragon Violet). (b) Immunofixation of serum and urine from serum and urine of case shown in (a) (Paragon system stained with Paragon Violet; anode at the top).

INTERPRETATION OF FIG. 10.27

The α_1-region band in the serum (top lane) stains weakly. The α_2-region is darker and broader than normal. There is a diffuse haze between the α_2-region band and the dense sharp band (presumed to be β_1-lipoprotein). This serum was not hemolysed. The β_1-region band (transferrin) is beneath the dark β_1-lipoprotein band. The β_2-region has two bands. The first is C3, the second is not identified. In the mid-γ-region of the serum there is a weakly staining band that is barely visible. The γ-region stains very weakly indicating that hypogammaglobulinemia is present. The corresponding urine has an artifactual restriction at the origin. Two small bands are present in the γ-region of the urine.

Immunofixation of the serum and urine shown in Fig. 10.27b demonstrates that the complex serum and urine patterns described above are the result of the presence of biclonal gammopathy and one with the further complication of monoclonal free light chains. IgA κ and IgA λ monoclonal proteins are in the serum while κ and λ MFLC are in the urine. Note how much more densely the Bence Jones proteins stain in the immunofixation than in the urine protein electrophoresis gel. This illustrates why merely performing serum protein electrophoresis on concentrated urine samples is an inadequate screen for MFLC. The recent availability of immunoassays for free light chains in serum and urine are already available to aid in the diagnostic process (see Chapter 7). Also, note that tiny amounts of the intact IgA monoclonal proteins are also present (albeit barely visible) in the urine, but their light chain counterparts are not seen at the dilution of urine used for this study (concentrated 100 times).

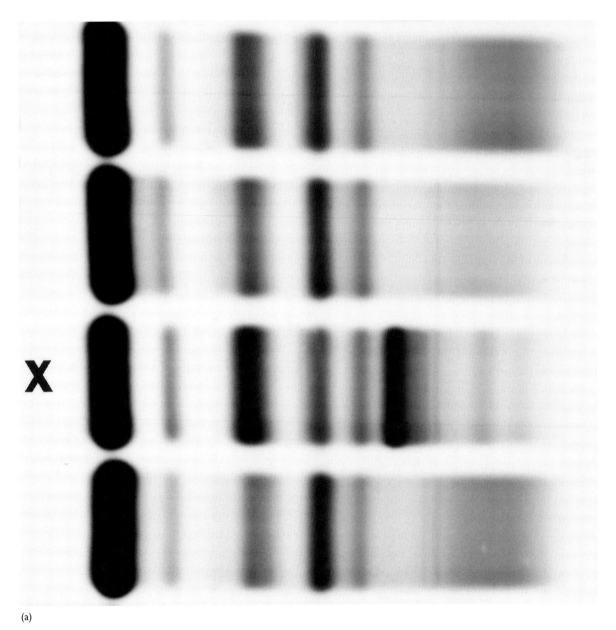

(a)

Figure 10.28 (a) Serum protein electrophoresis on four samples. Case X is from an 82-year-old man (his name is not Lirpa Loof!) (Paragon SPE2 stained with Paragon Violet). (b) Densitometric scan of serum from case X. (c) Immunofixation of serum from case X (Paragon system stained with Paragon Violet; anode at the top).

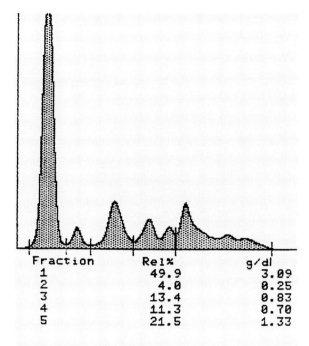

Fraction	Rel%	g/dl
1	49.9	3.09
2	4.0	0.25
3	13.4	0.83
4	11.3	0.70
5	21.5	1.33

Total g/dl   6.20

(b)

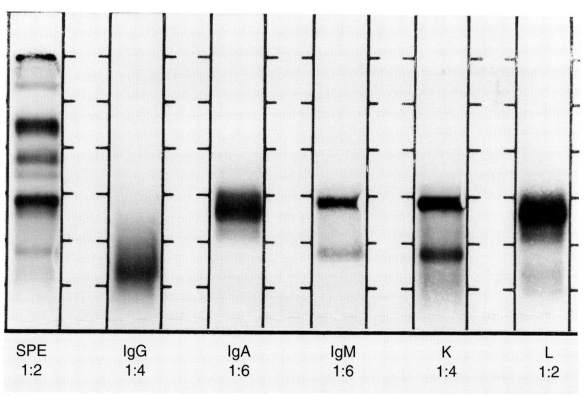

SPE	IgG	IgA	IgM	K	L
1:2	1:4	1:6	1:6	1:4	1:2

(c)

INTERPRETATION OF FIG. 10.28

This is a true story (though perhaps embellished with time). On April 1, not quite 20 years ago, I was presented with a sample demonstrating a tri-clonal gammopathy to interpret on a patient supposedly named Lirpa Loof. My residents were able to contain themselves for only a few minutes while I waxed poetic about the theoretical possibility of this happening, and its even greater likelihood in this era of acquired immunodeficiency syndrome (AIDS) (which was then a new disease). After my monologue, they gleefully pointed out that Lirpa Loof was April Fool backward, and showed me the serum gels from the three separate monoclonal patients they had mixed to produce the artifact.

The serum X from Fig. 10.28a is not a mixture of sera, and this 82-year-old man does not have AIDS. The serum shows three distinct monoclonal bands. The darkest band is in the fibrinogen region. In the mid-γ-region a small, but distinct band is present. Just cathodal to the mid-γ-band is a slightly fainter, but also distinct band. The densitometric scan in Fig. 10.28b shows these three irregularities. In Fig. 10.28c, the immunofixation demonstrates that the large band in the fibrinogen region is an IgA λ monoclonal gammopathy. The breadth of the IgA and associated λ light chain band probably reflects the glycosylation of many IgA molecules. Interestingly, migrating at the same location is a smaller IgM κ monoclonal gammopathy, a band I was not able to appreciate on the serum protein electrophoresis or on the densitometric scan. This same combination (IgM κ) is responsible for the mid-γ-region band seen on the electrophoresis in Fig. 10.28a and on the densitometric scan. It is possible that this is a multimer of the IgM κ band seen in the β-region. However, it could be a product of an entirely separate clone. The slowest band is due to an IgG λ monoclonal gammopathy. Note that the light chain does not stain as well as the IgG on this immunofixation pattern. The patient did not have a lymphoproliferative process.

If my former residents are reading this, all I can say is Lirpa Loof lives!

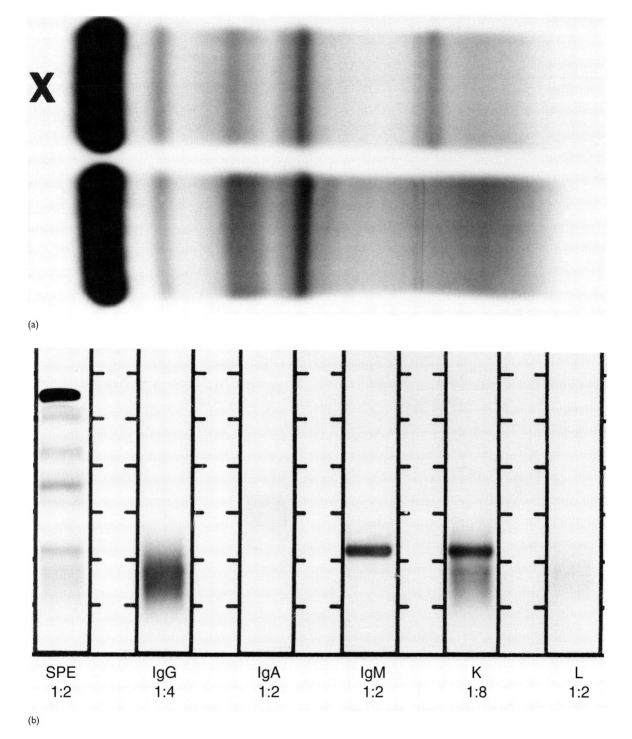

(a)

(b)

Figure 10.29 (a) Serum protein electrophoresis on two samples is shown. Case X is a College of American Pathologists (CAP) survey sample EC-08 (Panagel system stained with Paragon Violet). (b) Immunofixation of Case X. (Paragon system stained with Paragon Violet; anode at the top).

INTERPRETATION FOR FIG. 10.29

The electrophoretic pattern shows a small fast γ-restriction. Note that the normal sample below has a tiny origin artifact which should not be mistaken for a monoclonal gammopathy. The immunofixation of the College of American Pathologists (CAP) sample identifies this band as an IgM κ monoclonal gammopathy. As discussed in Chapter 2, as many as two-thirds of the laboratories that used the other methods missed this small, but obvious monoclonal gammopathy. Earlier examples demonstrated that small serum monoclonal gammopathies may be associated with excretions of large amounts of MFLC found most readily in the urine by electrophoresis and immunofixation. Further, as discussed in Chapter 6, small IgM monoclonal gammopathies are associated with peripheral neuropathies.

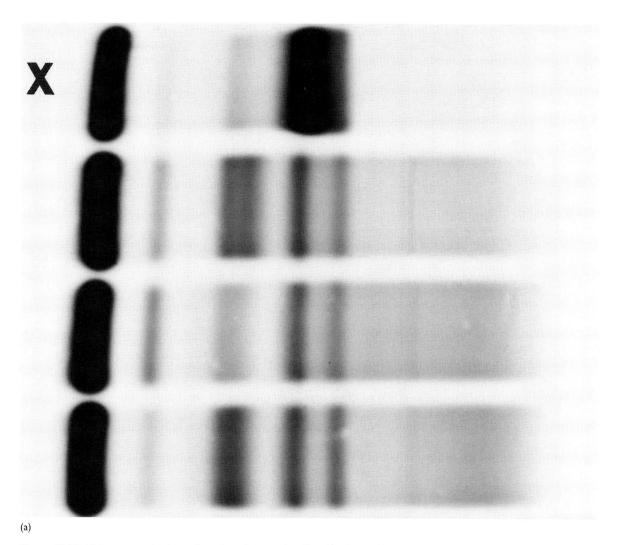

(a)

Figure 10.30 (a) Serum protein electrophoresis on four samples. Case X is from a 46-year-old woman with anemia, elevated calcium and lytic bone lesions (Paragon SPE2 stained with Paragon Violet).

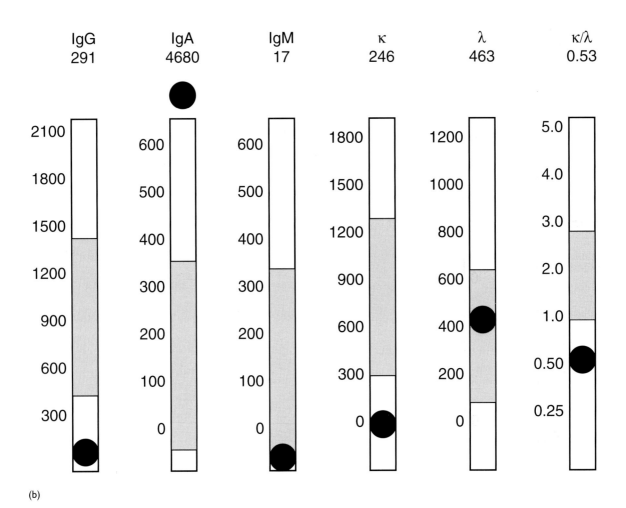

(b)

Figure 10.30 (*contd*) (b) Immunoglobulin measurements on the serum from case X on the Beckman Nephelometer.

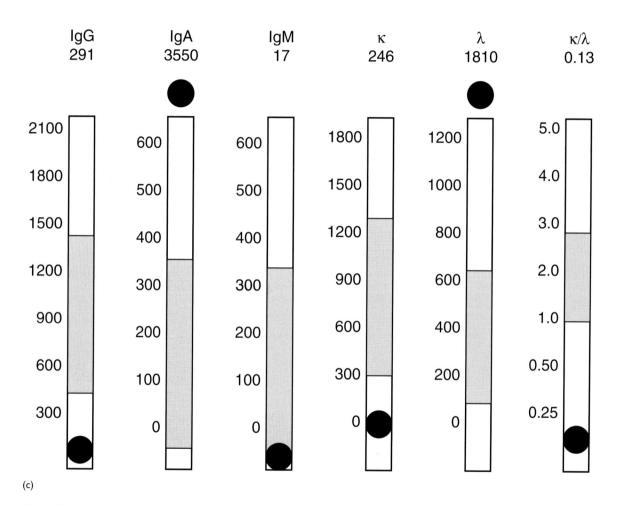

| IgG | IgA | IgM | κ | λ | κ/λ |
| 291 | 3550 | 17 | 246 | 1810 | 0.13 |

(c)

Figure 10.30 (*contd*) (c) Immunoglobulin measurements same data for IgG, IgM, and κ, but IgA and λ remeasured on the serum diluted 1:10 prior to analysis.

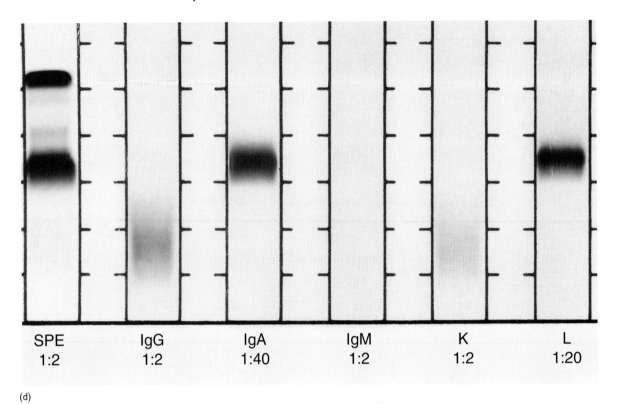

SPE	IgG	IgA	IgM	K	L
1:2	1:2	1:40	1:2	1:2	1:20

(d)

Figure 10.30 (*contd*) (d) Immunofixation of serum from case X (Paragon system stained with Paragon Violet; anode at the top).

INTERPRETATION FOR FIG. 10.30

A broad, massive band is present in the β-region. The γ-region is markedly decreased compared with the other samples on this gel. This is certainly a massive monoclonal gammopathy. The immunoglobulin measurements in Fig. 10.30b disclose an extraordinary increase in the amount of IgA present that correlates well with the amount of protein seen on the serum protein electrophoresis gel. Please note that the light chain measurements are presented as the total concentration of the immunoglobulin they are attached to (i.e. calculated as the molecular weight of the molecule to which they are attached rather than as the weight of the free light chain). At the time this study was performed, that was the standard. More recently, only the molecular weight of the light chain is considered in calculating the quantity of light chain present. These numbers will likely differ from current numbers you may see in your patients. Therefore, with the study numbers, in theory κ and λ should add up to the total of IgG + IgA + IgM. In this case they do not. Indeed, there seems to be no corresponding light chain for the massive IgA. Could this be a case of α heavy chain disease? No, but situations like this are often mistaken for heavy chain disease despite the fact that the history does

not fit. α heavy chain disease occurs among individuals in their second and third decades in the Middle East and Mediterranean regions. It presents as a gastrointestinal disease and does not cause lytic bone lesions or elevated calcium values. This is a woman in the latter half of her fifth decade. So what is wrong? Even the κ/λ ratio is only barely abnormal. This is a demonstration of the problems of using nephelometry alone to detect monoclonal proteins. As discussed in Chapter 9, some monoclonal proteins do not react well with antisera standardized against polyclonal immunoglobulins. The immunofixation in Fig. 10.30d clearly demonstrates an IgA λ monoclonal gammopathy. When the immunoglobulin quantifications were repeated prediluting the sample 1:10 as shown in Fig. 10.30c, more of the monoclonal λ light chains were detected. However, further dilutions did not allow better approximation of the amount of IgA present. When a patient appears to have a heavy chain disease because of the *lack* of reactivity of a light chain either by nephelometry or by immunofixation, be careful. Use another technique such as immunosubtraction, or send it to a laboratory that performs immunoselection to make a definitive diagnosis. This patient has classic multiple myeloma.

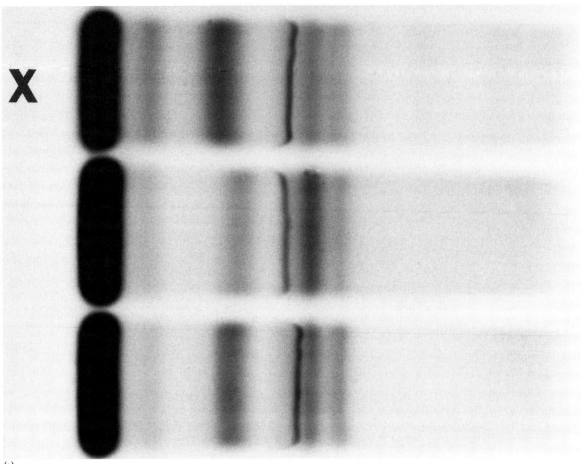

(a)

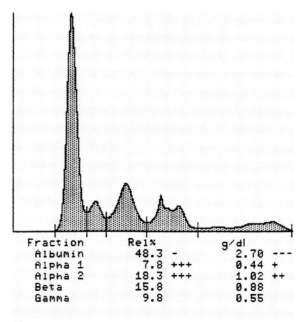

Fraction	Rel%		g/dl	
Albumin	48.3	-	2.70	---
Alpha 1	7.8	+++	0.44	+
Alpha 2	18.3	+++	1.02	++
Beta	15.8		0.88	
Gamma	9.8		0.55	

(b)

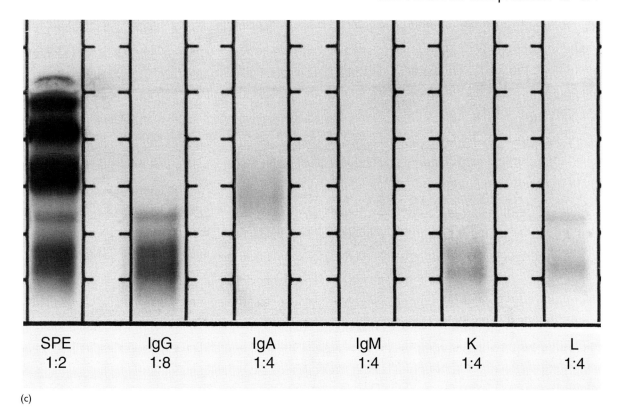

SPE IgG IgA IgM K L
1:2 1:8 1:4 1:4 1:4 1:4

(c)

Figure 10.31 (a) Serum protein electrophoresis on three samples. Case X is from a 67-year-old man with recurrent pneumonia (Paragon SPE2 stained with Paragon Violet). (b) Densitometric scan of serum from case X. (c) Immunofixation of serum from case X (Paragon system stained with Paragon Violet; anode at the top).

INTERPRETATION OF FIG. 10.31

Although the albumin band on the photograph of case X looks similar to the other two samples on the gel, the densitometric scan in Fig. 10.31b documents a decrease in albumin concentration. Both the α_1- and α_2-regions are increased, consistent with an acute inflammatory response. The transferrin band stains fainter on the gel specimen (Fig. 10.31a) than the transferrin bands in the two samples below it. The C3 band is slightly broader than the two samples below it (possibly indicating a subacute inflammation). The γ-region stains more weakly than the two samples below it

and a couple of faintly stained bands are barely discernable. Immunofixation was ordered on this sample. It demonstrates that the slightly broader C3 band was really due to a tiny IgG λ band that migrates in the C3 region. There are also two slower-moving IgG κ bands and a third slow-migrating IgG λ band. This oligoclonal expansion in the context of a borderline hypogammaglobulinemia can be seen in patients with B-cell lymphoproliferative disorders (discussed in Chapter 6). A call to the clinician revealed that this patient has chronic lymphocytic leukemia with a leukocyte count of 70 000 (virtually all mature lymphocytes).

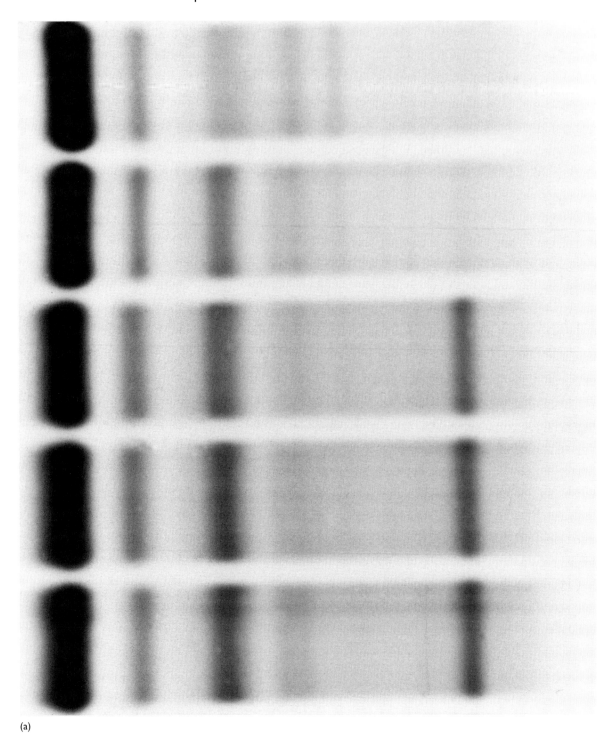

(a)

Figure 10.32 (a) Serum protein electrophoresis on a 79-year-old man with endocarditis. All sera on this gel are from this patient. Top to bottom: January 15, 18, 27, 29, and February 3 (Paragon SPE2 stained with Paragon Violet). (b) Immunofixation of serum from January 27 sample (Paragon system stained with Paragon Violet; anode at the top). (c) Immunofixation of urine from January 27 sample (Paragon system stained with Paragon Violet; anode at the top).

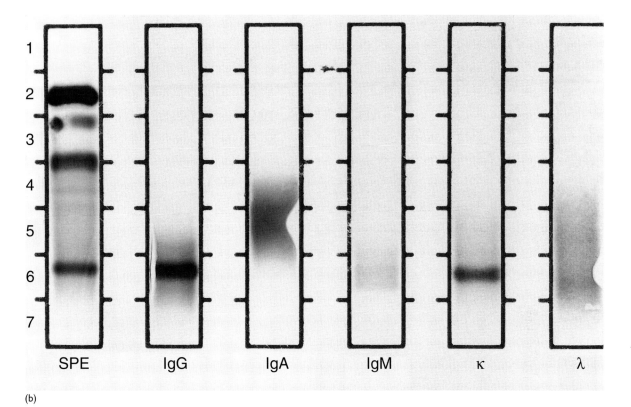

(b)

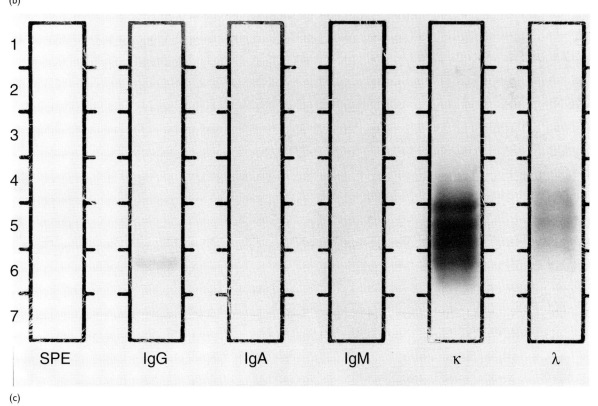

(c)

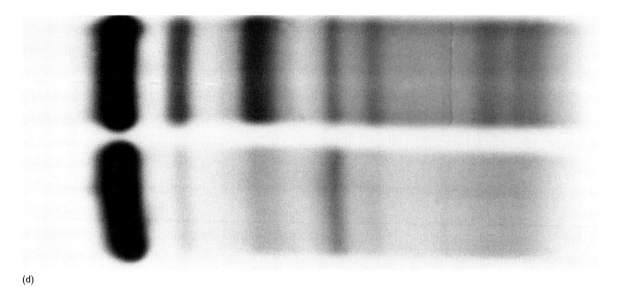

(d)

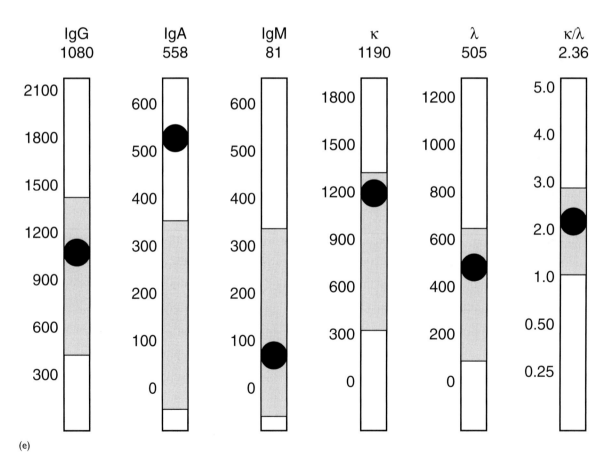

(e)

Figure 10.32 (*contd*) (d) Patient serum from March 5 on top and control serum on bottom (Paragon SPE2 stained with Paragon Violet). (e) Immunoglobulin measurements from January 27 sample.

INTERPRETATION OF FIG. 10.32

The earliest electrophoretic sample on the gel from Fig. 10.32a shows a barely discernable mid-γ band. The second sample shows an increase in α_1- and α_2-globulin and the presence of a faint band in the mid-γ-region. The third through the bottom sera all have anodal slurring of albumin (presumably due to the antibiotic therapy), an acute-phase reaction and a prominent mid-γ-region band. The immunofixation performed on the serum from January 27 (Fig. 10.32b) shows an obvious IgG κ monoclonal gammopathy. The urine immunofixation from the same date (Fig. 10.32c) shows a very faint IgG band which corresponds to the band in the serum and a ladder pattern with both kappa and lambda. No MFLC (Bence Jones protein) is seen.

With time, this pattern evolved into an oligoclonal pattern (Fig. 10.32d), demonstrating that the original monoclonal band was likely an early response of a prominent B-cell clone to the infectious agent causing the endocarditis. Transient monoclonal gammopathies have been reported with endocarditis and other infectious diseases and are discussed in Chapter 6. Typically, they do not have accompanying MFLC. Also, in cases with reactive clonal bands, the κ/λ ratio is often in the normal range, as it was in this case (Fig. 10.32e).

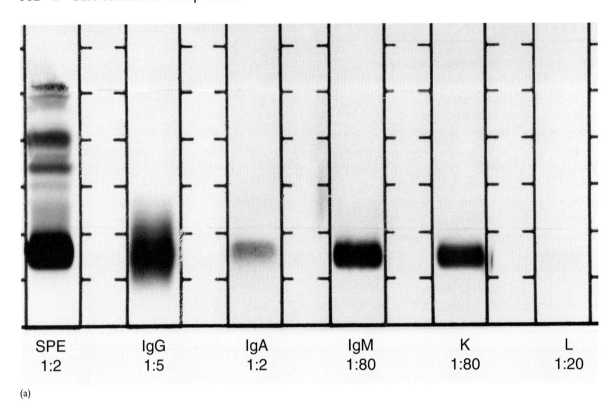

SPE 1:2   IgG 1:5   IgA 1:2   IgM 1:80   K 1:80   L 1:20

(a)

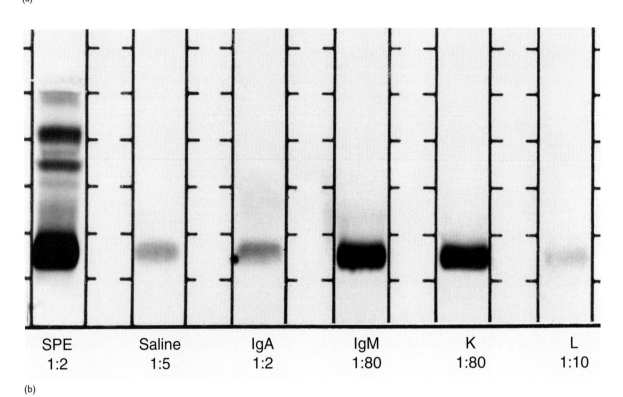

SPE 1:2   Saline 1:5   IgA 1:2   IgM 1:80   K 1:80   L 1:10

(b)

INTERPRETATION FOR FIG. 10.33

The serum protein electrophoresis lane (fixed in acid) in Fig. 10.33a demonstrates a massive band at the origin. Inspection of the immunofixation reaction (dilutions of patient's serum used for the reactions are recorded below) looks as though there is an IgM κ monoclonal gammopathy. However, there is also a distinct restriction in the IgA lane that has the same migration Also, on the gel I can see a faint band in the same location of the λ lane (this may not show up in the final figure you are looking at). Is this a double gammopathy, or is this just a protein precipitating at the origin of the gel? I suspected that it was a cryoglobulin that was precipitating on the cool gel. Except for the reactions in the IgM and κ lanes (which were diluted a whopping 1:80), the density of the precipitate seemed to correlate with the dilution of the patient's serum. That is, a stronger precipitate was seen with IgA at 1:2 than with λ at 1:20 (even if you cannot see it, take it on trust for a moment). To determine if the bands in the IgA and λ regions

are due to a cryoglobulin or really represent a second gammopathy, we repeated the immunofixation using the a dilution of 1:2 of the patient's serum in the IgG lane, but instead of using the anti-IgG, we just placed saline on this lane. Also, I was curious what would happen in the λ lane with a 1:10 dilution of the patient's serum. As shown in Fig. 10.33b, the same band appears in the lane where saline was used. The density of staining correlates with the dilution in these lanes λ ≤ saline ≤ IgA indicating that this is the precipitation of this patient's Type I cryoglobulin (massive IgM κ, no rheumatoid factor activity) at the origin. How do we know that the IgM κ is not just cryoprecipitation? The massive dilution of IgM and κ (1:80) should produce no artifactual band (witness the result of the 1:20 dilution of patient's serum with anti-λ). Therefore, when an immunofixation of a serum sample shows an origin precipitate as in Fig. 10.33a, I repeat the immunofixation replacing antisera against one component with buffer or saline.

Figure 10.33 (a) Immunofixation on serum from an 82-year-old man with dizziness, paresthesias, and skin ulcers surrounded by erythematosus plaques on his legs (Paragon system stained with Paragon Violet; anode at the top). (b) Repeat of the immunofixation on the same sample, this time, instead of overlaying the sample with anti-IgG, saline was placed in that lane (Paragon system stained with Paragon Violet; anode at the top).

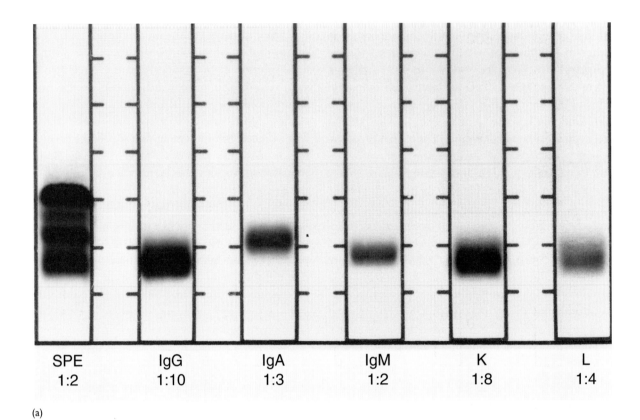

SPE
1:2

IgG
1:10

IgA
1:3

IgM
1:2

K
1:8

L
1:4

(a)

Figure 10.34 (a) Immunofixation on serum from a 49-year-old man with shoulder pain. (Paragon system stained with Paragon Violet; anode at the top).

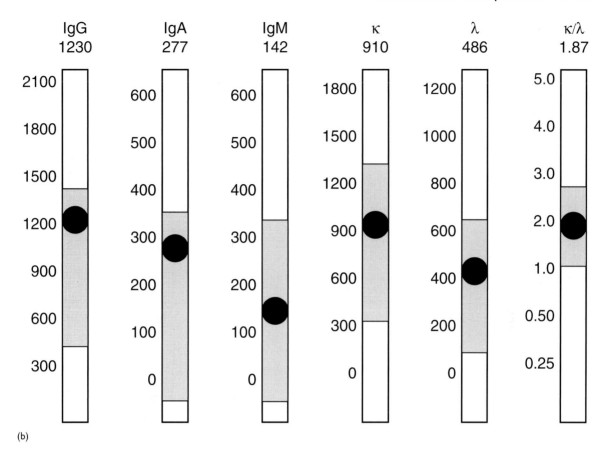

(b)

Figure 10.34 *(contd)* (b) Immunoglobulin measurements from the same serum.

INTERPRETATION FOR FIG. 10.34

The immunofixation in Fig. 10.34a shows a dense band in each of the immunoglobulin classes. The pattern resembles, somewhat, the origin artifacts that are often seen in patients with cryoglobulinemia. Examination of the SPE lane discloses the major problem in this case. There has been grossly inadequate migration. (Someone turned off the power too soon.) The poor separation of the protein led to this confusing picture. The immunoglobulin quantifications are consistent with a normal sample, but the immunofixation needs to be repeated.

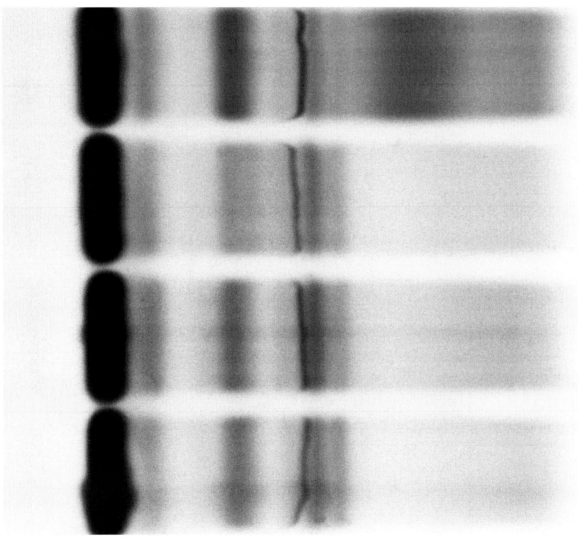

(a)

Figure 10.35 (a) Serum protein electrophoresis on four samples is shown. The sample in the top lane is from a patient at the oncology clinic (Paragon SPE2 stained with Paragon Violet). (b) Immunofixation of serum from case in top lane in (a) (Paragon system stained with Paragon Violet; anode at the top). (c) Immunofixation of serum from same case using antisera against IgG subclasses.

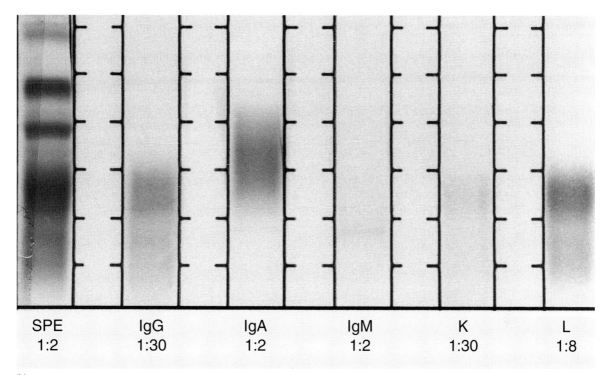

SPE	IgG	IgA	IgM	K	L
1:2	1:30	1:2	1:2	1:30	1:8

(b)

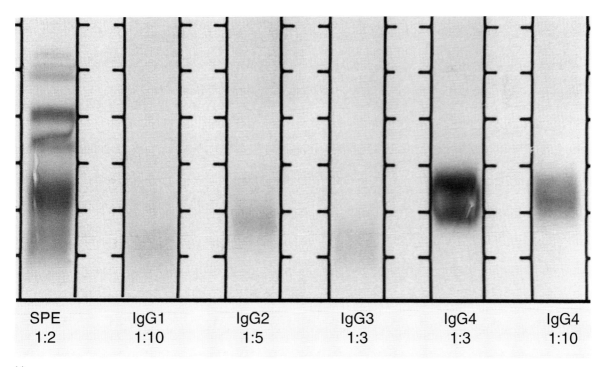

SPE	IgG1	IgG2	IgG3	IgG4	IgG4
1:2	1:10	1:5	1:3	1:3	1:10

(c)

INTERPRETATION OF FIG. 10.35

The top lane of the electrophoretic pattern in Fig. 10.35a shows a broad, but distinct band in the slow β- to fast γ-region. Although one might suspect that this increase would be associated with an elevated serum IgA, it was in the mid-normal range (233 mg/dl). However, the IgG was elevated (3070 mg/dl, about twice the normal upper limit). The immunofixation shown in Fig. 10.35b demonstrates that both κ and λ lanes show a broad band similar to that seen in the serum protein electrophoresis lane. Although the κ lane stains more weakly than the λ lane in this region, note that the κ is diluted 1:30, whereas the λ is diluted only 1:8. This broad band is not a cryoglobulinemia because no such bands occur in the IgM lane (diluted 1:2), although a tiny origin artifact is seen at this concentration. A logical supposition for this pattern is that there is a polyclonal increase in one subclass of IgG. The IgG4 subclass was elevated but the other subclasses were not. The immunofixation, in Fig. 10.35c, performed using subclass antisera demonstrates that the IgG4 subclass corresponds to the same migration. Follow-up on the patient revealed he had an epithelial malignancy and chronic respiratory disease, not a monoclonal gammopathy. This case contributed by Dr A. C. Parekh.

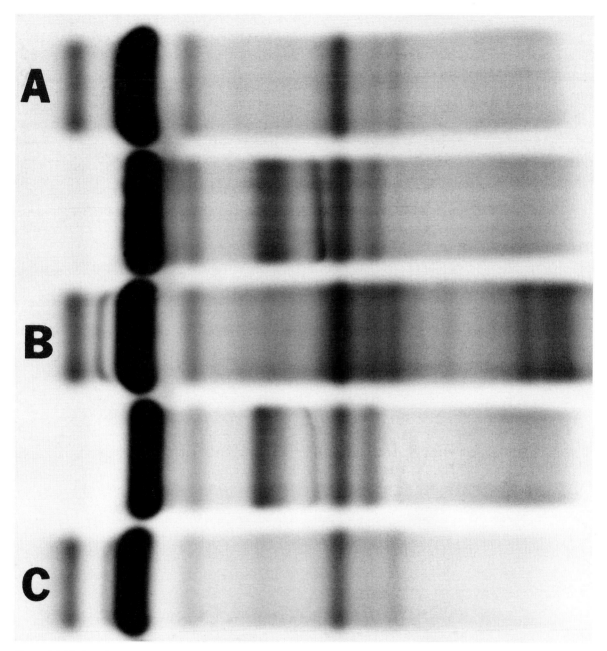

Figure 10.36 Case A: cerebrospinal fluid (CSF) concentrated 80-fold and serum (immediately below) diluted 1:3 from a 28-year-old woman. Case B: CSF concentrated 80-fold and serum (immediately below) from a 42-year-old man. Case C: CSF only is shown from a 12-year-old girl (no serum was sent) (Paragon SPE2 system stained with Paragon Violet).

INTERPRETATION FOR FIG. 10.36

I no longer recommend using serum protein electrophoresis to detect oligoclonal bands. Either isoelectric focusing or immunofixation-based techniques are preferred for sensitivity and for identification of the bands as immunoglobulins. However, many clinical laboratories are still using these techniques. Therefore, I've included this example.

Case A: the cerebrospinal fluid (CSF) is negative for oligoclonal bands. Case B: the CSF is positive for oligoclonal bands; the corresponding serum is negative for these bands. Case C: the CSF is negative for oligoclonal bands. The lack of a corresponding serum does not interfere with interpretation in a negative CSF. However, if there had been oligoclonal bands in this CSF, we would have requested that a serum be sent to rule out systemic oligoclonal bands that had diffused across the blood–brain barrier.

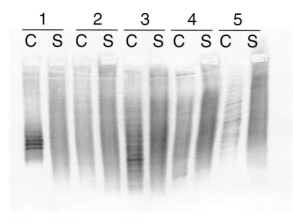

Figure 10.37 Five pairs of CSF and serum samples are present for review. These samples have been processed without concentration on the Helena SPIFE IgG Isoelectric Focusing (IEF) technique (Helena Laboratories, Beaumont, Texas).

INTERPRETATION FOR FIG. 10.37

This technique uses both IEF and immunologic identification of IgG to allow detection of oligoclonal bands in unconcentrated CSF. However, to achieve this level of sensitivity, after electrophoresis, the samples must be blotted onto nitrocellulose paper for the immunofixation reaction with peroxidase-conjugated anti-IgG. In this case, although there was minor distortion of samples 4 and 5 during this transfer, it does not affect the interpretation. The samples are applied at the top of the gel, therefore, all lanes have one or two artifactual bands at the top that should be ignored during interpretation.

In Case 1, the CSF (C) has several obvious oligoclonal bands and the corresponding serum (S) is negative. This indicates a positive test for CSF oligoclonal bands.

In Case 2, the CSF and serum are both negative for oligoclonal bands.

In Case 3, many oligoclonal bands are readily evident in CSF, but serum contains only a few barely discernible bands. This case is positive for CSF oligoclonal bands.

Case 4 provides a more difficult example. Here one finds several weakly staining oligoclonal bands in the CSF, most of which are present in the serum (where they stain more strongly). This likely reflects the passage of serum oligoclonal bands into the CSF. With this technique, it is harder to detect a slight traumatic tap than with the serum protein electrophoresis technique where one can examine the α_2-region molecules (see Chapter 8). In this case, however, there is another line in the CSF (indicated) that doesn't match up with a serum line. Nonetheless, because the overwhelming pattern shows the same bands, I interpret this as a mirror pattern – insufficient to support the diagnosis of multiple sclerosis.

Dr. Jerry A. Katzmann (Mayo Clinic) requires four or more distinct bands in the CSF that do not have parallel bands in the serum in order to interpret the CSF sample as being positive for oligoclonal bands. Further, he and his colleagues have improved the sensitivity of detection of CSF oligoclonal bands from 60 per cent with a high-resolution method to 90 per cent with the Helena SPIFE IgG Isoelectric Focusing (IEF) technique (personal communication).

Case 5 demonstrates several oligoclonal bands in the CSF and like Case 3 has a few barely discernible bands. This case is positive for oligoclonal bands.

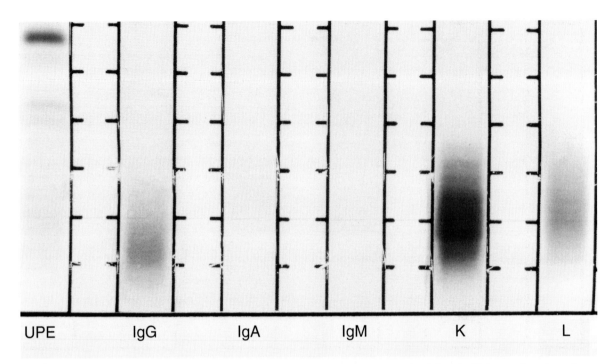

UPE   IgG   IgA   IgM   K   L

Figure 10.38 Urine (concentrated 100-fold) from a 56-year-old man (Paragon system stained with Paragon Violet; anode at the top).

INTERPRETATION FOR FIG. 10.38

The urine protein electrophoresis lane shows mainly albumin with a couple of smaller bands in the α_2-region. A diffuse IgG band is seen. No IgA or IgM is detectable. There is a classic ladder pattern in κ and a fainter ladder pattern in λ. The interpretation in this case is 'Negative for monoclonal free light chains (Bence Jones protein)'. Ladder patterns merely reflect the limited heterogeneity of polyclonal free light chain migration. Terms such as oligoclonal or minimonoclonal are confusing and ambiguous.

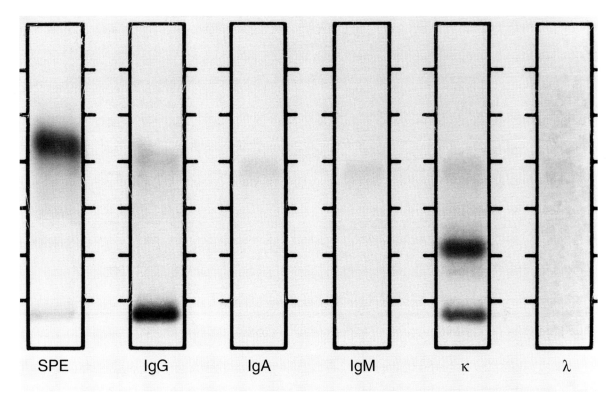

Figure 10.39 Urine (concentrated 100-fold) from a 68-year-old man with a known serum IgG κ monoclonal gammopathy from our files (Paragon system stained with Paragon Violet; anode at the top).

INTERPRETATION FOR FIG. 10.39

The urine protein electrophoresis lane has a single dense band in the α–β-region, no albumin band is visible. There is also a tiny band in the slow γ-region. The α–β-region band is hemoglobin and the urine was red. This band causes a faint artifactual staining in all of the immunoglobulin antisera lanes. The slow γ-region band is the counterpart of the IgG κ monoclonal protein seen previously in this patient's serum. However, in addition, there is a free κ MFLC that migrates in the fast γ-region. The interpretation noted the MFLC and recommended a 24-h protein quantification of this component. I also reported the presence of the IgG κ monoclonal protein.

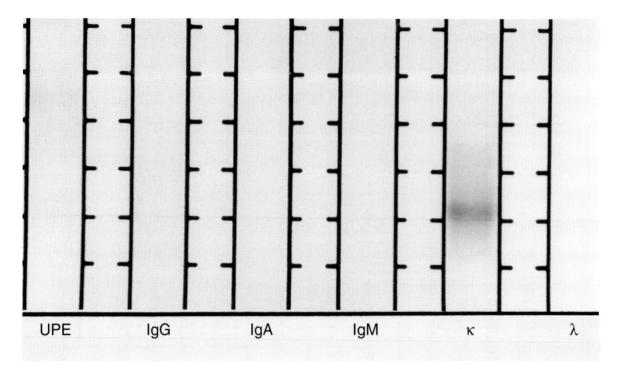

UPE   IgG   IgA   IgM   κ   λ

Figure 10.40 Urine (concentrated 100-fold) from a 53-year-old woman (rule out monoclonal free light chain) (Paragon system stained with Paragon Violet; anode at the top).

INTERPRETATION FOR FIG. 10.40

Two very faint bands are barely discernable in the albumin and α-regions of the urine protein electrophoresis lane. No bands are seen in the IgG, IgA, IgM or λ lanes. However, the κ lane has a small but distinct band superimposed on a diffuse hazy area with some suggestion of a ladder pattern. Some individuals would say this is a variant of a ladder pattern with one prominent band. Others would say that this must be a MFLC because it stains out of proportion to the diffuse hazy background. I do not call this a MFLC or a ladder pattern, because I am not certain about its significance. I interpreted this pattern as 'There is a small κ restriction present. The significance of such tiny κ restrictions is not known. Recommend serum immunofixation now and follow the urine at 3- to 6-month intervals to determine if the process evolves or regresses.' A new alternative not available when this case was seen is the use of free light chain assays in the serum and urine. As discussed in Chapter 7, almost all cases with clinically significant MFLC will have an abnormal ratio of free κ/λ in serum. Beyond testing, however, I often call the clinicians in these unusual cases. In this case he informed me that the patient had joint pains which resembled osteoarthritis, and he wished to rule out a MFLC (Bence Jones protein). I discussed our findings and he sent a serum for study. It was negative. We continue to follow this patient. This case emphasizes the need to let the clinician know our limitations in the laboratory. When I am not certain about the interpretation, I emphasize the importance of following the process.

Index

Note – Page numbers in **bold** type refer to figures and *italic* type indicates tables.